OF THE ELEMENTS

		8A
		2
		Helium
		He

3A	4A	5A	6A	7A
5	6	7	8	9
Boron	Carbon	Nitrogen	Oxygen	Fluorine
B	C	N	O	F
10.81	12.011	14.0067	15.9994	18.9984

		3A	4A	5A	6A	7A	8A
		13	14	15	16	17	18
1B	2B	Alumi-num	Silicon	Phos-phorus	Sulfur	Chlorine	Argon
		Al	Si	P	S	Cl	Ar
		26.9815	28.0855	30.9738	32.06	35.453	39.948
29	30	31	32	33	34	35	36
Copper	Zinc	Gallium	Germa-nium	Arsenic	Sele-nium	Bromine	Krypton
Cu	Zn	Ga	Ge	As	Se	Br	Kr
63.546	65.38	69.72	72.59	74.9216	78.96	79.904	83.80
47	48	49	50	51	52	53	54
Silver	Cad-mium	Indium	Tin	Anti-mony	Tellu-rium	Iodine	Xenon
Ag	Cd	In	Sn	Sb	Te	I	Xe
107.868	112.41	114.82	118.69	121.75	127.60	126.9045	131.29
79	80	81	82	83	84	85	86
Gold	Mercury	Thallium	Lead	Bismuth	Polo-nium	Astatine	Radon
Au	Hg	Tl	Pb	Bi	Po	At	Rn
196.9665	200.59	204.383	207.2	208.9804	(209)	(210)	(222)

63	64	65	66	67	68	69	70	71
Euro-pium	Gado-linium	Terbium	Dyspro-sium	Hol-mium	Erbium	Thulium	Ytter-bium	Lutetium
Eu	Gd	Tb	Dy	Ho	Er	Tm	Yb	Lu
151.96	157.25	158.9254	162.50	164.9304	167.26	168.9342	173.04	174.967
95	96	97	98	99	100	101	102	103
Amer-icium	Curium	Berke-lium	Califor-nium	Einstei-nium	Fermium	Mend-levium	Nobel-ium	Lawren-cium
Am	Cm	Bk	Cf	Es	Fm	Md	No	Lr
(243)	(247)	(247)	(251)	(252)	(257)	(258)	(259)	(260)

*Basics for
chemistry*

DAVID A. UCKO
Science Director
Museum of Science and Industry,
Chicago

Basics for chemistry

ACADEMIC PRESS

New York/London/Toronto/Sydney/San Francisco

A SUBSIDIARY OF HARCOURT BRACE JOVANOVICH, PUBLISHERS

Cover illustration by Jerry Wilke

ACADEMIC PRESS, INC.
111 Fifth Avenue, New York, New York 10003
United Kingdom edition published by
Academic Press, Inc. (London) Ltd.
24/28 Oval Road, London NW1

ISBN: 0-12-705960-1
Library of Congress Catalog Card Number:
80-67631

Printed in the United States of America

Designed by Betty Binns Graphics/Martin Lubin

Preface

Many students enter college without a good background in problem-solving, quantitative skills, and scientific terminology. This book provides students with these tools, and thereby gives them a fighting chance to succeed in General Chemistry. It is designed for a one-semester or one-quarter preparatory chemistry course and is based on my teaching experience at both Antioch College and Hostos Community College.

PROBLEM-SOLVING APPROACH

Basics for Chemistry is carefully tailored to meet the specific needs of underprepared students. It teaches students how to approach problems, and features a unique format for problem-solving that encourages success and confidence. Students are taught to organize the information needed to solve a problem into three categories: the **given**, the **unknown**, and the **connection**. This structured format helps the students set up the calculation properly. In most cases, the problem is then solved using dimensional analysis.

Hundreds of problems are worked out in this way in the text. And, the problems are not only presented in a step-by-step manner, but each step is thoroughly explained. The estimation of the result before each calculation is emphasized in every problem to discourage the blind manipulation of numbers.

SI units are introduced, and supplemented by customary units in those cases where they are still commonly used.

LEVEL OF PRESENTATION

Concepts are presented only as they are needed, and developed from the simple to the complex. For example, the atom is introduced first, then elements and periodicity, then chemical bonding, and finally, chemical compounds and chemical reactions. Within each topic, information and examples are presented in a carefully organized way.

Few assumptions are made about the student's prior knowledge of science or mathematics. New terms, no matter how basic, are clearly explained and illustrated. Mathematical concepts such as significant figures, exponential notation, rounding off, and estimating are treated within the body of the text rather than in an appendix.

The language is kept simple and direct. Pronunciation aids are provided where appropriate. A large number of line drawings, photographs, charts, and tables enhance the written text.

CHAPTER ORGANIZATION

Each chapter begins with an overview of the material to be covered and a set of learning objectives. There is one objective for each chapter section.

Each time a new type of problem is introduced, it is explained step-by-step and is then followed by worked out **Examples** of similar problems as well as by **Practice Problems** that supply the answer but leave the calculation to the student. A **Summary** at the end of every chapter reinforces major concepts.

End-of-chapter exercises consist of three parts. The **Key-Word Matching Exercise** tests understanding of important terms introduced in the chapter. The **Exercises By Topic** are questions and problems relating to each section of the chapter. The **Additional Exercises** provide further practice and more challenging problems. Also, a set of **Review Exercises** appears after each group of three or four chapters. These questions and problems reinforce longer-term learning.

Progress Charts help the students organize and monitor their study of each chapter. To make studying easier, the same number that identifies a chapter section is used to label that section's objective, summary paragraph, set of exercises, and check-off on the Progress Chart.

SPECIAL FEATURES

In addition to using interesting applications from many fields to help motivate students, each chapter also features the following:

Do-it-yourself A simple experiment or demonstration that can be done at home.

Box A practical sidelight or a look at a new development related to chemistry.

Biography A brief discussion of the life and work of a key scientific figure.

Descriptive material about the uses of elements and compounds, important industrial processes, energy sources and uses, and air and water pollution, also appears in appropriate places throughout the book.

The first chapter teaches students good study habits, and explains how to get the most out of this book. Appendices deal with overcoming fear of science and math, using a calculator, and the review of basic mathematics. Answers to numerical exercises, and a Glossary are also included.

Because of its unique problem-solving approach and interesting features, I believe that *Basics for Chemistry* is a significant advance in helping underprepared students learn the basics of chemistry. The following supplements make the text even more effective:

Study Guide for Basics for Chemistry Contains many additional exercises, self-tests, chapter outlines, and examples of common errors.

Instructor's Manual for Basics for Chemistry Contains detailed solutions to all numerical problems in the text.

Overhead Transparency Masters Selected illustrations for classroom projection.

ACKNOWLEDGMENTS I am grateful for the many helpful comments and suggestions made by William Bailey (Broward Community College), David Brouse (Lansing Community College), Tom Fredericks (Cypress College), Thomas Frey (California Polytechnic University, San Luis Obispo), William Hausler (Madison Area Technical College), Judith M. Jones (Valencia Community College), Paul Lauren (Suffolk College), Martha Mackin (Truman College), Ted Musgrave (Colorado State University), Gordon A. Parker (University of Toledo), Otis Rothenberger (Manatee Junior College), and Charles R. Ward (University of North Carolina, Wilmington). Special thanks to Dilip Balamore (Lehman College) for suggested additions to the text.

I would also like to thank Kathy Civetta, Richard Christopher, and the rest of the staff at Academic Press. Finally, I wish to thank my wife Barbara for her assistance and her forebearance.

Contents

ONE

Introduction

By studying this book, you can gain the skills and confidence that you need to go on in chemistry. You will be introduced to the language and basic concepts of chemistry. You will also acquire essential tools for further study, such as the techniques for solving problems. Perhaps most importantly, studying chemistry can change the way you look at and think about almost everything.

Chapter One is an introduction to chemistry both as a part of science and as a college course. The first sections of this chapter describe the nature of chemistry, and the nature of science in general. Later sections describe methods of studying that will help you to learn chemistry and to make the best use of this book. To demonstrate an understanding of Chapter One, you should be able to:

1 Define chemistry and describe what chemists do.

2 Explain the nature of science and the scientific method.

3 Describe the background of chemistry.

4 State ways of improving your study habits.

5 Describe the features of this textbook and how to make the best use of them.

1.1
CHEMISTRY
DEFINED

Although you might not realize it, you are already familiar with chemistry. Chemistry deals with substances, the materials from which things are made. Everything you see depends on chemistry. So does much that you cannot see, such as the workings of your body. Even a simple cup of coffee contains a large number of chemical substances which are dissolved in hot water. Some of them are listed in Table 1-1. Even though the names in this list may seem strange to you, these and other chemical substances are actually an ordinary part of your life.

You are also already familiar with some of the changes that sub-

TABLE 1-1 *Chemical composition of coffee*[a]

caffeine	methyl formate	acetone	butyl alcohol
essential oils	ethyl alcohol	methyl acetate	methylfuran
methyl alcohol	dimethyl sulfide	furan	isoprene
acetaldehyde	propionaldehyde	diacetyl	

[a]A partial list.

stances undergo. For example, when you burn a log in a fire, you convert wood, one type of substance, into ash, another substance. In the process, you also produce heat and light.

Chemistry can be defined as *a study of the composition, structure, and properties of substances, and the changes that they undergo.* There are now over 5 million distinct chemical substances known and about 350,000 new ones are added to the list each year. Most of these millions of chemical substances are rare, but about 65,000 are in common use.

Chemistry can also be defined in terms of what chemists do. For example, *synthetic chemists* combine different types of materials to create new ones that have interesting properties or practical uses. The chemists who created nylon were synthetic chemists. *Physical chemists* study and predict the properties and interactions of substances. Physical chemists provide basic information that is used by other chemists. They also propose fundamental explanations for why substances act as they do. *Analytical chemists* separate materials into various components and determine their composition. For example, analytical chemists determined the makeup of rock samples that were brought back from the moon.

Chemists also specialize according to the types of materials they work with. *Biochemists* study substances found in living things, while *inorganic chemists* deal with substances derived from minerals. The largest group of chemists, *organic chemists,* work with substances that contain a special ingredient called carbon. (See Figure 1-1.)

Most chemists work in industry. Some of the major chemical industries are listed in Table 1-2. They provide a tremendous variety of products and services. Starting with raw materials from the earth, such

TABLE 1-2 *Examples of chemical products and industries*

detergents	industrial chemicals	petroleum refining
drugs	metals	photographic films
electronics	nuclear energy	plastics
fertilizer	paper	rubber
food processing	personal care products	textile fibers
glass and ceramics	pesticides	

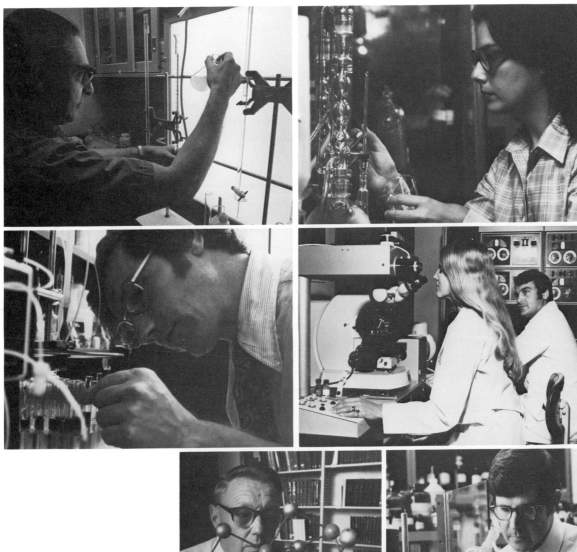

FIGURE 1-1

Chemists at work. Chemists perform a wide variety of jobs in many different settings. (Photos courtesy of Du Pont Company and NASA.)

as petroleum or minerals, these industries produce basic chemical substances. Some substances, like baking soda, are used directly, while others are used in further manufacturing. For example, ammonia is used for making fertilizers and explosives. Chemical industries provide finished products like plastics, fibers, glass, paper, metals, and drugs. (See Figure 1-2.)

In addition to making chemical products, chemists also provide important services in fields outside of the chemical industries. They monitor and control pollutants in the air and water, and help develop

---

Box 1

Chemistry as an occupation

Most chemists work in industry. In fact, about 60% of all chemists are employed by companies that produce either chemicals, drugs, personal care products, coatings, petroleum, rubber, paper, metals, minerals, or food. The five companies that sell the most chemical products are Du Pont, Dow Chemical, Exxon, Union Carbide, and Monsanto. About 20% of all chemists teach, and another 10% work for government agencies.

The type of work performed by most chemists, whether they work for industry, government, or a university, is called R & D, or research and development. Research is the process of gathering scientific information. The purpose of basic research is to learn more about an aspect of science that the researcher finds interesting. Exploring the solar system with spacecraft is an example of basic research. Many chemistry professors, in addition to teaching, also carry out chemical research, or supervise the research projects of their students. Development is applied science. It involves taking the results of research and applying them toward the solution of a practical problem. Making a glue stronger or a paint more resistant to chipping are examples of development.

After R & D, the next most common activity of chemists is management. Most managers in chemical industry or government started out in R & D, and then were promoted. Chemists also become involved in marketing and production.

The American Chemical Society (ACS) is the professional organization for chemists in the United States. It has well over 100,000 members. The ACS publishes books and journals, holds national and regional meetings, and provides various other services for its members. You might consider joining the ACS as a student affiliate to obtain journal subscriptions and other membership advantages at a discount rate. If you are interested, speak to your instructor or write directly to: American Chemical Society, 1155 16th Street N.W., Washington, D.C. 20036.

FIGURE 1-2

Chemical production. The chemical industry produces a wide range of important substances. Shown here is a major chemical manufacturer's storage area for raw materials used in the production of industrial chemicals.

new sources of energy. In police laboratories, chemists analyze materials involved in crime. Clinical chemists perform tests to analyze samples taken from patients in hospitals. Chemists are also involved in agriculture, art, engineering, nutrition, oceanography, and many other fields. In short, chemists play important roles throughout our society.

1.2 THE NATURE OF SCIENCE

Chemistry is part of **science**. The word science comes from the Latin word *scientia*, which means knowledge. Today the word science refers to the systematic organization of facts that describe our world. This book deals mainly with the part of science that is called chemistry.

But the word science also refers to the process by which knowledge is gathered. The process of science is based on experiments. Experiments consist of observations and measurements carried out under conditions that are controlled for the purpose of study. For example, if we were studying how rapidly different dyes faded in sunlight, we would want to make sure that each dye received the same amount of sunlight, that each was kept at the same temperature, and so on.

Scientific laws are statements about the way nature behaves. They summarize experimental observations that show a common pattern. For example, we observe that objects fall when we drop them. These observations are summarized by the law of gravity, which states that

Smithsonian Institution: Photo #10045-N

Roger Bacon was a 13th century English scholar with an interest in science that was far ahead of his time. Bacon was once believed to have invented gunpowder, although we now know that it was made centuries earlier in China. He described the structure of the eye, and experimented with lenses and mirrors.

Impressed by the writings of Aristotle and other philosophers, Bacon attempted to compile a universal summary of knowledge. However, even as early as the 13th century this task was impossible.

At the age of 38, Bacon became a Franciscan monk. Because he criticized some practices of the order, Bacon's writings were condemned by the Church and he was imprisoned. His greatest work, *Opus Majus,* was not published until nearly 500 years after his death.

Like many others living in the Middle Ages, Bacon believed in alchemy. Alchemists sought to transform common metals into gold, cure any illness, and bring about eternal life. Of course no alchemist could work such miracles, but Bacon left behind recipes for chemicals, and recorded useful information about minerals and glazes for pottery.

Bacon's most important contribution to science is his recognition of the importance of experiments. Believing that experience was the only basis of certainty, Bacon said:

There are two methods in which we acquire knowledge—argument and experimentation. Argument allows us to draw conclusions, and may cause us to admit the conclusion, but it gives no proof, nor does it remove doubt, and cause the mind to rest in conscious possession of truth, unless the truth is discovered by means of experience.

Bacon's understanding of the significance of experiments makes him one of the founders of modern science.

FIGURE 1-3

The scientific method. The process of science involves the interplay between observation and measurement (experiment) and proposed explanation (hypothesis).

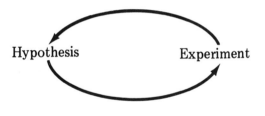

all objects attract each other. A scientific law can be shown to be false or limited in scope if something is found that cannot be explained by the law.

In order to explain experimental observations, scientists form hypotheses. A **hypothesis** is a model or framework for organizing facts. It describes how and why things behave as they do. The **scientific method** is the name given to the process of creating and testing hypotheses. (See Figure 1-3.) A hypothesis is useful if we can make predictions based on it that we can check by further experiments. The hypothesis must be scrapped or modified if it does not agree with the results of our new experiments.

Let us see how the scientific method might be applied. Suppose that we want to know why a certain television set, which has been plugged in and switched on, is not working. The screen is blank; there is no picture or light. What is the cause of the blank screen? One preliminary hypothesis might be: the trouble is in the wall outlet into which the set is plugged. But we need an experiment to test this hypothesis. We could test the outlet by plugging a lamp into it. If the

Do-it-yourself 1

The burning candle

Careful observation is a key part of the scientific process. You can test your own powers of observation by studying a burning candle.

To do: Light a candle. On a sheet of paper, list as many things as you can observe about the burning candle in 10 minutes.

If you wish, have a friend list his or her observations at the same time as you do. Compare your lists at the end of the 10 minutes.

In 1860, the English scientist Michael Faraday delivered a series of lectures for children about the candle. He described and explained how the wax burns, the shape and regions of the flame, the need for air, the products of burning, and many other observations. Your library may have a copy of Faraday's book *The Chemical History of a Candle*, which contains these lectures.

lamp works, we can then reasonably say that the outlet is okay. When we carry out this experiment, we find that the lamp does indeed work. (If it didn't work, we would want to use a different lamp, or another method to test the line, just to be sure the outlet was dead.) We conclude that the outlet works and that the problem is probably within the television set. Our next hypothesis might be that one of the tubes is not working. We would continue the process by testing this new hypothesis. If this new hypothesis is proven wrong, we could form still another hypothesis, test it, and so on, until at last we find a hypothesis that is supported by testing.

We can also view medical diagnosis as an application of the scientific method. In order to make a diagnosis, a physician first observes the state of the patient's health and may perform certain tests. Next, the physician proposes a possible explanation or hypothesis, such as the presence of a particular disease. Checking this hypothesis involves treatment of the condition. (For example, a particular drug may be administered.) The patient is observed again to determine whether the proposed explanation, and the therapy based on it, were correct. If the symptoms disappear, the hypothesis is supported. If they do not, then another hypothesis may be necessary, along with a different treatment.

A **scientific theory** is a hypothesis that has been tested and confirmed many times under a wide range of conditions. It must unify experience, be simple, and make meaningful predictions. Scientists generally try to find support for their own theories and to disprove rival theories by doing experiments.

The scientific method is a powerful tool. But it is not a guaranteed means for scientific discovery. Creativity, inventiveness, imagination, and sometimes even luck, are also needed.

**1.3
THE BACKGROUND
OF CHEMISTRY**

Chemistry has several origins. Ancient civilizations developed practical arts, such as working with metals, pottery and dyes, and making wine. While these arts were being developed in the Middle East several thousand years ago, the ancient philosophers of Greece, India, and China, first attempted to describe the composition of substances.

These two activities led to alchemy, which combined experience from the practical arts with philosophy and magic. Alchemists sought to turn metals, such as lead, into gold. (See Figure 1-4.) They also tried to make the "elixir of life," a magical drink that would prevent sickness and prolong life. Although alchemists failed in these attempts, they developed laboratory techniques and prepared several chemicals that are commonly used today. Figure 1-5 shows an alchemical laboratory.

Modern chemistry began little more than 200 years ago when the scientific method was applied to chemical experimentation. The theories presented in this book are based on work carried out since that

FIGURE 1-4

(Copyright © 1977 by Sidney Harris/"What's so Funny about Science.")

"FRANKLY I'D BE SATISFIED NOW IF I COULD EVEN TURN GOLD INTO LEAD."

time. They explain present chemical data, and are well accepted. But as new information is obtained, these theories may change.

Chemistry is closely related to the other branches of modern science. Biology, the study of living things, is based on principles of chemistry. Chemistry depends in part on physics, which is the study of the fundamental laws of nature. Geology, or earth science, is founded on both chemistry and physics. Thus, the science of chemistry holds a key position among the other basic sciences.

FIGURE 1-5

An alchemical laboratory. Alchemists tried to make the "elixir of life" and to turn lead into gold. ("The Alchemist" by David Teniers the Younger. Reproduction courtesy of the Fisher Collection of Alchemical and Historical Pictures.)

**1.4
STUDYING
CHEMISTRY**

Knowing how to study is the key to learning chemistry. Most students have a limited amount of time. Effective study involves making the best use of that time.

Organization is an important factor in studying. Set aside a regular time and place to study. Choose a time when you are most alert. Do not try to study in bed before going to sleep. Studying is hard work! Pick a work area that has good lighting and that is private. Sitting upright at a desk, in a well-lit room, can greatly improve your studying efficiency. If your room is noisy or distracting, you may need to isolate yourself in a library.

Concentrate fully on what you study. Listening to the radio while you study can interfere with your concentration, and will probably make it necessary to spend more time studying. The goal of studying is to learn the material for your course as efficiently as possible. Although it may be difficult, try not to think about personal problems while you study.

Study to learn ideas, not just to cover a certain number of assigned pages. As you read, think about these ideas, relate them to previously studied ideas and even to your own experience. As you learn new material, try to organize it into a pattern, fitting it into the framework of the course. Organizing ideas is just as important as organizing your time.

Repetition is also basic to learning. Studying for many short periods is better than having a long study session, because it enables you to review past material. Therefore try to study chemistry for one or two hours each day rather than all day on the weekend. If you make a conscious effort at first, good study patterns will become a habit with time.

Practice solving the problems that are worked out in the text. You cannot learn something just by seeing how someone else has done it. *You must do it yourself.* After studying an example in the text, copy the question, close the book, and try to solve the problem on your own. If you can solve it, then try additional problems to reinforce your skills. If you have difficulty, go back to the text and find out where you went wrong. Then try again. Solving problems on your own is an excellent learning device.

You may discover other useful study aids. For example, some students find it helpful to make flash cards, which have words or symbols on one side and the corresponding meaning on the other side. This tool is often effective for learning new vocabulary words. Since chemistry uses its own special language, flash cards can be very helpful.

Finally, your class notes are an important study aid. Learn to take brief notes and follow the lesson at the same time. Review the notes after class, while they are still fresh in your mind. Mark any parts that you do not completely understand. Then ask the instructor for some help before the next class.

If you have already had problems with science or math and are now worried that you will have difficulty learning chemistry, read Appendix A at the end of this book. It contains valuable advice that will help you overcome this fear and be successful in this course.

**1.5
STUDYING
THIS TEXTBOOK**

Just as studying chemistry is a skill, so is knowing how to get the most out of your chemistry textbook. You should first learn how this book is organized. It is divided into 18 chapters, each covering some particular aspect of chemistry. Each of the chapters is further divided into sections, such as the one that you are now reading (Section 1.5). The sections contain relatively small, digestible chunks of information. The Table of Contents in the front of this book provides a convenient listing of all the chapters and their sections.

By following the steps that are described here, along with the recommendations that were made in the previous section, you will be able to learn chemistry on a chapter-by-chapter basis.

1 *Preview the chapter.* Read the introduction and the set of objectives at the beginning of the chapter. Each objective corresponds to one section in the chapter. The objectives describe the skills that you will have gained when you have successfully completed your studying. Survey the section headings and look over the pages of the chapter. Also read the summary at the end of the chapter, and skim the end-of-chapter exercises. The purpose of this preview is to give yourself a general picture of the material that is covered and its overall organization. By providing a framework for your study, previewing allows you to fit the pieces together more readily as you read each section.

2 *Study each section.* This process consists of three parts, repeated for each section.

(a) *Turn the section heading into a question.* For example, if the heading is "Studying this textbook," the question in your mind should be "How do I study this textbook?" This procedure focuses your attention so that your mind will be actively involved in reading the section.

(b) *Carefully read the section, answering the question that you have just created.* As you read, look up words that you don't know and note in the margin any points that you do not understand. Carefully follow all calculations step by step. On a separate piece of paper, repeat each Example in order to test your understanding of the material. Reread the section until you are sure that you understand it.

Make notes in the margin when you have questions, then cross out these marginal notes as you clear up the difficulties. If you still have questions, ask your instructor, or other students, for help until all of your questions are answered. Check the objective at the beginning of the chapter that has the same number as the section that you are studying. For example, the objective for this section (number 5) is "Describe the features of this textbook and how to make the best use of them." If you did a good job, you will have gained the skill or skills described.

(c) *Review the section and pick out its major points.* Put the book down and answer from memory the question that you originally created from the section heading. State the basic concepts of the section in your own words; then write these concepts down in a notebook.

Open the textbook again and look over the section, underlining one or two key phrases. (Do not underline too many.) If there is a Practice Problem at the end of the section, solve it yourself and then compare your answer with the one given. If you have difficulties, read the section again, concentrating on the worked-out examples.

3 *Review the entire chapter.* After you complete each section as just described, reinforce your understanding of the chapter by reviewing. Look over any notes you have taken as well as the underlinings you have made in the book. Without looking at the book, state the major concepts of the chapter, section by section. Go over the list of objectives at the beginning of the chapter to see if you are now confident that you can satisfy them. Study the chapter summary.

4 *Answer the chapter exercises.* If possible, do the assigned exercises without looking back at the chapter. Then, to develop your confidence, try additional exercises and check your answers with those given in Appendix J. Do the Key-Word Matching Exercise, and be sure that you understand the words and can explain them.

Each chapter contains a Progress Chart. By checking off each step as it is completed, you can keep track of your progress through the chapter. The chart will also help you to organize your studying.

A large part of the study procedure just described involves reflection: questioning as you read, thinking about and restating ideas after you read them. Reflection helps to insure real understanding of the material. To make sure that you retain this understanding, review the chapter periodically and complete the Review Exercises. These exercises are designed to reinforce long-term memory as well as to help relate the material from several chapters.

This textbook contains still more learning aids. Look through the Appendices. They can give you much useful information, including a review of mathematics. The Glossary defines key terms from the text. Use the index to look up words or concepts. The Study Guide that accompanies this book contains many more questions and examples for further practice.

Summary   1.1 Chemistry is defined as a study of the composition, structure, and properties of substances, and the changes that they undergo. Over 5 million distinct chemical substances are known. Chemists prepare and study these substances. Although most chemists work in the chemical industry, many perform important services in other fields.

1.2 Modern science is the systematic organization of facts that describe our world. It is based on experiments, which are carefully controlled observations and measurements. The scientific method is the process of creating and testing hypotheses, which are explanations of experimental observations. A theory is a hypothesis that has been tested and confirmed many times under a wide range of conditions.

1.3 The roots of chemistry are the ancient practical arts and ancient philosophy. Chemistry as we know it began little more than 200 years ago when the scientific method was applied to chemical experimentation.

1.4 Studying effectively is the key to learning chemistry. Organizing your time is important, as is organizing the ideas that you learn. Repetition, practice, and daily study are other important habits.

1.5 The following steps will help you to study each chapter of this textbook. First, preview the chapter to get a general picture of the material covered. Then study each section, turning the heading into a question, and reading the section to answer the question. Review the section from memory and pick out its major points. Finally, after you have studied each section, review the entire chapter and answer the chapter exercises.

PROGRESS CHART   Check off each item when completed.

_____   Previewed chapter

Studied each section:

_____   **1.1** Chemistry defined

_____   **1.2** The nature of science

_____   **1.3** The background of chemistry

_____   **1.4** Studying chemistry

_____   **1.5** Studying this textbook

_____   Reviewed chapter

TWO

Matter, energy, and measurement

In Chapter Two we introduce the concepts of matter and energy. These concepts are fundamental to chemistry. When we deal with matter and energy, we use measurements to describe how much matter or how much energy. The system of measurement that we use most often in chemistry is called the metric system and is described in this chapter. The techniques of exponential notation, significant figures, rounding-off, and estimating, are also explained here as ways of handling measurements. To demonstrate an understanding of Chapter Two, you should be able to:

1 Define matter and describe the three states of matter.
2 State the meaning of the most common metric system prefixes.
3 Write numbers in exponential notation.
4 Define mass and identify units of mass.
5 Define length and identify units of length.
6 Define volume and identify units of volume.
7 Define energy and state its relationship to matter.
8 Identify the number of significant figures in a measurement.
9 Round off measured values, using the rules for significant figures.
10 Perform calculations using exponential notation.
11 Estimate the results of numerical calculations.

**2.1
MATTER**

Chemistry deals with **matter.** The concept of matter is so basic that it is difficult to define. We know, however, that matter is the "stuff" that surrounds us. It is, in fact, anything that occupies space. As shown in Figure 2-1, matter exists in three forms or states: solid, liquid, and gas. (A fourth possible state, called plasma, is related to the gaseous state and exists only under special conditions.)

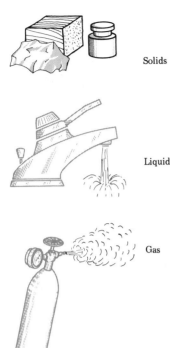

Solids

Liquid

Gas

FIGURE 2-1

The three states of matter. Solids have a definite volume and shape. Liquids also have a definite volume but not a definite shape. Gases have neither a definite volume nor a definite shape.

TABLE 2-1 *The three states of matter and their properties*

Property	Solid	Liquid	Gas
shape	definite	not definite	not definite
volume	definite	definite	expands without limit
mass	definite	definite	definite
compressibility	not compressible	not compressible	can be compressed
effect of heating	very slight expansion	small expansion	large expansion

Solids take up definite amounts of space. They generally have rigid shapes that resist change. Solids can be compressed (pressed into less space) only very slightly. They expand only very slightly when heated. Some examples of solids are wood, rock, bone, gold, and table salt.

Liquids also take up fixed amounts of space. However, they do not have rigid shapes. They take the shapes of their containers, filling them up from the bottom. Liquids can be compressed only very slightly, and when heated they expand slightly more than solids do. Milk, water, blood, alcohol, and mercury are examples of liquids.

Gases do not take up definite amounts of space and have no definite shapes. Instead, they expand without limit to uniformly fill the space that is available. Gases can be compressed into very little space. For example, a large amount of air can be compressed into a scuba diver's tank. Air is a gas. Some other examples of gases are steam, oxygen, neon, and helium. (Note that gasoline, sometimes called ''gas,'' is actually not a gas but a liquid.)

Table 2-1 summarizes the most important properties of the three states of matter. In Chapters Eleven and Twelve we discuss these states further and describe what happens when matter changes from one state to another.

2.2 MEASUREMENT AND THE METRIC SYSTEM

When describing matter, chemists must specify the size or quantity. To do this, they use measurements. A **measurement** is a number that describes the amount of something. Lord Kelvin, a 19th century English scientist, said that ''when you can measure what you are speaking about, and express it in numbers, you know something about it.''

Measurements are based on a system of units. These units are agreed-upon standards of size. For example, each time you buy a quart of milk you get the same amount because the quart is an agreed-upon size and does not change.

The result of a measurement is always expressed in terms of a number and a unit. For example, we write the measured length of a soccer field as 110 yards. Here, the number is 110, and the unit is the yard. This particular unit, the yard, is part of the English system of measurement, which is still widely used in the United States. However, people in most other countries, and scientists in all countries, use the metric system.

The **metric system** is based on the number 10. Each unit for a given type of measurement is related to the other units by some exact multiple of 10. This system is convenient because our decimal system (described in Appendix B.3) is also based on the number 10. We do not have to memorize numbers like 5280 (the number of feet in a mile in the English

Box 2

The U.S. Customary System of Units

The United States and the countries Brunei, Burma, Liberia, and Yemen still use the English system of measurement, which in the United States is called the U.S. Customary System of Units.

Although you are probably very familiar with the U.S. Customary System, it is actually more complex than the metric system. For example, there are three different sets of units for measuring weight in this U.S. system. **Avoirdupois** units are used most often. The standard (or base) avoirdupois unit is the pound. It is divided into 16 ounces. Each ounce is divided into 16 drams, and each dram into 27 11/32 grains. **Troy** units are used for weighing precious metals. A troy pound is slightly smaller than an avoirdupois pound, and is divided into 12 troy ounces. Gold and silver prices are quoted in dollars per troy ounce of the metal. The troy ounce is divided into 20 pennyweights, and each pennyweight contains 24 grains. **Apothecaries** units are used for weighing drugs. The apothecaries pound is equal to a troy pound and is divided into 12 ounces. But unlike either the troy ounce or the avoirdupois ounce, the apothecaries ounce is divided into eight drams, each dram is divided into three scruples, and each scruple into 20 grains. An ordinary aspirin tablet contains 5 grains of aspirin.

There are four sets of units for measuring volume. The gallon is the basic unit for liquid volume. It is divided into four quarts, each quart is divided into 2 pints, and each pint is divided into 16 fluid ounces. In apothecaries units for liquid volume, the fluid ounce is further divided into 8 fluid drams, and each fluid dram into 60 minims. The bushel is the basic unit for measuring dry volume. Volume can also be expressed in cubic inches or cubic feet.

There is only one set of units for measuring length in the U.S. Customary System. The base unit is the yard. The foot is 1/3 yard, and the inch is 1/36 yard. A mile is 1760 yards, or 5280 feet.

As you have probably noticed, there is no consistent relationship between subdivisions of units in the U.S. system. Metric units, on the other hand, are consistently based on multiples of 10, making the division of the basic units simpler. Partly for this reason, the metric system has become the international system of measurement for all scientists.

TABLE 2-2 *Metric system prefixes*

	Prefix[a]	Symbol	Pronunciation	Meaning
	giga	G	jig' ah	1,000,000,000 or 10^9
	mega	M	meg' ah	1,000,000 or 10^6
larger	**kilo**	k	kill' oh	1000 or 10^3
	hecto	h	hek' toh	100 or 10^2
	deka	da	dek' ah	10 or 10^1
base unit				1 or 10^0
	deci	d	dess' ee	0.1 or 10^{-1}
	centi	c	sen' tee	0.01 or 10^{-2}
	milli	m	mill' lee	0.001 or 10^{-3}
smaller	**micro**	μ	my' kroh	0.000001 or 10^{-6}
	nano	n	nan' oh	0.000000001 or 10^{-9}
	pico	p	pee' ko	0.000000000001 or 10^{-12}

[a]The most common prefixes are in black type.

system) in order to convert from one unit to another in the metric system.

Every measurement in the metric system is expressed in terms of an official standard called the *base unit*. A prefix, consisting of a letter or symbol, can be placed in front of the base unit to change its size by a certain multiple of 10. Table 2-2 lists some of these prefixes. The most common ones are:

kilo- one thousand (1000) times larger

milli- one thousand times smaller (1/1000 or 0.001).

Do-it-yourself 2

Metric muffins

1 egg
235 mL milk
60 mL salad oil
475 cm³ (330 g) flour
60 cm³ (60 g) sugar
15 cm³ (10 g) baking
powder
5 cm³ (5 g) salt

The metric system is not yet commonly used in the United States. If and when it is, you may find a list of ingredients such as the one on the left in a cookbook.

To do: Heat the oven to 205°C. Grease the bottoms of 12 medium-size (7 cm diameter) muffin cups.

Beat the egg, and stir in the milk and oil. Add the other ingredients and mix until the flour is just moistened. The batter should stay lumpy.

Fill the muffin cups 2/3 full. Bake for 20 to 25 minutes, or until the muffins are golden brown. Remove them immediately from the muffin cups.

Although the metric measurements may appear strange to you, this recipe makes perfectly ordinary muffins.

The modern version of the metric system is known as the *International System of Units,* abbreviated SI (see Appendix E for further details). At the present time, not all of the SI base units are widely used in chemistry. In this book we emphasize the units that you will come across most often. When measurements are expressed in the English system, the SI value generally will be given in parentheses.

2.3
EXPONENTIAL
NOTATION

Exponential or **scientific notation** is a shorthand way of writing both large and small numbers, such as the values of measurements. It is based on expressing a number as a power of 10. A power of 10 is 10 multiplied by itself a certain number of times, as shown in Table 2-3.

TABLE 2-3 *Powers of 10*

Power of 10	Meaning
10^3	$10 \times 10 \times 10$ or 1000
10^2	10×10 or 100
10^1	10
10^0	1
10^{-1}	$1/10^1$ or 1/10 or 0.1
10^{-2}	$1/10^2$ or 1/100 or 0.01
10^{-3}	$1/10^3$ or 1/1000 or 0.001

The raised number, called the *exponent,* shows how many times 10 is multiplied by itself. For example, 10^4 (called "ten to the fourth power") is $10 \times 10 \times 10 \times 10$, or 10,000. Also, 10^5 (called "ten to the fifth power") is $10 \times 10 \times 10 \times 10 \times 10$, or 100,000. Notice that 10^5 is ten times greater than 10^4. Increasing the exponent by one increases the number itself by 10 times. Thus 10^8 is 10 times greater than 10^7, and 10^7 is in turn ten times greater than 10^6. The numbers 10^2 and 10^3 have special names; 10^2 is called "ten squared" and 10^3 is called "ten cubed." The symbol 10^0 is given the value 1.

Sometimes we have to write small numbers such as 1/1000 or 1/10,000. We could express these numbers as $1/10^3$ or $1/10^4$ respectively. This notation, however, is a little clumsy. Instead, we may express $1/10^3$ as 10^{-3} (called "ten to the minus third power"). Similarly $1/10^4$ may be written 10^{-4} (ten to the minus fourth power). Notice that 10^{-3} is 1/1000 and 10^{-4} is 1/10,000. Therefore, 10^{-4} is 10 times smaller than 10^{-3}. When the exponent becomes one unit more negative, the number itself becomes 10 times smaller.

As we saw in the previous section, the metric system has several prefixes. The last column of Table 2-2 shows the values of metric prefixes written as powers of 10.

ASIDE . . .

If you started counting today, you would not be able to reach the number 10^{10} within your lifetime.

We express a number in exponential notation by writing a number between 1 and 10, multiplied by a power of 10. The number between 1 and 10 is called the *coefficient*. The population of the United States is about 2.2×10^8; the distance to the sun is 1.5×10^8 km; the number of seconds in a year is roughly 3.15×10^7. All of these numbers are examples of exponential notation.

Number expressed in
exponential notation: coefficient $\times 10^{\text{exponent}}$

                                             power of 10

We obtain the coefficient by moving the decimal point in our original number until we get a value between 1 and 10. For example, if our original number is 1450, the decimal point must be shifted three places:

$$1\underset{3}{.}4\underset{2}{.}5\underset{1}{.}0. \qquad \text{shift decimal point 3 places to the left}$$

This shift gives us 1.450, a number between 1 and 10. The number of places that we move the decimal becomes the exponent of the power of 10. The exponent is positive if we move the decimal point to the *left*, that is, if the original number is greater than one. Therefore, because we move the decimal point three places to the left in this example, the power of 10 is 10^3 and we write

$$1.45 \times 10^3$$
coefficient   power of 10

Now consider the number 0.00029, which is less than one. In order to make a coefficient between 1 and 10, we must shift the decimal point to the *right* 4 places:

$$0 \underset{1\ 2\ 3\ 4}{.0\ 0\ 0\ 2}9 \qquad \text{shift decimal point 4 places to the right}$$

In this case, the exponent is negative, -4, and the number is written:

$$2.9 \times 10^{-4}$$

Just remember that a number greater than one must have a positive power of 10 and that a number less than one must have a negative power of 10. Table 2-4 shows additional examples.

To convert a number from exponential notation to regular decimal form, we simply shift the decimal point in the coefficient. The sign of the exponent shows us in which direction to move the decimal point.

TABLE 2-4 *Examples of numbers in exponential notation*

Number	Notation
5,300,000	5.3×10^6
39,800	3.98×10^4
1480	1.48×10^3
56	5.6×10^1
0.33	3.3×10^{-1}
0.0067	6.7×10^{-3}
0.0000051	5.1×10^{-6}

A positive exponent ($+1$ or greater) means that the number is larger than one. Therefore, we move the decimal point in the coefficient to the *right* to make a larger number. On the other hand, a negative exponent (-1 or less) means that the number is less than one. In this case, we move the decimal point to the *left* to make the coefficient a smaller number.

The numerical value of the exponent shows us how many places to move the decimal point in the coefficient. Suppose that we want to convert the following number to decimal form: 3.75×10^4. Because the exponent is 4 we move the decimal point four places. And because the exponent is positive, the decimal point is shifted four places to the right. The number in decimal form is 37,500.

Consider the following number expressed in exponential notation: 8.1×10^{-3}. To convert this number to regular decimal form, we move the decimal point in the coefficient *three* places to the *left* because the exponent is -3. The number becomes 0.0081. For practice, cover up the first column in Table 2-4 and try to write the numbers in the second column in decimal form.

Practice problem   Write the following numbers in exponential notation:

a 3785

b 0.093

c 19.2

d 552,000

e 0.00075

ANSWERS   **a** 3.785×10^3 **b** 9.3×10^{-2} **c** 1.92×10^1
d 5.52×10^5 **e** 7.5×10^{-4}

Practice problem   Express the following numbers in regular decimal form:

a 4.42×10^2

b 8.9×10^{-3}

c 3.01×10^{-1}

d 6.9×10^4

e 7.7×10^0

ANSWERS   **a** 442   **b** 0.0089   **c** 0.301   **d** 69,000   **e** 7.7

**2.4
MASS**

Mass is defined as the amount of matter that an object contains. Everything around you, this book, your clothing, even the air, has mass. You can also think of mass in another way, based on a scientific law discovered during the 17th century by Sir Isaac Newton. He noted that when an object is standing still, a push or pull is needed to make it move. Newton found that the mass of an object at rest (standing still) is a measure of its tendency to stay at rest. The greater the mass of an object at rest, the greater is its tendency to remain at rest. Consider what happens when you kick an empty tin can. You can easily move the tin can because its mass is small. But if you try this experiment with a brick, the results may be painful. Since a brick has a greater mass than a tin can, a much greater kick is needed to make it move.

We commonly use the word weight when we actually mean mass. Weight is the earth's attraction for an object, due to gravity. The weight of an object may change with the object's location, but its mass always remains the same. Astronauts on the moon, for example, have the same mass as they do on earth but weigh only one-sixth as much because of the smaller pull of gravity that exists on the moon.

A balance is used to measure the mass of an object. As shown in Figure 2-2, a balance looks like a seesaw. The balance consists of a beam balanced on a sharp edge (part a in Figure 2-2). When an object of unknown mass is placed on one side, the beam is no longer balanced (part b). Known masses are then added to the other side until the beam balances again (part c). The mass of the original object equals the sum of the known masses, because the masses on both sides must be equal when the beam balances.

Figure 2-3 shows a common laboratory balance. To use this type of balance, we slide masses along the marked beams from left to right in order to balance an object of unknown mass placed in the tray at the left side.

The unit of mass in the metric system is the **gram.** Its symbol is g, but it is sometimes written as gm. One gram (1 g) is the mass of a small amount of matter, a little more than that of a standard-size paper clip.

FIGURE 2-2

The operation of a balance. (a) The beam is initially balanced. (b) The object whose mass is to be measured (X) is placed on one side, making the beam unbalanced. (c) Known masses are added to the other side until balance is restored. The masses on both sides are now equal. The mass of object X is 5 g + 1 g + 1 g = 7 g.

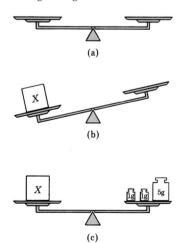

(a)

(b)

(c)

Smithsonian Institution: Photo #9585-B

You probably have heard the story of Isaac Newton and the apple tree. Seeing an apple fall from a tree, Newton is said to have realized that the force that pulls an apple to the ground is the same one that attracts the moon and keeps it in orbit around the Earth. He proposed that both "earthly" and "heavenly" objects obey the same law—the law of universal gravitation.

Newton developed the three fundamental laws of motion. They are:

1. An object in motion tends to stay in motion.

2 A force (a push or pull) is needed to change the motion of an object.

3 Every action causes an equal but opposite reaction.

According to Newton, the mass of an object is a measure of its inertia, which is its tendency to stay at rest or in uniform motion.

Newton also invented the calculus. Today, anyone who wants to be a physical or social scientist studies this branch of mathematics. In addition, Newton invented the reflecting telescope, and showed that white light is a combination of the colors of the rainbow.

Newton's brilliance did not extend to financial matters, however. He lost a great deal of money speculating in the stock of the South Sea Company. Newton explained, "I can calculate the motions of heavenly bodies, but not the madness of people."

Newton was knighted by Queen Anne of England in 1705 for his service as Master of the Mint as well as for his scientific accomplishments. In his later years, Sir Isaac treated physics and mathematics as hobbies and devoted himself to alchemy, theology, and history.

Newton is often called the greatest of all scientists. The poet Alexander Pope praised his work in these famous lines:

Nature and nature's laws lay hid in night:
God said, Let Newton be! and all was light.

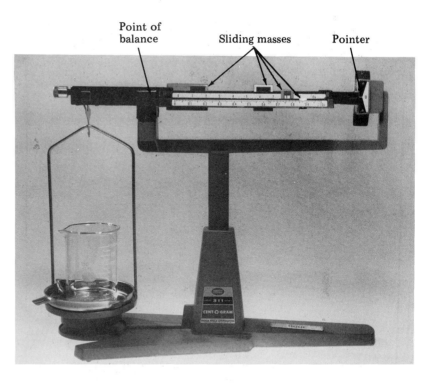

Point of balance   Sliding masses   Pointer

FIGURE 2-3

A laboratory balance. The mass of the object in the pan is found by sliding masses along the scale until the beam is balanced. The beaker shown has a mass of 68.87 g. (Photo by Al Green.)

The official standard or base unit for mass is actually a larger unit, the kilogram. Since kilo- means "1000 times larger than," a **kilogram** is 1000 grams, or 10^3 g.

kilo	gram
1000 times	unit of mass

$$1 \text{ kg} = 10^3 \text{ g}$$

The symbol for the kilogram, kg, is made by combining the symbol for kilo-, which is k, with the symbol for gram, g.

Americans generally measure weight in pounds, abbreviated lb. An object that weighs one pound (1 lb) on earth has a mass of 454 g. Another way of stating this relationship is to say that one kilogram (1 kg) weighs 2.20 pounds. The astronauts who walked on the moon wore spacesuits having masses of about 150 kg each. On Earth, these spacesuits weighed 330 lb. On the moon, however, the suits weighed only 55 lb, or one-sixth as much. Table 2-5 illustrates the relationship between mass and weight by comparing the mass in kilograms and the weight in pounds of parts of a human body.

Just as large amounts of mass are expressed in kilograms, which are 1000 times larger than the gram, smaller amounts of mass are expressed in **milligrams**, or thousandths of a gram. The small amounts

TABLE 2-5 *Mass and weight of body components of a 120-pound female*

Body component	Mass (kilograms)	Weight (pounds)
muscle	24	52
fat	8	17
skeleton	8	17
skin	5	11
blood	4	10
miscellaneous	6	13
	55	120

TABLE 2-6 *Metric units for mass*

Unit	Symbol	Meaning
kilogram	kg	1000 grams or 10^3 g
gram	g	base unit (454 grams weighs 1 pound)
milligram	mg	1/1000 (or 0.001) gram or 10^{-3} g

of mass that we commonly measure in the laboratory are expressed in milligrams.

$$\underset{\text{1/1000 times}}{\text{milli}} \mid \underset{\text{unit of mass}}{\text{gram}}$$

$$1\ \text{mg} = 10^{-3}\ \text{g}$$

The symbol for milligram, mg, is formed by combining the symbol for milli-, which is m, with the symbol g for gram. Since the milligram is 1000 times smaller than the gram, there are 1000 mg in one gram. Table 2-6 summarizes the basic information about these most common metric units for mass.

**2.5
LENGTH**

Length is a measure of the distance between two points. We use length to express the size of matter. The basic metric unit of length is the **meter**, and its symbol is m. In terms of English units, one meter (1 m) is slightly longer than 3 feet (ft) or 1 yard (yd). Table 2-7 lists a sampling of approximate lengths in terms of the meter.

TABLE 2-7 *Approximate lengths in meters*

Measurement	Length, height, size, or distance		
distance from the earth to the sun	150,000,000,000 m	1.5×10^{11} m	150 Gm
diameter of the earth	13,000,000 m	1.3×10^{7} m	13 Mm
height of Mt. Everest	10,000 m	1.0×10^{4} m	10 km
height of Empire State Building	400 m	4.0×10^{2} m	400 m
human height	1.7 m	1.7×10^{0} m	1.7 m
length of a cockroach	0.040 m	4.0×10^{-2} m	40 mm
size of a grain of sand	0.00010 m	1.0×10^{-4} m	100 μm
size of a red blood cell	0.0000065 m	6.5×10^{-6} m	6.5 μm
size of a poliovirus	0.000000025 m	2.5×10^{-8} m	25 nm

In scientific work, very small distances often must be measured. Units smaller than the meter are therefore useful. One such unit, the **centimeter**, symbolized cm, is 1/100, 0.01, or 10^{-2} meter, since centi- means 100 times smaller.

$$\begin{array}{c|c} \text{centi} & \text{meter} \\ \text{1/100 times} & \text{unit of length} \\ \hline 1 \text{ cm} & = 10^{-2} \text{ m} \end{array}$$

There are 100 centimeters in a meter. A centimeter is less than half an inch, or about the width of a standard-size paper clip.

For even smaller distances, we use the **millimeter**. Just as a milligram is one thousand times smaller than a gram, a millimeter, symbolized mm, is one thousand times smaller than a meter:

$$\begin{array}{c|c} \text{milli} & \text{meter} \\ \text{1/1000 times} & \text{unit of length} \\ \hline 1 \text{ mm} & = 10^{-3} \text{ m} \end{array}$$

One meter contains 1000 millimeters. Since the millimeter is 10 times smaller than the centimeter, there are 10 millimeters in a centimeter:

10 mm = 1 cm

Figure 2-4 compares a portion of a yardstick to part of a meter stick. A meter stick is a "ruler" one meter long, and is used to measure metric lengths. It is divided into 100 small divisions, called centimeters,

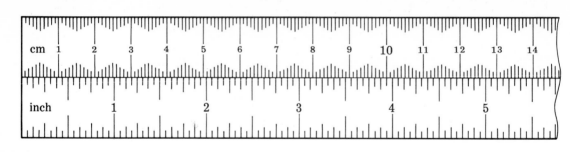

FIGURE 2-4

Portions of a meter stick and a yard stick (shown actual size). The large divisions of the meter stick are centimeters (0.01 m); the smallest divisions are millimeters (0.001 m). Notice that 1 in corresponds to approximately 2.5 cm.

each of which is further divided into 10 smaller divisions, called millimeters. Notice that 1 inch (in) is slightly larger than 2.5 centimeters:

$$1 \text{ in } = 2.54 \text{ cm}$$

Units that are even smaller than the millimeter are sometimes useful. For example, the smallest particles of matter are separated by distances that are expressed in terms of the picometer, pm, which is 10^{-12} meter. Another unit, the angstrom, Å, is sometimes used in chemistry, although it is not an official SI unit:

$$1 \text{ Å } = 10^{-10} \text{ m } = 100 \text{ pm}$$

Table 2-8 summarizes the most common metric units of length.

TABLE 2-8 *Metric units for length*

Unit	Symbol	Meaning
meter	m	base unit (equal to 39.37 inches)
centimeter	cm	1/100 (or 0.01) meter or 10^{-2} m
millimeter	mm	1/1000 (or 0.001) meter or 10^{-3} m

2.6
VOLUME

Matter takes up space. **Volume** is a measure of the amount of space occupied. The volume of a rectangular object like the box shown in Figure 2-5 is found by multiplying its length times its height times its width (volume = length × height × width). If each of these dimensions is exactly 1 meter, the volume of the box is 1 m × 1 m × 1 m, or 1 m³, namely one cubic meter. The **cubic meter** is the SI unit of volume.

Since the cubic meter is rather large for chemical measurements, smaller units of volume are more commonly used. For example, if a box has dimensions of 1 decimeter (0.1 meter) for its length, height, and width, its volume is 1 dm × 1 dm × 1 dm, or 1 dm³, called one cubic decimeter (equal to 10^{-3} m³). This volume is called the **liter**, and

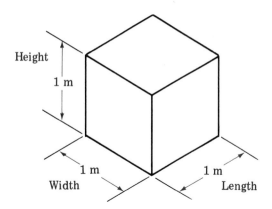

FIGURE 2-5

A cubic meter. The volume of this cube is 1 m × 1 m × 1 m = 1 m³.

its symbol is L or l. The liter is a commonly used unit in chemistry. One liter is a little larger than one quart (qt) in the English system:

$$1 \text{ dm}^3 = 1 \text{ L} = 1.06 \text{ qt.}$$

Because 1 decimeter equals 10 centimeters, we can think of the cubic decimeter or liter as a box having the dimensions 10 cm × 10 cm × 10 cm (see Figure 2-6). The volume of this box equals 1000 cm³, or one thousand cubic centimeters. Thus, the **cubic centimeter** is one-thousandth of a liter, and is also called a **milliliter**. The symbol for milliliter is mL or ml.

milli	liter
1/1000 times	unit of volume

$$1 \text{ mL} = 10^{-3} \text{ L}$$

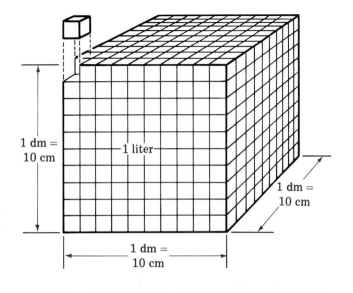

FIGURE 2-6

The relationship between a milliliter and a liter. One liter (dm³) contains 1000 milliliters (cm³).

1 dm = 10 cm

1 liter

1 dm = 10 cm

1 dm = 10 cm

TABLE 2-9 *Metric units for volume*

Unit	Symbol	Meaning
cubic meter	m^3	base unit
cubic decimeter	dm^3	1/1000 (or 0.001) m^3 or 10^{-3} m^3
liter	L (or l)	same as dm^3
cubic centimeter	cm^3	1/1000 (or 0.001) dm^3 or 10^{-3} dm^3
milliliter	mL (or ml)	1/1000 (or 0.001) L or 10^{-3} L

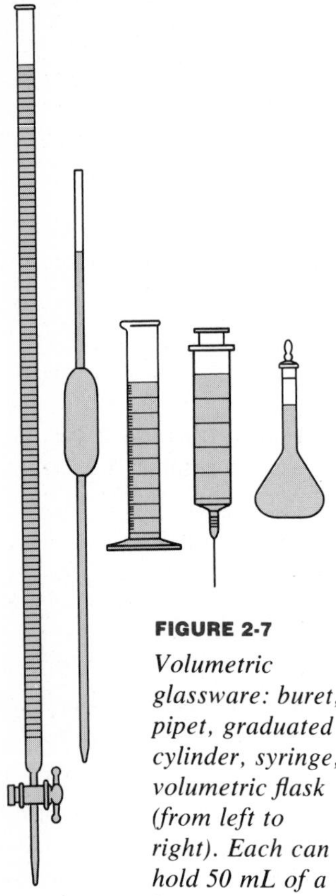

(One milliliter of liquid fills about one-fifth of a teaspoon. A cup of coffee has a volume of about 200 mL.) Thus,

$$1 \text{ L} = 1 \text{ dm}^3 = 1000 \text{ cm}^3 = 1000 \text{ mL}.$$

The cubic centimeter and milliliter have the same volume, which is 1000 times smaller than the cubic decimeter or liter. Table 2-9 summarizes these relationships.

Volumes of liquids can be measured in several ways. Figure 2-7 shows some of the equipment that is used to measure liquids. A buret is used to dispense small known volumes of liquid. The flow of liquid is controlled by a valve, called a stopcock, at the buret's tip. A pipet is used to make a single transfer of a measured volume of liquid from one container to another. A graduated cylinder is used to measure the approximate volume of a liquid before pouring it into another container. A syringe, similar to a pipet, is used to inject a known volume of liquid. A volumetric flask holds a fixed volume of liquid when it is filled to the mark on its neck. Each of these pieces of glassware is available in different sizes, depending on its intended use.

FIGURE 2-7

Volumetric glassware: buret, pipet, graduated cylinder, syringe, volumetric flask (from left to right). Each can hold 50 mL of a liquid.

2.7 ENERGY

Energy is a basic scientific concept that is closely related to matter. Energy is simply the ability to do work, to make matter move. As we will see, chemistry deals with energy as well as matter.

Energy exists in different forms. **Potential energy** is the energy that an object has due to its position or condition. A raised stone has potential energy; it can be allowed to drop to earth and do work. Water on top of a hill can be allowed to flow down; the energy released can run a mill. A wound clock spring has potential energy which is released as it unwinds.

Often when stored-up energy is released, it becomes kinetic energy. **Kinetic energy** is the energy an object has due to its motion. Fast-blowing wind has kinetic energy. A windmill taps this energy. The speed of the moving air is slowed down, and the energy released is used for work.

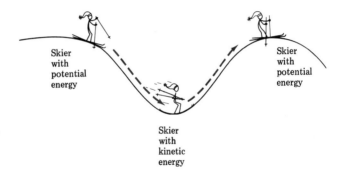

FIGURE 2-8

Energy conversion. This skier converts potential energy into kinetic energy and then back into potential energy.

Potential and kinetic energy can be changed into each other. As illustrated in Figure 2-8, a skier at the top of a hill has potential energy. As the skier moves down the hill, losing height and gaining speed, potential energy changes into kinetic energy. At the bottom of the hill, the skier has no potential energy, but a great deal of kinetic energy. The skier can use this kinetic energy to coast up another hill. The kinetic energy will then be converted back into potential energy, as the skier reaches the top of the second hill.

Chemists are especially concerned with a third form of energy called **internal energy**. Internal energy is the energy present within matter, rather than the energy (kinetic or potential) of an object as a whole. This form of energy will be discussed in later chapters.

The unit of energy in the SI system is the **joule** (pronounced jool), symbolized J. One joule is a rather small amount of energy. For example, your heart uses about 1 J with each beat. An electric toaster may use 1000 J each second that it is on. To describe larger amounts of energy, the **kilojoule** is used. The kilojoule, symbolized kJ, is 1000 times larger than the joule.

kilo	joule
1000 times	unit of energy

$$1 \text{ kJ} = 10^3 \text{ J}$$

You are probably familiar with another unit of energy, the **calorie**, symbolized cal. The calorie is over four times larger than the joule. One joule is therefore about one-fourth the size of a calorie:

$$1 \text{ J} = 0.239 \text{ cal}$$

The **kilocalorie**, symbolized kcal, is a unit 1000 times larger than the calorie.

$$1 \text{ kcal} = 1000 \text{ cal}$$

29

Nutritionists describe the energy value of foods using kilocalories, except that they refer to the kilocalorie as a Calorie (Cal) with a capital C. For example, the recommended caloric intake for women between the ages of 19 and 22 is 2100 Cal (2100 kcal) per day. In terms of joules, the amount of energy is 8786 kJ.

An important property of energy is that the total amount of energy stays the same, even though energy may change from one form to another. We say that energy is *conserved* and call this statement the **law of conservation of energy**. Here is an example of a conservation law. Imagine four people gathering together to gamble. Each brings some money, and the total amount is $600. After the gambling session is over, the gamblers leave with different amounts of money than they arrived with. The total amount among them, however, still remains $600. We say that the total amount of money is conserved in a gambling session.

The law of conservation of energy holds true under most conditions. However, at the beginning of this century, Albert Einstein predicted that under certain conditions energy and mass can be converted into each other. His prediction was dramatically shown to be correct when the first atomic bomb was exploded. Under these conditions, a small amount of matter is converted into a large amount of energy. For example, the energy needed to power New York City for one day (about 10^{15} J) could be obtained from only 5 g of matter if the mass were completely converted into energy. The relationship between energy and matter is discussed further in the chapter on Nuclear Chemistry (Chapter Seventeen).

2.8
SIGNIFICANT
FIGURES

Whenever we make a measurement of mass, length, or volume, the result is never exact. The result depends on how carefully we do the measuring and on the type of device we use. For example, suppose that you step on a digital scale that reads to the nearest pound. If the scale reads 154 lb this does not necessarily mean that your weight is exactly 154 lb. Instead, because the scale measures to the nearest pound and does not show fractions of a pound, the number 154 means that your weight has some value between 153.5 and 154.5 lb. Had your weight been less than 153.5 lb, the scale would have read 153 lb. Had your weight been greater than 154.5 lb, the scale would have read 155 lb.

The number of *meaningful* digits in a measured value is called the number of **significant figures.** Figure 2-9 shows how the same volume can be expressed with different numbers of significant figures, depending on how the measurement is made. Graduated cylinder A has markings for every 10 mL. On this scale, we can see that the measured volume is more than 70 mL and less than 80 mL. Since the level of the liquid lies about halfway between these two values, the volume is read as 75 mL. This measurement has two significant figures; the 7 is certain,

FIGURE 2-9

Significant figures. In graduated cylinder A, the level of liquid is about half-way between 70 mL and 80 mL. The volume is read as 75 mL (2 significant figures). In graduated cylinder B, the same level appears half-way between 74 mL and 75 mL. This volume is read as 74.5 mL (3 significant figures).

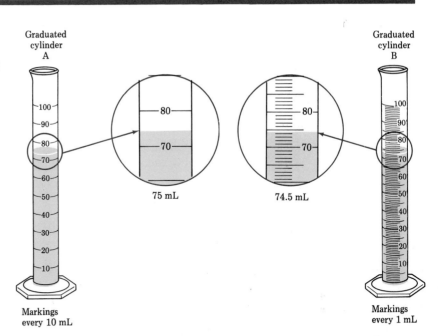

TABLE 2-10

Counting significant figures

Measured value	Number of significant figures
1	
3 cm	1
12 34	
45.87 mL	4
123	
0.223g	3
1234	
4529 m	4
12	
0.70 kg	2
12 3	
30.6 cm	3
12 34	
65.00 m³	4
1	
0.005 mg	1
1 2	
3.8 × 10³ L	2
1 23	
1.05 × 10⁻¹⁰ mm	3

but the 5 is an estimate. Using graduated cylinder B, which has markings for every milliliter, we can see that this same volume is between 74 mL and 75 mL, again about halfway between the two values. In this case, the volume is read as 74.5 mL. This measurement has three significant figures. The 7 and 4 are certain, and the 5 is an estimate. Significant figures are all the numbers that we are sure about, plus one more that we estimate.

Table 2-10 presents examples of measurements, and the number of significant figures in each case. Note that initial zeroes, used as ''place holders'' before the first nonzero digit, are not counted as significant figures. For example, the number 0.005 contains only one significant figure. Zeroes that follow other digits are significant if they come after the decimal point. Such zeroes show how reliable the measurement is. For example, there are four significant figures in the number 65.00. The two zeroes after the decimal point show that the measurement is reliable to the hundredths place. But zeroes that follow other digits and come *before* the decimal point may or may not be significant. For example, the number 500 may have one, two, or three significant figures depending on how the measurement was made.

Measured value	Meaning	Number of significant figures
500	between 450 and 550	1
500	between 495 and 505	2
500	between 499.5 and 500.5	3

To show whether a zero is a significant number or merely a place holder, we write the measurement in exponential notation. For example, the three values in the preceding list are written as 5×10^2, 5.0×10^2, and 5.00×10^2, respectively. Note that writing the number as 500. (with a decimal point after the last zero) also indicates three significant figures.

Keep in mind that not all numbers represent measurements. For example, there are *exactly* 10 millimeters in a centimeter, 12 objects in a dozen, and so on. These numbers are exact because they result from definitions rather than from measurements. We need to be concerned with significant figures only when we deal with measurements. Measurements are inexact and require estimation of the last digit.

In summary, here are the rules for determining the number of significant figures:

1 The digits in a number that are significant are those that are known with certainty plus the first digit whose value is estimated.

2 Zeroes following a number after the decimal point are counted as significant figures. (Example: 3.150 has four significant figures.)

3 Zeroes in front of a number are not counted as significant figures. (Example: 0.041 has two significant figures.)

4 Zeroes within a number are counted as significant figures. (Example: 20.6 has three significant figures.)

Practice problem   State the number of significant figures in each of the following quantities:

a 15.20 m

b 0.23 L

c 10.05 g

d 6.02×10^2 km

e 0.003 cm³

ANSWERS **a** 4 **b** 2 **c** 4 **d** 3 **e** 1

2.9
ROUNDING-OFF
NUMBERS

When we perform calculations involving measured values, it is often necessary to round-off numbers. Otherwise, the answer may contain more digits than are significant. Rounding-off involves reducing the number of digits according to certain rules. These rules are based on the value of the digits being removed. For example, when we fill out income tax forms, the Internal Revenue Service permits us to round-off all figures to whole dollar amounts, thus eliminating cents. According to their guidelines, any amount less than (but not including) $0.50 can be dropped from a larger number. Thus $89.37 is rounded to $89. But for this privilege, we must raise any amount to the next

FIGURE 2-10

Calculator. This hand-held calculator displays a number in exponential notation: 1.2367954×10^{-2}. In general, the result displayed on a calculator must be rounded off to the proper number of significant figures. (Photo courtesy of Texas Instruments.)

highest dollar if its cents are 50 or more. As a result, $89.57 becomes $90.

Rounding-off is particularly important when using a calculator. As shown in Figure 2-10, most calculators display an answer with eight digits. In nearly every case, not all of these digits are significant and the answer must therefore be rounded-off. (Appendix D contains additional information about using a calculator.)

In general, the value of the last digit that is kept, such as dollars in the income tax example, depends on the value of the digit just to its right, which is dropped. Even if several digits are being eliminated by rounding-off, they are removed together rather than one at a time. Suppose we wish to round off the following number to four significant figures (to the tenths place).

first digit being removed

$$378.5134$$

last digit being kept

Always start counting significant figures from the left. Whether or not the last digit being kept changes is determined by these rules:

Value of first digit being removed	Effect on last digit being kept
Rule A 0, 1, 2, 3, 4 (less than 5)	**stays the same**
Rule B 5, 6, 7, 8, 9, (5 or greater)	**increases by 1**

Thus our number is rounded to 378.5 by Rule A. Other examples are given in Table 2-11. (A third rule, which modifies Rule B may be added for exact work: If the first digit being removed is equal to 5, the last

TABLE 2-11 *Examples of rounding-off numbers*

Original number[a]	Significant figures desired	Rule used	Final number
1.6_7_	2	B	1.7
521._1_	3	A	521
4.56_5_	3	B	4.57
351._5_3	3	B	352
1._5_68	1	B	2
16.0_4_3	3	A	16.0
8.2_3_5 $\times 10^3$	2	A	8.2 $\times 10^3$
7540._9_16	4	B	7541

[a]The first number being dropped is underlined.

TABLE 2-12

Examples of computations with significant figures

Computation	Result
18.2 + 5.35 + 20.	44
15.02 + 0.003 + 700.1	715.1
7109.3 − 40.352	7068.9
1.521 − 0.81	0.71
38.2 × 0.95	36
17.32 × 1.66	28.8
182/32.800	5.55
0.881/5.2	0.17
$\dfrac{39.3 + 17.21}{190.}$	0.297
(58.1 × 0.82) − 1.19	46

digit being kept is rounded off to the nearest *even* digit. For example, the following numbers are rounded off to two significant figures using this rule: 8.55 becomes 8.6; 8.65 becomes 8.6; 8.75 becomes 8.8; 8.85 becomes 8.8.)

The number of digits that remain in the answer to an arithmetical calculation depends on the measured values involved. Certain rules apply to addition and subtraction, and different ones apply to multiplication and division. When adding or subtracting measurements, the final answer can have only as many *decimal places* as the measurement with fewest decimal places. (Decimal places are the numbers after the decimal point.) Consider addition:

$$
\begin{array}{r}
14.57 \\
1835.6 \\
0.875 \\
\hline
1851.045
\end{array}
$$

1851.0   rounded answer

In this example, the number with the fewest decimal places is 1835.6. Because this number has only one decimal place, the answer cannot be stated with any more certainty than one decimal place and must be rounded-off as indicated.

Now consider subtraction:

$$
\begin{array}{r}
63.523 \\
- \quad 1.41 \\
\hline
62.113
\end{array}
$$

62.11   rounded answer

In this example, the number 1.41 (with two decimal places) has fewer decimal places than the number 63.523 (which has three decimal places). Therefore, the answer is limited to two decimal places.

When we round-off results from multiplication or division, our answer should have the same number of *significant figures* (not decimal places) as the measurement with the fewest significant figures. Consider multiplication:

$$
\begin{array}{r}
5.87 \\
\times \quad 2.3 \\
\hline
1761 \\
1174 \\
\hline
13.501
\end{array}
$$

14   rounded answer

In this example, the measurement with fewer significant figures is 2.3. Because this number has two significant figures, the answer must also have two significant figures.

Now consider division:

$$
\begin{array}{r}
.058 \\
6\overline{)0.348} \\
\underline{30} \\
48 \\
\underline{48} \\
\end{array}
$$

0.06   rounded answer

In this example, the measurement 6 (the divisor) has only one significant figure. The answer must also have only one significant figure. Further examples of computations are shown in Table 2-12.

Practice problem   Round-off each of the following numbers to three significant figures.

a 187.6   d 39,215
b 5281   e 8.346×10^5
c 1.0334

ANSWERS   a 188   b 5.28×10^3 c 1.03   d 3.92×10^4 e 8.35×10^5

Practice problem   Perform the following calculations and express the results with the proper number of significant figures.

a $77.2 + 0.531 + 13.27$ d $1.594/6.23$
b 68.3×0.92
c $0.815 - 0.2346$ e $\dfrac{2.151 - 1.30}{0.591}$

ANSWERS   a 91.0   b 63   c 0.580   d 0.256   e 1.4

**2.10
CALCULATIONS
USING
EXPONENTIAL
NOTATION**
Many calculations are simplified when numbers or measurements are written in exponential notation. To do such computations, we must operate separately on the coefficients and on the exponents of the numbers. When multiplying numbers, for example, the rule is to *multiply* the coefficients and to *add* the exponents.

$$(3 \times 10^4) \times (2 \times 10^5)$$

$$= (3 \times 2) \times 10^{(4+5)}$$

multiply coefficients   add exponents

$$= 6 \times 10^9$$

If one or more exponents have negative signs, they must be added algebraically (see Appendix B.1).

$$(4 \times 10^{-6}) \times (2 \times 10^3)$$

$$= (4 \times 2) \times 10^{(-6+3)}$$

$$= 8 \times 10^{-3}$$

Division is carried out by the reverse procedures. We *divide* the coefficients (numerator ÷ denominator) and *subtract* the exponents (numerator − denominator). For example:

$$\frac{6 \times 10^4}{2 \times 10^3}$$

divide coefficients $\nearrow = \frac{6}{2} \times 10^{(4-3)}$ $\nwarrow$ subtract exponents

$$= 3 \times 10^1$$

In division, unlike multiplication, the order in which we carry out the division and subtraction is critical. Notice that the exponent in the denominator is always subtracted from the exponent in the numerator. The subtraction is done taking the rules for signed numbers into account.

$$\frac{2 \times 10^5}{2 \times 10^{-3}}$$

$$= \frac{2}{2} \times 10^{(5-[-3])}$$

$$= 1 \times 10^{(5+3)}$$

$$= 1 \times 10^8$$

To add or subtract numbers in exponential notation, we must first express all of the numbers in terms of the same power of 10. Then the sum is simply equal to the sum of the coefficients, expressed to this common power of 10. Consider this addition:

$$\begin{array}{r} 7 \ \times 10^3 \\ + \ 2.0 \times 10^4 \end{array}$$

The numbers cannot be added as they are written because they have different powers of 10. In such a case, it is often simplest to change the smaller power of 10 to the larger. For instance, 7×10^3 can be written as some coefficient times 10^4. Since the value of the power is being *increased* by 10 (multiplied by 10), the coefficient must be *decreased* by 10 (divided by 10) to keep the number unchanged:

$$7 \times 10^3 = 0.7 \times 10^4$$

The coefficient becomes 7/10, or 0.7, since the power becomes $10^3 \times 10$, or 10^4. When changing a power of 10, remember that the coefficient must always be changed by the same amount but in the *reverse* direction. In our example, since the power becomes larger, the coefficient becomes smaller. Now the addition can be performed.

$$\begin{array}{r} 0.7 \times 10^4 \\ + \ 2.0 \times 10^4 \\ \hline 2.7 \times 10^4 \end{array}$$

We treat subtraction similarly. First we express the numbers to the same power of 10. Then we find the difference between the coefficients. For example:

$$\begin{array}{r} 3.0 \times 10^6 \\ - \ 5 \ \ \times 10^5 \\ \hline \end{array} \quad \text{becomes} \quad \begin{array}{r} 3.0 \times 10^6 \\ - \ 0.5 \times 10^6 \\ \hline 2.5 \times 10^6 \end{array}$$

The answer is simply the difference between the coefficients, 2.5, expressed to the common power of 10, which is 10^6.

More complicated calculations can be carried out by using several different rules in the same calculation. For example:

$$\frac{8 \times 10^3 + 6.0 \times 10^4}{5.0 \times 10^{-9}}$$

$$= \frac{(0.8 \times 10^4) + (6.0 \times 10^4)}{5.0 \times 10^{-9}} \qquad \text{change powers of 10 for addition}$$

$$= \frac{6.8 \times 10^4}{5.0 \times 10^{-9}} \qquad \text{perform addition}$$

$$= \frac{6.8}{5.0} \times 10^{(4-(-9))} \qquad \text{perform division}$$

$$= 1.4 \times 10^{13} \qquad \text{answer}$$

The following list summarizes these rules.

Operation	Coefficients	Powers of 10
addition	add	must be the same
subtraction	subtract	must be the same
multiplication	multiply	add
division	divide	subtract

TABLE 2-13 *Calculations with exponential notation*

Calculation	Answer	Calculation	Answer
$(3.4 \times 10^{-2}) \times (1.7 \times 10^{-5})$	5.8×10^{-7}	$\dfrac{9.51 \times 10^{25}}{3.30 \times 10^{10}} - 1.1 \times 10^{15}$	1.8×10^{15}
$(6.8 \times 10^{12}) + (9.2 \times 10^{11})$	7.7×10^{12}	$(8.9 \times 10^3) + (9.2 \times 10^2) + (1.30 \times 10^4)$	2.3×10^4
$\dfrac{5.51 \times 10^{-3}}{8.93 \times 10^{-6}}$	6.17×10^2	$5.62 \times 10^{-10} + \dfrac{8.2 \times 10^{-14}}{1.9 \times 10^{-4}}$	9.9×10^{-10}
$(9.8 \times 10^5) - (1.20 \times 10^6)$	-2.2×10^5	$\dfrac{(4.3 \times 10^8) + (5.60 \times 10^9)}{(6.51 \times 10^{-3}) - (1.8 \times 10^{-4})}$	9.53×10^{11}
$\dfrac{(7.40 \times 10^3) - (2.1 \times 10^2)}{8.3 \times 10^{-4}}$	8.7×10^6	$\dfrac{1.00}{8.93 \times 10^{-19}}$	1.12×10^{18}

Table 2-13 contains more examples of computations using exponential notation. Perform them yourself for practice, checking your answers with those that are given. Note that the rules for handling significant figures apply to the coefficients of numbers written in exponential notation.

Practice problem   Perform the following calculations.

a $(8.91 \times 10^{-2}) \times (1.2 \times 10^5)$

b $(1.15 \times 10^8) + (6.2 \times 10^7)$

c $\dfrac{5.09 \times 10^5}{8.2 \times 10^{-2}}$

d $(9.82 \times 10^4) - (1.35 \times 10^3)$

e $\dfrac{(2.731 \times 10^{-2}) \times (1.52 \times 10^9)}{8.3 \times 10^{14}}$

ANSWERS   **a** 1.1×10^4 **b** 1.77×10^8 **c** 6.2×10^6 **d** 9.69×10^4
e 5.0×10^{-8}

**2.11
ESTIMATING
CALCULATIONS**

Estimating calculations is valuable because it helps you to catch mistakes. If you use a calculator, estimating is especially important because it makes you think about your answer. Estimating involves simplifying a calculation so that you can quickly produce an approximate result. Each quantity is rounded-off to only one significant figure and then written in exponential notation, as shown in Table 2-14. If your computation involves intermediate results, they too are rounded-off to one significant figure each.

Consider the following calculation:

$$\frac{8370 \times 0.092}{4.81}$$

TABLE 2-14

Estimating numbers

Number	Estimate
529	5×10^2
0.1372	1×10^{-1}
8499	8×10^3
4.32	4×10^0 or 4
0.0088	9×10^{-3}
25.9	3×10^1

It can be estimated as

$$\frac{(8 \times 10^3) \times (9 \times 10^{-2})}{5}$$

Now you can easily do the computation:

$$= \frac{(8 \times 9) \times 10^{(3-2)}}{5}$$

$$= \frac{72 \times 10^1}{5}$$

This intermediate result is rounded off to:

$$\frac{7 \times 10^2}{5}$$

$$= \frac{7}{5} \times 10^2$$

$$= 1 \times 10^2$$

The estimated answer for this calculation is 100. If you actually carried out the computation, you would get the result 160, or 1.6×10^2. If you had calculated a number greatly different from the estimate, say 16, you would know that you should check your computation.

Although estimating an answer is extra work, practice reduces the amount of time it takes. Study the following sample problem.

$$9.12 \times \frac{8291}{0.63} \times \frac{50.3}{230.}$$

The estimated result is:

$$9 \times \frac{8 \times 10^3}{6 \times 10^{-1}} \times \frac{5 \times 10^1}{2 \times 10^2}$$

$$= \frac{9 \times 8 \times 5 \times 10^{(3+1)}}{6 \times 2 \times 10^{(-1+2)}}$$

$$= \frac{360 \times 10^4}{12 \times 10^1}$$

$$= \frac{4 \times 10^6}{1 \times 10^2}$$

$$= 4 \times 10^4$$

TABLE 2-15   *Estimating calculations*

Calculation	Set-up for estimate	Estimated answer	Actual answer
732 + 1526	$(7 \times 10^2) + (2 \times 10^3)$	3×10^3	2258
$\dfrac{23.5}{0.0581}$	$\dfrac{2 \times 10^1}{6 \times 10^{-2}}$	3×10^2	404
852×21	$(8 \times 10^2) \times (2 \times 10^1)$	2×10^4	1.8×10^4
$382 \times \dfrac{0.16}{7.32}$	$(4 \times 10^2) \times \dfrac{2 \times 10^{-1}}{7}$	1×10^1	8.3
$\dfrac{0.498}{53} \times \dfrac{651}{239}$	$\dfrac{5 \times 10^{-1}}{5 \times 10^1} \times \dfrac{7 \times 10^2}{2 \times 10^2}$	4×10^{-2}	0.026
$\dfrac{2.86 \times 0.0182 \times 460.}{39.3}$	$\dfrac{(3) \times (2 \times 10^{-2}) \times (5 \times 10^2)}{4 \times 10^1}$	8×10^{-1}	0.610

The actual result is 2.6×10^4. More examples are given in Table 2-15. Try working them out yourself.

Practice problem   Write estimates for the following numbers.

a 93

b 0.0378

c 4521

d 1.25

e 0.0022

ANSWERS   a 9×10^1 b 4×10^{-2} c 5×10^3 d 1   e 2×10^{-3}

Practice problem   Estimate answers to the following calculations.

a 4251×837

b $\dfrac{76.8}{3.91}$

c $6956 - 5412$

d $\dfrac{18.2 \times 71.9}{0.553}$

e $\dfrac{391}{0.936 \times 12.2}$

ANSWERS   a 3×10^6 b 2×10^1 c 2×10^3 d 2×10^3 e 4×10^1

Summary

2.1 At this stage, we have learned that chemistry deals with matter. Matter is defined as anything that occupies space. It exists in three states: solid, liquid, and gas.

2.2 Measurements are used to describe matter and energy. A measurement consists of a number and unit that communicate the amount of something. Chemists use the metric system of measurement, which is based on the number 10. The most common metric prefixes are kilo-, meaning one thousand times larger, and milli-, meaning one thousand times smaller.

2.3 Large or small measurements and numbers can be conveniently expressed using exponential (scientific) notation. Exponential notation is based on writing a number in terms of a power of 10 multiplied by a coefficient between 1 and 10.

2.4 Mass is a measure of the amount of matter that an object contains. The base unit of mass in the metric system is the gram (symbolized g). Large masses are measured in terms of kilograms (kg) and smaller masses are measured in milligrams (mg).

2.5 Length is a measure of the distance between two points. The basic metric unit of length is the meter (symbolized m). Other common units are the kilometer (km), centimeter (cm), and millimeter (mm).

2.6 Volume is a measure of the amount of space occupied by matter. The units of volume used by chemists are the liter (L), and the milliliter (mL) or cubic centimeter (cm^3).

2.7 Energy is the ability or capacity to do work, that is, to force matter to move. Energy can exist in many different forms, and can be converted from one form to another. However, the total amount of energy always remains the same.

2.8 Measurements must always be expressed with the proper number of significant figures. They represent the number of meaningful digits in a measured value.

2.9 When performing calculations with measurements, it is often necessary to round-off numbers. This process involves reducing the number of digits according to certain rules. In addition and subtraction, the final answer is limited to the number of decimal places that are in the measurement with fewest decimal places. In multiplication and division, the answer should have the same number of significant figures as the measurement with fewest significant figures.

2.10 Many calculations are simplified when numbers or measurements are written in exponential notation. To do such computations, we must operate separately on the coefficients and the exponents of the numbers.

2.11 Estimating involves simplifying a calculation so that we can quickly produce an approximate result. This technique helps us to catch mistakes. In estimating, each number is first rounded-off to one significant figure and then written in scientific notation.

Exercises

KEY-WORD MATCHING EXERCISE For each term on the left, choose a phrase from the right-hand column that most closely matches its meaning.

1 matter		**a**	unit of mass
2 energy		**b**	based on 10
3 solid		**c**	space occupied
4 gram		**d**	distance between points
5 significant figure		**e**	energy cannot be destroyed
6 meter		**f**	gives amount of something
7 volume		**g**	unit of length
8 metric system		**h**	has mass
9 length		**i**	unit of volume
10 law of conservation of energy		**j**	definite shape and volume
11 measurement		**k**	expands to fill container
12 liter		**l**	moves matter
13 gas		**m**	amount of matter
14 exponential notation		**n**	definite volume, not shape
15 liquid		**o**	meaningful digit
16 mass		**p**	uses exponents

EXERCISES BY TOPIC

Matter (2.1)

17 What is matter? In what forms can it exist?

18 How can you tell if a substance is a solid, liquid, or gas?

19 Give two examples each of a solid, a liquid, and a gas (other than those given in the textbook).

20 For each of the following substances, state whether it exists as a solid, liquid, or gas (at room temperature): molasses, iron, vinegar, neon, gold.

21 Which state of matter **a** has a definite shape? **b** can be compressed? **c** fills a container from the bottom? **d** expands greatly upon heating?

Measurement and the metric system (2.2)

22 How is the metric system simpler than the English system?

23 If a "buck" is dollar, how much money is a "kilobuck?"

24 Arrange in order of increasing size: micro-, kilo-, centi-, mega-, deci-.

25 How many times larger is the prefix kilo- compared to milli-?

26 Why is measurement needed to describe matter or energy?

Exponential notation (2.3)

27 Why is exponential notation useful?

28 The sun is about 93 million miles from Earth. Express this distance using exponential notation.

29 Use exponential notation to express the number of pages in this book.

30 If a poliovirus is 0.0000009843 inches long, what is its length in exponential notation?

31 About 1.85×10^8 bottles of Coca Cola are sold in the world each day. Express this number in decimal form.

32 The smallest free-living organisms have a diameter of 4×10^{-6} inches. Write this number as a decimal.

33 Express in exponential notation: **a** 92.2 **b** 0.00555 **c** 389,400 **d** 15 **e** 0.025

34 Write the following numbers in decimal form: **a** 2.8×10^3 **b** 1.52×10^{-2} **c** 8×10^4 **d** 7.6×10^{11} **e** 3.1×10^{-5}

Mass (2.4)

35 Mass is a measurement of what?

36 How does mass differ from weight?

37 Describe how the mass of an object can be determined.

38 In each case, which is larger: **a** 10,000 g or 1 kg? **b** 0.010 g or 1 mg? **c** 10^{-7} g or 1 μg?

39 Using Table 2-5, which component of the human body has the greatest mass?

Length (2.5)

40 Which is longer, a 180 m race or a 100 yd race?

41 How many meters are equal to one centimeter?

42 How many times larger is a millimeter than a nanometer?

43 Arrange in size from the shortest to the longest: 100 mm, 10 km, 100 m, 1000 μm, 1Å, 10 pm.

Volume (2.6)

44 What is the relationship between a liter and **a** a cubic decimeter? **b** a cubic centimeter?

45 What fraction of a liter is a milliliter?

46 Describe three ways that you might measure the volume of a liquid.

47 How many microliters are in a milliliter?

48 Which contains more liquid: a quart of milk or a liter of milk?

Energy (2.7)

49 What is energy? How is it related to matter?

50 Give one example of **a** kinetic energy and **b** potential energy (other than those given in the book).

51 What is internal energy?

52 What does it mean to say that energy is conserved?

53 How are kinetic and potential energy related?

Significant figures (2.8)

54 Why is it important to be able to recognize significant figures?

55 An African goliath beetle, the world's heaviest insect, has a mass of 99.790 g. To how many significant figures is this mass expressed?

56 Identify the number of significant figures in each measurement: **a** 1.030 μg **b** 60.2 m **c** 0.005 L **d** 3.10×10^{-2} m **e** 5.1×10^{-3} dm^3

57 Express the number 1500 with **a** three significant figures **b** two significant figures.

58 How many significant figures are in each of the following quantities? **a** 10. kg **b** 1×10^2 L **c** 0.022 cm **d** 82 mg **e** 7.542 m

Rounding-off numbers (2.9)

59 Round off each number to two significant figures: **a** 1.09 **b** 358 **c** 6.66×10^{-2} **d** 0.04782 **e** 58.5

60 Round off each number to three significant figures: **a** 4.588 **b** 0.008355 **c** 34.51 **d** 0.06648 **e** 1256

61 Two of the world's heaviest people weighed 875.5 lb and 800.25 lb, respectively. Add their weights and round-off if necessary.

62 Find the difference in mass between two apples that have masses of 119 g and 103.85 g.

63 Perform the following calculations: **a** 16.4 + 2.55 **b** 5.4210 − 3.1

64 Perform the following calculations: **a** 1.112 × 0.2 **b** 4.56 ÷ 3 **c** 3.15 × 1.9 **d** 3926 ÷ 13

65 Express the results of these calculations with the proper number of significant figures: **a** $\dfrac{43.15 - 0.8}{10.3}$ **b** $(7.9 - 1.25) \times (12.2 + 109)$

Calculations using exponential notation (2.10)

66 The largest swimming pool has a length of 4.80×10^2 m and a width of 7.5×10^1 m. Find its area (length $\times$ width) in square meters.

67 Perform the following calculations: **a** $(5 \times 10^3) \times (3 \times 10^{-2})$ **b** $(2 \times 10^4) \times (3 \times 10^7)$ **c** $(4 \times 10^{-5}) \times (2 \times 10^{-3})$ **d** $(2 \times 10^{-4}) \times (2 \times 10^3)$

68 Do the following calculations:

a $\dfrac{12 \times 10^4}{2 \times 10^5}$ **b** $\dfrac{6 \times 10^4}{3 \times 10^3}$ **c** $\dfrac{7.5 \times 10^7}{3 \times 10^6}$ **d** $\dfrac{8 \times 10^{-2}}{2 \times 10^4}$ **e** $\dfrac{1.0 \times 10^4}{5 \times 10^{-2}}$

69 Carry out these calculations: **a** $8 \times 10^2 + 3 \times 10^3$
b $5 \times 10^4 + 4 \times 10^4$ **c** $3 \times 10^{-3} + 4.0 \times 10^{-2}$
d $6.2 \times 10^{-1} + 1.9 \times 10^{-1}$

70 Perform the following calculations: **a** $6.0 \times 10^7 - 4 \times 10^6$
b $4.0 \times 10^{-1} - 5.0 \times 10^{-2}$ **c** $5.0 \times 10^5 - 2.0 \times 10^6$ **d** $8.1 \times 10^8 - 4.92 \times 10^7$

71 Do these calculations:
a $\dfrac{3.0 \times 10^4 - 1.9 \times 10^3}{2.7 \times 10^6}$ **b** $\dfrac{6.1 \times 10^{-2} + 3.8 \times 10^{-3}}{9.8 \times 10^5}$

Estimating calculations (2.11)

72 The deepest cave in the world, located in France, is 3851 ft below the surface of the earth. Write an estimate for this depth.

73 Write estimates for the following numbers: **a** 945 **b** 0.0756 **c** 25.1 **d** 1.42 **e** 365,000

74 Estimate answers to the following calculations: **a** $3092 + 540$ **b** 65×432 **c** $\dfrac{52}{0.085}$ **d** $1152 - 879$.

75 Estimate: **a** $\dfrac{45 \times 0.87}{8.23}$ **b** $982 \times \dfrac{0.054}{33}$ **c** $\dfrac{125 - 8.7}{1452}$

76 Estimate answers to the calculations in Exercise 71.

ADDITIONAL EXERCISES

77 What is the weight in pounds of one gram?

78 Why are units of volume expressed as cubes: cm^3, m^3, and so on?

79 It took 1.03×10^2 hours for the Apollo XI spacecraft to reach the moon, 3.84×10^8 m away from Earth. Calculate the average speed in m/hr for this spacecraft.

80 The South China Sea, the world's largest sea, has an area of $1,148,500$ mi^2. Write an estimate for this area.

81 Perform the following calculations: **a** $51.003 + 2.009$ **b** $7.59 - 1.1$
c $\dfrac{4 \times 10^{-6}}{2 \times 10^{-3}}$

82 Why is your weight less, but your mass the same, on the moon?

83 A cockroach was measured to be 20 mm long. What is its length in centimeters?

84 A 1 dm^3 graduated cylinder is filled with 500 mL of water. How much of the cylinder is filled?

85 How many milliliters are contained in a 2 L bottle?

86 An Australian mouse has an average weight of only 0.141 oz. To how many significant figures is this weight given?

87 A measured length is found to be between 145 and 155 m. Express this length with the proper number of significant figures.

88 Round off the average human height, 1.70 m, to one significant figure.

89 A box has the dimensions 1.63×10^{-1} m, 2.5×10^1 m, and 4.85×10^1 m. Find its volume.

90 Perform the following calculations: **a** $\dfrac{7.12 \times 10^3 - 6.0 \times 10^2}{1.99 \times 10^{-3}}$

b $(17.9 + 5) \times (16 - 1.8)$

91 Estimate the answers for Exercise 90.

92 A bullet cap explodes in 0.000001 second. Express this time in exponential notation.

93 State whether each of the following objects has mainly kinetic or potential energy: **a** a moving car **b** a rock on a ledge **c** a wound-up mechanical toy **d** a rolling ball

94 Why is it especially important to keep track of significant figures when you use a calculator?

95 Which measurement does not belong in the following list? 100 mm 0.10 m 10,000 μm 10 cm

96 The mass of Jupiter is much greater than the mass of Earth. How would an astronaut's mass and weight on Earth compare with his or her mass and weight on Jupiter?

97 Express the following numbers in exponential notation: **a** 0.00345 **b** 1,108,000 **c** 13.52 **d** 880. **e** 0.1

98 A scientist friend tells you that she won 2 megadollars in a lottery. How much money did she win?

99 The tallest tree, which is located in Redwood Grove, California, is 109,860 mm tall. Round off this measurement to 2 significant figures.

100 Which of the following units are used to express **a** mass? **b** energy? **c** length? **d** volume?: cm cm^3 Å kJ m mL g dm^3 cal nm

101 The Statue of Liberty weighs 450,000 pounds. Express this number in exponential notation using two significant figures.

PROGRESS CHART   Check off each item when completed.

———————   Previewed chapter

Studied each section:

———————   **2.1**   Matter

———————   **2.2**   Measurement and the metric system

———————   **2.3**   Exponential notation

———————   **2.4**   Mass

———————   **2.5**   Length

———————   **2.6**   Volume

———————   **2.7**   Energy

———————   **2.8**   Significant figures

———————   **2.9**   Rounding-off numbers

———————   **2.10** Calculations using exponential notation

———————   **2.11** Estimating calculations

———————   Reviewed chapter

———————   Answered assigned exercises:

      #'s ————————————————————————

THREE

Problem solving and measurement applied

In this chapter, we introduce a general method for solving problems. You will see that, in problem solving, the units of measurement you learned in Chapter Two must often be converted to other units. To do this, we use what we call the conversion factor method. Using this method, we explain conversions between metric units, and between the English and metric systems. We also discuss temperature conversions and the nature of heat. Finally, the concept of density is introduced, and several density problems are solved.

To demonstrate an understanding of Chapter Three, you should be able to:

1 Identify the steps involved in problem solving.

2 Explain the conversion factor method for solving problems.

3 Perform metric conversions using a single conversion factor.

4 Perform metric conversions that involve several conversion factors.

5 Convert between quantities in the metric and English systems.

6 Compare the Kelvin, Celsius, and Fahrenheit temperature scales, and convert readings from one scale to another.

7 Explain the relationships between heat, energy, and temperature.

8 Define density and solve problems based on density.

**3.1
PROBLEM SOLVING**

Many chemical concepts and relationships are expressed in mathematical form. While learning or actually "doing" chemistry, you will be required to solve many problems. Problem solving is therefore an essential skill that you will need to develop. This section contains important hints that will help you to work out problems in a systematic way. Read these hints now, and then again each time that you solve a problem, until the steps become automatic. Then review this section occasionally to refresh your memory.

The first step in problem solving is to *understand the problem completely*. Read through the problem carefully. Look up any words that you don't know (either in a dictionary or in the glossary in Appendix I). It is often helpful to restate the problem in your own words. If you cannot restate the problem, it may mean that you do not understand it clearly. Pick out the key parts of the problem. First, identify the **unknown,** what you are being asked to find. Next, identify the **given** information, the data or facts that you will use to solve for the unknown. Be cautious here. Details that are not needed for the solution are sometimes given in the problem.

It is helpful to outline a problem as follows, listing the unknown and the given information:

UNKNOWN	GIVEN

For example, consider the following problem.

A Texan plans to visit Saudi Arabia to talk about oil drilling. The currency of that country, the riyal, can be exchanged at the rate of three riyals per dollar. The Texan wants to exchange 100 dollars. How many riyals is this amount worth?

This problem involves conversion, changing dollars to riyals. You can restate the problem as "Convert 100 dollars to riyals" and outline it as shown.

UNKNOWN	GIVEN
riyals	100 dollars

Once you have identified the unknown and the given information, you must find a **connection** between them. Such a connection will enable you to set up the problem for solution. If the connection is not given in the problem, it may be based on a chemical principle. Review the relevant information in this book or in your notes if necessary.

Another possible starting point is to examine a similar or related problem that you solved previously. Methods used in the old problem may give clues to the solution of the new problem. If the problem is entirely new, however, and you cannot see an obvious approach to the solution, try looking at the problem in different ways. If you can't see how to go from the given information to the unknown, try the reverse. Start with the unknown and try to relate it to the given information.

If the problem is complex, you may have to break it down into smaller parts. Many problems involve more than one relationship. Such problems must be solved in several steps, with the answer from one step becoming the starting point for the next step.

Once you have found the needed connection, add it to the outline of your problem:

UNKNOWN	GIVEN	CONNECTION

Now you are ready to set up the problem.

In our sample problem, the connection between the unknown and given information is stated in the problem itself: 3 riyals per dollar. Since "per" means "divided by," you can write this relation as a fraction, 3 riyals/1 dollar, and place it in the connection column.

UNKNOWN	GIVEN	CONNECTION
riyal	100 dollars	$\dfrac{3 \text{ riyals}}{1 \text{ dollar}}$

Once you have related the unknown to the given information, you must translate this understanding into a mathematical form, which we call a *set-up,* that will provide the solution. It is essential that you carry out this step carefully, and think about what you are doing. The plan may not be obvious right away, but do not give up. This step is often the hardest part of problem solving and is usually the most important. Scientists sometimes spend many hours, days, or even months thinking about difficult problems.

The set-up generally consists of an arithmetical computation or an algebraic equation or both. You might want to refer to Appendices B and C for a review of basic arithmetic and algebra. If you use a ready-made formula from the textbook, make sure that it applies to the problem. Also be sure that you understand how to substitute the information given in the problem for the symbols in the formula.

Once you have set up the problem, check to see whether or not your method will actually provide the answer required. For example, if you have written an algebraic equation, is the unknown in the equation the same as the unknown in the problem? If not, can you solve for the unknown in the problem using the answer from the equation? If you can answer "yes" to one of these questions, then the problem is set up correctly. If you *cannot* answer "yes" to one of these questions, then the set-up is incorrect.

Returning again to our sample problem, you could set up the

following computation to convert 100 dollars to riyals:

$$100 \text{ dollars} \times \frac{3 \text{ riyals}}{1 \text{ dollar}}$$

At this point, it may not be clear to you why this particular computation will provide the solution. However, the set-up will be explained in the following section, when we present the conversion factor method (an almost foolproof approach for handling many types of problems).

The next step, *solving the problem,* has two parts. First, before you actually do any calculation, *estimate* the result. Estimating (described in Section 2.11) will give you an approximate answer that is useful in avoiding mistakes.

Next, perform the necessary algebra and arithmetic. Check and justify each step as you carry it out. Take each step in proper order. Follow the rules for significant figures and for rounding-off calculations, as described in Sections 2.9 and 2.10.

Now *check* your answer by comparing the calculated result with your estimated result. If they are not reasonably close, check both your estimate and your calculation for errors. If, on the other hand, the two results are similar, you probably have the correct answer.

In our sample problem, the calculation step is the same as the estimation step, since the computation consists of numbers that need not be rounded off.

$$100 \text{ dollars} \times \frac{3 \text{ riyals}}{1 \text{ dollar}} = 300 \text{ riyals}$$

Even after you have obtained the solution, you are not yet finished with the problem. You must check to see that the solution makes sense and that it answers the original question. Since 3 riyals are exchanged for each dollar in this example, your answer in riyals must be larger, in fact 3 times larger, than the given number of dollars. Make sure that you have made use of all *relevant* information (not necessarily *all* of the information given).

Check to see that the units of your answer are correct and are written along with the answer. Be sure that the answer is expressed with the proper number of significant figures. If possible, check your answer. Solutions to most of the numerical problems in this book are given in Appendix J. Even if the correct answer is not available, you may be able to check your answer. Try to work out the problem again, using a different approach, or by working backwards from your answer.

When you are satisfied that you have solved the problem correctly, *review* it completely. Study your method of solution to reinforce your understanding of how you solved the problem. This step may make the next problem easier to handle.

The following summarizes the steps involved in problem solving:

1 Understand.

2 Make connections.

3 Set up.

4 Solve: estimate, calculate.

5 Check and review.

Practice is essential to problem solving. By working out many examples in this step-by-step manner, you will develop skill and confidence in your ability to tackle new problems. Don't feel discouraged if you have difficulties at first. Concentrate on what you are doing and don't give up! Get help from others if you need it.

**3.2
CONVERSION
FACTOR METHOD**

Many chemistry problems can be solved by an approach that we call the **conversion factor method**. This method is sometimes called dimensional analysis. It is based on the units (or dimensions) of the quantities involved in a calculation. The conversion factor is simply a fraction that has different units in the numerator (top) and in the denominator (bottom). But these units must have a special relationship to each other. They may be different ways of describing the same thing, such as 12 and one dozen, or they may be other mathematically equivalent quantities like 100 cents and one dollar, as shown in Figure 3-1. A conversion factor may also represent a relationship between two very different quantities, such as the conversion factor 55 miles per hour.

Because it is a fraction, a conversion factor can be written in two different ways, depending on which quantity goes into the numerator and which goes into the denominator. For example, the relation between dollars and cents may be written as:

$$\frac{100 \text{ cents}}{1 \text{ dollar}} \quad \text{or} \quad \frac{1 \text{ dollar}}{100 \text{ cents}}$$

These factors can be used to convert a certain amount of money from dollars to cents, or from cents to dollars. The type of conversion that you want to make determines which factor you use.

Similarly, the speed of a car could be expressed as the following conversion factors:

$$\frac{55 \text{ miles}}{1 \text{ hour}} \quad \text{or} \quad \frac{1 \text{ hour}}{55 \text{ miles}}$$

The second factor is more unusual than the first, but it contains similar information. Notice that the phrase "miles per hour" means miles per 1 hour. The number 1 is understood in common speech. However, be

100 cents = 1 dollar

FIGURE 3-1

A conversion factor. The relationship between two equivalent quantities can be expressed as a conversion factor. For example, the relationship between 100 cents and 1 dollar can be expressed by the conversion factor $\frac{100 \text{ cents}}{1 \text{ dollar}}$ or $\frac{1 \text{ dollar}}{100 \text{ cents}}$

sure to write the 1 when you use a conversion factor. In nearly all cases, the 1 in a conversion factor is exact and does not limit the number of significant figures in a calculation. Also, remember that "per" means "divided by" and is commonly used to relate different quantities.

To solve a problem using the conversion factor method, set up the calculation as:

unknown quantity = given quantity × conversion factor

For example, if you want to know the value in dollars of 543 cents, begin by writing down the unknown, dollars. Then set equal to the unknown the given information, 543 cents, multiplied by a conversion factor:

dollars = 543 cents × conversion factor

The conversion factor allows you to change the units of the given quantity into the units of the unknown quantity.

Now you must determine which of the possible conversion factors to use. Look at the two possibilities for this example:

A (incorrect) *B* (correct)

$$\text{dollars} = 543 \text{ cents} \times \frac{100 \text{ cents}}{1 \text{ dollar}} \qquad \text{dollars} = 543 \text{ cents} \times \frac{1 \text{ dollar}}{100 \text{ cents}}$$

A will not work because it will not give an answer in dollars. Instead, the answer will have the meaningless units (cents)^2/dollar. In *B* however, the cents in 543 cents will cancel the cents in the denominator of the conversion factor, giving an answer in dollars.

$$\text{dollars} = 543 \;\cancel{\text{cents}} \times \frac{1 \text{ dollar}}{100 \;\cancel{\text{cents}}}$$

$$= \frac{543 \times 1 \text{ dollar}}{100} = 5.43 \text{ dollars} = \$5.43$$

Units cancel when they appear in both the numerator (on top) of one fraction (543 cents is understood to mean 543 cents/1) and in the denominator (on the bottom) of another fraction. Thus, when the unit you wish to cancel is in the numerator of the given quantity (as in 543 cents/1), you arrange the conversion factor so that the unit to be cancelled is in its denominator. The *desired* unit should be in the numerator of the conversion factor.

This type of conversion problem can also be solved using a proportion:

$$\frac{x}{1 \text{ dollar}} = \frac{543 \text{ cents}}{100 \text{ cents}}$$

Note that cross-multiplying the proportion gives an equation that is identical to the one arrived at by the conversion factor method.

$$x = \frac{543 \text{ cents} \times 1 \text{ dollar}}{100 \text{ cents}}$$

Appendix C.2 provides further information about the use of proportions. In this book, we will use the conversion factor method for most problems.

One of the major advantages of the conversion factor method is that it provides an automatic means of checking your result. The answer must come out with the proper units. If it does not, you know that you have set up the problem wrong, or used an incorrect conversion factor, or cancelled units incorrectly.

3.3
METRIC
CONVERSIONS

Converting one metric unit to another is very simple because each unit differs from another only by a power of 10. (This straightforward type of conversion will give you practice in using the methods just described, as well as in handling metric units.) All you need to know is the meaning of the metric prefixes and how to use the conversion factor method. To form the conversion factors, simply write fractions based on the meaning of the unit, expressed in exponential notation. For example, suppose that you want to convert grams to milligrams. We know that $1 \text{ mg} = 10^{-3} \text{ g}$. Therefore, the conversion factors are:

$$\frac{1 \text{ mg}}{10^{-3} \text{ g}} \quad \text{and} \quad \frac{10^{-3} \text{ g}}{1 \text{ mg}}$$

Note that both terms in a metric conversion factor are exact and could be written with any number of significant figures. The number of significant figures in the answer depends upon the number of significant figures in the given information, not in the metric conversion factor.

Consider the following problem:

The average woman needs about 0.018 g of iron each day from her diet. How would we state this requirement in milligrams?

According to our guidelines for solving problems, the first step is to understand the problem clearly. In this case, you might first restate the problem as: "Convert 0.018 g to milligrams," and then outline it as shown:

UNKNOWN	GIVEN
milligrams	0.018 g

You are given a quantity in grams and must find its value in milligrams.

The second step is to relate the unknown to the given information, to make a connection. Knowing that *milli-* means 10^{-3} allows you immediately to write the two possible conversion factors.

UNKNOWN	GIVEN	CONNECTION
milligrams	0.018 g	$\dfrac{1 \text{ mg}}{10^{-3} \text{ g}}$ or $\dfrac{10^{-3} \text{ g}}{1 \text{ mg}}$

The third step is to set up the problem. According to the conversion factor method, the unknown must be equal to the given information multiplied by some conversion factor:

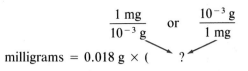

$$\frac{1 \text{ mg}}{10^{-3} \text{ g}} \quad \text{or} \quad \frac{10^{-3} \text{ g}}{1 \text{ mg}}$$

$$\text{milligrams} = 0.018 \text{ g} \times (\qquad ? \qquad)$$

Because you want the answer to be in milligrams, you must cancel the unit *grams* in 0.018 g. To do this, you must choose the conversion factor that has grams in the denominator: $1 \text{ mg}/10^{-3} \text{ g}$. The *desired* unit, milligrams, is in the numerator. The set-up then becomes:

$$\text{milligrams} = 0.018 \text{ g} \times \frac{1 \text{ mg}}{10^{-3} \text{ g}}$$

The fourth step is to estimate and then calculate the solution. (Although estimating may seem to be a waste of time in such a simple problem, it is a good habit to develop and will be especially valuable in more complex problems.)

ESTIMATE $\quad 2 \times 10^{-2} \cancel{g} \times \dfrac{1 \text{ mg}}{10^{-3} \cancel{g}}$

$$= \frac{2 \times 10^{-2} \times 1 \text{ mg}}{10^{-3}}$$

$$= \frac{2 \times 10^{-2}}{10^{-3}} \text{ mg}$$

$$= 2 \times 10^{[-2 - (-3)]} \text{ mg}$$

$$= 2 \times 10^{(-2 + 3)} \text{ mg}$$

$$= 2 \times 10^{1} \text{ mg}$$

$$= 20 \text{ mg}$$

CALCULATION $\quad 0.018 \text{ g} \times \dfrac{1 \text{ mg}}{10^{-3} \text{ g}}$

$$= 1.8 \times 10^{-2} \cancel{g} \times \frac{1 \text{ mg}}{10^{-3} \cancel{g}}$$

$$= \frac{1.8 \times 10^{-2} \times 1 \text{ mg}}{10^{-3}}$$

$$= \frac{1.8 \times 10^{-2}}{10^{-3}} \text{ mg}$$

$$= 1.8 \times 10^{1} \text{ mg}$$

$$= 18 \text{ mg}$$

Because the calculated result is very close to the estimated value, you can assume that it is correct.

Finally, you should check the problem. You should see that the answer makes sense: the value in milligrams, 18 mg, must be 1000 times greater than the value in grams, 0.018 g, because the milligram is 1000 times smaller than the gram. The units are also correct; you started with grams and ended with milligrams.

Note that the only change that took place in this problem was a shifting of the decimal point and a change in units of the given value. This result occurs because metric units differ from each other only by some multiple of ten. Since you divided by 10^{-3}, which is the same as multiplying by 10^3, the decimal point was moved three places to the right.

Because metric conversions are based on the meaning of the prefixes of the units, the method of conversion is similar for every type of unit. For example, the method that we just used to convert grams to milligrams could also be used to convert meters to millimeters or to convert liters to milliliters. In each of these cases, the change is from the base unit to a unit that is 1000 times smaller.

Example If a person's mass is 56 kg, what is this mass in grams?

UNKNOWN	GIVEN	CONNECTION
grams	56 kg	$\dfrac{10^3 \text{ g}}{1 \text{ kg}}$ or $\dfrac{1 \text{ kg}}{10^3 \text{ g}}$

$$\text{grams} = 56 \text{ kg} \times \frac{10^3 \text{ g}}{1 \text{ kg}}$$

ESTIMATE $6 \times 10^1 \text{ kg} \times \dfrac{10^3 \text{ g}}{1 \text{ kg}}$

$= 6 \times 10^4 \text{ g}$

CALCULATION $5.6 \times 10^4 \text{ g} = 56{,}000 \text{ g}$

Example What is the volume in liters of a 500. mL volumetric flask?

UNKNOWN	GIVEN	CONNECTION
liters	500. mL	$\dfrac{10^{-3} \text{ L}}{1 \text{ mL}}$ or $\dfrac{1 \text{ mL}}{10^{-3} \text{ L}}$

$$\text{liters} = 500.\ \cancel{\text{mL}} \times \frac{10^{-3}\ \text{L}}{1\ \cancel{\text{mL}}}$$

ESTIMATE $\quad 5 \times 10^2\ \cancel{\text{mL}} \times \dfrac{10^{-3}\ \text{L}}{1\ \cancel{\text{mL}}}$

$$= 5 \times 10^{-1}\ \text{L}$$

CALCULATION $\quad 0.500\ \text{L} = 5.00 \times 10^{-1}\ \text{L}$

Example   If a person's hand is 17.0 cm long, what is its length in meters?

UNKNOWN	GIVEN	CONNECTION
meters	17.0 cm	$\dfrac{1\ \text{cm}}{10^{-2}\ \text{m}}$ or $\dfrac{10^{-2}\ \text{m}}{1\ \text{cm}}$

$$\text{meters} = 17.0\ \cancel{\text{cm}} \times \frac{10^{-2}\ \text{m}}{1\ \cancel{\text{cm}}}$$

ESTIMATE $\quad 2 \times 10^1\ \cancel{\text{cm}} \times \dfrac{10^{-2}\ \text{m}}{1\ \cancel{\text{cm}}}$

$$= 2 \times 10^{-1}\ \text{m}$$

CALCULATION $\quad 1.70 \times 10^{-1}\ \text{m} = 0.170\ \text{m}$

Example   A neighboring town is 15.2 km away. What is this distance in meters?

UNKNOWN	GIVEN	CONNECTION
meters	15.2 km	$\dfrac{1\ \text{km}}{10^{3}\ \text{m}}$ or $\dfrac{10^{3}\ \text{m}}{1\ \text{km}}$

$$\text{meters} = 15.2\ \cancel{\text{km}} \times \frac{10^{3}\ \text{m}}{1\ \cancel{\text{km}}}$$

ESTIMATE $\quad 2 \times 10^1\ \cancel{\text{km}} \times \dfrac{10^{3}\ \text{m}}{1\ \cancel{\text{km}}}$

$$= 2 \times 10^4\ \text{m}$$

CALCULATION $\quad 1.52 \times 10^4\ \text{m} = 15,200\ \text{m}$

Practice problem   A box has a mass of 30.5 mg. Find its mass in grams.

ANSWER $\quad 0.0305\ \text{g}$ (or $3.05 \times 10^{-2}\ \text{g}$)

Practice problem   A laboratory flask holds 0.500 L. What is its volume in milliliters?

ANSWER   500. mL

3.4
MULTIPLE
CONVERSIONS

In many problems, you will find it necessary to use more than one conversion factor to change the units of the given quantity to the units of the desired answer. You may need several conversion factors when, for example, you are converting between two different metric units that are each some multiple of the base unit (such as centimeters and millimeters). By defining each unit in terms of the base unit (the meter) you can easily find the connection. Thus to find the number of millimeters in a centimeter, the outline is:

UNKNOWN	GIVEN	CONNECTION
millimeters	1 cm	$\dfrac{1\ cm}{10^{-2}\ m}$ or $\dfrac{10^{-2}\ m}{1\ cm}$
		$\dfrac{1\ mm}{10^{-3}\ m}$ or $\dfrac{10^{-3}\ m}{1\ mm}$

To set up the problem, notice that one set of conversion factors connects centimeters and meters, while the other set connects meters and millimeters. You can then convert centimeters to millimeters through the base unit, the meter, which both units have in common.

$$cm \quad \rightarrow \quad m \quad \rightarrow \quad mm$$
given unit   base unit   desired unit

You now must select the proper forms of the two conversion factors:

$$\frac{1\ cm}{10^{-2}\ m} \quad or \quad \frac{10^{-2}\ m}{1\ cm} \qquad \frac{1\ mm}{10^{-3}\ m} \quad or \quad \frac{10^{-3}\ m}{1\ mm}$$

millimeters = 1 cm × (?) × (?)

The first conversion factor must convert the given centimeters to meters, and therefore must have cm in the denominator:

$$\text{millimeters} = 1\ \cancel{cm} \times \frac{10^{-2}\ m}{1\ \cancel{cm}} \times (\,?\,)$$

The second factor now must cancel the unit *meter* and therefore must have m in the denominator:

$$\text{millimeters} = 1\ \cancel{cm} \times \frac{10^{-2}\ \cancel{m}}{1\ \cancel{cm}} \times \frac{1\ mm}{10^{-3}\ \cancel{m}}$$

Since the calculation cannot be further simplified to estimate an answer, the result is found immediately:

CALCULATION $\quad \dfrac{10^{-2} \text{ mm}}{10^{-3}}$

$$= 10^{[-2 \,-\, (-3)]} \text{ mm} = 10^{(-2 \,+\, 3)} \text{ mm}$$

$$= 10^{1} \text{ mm} = 10 \text{ mm}$$

Check the result: Millimeters are smaller than centimeters, so the answer makes sense.

It might have been clear to you, from the definitions of millimeter and centimeter, that 10 mm = 1 cm. If that was the case, you could have solved the problem in just one step:

$$\text{millimeters} = 1 \, \cancel{\text{cm}} \times \frac{10 \text{ mm}}{1 \, \cancel{\text{cm}}} = 10 \text{ mm}$$

As you become more familiar with metric units, such relationships will become more and more obvious. Nevertheless, the technique of using several conversion factors is extremely useful and will be applied often throughout this book.

Example A micropipet is used to transfer very small volumes of liquid. What is the volume in milliliters that can be delivered by a 5.00 μL pipet?

UNKNOWN	GIVEN	CONNECTION	
milliliters	5.00 μL	$\dfrac{1 \, \mu\text{L}}{10^{-6} \text{ L}}$ or $\dfrac{10^{-6} \text{ L}}{1 \, \mu\text{L}}$	
		$\dfrac{1 \text{ mL}}{10^{-3} \text{ L}}$ or $\dfrac{10^{-3} \text{ L}}{1 \text{ mL}}$	

$$\text{milliliters} = 5.00 \, \cancel{\mu\text{L}} \times \frac{10^{-6} \, \cancel{\text{L}}}{1 \, \cancel{\mu\text{L}}} \times \frac{1 \text{ mL}}{10^{-3} \, \cancel{\text{L}}}$$

ESTIMATE $\quad 5 \times 10^{-3}$ mL

CALCULATION $\quad 5.00 \times 10^{-3}$ mL

Example Using X-rays, very small distances can be measured. Convert the distance 30.5 pm to centimeters. (Note: 1 picometer (pm) $= 10^{-12}$ m.)

UNKNOWN	GIVEN	CONNECTION
centimeters	30.5 pm	$\dfrac{1\ \text{pm}}{10^{-12}\ \text{m}}$ or $\dfrac{10^{-12}\ \text{m}}{1\ \text{pm}}$
		$\dfrac{1\ \text{cm}}{10^{-2}\ \text{m}}$ or $\dfrac{10^{-2}\ \text{m}}{1\ \text{cm}}$

$$\text{centimeters} = 30.5\ \cancel{\text{pm}} \times \frac{10^{-12}\ \cancel{\text{m}}}{1\ \cancel{\text{pm}}} \times \frac{1\ \text{cm}}{10^{-2}\ \cancel{\text{m}}}$$

ESTIMATE $\quad 3 \times 10^{1}\ \cancel{\text{pm}} \times \dfrac{10^{-12}\ \cancel{\text{m}}}{1\ \cancel{\text{pm}}} \times \dfrac{1\ \text{cm}}{10^{-2}\ \cancel{\text{m}}}$

$$= 3 \times 10^{-9}\ \text{cm}$$

CALCULATION $\quad 3.05 \times 10^{-9}\ \text{cm}$

Example A book has a mass of 0.320 kg. Find its mass in milligrams.

UNKNOWN	GIVEN	CONNECTION
milligrams	0.320 kg	$\dfrac{1\ \text{kg}}{10^{3}\ \text{g}}$ or $\dfrac{10^{3}\ \text{g}}{1\ \text{kg}}$
		$\dfrac{1\ \text{mg}}{10^{-3}\ \text{g}}$ or $\dfrac{10^{-3}\ \text{g}}{1\ \text{mg}}$

$$\text{milligrams} = 0.320\ \cancel{\text{kg}} \times \frac{10^{3}\ \cancel{\text{g}}}{1\ \cancel{\text{kg}}} \times \frac{1\ \text{mg}}{10^{-3}\ \cancel{\text{g}}}$$

ESTIMATE $\quad 3 \times 10^{-1}\ \cancel{\text{kg}} \times \dfrac{10^{3}\ \cancel{\text{g}}}{1\ \cancel{\text{kg}}} \times \dfrac{1\ \text{mg}}{10^{-3}\ \cancel{\text{g}}}$

$$= 3 \times 10^{5}\ \text{mg}$$

CALCULATION $\quad 3.20 \times 10^{5}\ \text{mg}$

Example A room is 485 cm long. What is its length in kilometers?

UNKNOWN	GIVEN	CONNECTION
kilometers	485 cm	$\dfrac{1\ \text{cm}}{10^{-2}\ \text{m}}$ or $\dfrac{10^{-2}\ \text{m}}{1\ \text{cm}}$
		$\dfrac{1\ \text{km}}{10^{3}\ \text{m}}$ or $\dfrac{10^{3}\ \text{m}}{1\ \text{km}}$

$$\text{kilometers} = 485 \ \text{cm} \times \frac{10^{-2} \ \text{m}}{1 \ \text{cm}} \times \frac{1 \ \text{km}}{10^3 \ \text{m}}$$

$$\text{ESTIMATE} \quad 5 \times 10^2 \ \text{cm} \times \frac{10^{-2} \ \text{m}}{1 \ \text{cm}} \times \frac{1 \ \text{km}}{10^3 \ \text{m}}$$

$$= 5 \times 10^{-3} \ \text{km}$$

CALCULATION 4.85×10^{-3} km

Practice problem   According to a book on nutrition, you should eat 350. mg of magnesium each day. Find this mass in kilograms.

ANSWER 3.50×10^{-4} kg

Practice problem   You are told to measure 3.50 dL of a liquid. What is this volume in milliliters?

ANSWER   350. mL

3.5 METRIC–ENGLISH CONVERSIONS   In the United States, where English units are still commonly used, it is often necessary to convert units between the metric and English systems. This conversion involves more than simply changing the position of the decimal point. In these cases, you cannot write down a conversion factor based on the definitions of the metric prefixes. Instead, you must refer to a table, such as Table 3-1, which states relationships between units in these different systems. Many products

TABLE 3-1   *English units and their metric conversion factors*

	English units	English-metric conversion factors[a]
mass or weight	1 pound (lb) = 16 ounces (oz)	$\dfrac{1 \ \text{lb}}{454 \ \text{g}}$ or $\dfrac{454 \ \text{g}}{1 \ \text{lb}}$
volume	1 quart (qt) = 2 pints (pt)	$\dfrac{1 \ \text{qt}}{946 \ \text{mL}}$ or $\dfrac{946 \ \text{mL}}{1 \ \text{qt}}$
	1 gallon (gal) = 4 quarts (qt)	
length	1 foot (ft) = 12 inches (in)	$\dfrac{1 \ \text{in}}{2.54 \ \text{cm}}$ or $\dfrac{2.54 \ \text{cm}}{1 \ \text{in}}$
	1 yard (yd) = 3 feet (ft)	

[a]Alternate conversion factors are:

$\dfrac{1 \ \text{kg}}{2.20 \ \text{lb}}$ or $\dfrac{2.20 \ \text{lb}}{1 \ \text{kg}}$ (mass or weight) $\dfrac{1 \ \text{L}}{1.057 \ \text{qt}}$ or $\dfrac{1.057 \ \text{qt}}{1 \ \text{L}}$ (volume) $\dfrac{1 \ \text{m}}{39.37 \ \text{in}}$ or $\dfrac{39.37 \ \text{in}}{1 \ \text{m}}$ (length)

FIGURE 3-2

*A product label. Many
products now have labels that
express the quantity of the
product in both English and
metric units. (Photo by
Christine Osinski with the
permission of General Foods.)*

in the U.S. now carry labels with both metric and English units (see
Figure 3-2).

Consider the following example:

A large bottle contains 828 mL of soda. What is the volume of soda
in quarts?

The unknown, quarts, can be found from the given volume in millimeters by using the relationship (from Table 3-1) that 1 qt has the same
volume as 946 mL.

UNKNOWN	GIVEN	CONNECTION
quarts	828 mL	$\dfrac{1 \text{ qt}}{946 \text{ mL}}$ or $\dfrac{946 \text{ mL}}{1 \text{ qt}}$

You set up the problem by the conversion factor method:

$$\frac{1 \text{ qt}}{946 \text{ mL}} \quad \text{or} \quad \frac{946 \text{ mL}}{1 \text{ qt}}$$

$$\text{quarts} = 828 \text{ mL} \times (\qquad ? \qquad)$$

Because you want to cancel the unit mL, the factor must have mL in
the denominator. Therefore, you must write the equation as:

$$\text{quarts} = 828 \ \cancel{\text{mL}} \times \frac{1 \text{ qt}}{946 \ \cancel{\text{mL}}}$$

The estimated volume is:

ESTIMATE $8 \times 10^2 \ \cancel{\text{mL}} \times \dfrac{1 \text{ qt}}{9 \times 10^2 \ \cancel{\text{mL}}} = 1 \text{ qt}$

The actual, or calculated, volume is 0.875 qt. You should have expected an answer that was slightly less than one quart because the given volume, 828 mL, was less than 946 mL, the number of milliliters in one quart.

Box 3

The metric system and the U.S.

In his first message to Congress in 1790, George Washington called for a system of weights and measures for the new nation that was simpler than the English system. Around this time, the French created and adopted the metric system of measurement. Unlike the English system, related units in the metric system differ by an exact multiple of 10. This relationship makes it easy to perform conversions within the metric system. The U.S., however, decided to stay with the more familiar English units.

The U.S. signed the International Treaty of the Meter in 1875, and supported the metric system in principle but not in practice. Over the years, most countries except the U.S. have switched to the metric system. Our continued use of the English system, which we call the U.S. Customary System, has made international trade somewhat more difficult. For example, tools made in the United States cannot be easily used abroad because they do not fit metric-sized components. Anyone who has tried to tinker with a Toyota knows that U.S. wrenches won't fit metric nuts and bolts!

To overcome these difficulties, President Gerald Ford signed the Metric Conversion Act in 1975. It created a U.S. Metric Board to carry out the planning, coordination, and public education needed to encourage voluntary conversion. However, the path toward metrification is not an easy one. Many people would prefer to stay with the system that they know best. Also, although the English units are harder to convert and more complex than the metric units, they are based on human experience and in this sense are less artificial. The foot, for example, is a natural human measure.

Even though conversion to the metric system may further international trade, many businesses are slow to change because the necessary new machinery is expensive to acquire. They feel that the extra expense may not be worth it.

Since we have not yet given up the U.S. Customary System but are beginning to adopt the metric system, we presently have a dual system of measurement. Cereal boxes and soda bottles, for example, have their ingredients listed in both metric and English units. This arrangement can be confusing, and sometimes dangerous. In at least one instance, a nurse nearly gave a lethal dose of a drug to a patient because she confused metric and English units.

Today, organized groups in the U.S. are fighting metric conversion. One group, called ''Americans for Customary Weight and Measure,'' sells a t-shirt that reads ''Stop metric madness.''

Converting length measurements from the metric to the English system, and vice versa, is just as straightforward as converting volume measurements. Mass conversions are slightly different. In a mass conversion, you must keep in mind that the English unit, the pound, is not a unit of mass but of weight. The relationship given in Table 3-1 holds true only on Earth (as pointed out in Section 2.4) and even here weight undergoes slight changes with altitude.

English-metric conversions may also require the use of several conversion factors, a method described in the previous section. For example, suppose that your waistline is 32 in and you want to express this measurement in millimeters. The conversion factor given in Table 3-1 is 1 in/2.54 cm or 2.54 cm/1 in. Since you want to express the answer in millimeters, an additional conversion is required. You must convert from centimeters to millimeters. To do this conversion, you can use the relationship given in the previous section, namely that 1 cm = 10 mm.

The problem is outlined as shown:

UNKNOWN	GIVEN	CONNECTION
millimeters	32 in	$\dfrac{1 \text{ in}}{2.54 \text{ cm}}$ or $\dfrac{2.54 \text{ cm}}{1 \text{ in}}$
		$\dfrac{1 \text{ cm}}{10 \text{ mm}}$ or $\dfrac{10 \text{ mm}}{1 \text{ cm}}$

The set-up is as follows:

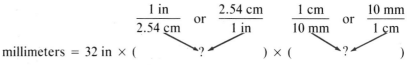

$$\text{millimeters} = 32 \text{ in} \times (\quad ? \quad) \times (\quad ? \quad)$$

The first factor must have inches in the denominator:

$$\text{millimeters} = 32 \text{ in} \times \frac{2.54 \text{ cm}}{1 \text{ in}} \times (\,?\,)$$

The second factor must have centimeters in the denominator:

$$\text{millimeters} = 32 \text{ in} \times \frac{2.54 \text{ cm}}{1 \text{ in}} \times \frac{10 \text{ mm}}{1 \text{ cm}}$$

An estimated answer is:

$$\text{ESTIMATE} \quad 3 \times 10^1 \text{ in} \times \frac{3 \text{ cm}}{1 \text{ in}} \times \frac{10^1 \text{ mm}}{1 \text{ cm}}$$

$$= 9 \times 10^2 \text{ mm}$$

The actual result is 8.1×10^2 mm, or 810 mm. As you should have expected, the size of your waistline expressed in millimeters is a considerably larger number than its value in inches.

Example What is the volume in milliliters of a pint (1.00 pt) of milk?

UNKNOWN	GIVEN	CONNECTION	
milliliters	1.00 pt	$\dfrac{2 \text{ pt}}{1 \text{ qt}}$ or $\dfrac{1 \text{ qt}}{2 \text{ pt}}$	
		$\dfrac{1 \text{ qt}}{946 \text{ mL}}$ or $\dfrac{946 \text{ mL}}{1 \text{ qt}}$	

$$\text{milliliters} = 1.00 \text{ pt} \times \frac{1 \text{ qt}}{2 \text{ pt}} \times \frac{946 \text{ mL}}{1 \text{ qt}}$$

ESTIMATE $\quad 1 \text{ pt} \times \dfrac{1 \text{ qt}}{2 \text{ pt}} \times \dfrac{9 \times 10^2 \text{ mL}}{1 \text{ qt}}$

$\quad\quad\quad = 5 \times 10^2 \text{ mL}$

CALCULATION $\quad 4.73 \times 10^2 \text{ mL} = 473 \text{ mL}$

Example A box of breakfast cereal weighs 15.0 oz. Find its mass in grams.

UNKNOWN	GIVEN	CONNECTION	
grams	15.0 oz	$\dfrac{16 \text{ oz}}{1 \text{ lb}}$ or $\dfrac{1 \text{ lb}}{16 \text{ oz}}$	
		$\dfrac{1 \text{ lb}}{454 \text{ g}}$ or $\dfrac{454 \text{ g}}{1 \text{ lb}}$	

$$\text{grams} = 15.0 \text{ oz} \times \frac{1 \text{ lb}}{16 \text{ oz}} \times \frac{454 \text{ g}}{1 \text{ lb}}$$

ESTIMATE $\quad 2 \times 10^1 \text{ oz} \times \dfrac{1 \text{ lb}}{2 \times 10^1 \text{ oz}} \times \dfrac{5 \times 10^2 \text{ g}}{1 \text{ lb}}$

$\quad\quad\quad = 5 \times 10^2 \text{ g}$

Example   What is the distance in feet that is covered by a runner in a 500. meter race?

UNKNOWN	GIVEN	CONNECTION	
feet	500. m	$\dfrac{1 \text{ cm}}{10^{-2} \text{ m}}$ or	$\dfrac{10^{-2} \text{ m}}{1 \text{ cm}}$
		$\dfrac{1 \text{ in}}{2.54 \text{ cm}}$ or	$\dfrac{2.54 \text{ cm}}{1 \text{ in}}$
		$\dfrac{1 \text{ ft}}{12 \text{ in}}$ or	$\dfrac{12 \text{ in}}{1 \text{ ft}}$

$$\text{feet} = 500. \text{ m} \times \frac{1 \text{ cm}}{10^{-2} \text{ m}} \times \frac{1 \text{ in}}{2.54 \text{ cm}} \times \frac{1 \text{ ft}}{12 \text{ in}}$$

ESTIMATE $5 \times 10^2 \text{ m} \times \dfrac{1 \text{ cm}}{10^{-2} \text{ m}} \times \dfrac{1 \text{ in}}{3 \text{ cm}} \times \dfrac{1 \text{ ft}}{1 \times 10^1 \text{ in}}$

$$= 2 \times 10^3 \text{ ft}$$

CALCULATION 1.64×10^3 ft = 1640 ft

Example   An automobile gasoline tank holds 16 gallons. What is its capacity in liters?

UNKNOWN	GIVEN	CONNECTION	
liters	16 gallons	$\dfrac{1 \text{ gal}}{4 \text{ qt}}$ or	$\dfrac{4 \text{ qt}}{1 \text{ gal}}$
		$\dfrac{1 \text{ qt}}{946 \text{ mL}}$ or	$\dfrac{946 \text{ mL}}{1 \text{ qt}}$
		$\dfrac{1 \text{ mL}}{10^{-3} \text{ L}}$ or	$\dfrac{10^{-3} \text{ L}}{1 \text{ mL}}$

$$\text{liters} = 16 \text{ gal} \times \frac{4 \text{ qt}}{1 \text{ gal}} \times \frac{946 \text{ mL}}{1 \text{ qt}} \times \frac{10^{-3} \text{ L}}{1 \text{ mL}}$$

ESTIMATE $2 \times 10^1 \ \text{gal} \times \dfrac{4 \ \text{qt}}{1 \ \text{gal}} \times \dfrac{9 \times 10^2 \ \text{mL}}{1 \ \text{qt}} \times \dfrac{10^{-3} \ \text{L}}{1 \ \text{mL}}$

$= 7 \times 10^1 \ \text{L} = 70 \ \text{L}$

CALCULATION 61 L

Practice problem What is the mass in kilograms of a person whose weight is 150. lb?

ANSWER 68.1 kg

Practice problem Find the height in meters of a person who is 5 ft 6 in (5.50 ft) tall.

ANSWER 1.68 m

**3.6
TEMPERATURE
CONVERSION**

Temperature is a measure of how hot or cold something is. Temperature is determined by means of a thermometer. A typical laboratory thermometer consists of a thin glass tube that is partially filled with mercury. An increase in the temperature at the bottom of the tube causes the mercury to expand in volume and to rise. Thermometers are calibrated so that the lengths of the mercury column correspond to certain temperatures. The temperature scale is etched onto the glass.

Three different scales are commonly used to describe temperature: Fahrenheit (°F), Celsius (pronounced sell′ see-us) (°C), and Kelvin (K). They are shown in Figure 3-3. The reference points of freezing and boiling water allow us to compare the sizes of the degrees of temperature on all three scales. On the Kelvin scale water boils at 373° and freezes at 273°. On the Celsius scale water boils at 100° and freezes at 0°. On the Fahrenheit scale water boils at 212° and freezes at 32°. Notice that there are 100 degrees between the boiling and freezing points on the Kelvin scale and also on the Celsius scale. In contrast, notice that there are 180 degrees between boiling and freezing points on the Fahrenheit scale. You can then see that the size of a degree is the same in the Celsius and Kelvin scales, and that it is larger than a degree in the Fahrenheit scale. Each degree on the Celsius and Kelvin scales is $180°/100° = 1.8$ times as large as each Fahrenheit degree.

Zero degrees means something different on each of the three scales. The Kelvin scale is the only one that calls the lowest possible temperature 0. Nothing can be colder than 0 K. This temperature is called absolute zero. For this reason, the Kelvin scale is sometimes called the absolute temperature scale. Because the lowest possible temperature is zero on this scale, a Kelvin temperature can never be negative. Degrees in the Kelvin scale are shown without the superscript degree (°) symbol, and the name kelvin now stands for the older term "degrees

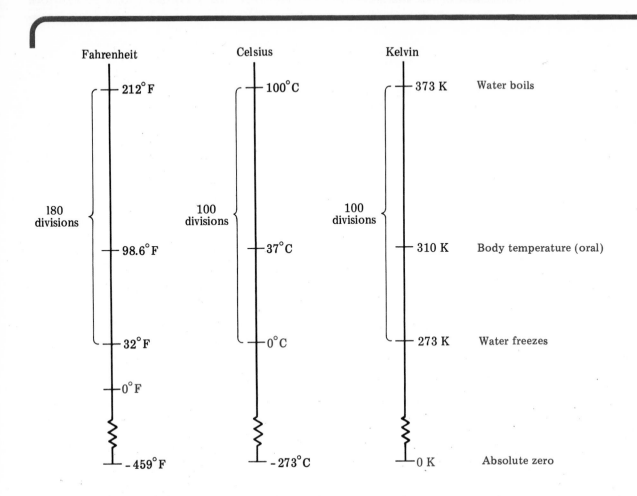

FIGURE 3-3

*Temperature scales:
Fahrenheit, Celsius, Kelvin.
Notice that the size of a
degree is smaller on the
Fahrenheit scale than in the
Celsius and Kelvin scales.
Also, note the different
locations of 0°.*

Kelvin.'' This scale is used in the SI system and is the one that you
will always use for chemical calculations.

You will usually use the Celsius scale to measure temperature in
the laboratory but you will perform calculations using the Kelvin scale.
Therefore, you must be able to convert from one set of temperature
units to the other. To do this conversion, you must use a mathematical
equation that gives the relationship between the Celsius and Kelvin
scales:

$$K = °C + 273$$

This simple formula states that to find the Kelvin temperature (K), you
must add 273 to the Celsius temperature (°C). (Appendix C.3 presents
a general introduction to formulas.) The formula is based on the fact
that the size of a degree is the same for both of these scales and that
each reading in kelvins is about 273 degrees higher than each Celsius
reading (because of the way zero is defined on the two scales). For

example, a cup of coffee with a temperature of 62°C has a reading on the Kelvin scale of

K = 62 + 273 = 335 K

To do the reverse type of calculation (converting from kelvins to degrees Celsius), you simply rearrange the formula, subtracting 273 from both sides.

°C = K − 273

Now, the Celsius reading is found by subtracting 273 from the temperature in kelvins. Thus, a freezer with a Kelvin temperature of 250 K, has a Celsius temperature of

°C = 250 − 273 = −23°C

If you remember that values on the Kelvin scale are always larger than corresponding values on the Celsius scale, you will know whether to add or to subtract 273.

Conversion between Fahrenheit and Celsius temperatures involves two changes. Not only do the locations of 0° differ (by 32°F) on these scales, but the size of each degree also differs (by the factor 180°F/100°C). A general formula that takes both these changes into account is:

°F − 32 = 1.8 (°C)

To find Fahrenheit degrees from a Celsius reading, you simply rearrange the formula so that only °F appears on the left by adding 32 to both sides of the equation:

°F = 1.8 (°C) + 32

You then substitute the Celsius reading. For example, substituting 33°C (the temperature on a hot day) into the formula gives:

°F = 1.8 (33) + 32

= 59 + 32

= 91°F

(Note that because the 32° difference is exact, the number of significant figures in the answer is not limited by this term. Similarly, the 1.8 in the formula is an exact number.)

To find the Celsius temperature from a Fahrenheit reading, rearrange the formula by dividing both sides by 1.8. Now °C appears by itself on one side of the equation:

$$°C = \frac{°F - 32}{1.8}$$

Thus, your normal (oral) body temperature, 98.6°F, would be converted to the Celsius scale as follows:

$$°C = \frac{98.6 - 32}{1.8}$$

$$= \frac{66.6}{1.8}$$

$$= 37.0°C$$

Example   A refrigerator keeps food at 10°C. What is this value in kelvins?

UNKNOWN	GIVEN	CONNECTION
kelvins	10°C	K = °C + 273

kelvins = 10 + 273 = 283 K

Note that in this and the following temperature conversion problems, the Connection is not a conversion factor but a formula. The use of a simple conversion factor is limited to quantities that can be converted by a multiplication or a division. Temperature conversions, on the other hand, involve an addition or subtraction (because of the different definition of zero degrees on each scale).

Example   What is the temperature inside this same refrigerator in degrees Fahrenheit?

UNKNOWN	GIVEN	CONNECTION
degrees Fahrenheit	10°C	°F = 1.8 (°C) + 32

degrees Fahrenheit = 1.8 (10) + 32°F

$$= (18 + 32) \, °F$$

$$= 50°F$$

Example   Helium freezes at 4 K. What is this temperature in degrees Celsius?

UNKNOWN	GIVEN	CONNECTION
degrees Celsius	4 K	°C = K − 273

degrees Celsius = (4 − 273)°C = −269°C

Example   What is the Celsius temperature if your thermometer reads 65°F?

UNKNOWN	GIVEN	CONNECTION
degrees Celsius	65°F	$°C = \dfrac{°F - 32}{1.8}$

$$\text{degrees Celsius} = \frac{65 - 32}{1.8} = \frac{33}{1.8} = 18°C$$

Practice problem   According to a thermometer, the room temperature is 21.5°C. What is this reading on the Fahrenheit scale?

ANSWER   70.7°F

Practice problem   A certain liquid, called aniline, freezes at −6.30°C. Convert this reading to kelvins.

ANSWER   267 K

3.7 HEAT   Heat and temperature are related concepts that are sometimes confused. As we saw in the previous section, temperature is a measure of how hot or cold an object is. **Heat**, on the other hand, is the energy that is transferred from a hotter object to a cooler one. For example, if you touch a hot stove, energy is transferred from the stove to your finger, probably causing a burn. Energy is always transferred from an object at a higher temperature (the stove) to an object at a lower temperature (your finger).

Heat is related to the concept of **work**. Like heat, work is a form of transferred energy. Work, however, involves moving an object by a push or pull. When you push a cart across the floor, you transfer energy from your body to the cart in the form of work. Work can be converted into heat. For example, if you rub a piece of wood with sandpaper, you may notice that the paper and wood become warmer. Heat can also be converted into work (although not completely). For instance, in an automobile, heat from the burning of gasoline is con-

Smithsonian Institution: Photo #58272

When it came to measurement, James Prescott Joule was a fanatic. Even while on his honeymoon, he stopped at a waterfall and carefully measured the temperatures at the top and bottom. He did this to test his theory that the kinetic energy of the falling water was released as heat, making the water at the bottom slightly warmer.

From this and other experiments, Joule determined the relationship between heat and mechanical energy. He found that the "loss" of mechanical energy through work is always balanced by a "gain" of thermal energy in the form of heat. This key relationship is one form of the law of conservation of energy.

Because he was a brewer and not a professor of science, Joule had trouble getting people to listen to his ideas. The only way he could publish his work was to give a public lecture and then have it reported by the Manchester newspaper for which his brother was music critic. But Joule kept trying and eventually managed to speak at a scientific meeting. Fortunately, William Thomson (1824–1907), later known as Lord Kelvin, happened to be in the audience. He became interested in Joule's work and helped him achieve recognition.

Joule was one of the greatest experimental scientists of the 19th century. Today, the SI unit of energy is called the joule in his honor.

verted into the motion of the car. It was through a series of experiments involving the conversion between work and heat that the English scientist James Prescott Joule discovered the law of conservation of energy.

Heat, as well as work, can be measured in units of energy such as the joule (named after James Prescott Joule) and the calorie. In fact, the calorie is defined as the amount of energy needed to raise the temperature of one gram of water by one degree Celsius (from 14.5°C to 15.5°C). Conversions between joules and calories can be carried out readily by using a conversion factor based on the relationship:

$$1 \text{ J} = 0.239 \text{ cal}$$

Engineers use still another unit for heat and energy, the British thermal unit (Btu), which is approximately equal to a kilojoule:

$$1 \text{ Btu} = 252 \text{ cal} = 1054 \text{ J}$$

Example   The amount of energy needed to melt 1 g of ice is 80. cal. Find this value in joules.

UNKNOWN	GIVEN	CONNECTION
joules	80. cal	$\dfrac{1 \text{ J}}{0.239 \text{ cal}}$ or $\dfrac{0.239 \text{ cal}}{1 \text{ J}}$

$$\text{joules} = 80. \ \cancel{\text{cal}} \times \frac{1 \text{ J}}{0.239 \ \cancel{\text{cal}}}$$

ESTIMATE $8 \times 10^1 \ \cancel{\text{cal}} \times \dfrac{1 \text{ J}}{2 \times 10^{-1} \ \cancel{\text{cal}}}$

$$= 4 \times 10^2 \text{ J}$$

CALCULATION $3.3 \times 10^2 \text{ J}$

The relationship between heat and temperature differs from one substance to the next. Some materials, like the metal in a frying pan, need relatively little heat to increase their temperature. Other substances, like water, need a great deal more heat to undergo the same change in temperature. The amount of heat required to raise the temperature of an object by one degree Celsius is known as that object's **heat capacity**. Consider, for example, one gram each of water, aluminum, iron and gold. If we transfer one calorie of heat to each of these substances, their temperatures will change by markedly different amounts. The gram of water will become 1°C hotter. The gram of aluminum will become 4.6°C hotter. The gram of iron will become 8.7°C hotter. The gram of gold will become 32°C hotter. Heat capacity is discussed further in Chapter Twelve.

Do-it-yourself 3

Counting your joules

Listed below are the energy values of common breakfast foods in kilojoules.

	kJ
orange juice, fresh (4 oz)	230
milk, whole (3/4 cup)	525
milk, skim (3/4 cup)	355
coffee or tea, black (cup)	8
coffee or tea, with sugar (cup)	85
coffee or tea, with cream and sugar (cup)	270
cereal (cup)	420
bread or toast (slice)	270
butter or margerine (pat)	210
jam (tablespoon)	210
muffin	525
danish pastry	835
doughnut, plain	630
doughnut, jelly	1050
egg, boiled	315
egg, fried	420
egg, scrambled	525
bacon, crisp (3 strips)	420
sausage, pork (2)	630
pancakes with butter and syrup (3)	1965

How rapidly you use up or burn the energy provided by food depends in part on your level of activity. The following list shows that the greater the level of activity, the faster energy is used up.

	Energy used (kcal/min)	
Activity	**Men**	**Women**
sleeping or reclining	1.1	1.0
light (walking)	3.7	3.0
moderate (cycling)	6.2	5.0
heavy (swimming)	10.0	8.0

To do: Determine the energy value of your breakfast, both in kilojoules and kilocalories (Calories). Now calculate how long it will take to burn your breakfast calories in light activity and in heavy activity. Remember that calories not used for energy by your body are stored as fat!

**3.8
DENSITY**

TABLE 3-2

Densities of common substances

Substance	Density (g/cm³)[a]
gold	19.3
mercury	13.5
lead	11.4
iron	7.9
diamond	3.5
aluminum	2.7
seawater	1.03
pure water	1.00
ice	0.92
gasoline	0.70
wood (pine)	0.50
air	0.0012

[a]At 20°C, except for pure water (3.98°C) and ice (0°C).

The relationship between mass and volume is an important property of substances. For example, a kilogram of iron takes up very little space; it is a little bigger than a fist. A kilogram of wood, however, is the size of a cantaloupe. The mass of iron is more concentrated, or more compact, than the mass of wood. In other words, a certain mass of iron takes up less volume than the same mass of wood. The difference in volume between equal masses of iron and wood is illustrated in Figure 3-4.

The ratio of the mass of an object to its volume is called **density**. Density of solids and liquids is usually expressed in grams per cubic centimeter, g/cm³.

$$\text{density} = \frac{\text{mass (g)}}{\text{volume (cm}^3)}$$

Thus, as shown in Table 3-2, iron (7.9 g/cm³) is much denser than wood (0.50 g/cm³). Because 1 cm³ = 1 mL, you will sometimes see the density of liquids written in the equivalent units, g/mL. Figure 3-5 shows an interesting application of liquid densities. (Specific gravity, used for comparing the density of liquids, is discussed in Chapter Twelve.) For gases, which have a very low ratio of mass to volume, density is often expressed as g/dm³ or g/L.

The volumes of many substances change with temperature. Most substances expand when heated and contract when cooled. (This effect of temperature is greatest for gases, much less for liquids, and still less for solids.) Because density depends on the volume of a substance, its value also changes with temperature. For example, mercury expands when heated. This expansion increases its volume and causes its

FIGURE 3-4

Wood and iron. The masses of wood and iron shown here are equal. Because wood is less dense than iron, the volume of the wood is greater than the volume of the iron. (Photo by Al Green.)

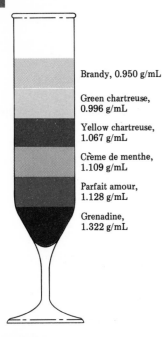

Brandy, 0.950 g/mL

Green chartreuse, 0.996 g/mL

Yellow chartreuse, 1.067 g/mL

Crème de menthe, 1.109 g/mL

Parfait amour, 1.128 g/mL

Grenadine, 1.322 g/mL

FIGURE 3-5

A pousse-café. This drink contains six different liqueurs. The liqueurs must be arranged in the order of their densities, with the most dense liqueur at the bottom of the glass.

density to decrease. The density of mercury is 13.60 g/mL at 0°C, 13.54 g/mL at 20°C, and 13.35 g/mL at 100°C. Therefore, the density values given in a table are usually accompanied by the temperature at which the density was measured.

Density is expressed as mass per unit volume of a substance, that is, mass per 1 cm³ or 1 dm³. It can be determined, however, by finding the ratio of mass to volume for any portion of convenient size. To make this calculation, we need two separate measurements: the object's mass (measured on a balance) and its volume. The volume is often measured in a graduated cylinder. It can be measured directly if the object is a liquid. If the object is a solid the volume is measured indirectly by placing the object in a graduated cylinder partially filled with liquid, and then observing the increase in the level of the liquid. Consider the following example:

A small amount, 5.5 cm³, of liquid benzene is found to have a mass of 4.84 g. Find the density of benzene.

The unknown, density, is easily calculated from the given information, mass and volume, using the definition of density.

UNKNOWN	GIVEN	CONNECTION
density	4.84 g = mass 5.5 cm³ = volume	$density = \dfrac{mass}{volume}$

The key step is to identify which quantity is the mass and which is the volume. Then substitute as shown.

$$density = \frac{mass}{volume} = \frac{4.84 \text{ g}}{5.5 \text{ cm}^3} = 0.88 \text{ g/cm}^3$$

Before calculating, your estimate should have shown the answer to be about 1. You should also have noticed that the density would be slightly less than 1, which indeed it is.

Density can be used as a conversion factor. If the volume is given, you can use density to find the mass. If the mass is given, you can use density to find the volume. Consider this example.

A gold coin has a mass of 31.1 g. What is its volume?

The unknown, volume, can be found using the given information, 31.1 g and the density of gold, which is given in Table 3-2.

UNKNOWN	GIVEN	CONNECTION
volume	31.1 g = mass	density = 19.3 g/cm³

ASIDE . . .

The density of this book is 1.4 g/cm³ and its mass is 1.6 kg. If it were made of ordinary air, its mass would be 1.4 g. If it were made of lead, its mass would be 13 kg.

When used as a conversion factor, the density of gold can be written as:

$$\frac{19.3 \text{ g}}{1 \text{ cm}^3} \quad \text{or} \quad \frac{1 \text{ cm}^3}{19.3 \text{ g}}$$

Set up the problem using the conversion factor method:

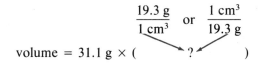

$$\text{volume} = 31.1 \text{ g} \times (\qquad ? \qquad)$$

The correct factor must have grams in the denominator to cancel the 31.1 g.

$$\text{volume} = 31.1 \cancel{g} \times \frac{1 \text{ cm}^3}{19.3 \cancel{g}}$$

Your estimate would be

$$\text{ESTIMATE} \quad 3 \times 10^1 \cancel{g} \times \frac{1 \text{ cm}^3}{2 \times 10^1 \cancel{g}}$$

$$= \frac{3}{2} \text{ cm}^3 = 2 \text{ cm}^3$$

The actual result is 1.61 cm³. You should have expected the volume to be greater than 1 cm³ since the given mass was 31.1 g and the density indicated that 1 cm³ has a mass of only 19.3 g.

Density also can be used as a conversion factor when we want to find the mass of a substance from its volume.

Example   A slab of wood that is used as a table top has a volume of 29,500 cm³ (2.95 × 10⁴ cm³). What is its mass?

The set-up for this problem is similar to that of the previous problem except that mass is now the unknown, and the volume is given.

UNKNOWN	GIVEN	CONNECTION
mass	2.95 × 10⁴ cm³ = volume	density = 0.50 g/cm³

You must refer to Table 3-2 for the density of wood. You can now use the conversion factor method:

$$\frac{0.50 \text{ g}}{1 \text{ cm}^3} \quad \text{or} \quad \frac{1 \text{ cm}^3}{0.50 \text{ g}}$$

$$\text{mass} = 2.95 \times 10^4 \text{ cm}^3 \times (\quad ? \quad)$$

The correct factor must have cm³ in the denominator to cancel the units of the 29,500 cm³.

$$\text{mass} = 2.95 \times 10^4 \text{ cm}^3 \times \frac{0.50 \text{ g}}{1 \text{ cm}^3}$$

The estimated mass is

$$\textsc{estimate} \quad 3 \times 10^4 \text{ cm}^3 \times \frac{5 \times 10^{-1} \text{ g}}{1 \text{ cm}^3}$$

$$= 15 \times 10^3 \text{ g} = 2 \times 10^4 \text{ g}$$

The actual value is 1.48 × 10⁴ g, or 14,800 g (also 14.8 kg or 32.6 lb).

Example   The mass of a cork that has a volume of 5.2 cm³ is 1.25 g. What is the density of the cork?

UNKNOWN	GIVEN	CONNECTION
density	1.25 g = mass 5.2 cm³ = volume	density $= \dfrac{\text{mass}}{\text{volume}}$

$$\text{density} = \frac{1.25 \text{ g}}{5.2 \text{ cm}^3} = 0.24 \text{ g/cm}^3$$

Example   Common glass has a density of 2.6 g/cm³. What is the mass of a piece of glass that has a volume of 200. cm³?

UNKNOWN	GIVEN	CONNECTION
mass	200. cm³ = volume	density = 2.6 g/cm³

$$\text{mass} = 200. \text{ cm}^3 \times \frac{2.6 \text{ g}}{1 \text{ cm}^3}$$

$$\textsc{estimate} \quad 2 \times 10^2 \text{ cm}^3 \times \frac{3 \text{ g}}{1 \text{ cm}^3}$$

$$= 6 \times 10^2 \text{ g}$$

CALCULATION 5.2×10^2 g = 520 g

Example   An experiment requires 18.2 g of alcohol. The density of alcohol is 0.79 g/cm³. What volume of alcohol is needed?

UNKNOWN	GIVEN	CONNECTION
volume	18.2 g = mass	density = 0.79 g/cm³

$$\text{volume} = 18.2 \text{ g} \times \frac{1 \text{ cm}^3}{0.79 \text{ g}}$$

ESTIMATE $2 \times 10^1 \text{ g} \times \dfrac{1 \text{ cm}^3}{8 \times 10^{-1} \text{ g}}$

$$= 3 \times 10^1 \text{ cm}^3 = 30 \text{ cm}^3$$

CALCULATION   23 cm³

Practice problem   50.0 cm³ of sugar has a mass of 79.5 g. Find the density of sugar.

ANSWER   1.59 g/cm³

Practice problem   What is the mass in grams of 15.5 cm³ of mercury?

ANSWER   211 g

Practice problem   What volume of air has a mass of 1.00 g? (See Table 3-2 for the density of air.)

ANSWER 7.7×10^2 cm³

Summary   **3.1** Problem solving is an essential skill. The first step in problem solving is to understand the problem completely. The second step is to find a connection between the unknown and the given information. The third step is to set up the computation. The fourth step, solving the problem, involves estimating and then calculating the answer. The last step is to review the problem by checking the answer and studying the method of solution.

3.2 Many chemistry problems can be solved by the conversion factor method. A conversion factor is a fraction that relates two quantities. The product of the appropriate conversion factor and the information given in a problem is often the solution, or partial solution, to that problem.

3.3 Converting metric units is simple because each unit differs from the others only by a power of 10. The conversion factor relates the desired unit to the given unit based on the meaning of their prefixes. Metric conversions change the position of the decimal point in the given quantity.

3.4 In many problems we must use several conversion factors to change the units of the given quantity to the units of the desired answer. Multiple conversions in the metric system can be carried out by defining each unit in terms of the base unit.

3.5 Changing units between the metric and English systems involves conversion factors that relate the units involved. Examples of conversions based on mass, length, and volume are presented in the section.

3.6 Temperature is a measure of how hot or cold something is. Temperature readings are obtained by means of a thermometer. Three different scales are commonly used to describe temperature: Fahrenheit (°F), Celsius (°C), and Kelvin (K). These scales differ in the size of the degree, or the location of 0°, or both. Simple formulas allow conversion from one scale to another.

3.7 Heat is a transfer of energy that takes place between two substances with different temperatures. The direction of heat transfer is always from the object with a higher temperature to the one with a lower temperature. Common units for heat are the calorie and the joule.

3.8 Density is the ratio of the mass of an object to its volume at a particular temperature. It is usually expressed in g/cm^3 for solids and liquids. Density can be used as a conversion factor to find the mass or volume of an object if the other quantity is given.

Exercises

For each term on the left, choose a phrase from the right-hand column that most closely matches its meaning.

1 conversion factor method   **a** ratio of mass to volume

2 temperature   **b** energy transfer

3 density   **c** degree of "hotness"

4 heat   **d** multiplication by ratio

EXERCISES BY TOPIC **Problem solving (3.1)**

5 Summarize the steps involved in problem solving.

6 What should you do if a calculated answer differs greatly from your estimate?

7 What should you do if you find it hard to solve a problem?

8 Outline the following problems: **a** convert 25 cents to dollars **b** convert 10 years into months.

9 Solve the problems in Exercise 8 using the steps given in Section 3.1.

Conversion factor method (3.2)

10 What are the advantages of solving problems by the conversion factor method?

11 If a rancher needs one acre of grazing land for 10 cattle, how many acres are needed for 55 cattle? Use the conversion factor method.

12 If a dollar is worth 200 Japanese yen, what is the value in dollars of 40 yen? Use the conversion factor method.

13 If 13 "ziggles" equal 2 "klopnys," what is the value of 29 "klopnys?" Use the conversion factor method.

14 Solve Exercise 12 using a proportion and compare your solution with that found by the conversion factor method.

Metric conversions (3.3)

15 If a one-ounce portion of cereal contains 3.0 g of protein, how many milligrams of protein does it contain?

16 Complete the following: **a** 4 cm = ___ m **b** 0.043 g = ___ mg **c** 15.5 m = ___ mm **d** 328 mL = ___ L **e** 0.98 kg = ___ g

17 The distance from the Earth to the sun is about 1.5×10^{11} m. What is the distance in cm?

18 An average African gorilla has a mass of 163 kg. What is this mass in grams?

19 How many meters are in a 10. km race?

20 What is the volume in liters of a 25.0 mL volumetric flask?

21 The lightest human adult had a mass of 2134 g. Convert this mass into milligrams.

22 If a soda bottle contains 0.207 L, what is the volume in mL?

Multiple conversions (3.4)

23 The average adult man has a mass of 68.04 kg. Convert this mass to milligrams.

24 What is the volume in mL of a 25.0 μL pipet?

25 A paper clip is 3.3 cm long. Express this length in millimeters.

26 Perform the following conversions: **a** 15.2 mg to kilograms **b** 35.1 nm to centimeters **c** 65 mL to microliters **d** 0.075 kg to micrograms

27 A particle of tobacco smoke has a diameter of 0.50 μm. Convert this value to centimeters.

28 At 50 mi/hr, a car travels 2234 cm in one second. How many kilometers does it travel in one second?

29 A can of soup has a mass of 3.05 × 10^5 mg. Find its mass in kilograms.

Metric-English conversions (3.5)

30 The world's heaviest man weighed 1069 lb. Express his mass in kilograms.

31 A recipe calls for 1.5 cups of milk. Convert this volume to milliliters. (Note: 1 qt = 4 cups.)

32 Which is worth more, a gold nugget with a mass of 50 g, or one weighing 2 oz?

33 Convert the following: **a** 16 lb to kilograms   **b** 5 qt to milliliters   **c** 17 in to meters   **d** 150. g to ounces   **e** 600. mL to quarts

34 A cook wants to pour 1.5 L of batter into a two-quart bowl. Will it fit?

35 What is the length in yards of a 100. m dash?

36 The fastest weight loss on record was 352 lb in eight months. Compute this loss in kilograms.

37 The smallest belt size is 13 in. Express this value in millimeters.

Temperature conversion (3.6)

38 Find the temperature in kelvins and in degrees Fahrenheit of a glass of water at 18°C.

39 Which temperature is higher, 16°C or 55°F?

40 A flu victim has a fever of 101.5°F. Convert this temperature to degrees Celsius.

41 The lowest recorded human body temperature is 16°C. Express this value in degrees Fahrenheit.

42 In 1969, French scientists measured a temperature of 5 × 10^{-7} K. Find this value in degrees Celsius and Fahrenheit.

43 Perform the following conversions:

a   15°F = ＿＿°C = ＿＿ K

b ＿＿°F =   29°C = ＿＿ K

c ＿＿°F = ＿＿°C =   325 K

d −30°F = ＿＿°C = ＿＿ K

e ＿＿°F = −10°C = ＿＿ K

44 Find the Celsius temperature when the thermometer reads −10°F during a blizzard.

Heat (3.7)

45 How is heat related to temperature?

46 A tablespoon of peanut butter has 95 kcal. Find its energy content in kilojoules.

47 An air conditioner has an energy rating of 6500. Btu/hr. Find its rating in kilocalories per hour.

48 A cup of whole milk has 160. Cal. What is this value in joules?

49 An egg has an energy value of 334 kJ. Find this value in kilocalories.

Density (3.8) [Refer to Table 3-2]

50 What is the mass of 50.0 cm³ of gold?

51 What volume does 250. g of wood occupy?

52 An unknown substance has a mass of 6.32 g and a volume of 0.8 cm³. From Table 3-2, what substance might it be?

53 An experiment requires 300. g of alcohol, the density of which is 0.791 g/cm³. What volume is needed?

54 If 500. mL of milk has a mass of 515 g, find the density of milk.

55 What is the mass of one liter of air?

ADDITIONAL EXERCISES

56 Make up a simple problem and solve it using the steps suggested in Section 3.1.

57 If a trucker maintains a speed of 50. mi/hr, how long will it take to travel from Dayton to New York, a distance of 650. miles.

58 One of the world's largest pools contains 2.38×10^7 L of water. What is its volume in milliliters?

59 Convert: **a** 10.2 g to kilograms **b** 155 mL to liters **c** 32.5 mm to meters **d** 5.23 mg to grams **e** 0.875 L to milliliters.

60 The concentration of lead in the air above Los Angeles was found to be 6.6 μg/m³. What is the mass of lead in milligrams per cubic meter?

61 A liquid occupies 2.32 dm³. What is its volume in milliliters?

62 A room is 12 ft by 15 ft. Find its area in square meters.

63 Find the length in meters of a football field 100. yd long.

64 What is the volume in quarts of a 250. mL flask?

65 The world's largest tomato had a mass of 3.09 kg. Find its weight in ounces.

66 Goats have the highest body temperature of all mammals, 39.9°C. Express this temperature in degrees Fahrenheit.

67 What is the mass of one gallon of gasoline?

68 The active ingredient in Kaopectate is a clay-like absorbant, kaolin. If there are 90. g of kaolin per (fluid) ounce of Kaopectate, and an 8-ounce bottle costs $1.50, what is the cost of kaolin per gram?

69 A "fifth" is a unit of volume for liquor equal to one-fifth of a gallon. How many milliliters are there in a "fifth"?

70 If a car can travel 30. miles per gallon of gasoline, how many kilometers can it travel per liter of gasoline?

71 The density of the substance iridium is 22.5 g/mL. What volume of iridium has the same mass as 1.00 mL of water?

72 Which is more fattening, a plain doughnut with 135 kcal or a hamburger with 565 J?

73 An auto engine runs at 250.°F. Convert this temperature to degrees Celsius.

74 A hockey puck weighs 0.38 lb. What is its mass in grams?

75 A standard champagne bottle holds 80. cL. Convert this volume to quarts.

76 The typical daytime temperature on Mars is −30.°C. What is this temperature in degrees Fahrenheit?

77 Hemoglobin, the oxygen carrier in your blood, has a diameter of 0.0000068 mm. Convert this value to nanometers.

78 The Earth's crust has an average density of 2.8 g/cm^3. What volume of the crust will weigh 1.00 lb?

79 A bowling lane is 60. ft long. What is its length in meters?

80 The energy released by a burning match is 4×10^3 J. Convert this value to calories.

81 The average American woman weighs 135 lb. What is this value in kilograms?

82 The human digestive tract is about 1.0×10^4 mm long. How long is it in meters? in inches?

83 The coldest spot on Earth, which is in Antarctica, reached a temperature of −126.9°F. Convert this value into degrees Celsius.

84 Sears Tower in Chicago is 1559 ft tall. What is its height in kilometers?

85 Oxygen becomes a solid at 54 K. Convert this temperature to degrees Celsius.

86 A Boeing 747 "Jumbo Jet" weighs 7.75×10^5 lb. Find its mass in kilograms.

87 The Pacific Ocean has a volume of 7.0×10^8 km^3. Find its volume in liters.

88 The average man's brain weighs 3.1 lb. What is its mass in grams?

89 At the top of Mt. Everest, water boils at 159.8°F. What is this temperature in degrees Celsius?

90 The tallest waterfall in the world, Angel Falls in Venezuela, is 3212 ft high. Find its height in meters.

91 A chocolate chip cookie has 50. kcal. What is its energy value in joules?

92 A queen honeybee is 1.5 cm long. Find its length in decimeters.

93 The temperature at the sun's surface is 9620.°F. Convert this value into kelvins.

94 You just bought a 1.5-carat diamond ring. If a carat is 200. mg, what is the weight of your diamond in ounces? (Note: 1 lb = 16 oz.)

95 An average liver cell has a diameter of 0.05 mm. Convert this value to inches and to centimeters.

96 A woman is 5 ft 6 in tall and weighs 125 lb. Express these measurements in centimeters and kilograms.

97 Convert the U.S. highway speed of 55 mi/hr into kilometers per hour. (Note: 1 mi = 5280 ft.)

PROGRESS CHART Check off each item when completed.

_____ Previewed chapter

Studied each section:

_____ **3.1** Problem solving

_____ **3.2** Conversion factor method

_____ **3.3** Metric conversions

_____ **3.4** Multiple conversions

_____ **3.5** Metric-English conversions

_____ **3.6** Temperature conversion

_____ **3.7** Heat

_____ **3.8** Density

_____ Reviewed chapter

_____ Answered assigned exercises:

#'s _____

FOUR

The atom

In Chapter Four we present the properties of atoms, which are the basis of chemistry. The size, charge, mass, names, and symbols of atoms are described here. Simplified atomic diagrams are presented, as well as modern concepts of the structure of the atom. This chapter concludes with a brief history of our model of the atom. To demonstrate an understanding of Chapter Four, you should be able to:

1 State the major features of the atom.

2 Identify the charges of the subatomic particles.

3 Compare the masses of the subatomic particles.

4 Describe the relationships between the following terms: atomic number, mass number, number of protons, number of electrons, and number of neutrons.

5 Write the names and symbols for the most common elements.

6 Recognize examples of isotopes and explain how isotopes are related to atomic weight.

7 Draw simplified diagrams of atoms based on electron shells.

8 Describe the most important aspects of the quantum theory of matter.

9 Identify the similarities and differences between s and p atomic orbitals.

10 Write the electron configurations of atoms based on atomic orbitals.

11 Explain the relationship between electron transitions and radiation.

4.1
ATOMS

The idea that matter consists of particles was first stated by the ancient Greek philosophers Democritus and Leucippus. They proposed that rather than being continuous, matter consists of tiny units that cannot be divided into smaller pieces. The Greeks called these units **atoms.**

ASIDE . . .

There are about 10^{14} atoms on the head of a pin. This number is over 10,000 times more than the number of people on earth.

In fact, the word atom comes from the Greek word meaning uncuttable. There was little experimental evidence to support the idea of atoms, however, until the early 19th century, when John Dalton proposed the basic concepts of modern atomic theory. And it was not until the 20th century that the inner structure of the atom was revealed. ("A Brief History of the Modern Atom" at the end of this chapter provides further details.) The concept of the atom now is fundamental to chemistry.

According to atomic theory, *all matter,* whether solid, liquid, or gas, *is composed of particles called atoms.* Atoms are the basic building blocks of nature. At the present time, 106 chemically different kinds of atoms are known. Just as the 26 letters of the alphabet create the entire English language, this limited number of atoms forms every material substance. *Chemistry is the study of atoms—how these basic units combine, and how substances made of atoms are changed into other substances.*

An atom is extremely small. It is too little to be seen by the naked eye and too light to be weighed on a balance. If we wanted to gather a few grams of atoms, enough to weigh a fraction of a pound, we would need a million times a million times a million of them.

We now know that the incredibly tiny atom is made up of even smaller pieces, the **subatomic particles**. Atoms consist of definite combinations of three types of subatomic particles: **protons** (p), **neutrons** (n), and **electrons** (e). The smallest atoms contain only a few subatomic particles each. The largest atoms consist of several hundred each.

Despite its small size, most of the atom is empty space. Its protons and neutrons are packed into a dense central core called the **nucleus**. The nucleus is only 1/10,000 the size of the entire atom. The electrons are outside of, and very far from, the nucleus. To give yourself some idea of the distance between the nucleus and the electrons of an atom, imagine that the period at the end of this sentence is an atom's nucleus. This atom's electrons would extend to the walls of the room you are in!

4.2 CHARGES OF SUBATOMIC PARTICLES

An essential property of subatomic particles is their **electrical charge**. Charges were discovered by ancient Greeks who found that amber (*electron* in Greek) acquired the ability to attract objects when rubbed with cloth or fur. In the 17th and 18th centuries additional pairs of objects were found to become charged when rubbed against each other. Glass rubbed with silk becomes charged; so does hard rubber rubbed with cat's fur. You may have noticed this effect when running a comb through your hair on a dry day. Your hair becomes charged and stands on end. The comb attracts lint and small pieces of paper.

In the 18th century, Benjamin Franklin experimented with electrical charge and saw that charged objects attracted or repelled other charged

Smithsonian Institution: Photo #9585-B

John Dalton is best known for his atomic theory. Although he borrowed the word "atom" from the Greek philosophers, his theory was soundly based on many years of chemical experimentation. He proposed that:

1 All matter is made of tiny, indivisible particles called atoms.

2 All the atoms of one element have the same mass and properties, but differ from atoms of other elements.

3 All substances are combinations of atoms; atoms remain unchanged when they combine.

4 Atoms of different elements combine in small whole-number ratios.

Dalton was the first to prepare a table of atomic weights, which he needed to interpret the chemical information on which his theory was based.

This Quaker schoolteacher also was interested in meteorology, the study of weather. Dalton recorded nearly 200,000 observations over 57 years. Perhaps for this reason he never found time for marriage. Dalton also studied his own color blindness. This condition is called daltonism in recognition of his work.

Because atoms cannot be seen with the naked eye, Dalton's atomic theory was not accepted immediately. As one critic said, "When you show me atoms as big as liver dumplings, then I will believe in them." But today the modern atomic theory, which differs only slightly from Dalton's theory, is the basis for our understanding of chemistry.

FIGURE 4-1

Electrical charges. Like charges, both positive (a) or both negative (b), repel each other. Opposite charges, positive and negative (c), attract.

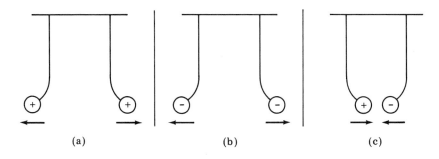

(a) (b) (c)

objects. He realized that there was a simple explanation. He said that charge is a fundamental property of nature that exists in two forms: **positive** ($+$) and **negative** ($-$). As shown in Figure 4-1, charges influence each other when they are brought together. Charges that are the same, either both positive or both negative (a and b in Figure 4-1), try to push each other apart. Opposite charges, positive and negative (c in Figure 4-1) try to pull themselves closer to each other. In other words, *like* charges repel, and *unlike* charges attract.

Charges can add or cancel. For example, if we place five positive charges together, they act five times more strongly than one. If, on the other hand, we place a positive charge and an equal negative charge together, the result at a distance is as if there were no charge at all. Combining two opposite charges is like adding $+1$ and -1 to get 0, or taking one step forward and then one step backward to end up where you started. An object with equal amounts of positive and negative charge is said to be **neutral**.

Experiments reveal that the proton is positively ($+$) charged, and the electron is negatively ($-$) charged. The neutron is neutral, having neither a positive nor a negative charge. One of the most important properties of the atom is that *the number of protons in an atom exactly equals the number of electrons*. That is, the number of positive charges in an atom equals the number of negative charges. Thus every atom is neutral, even though it consists of particles that have electrical charges.

4.3
MASSES OF SUBATOMIC PARTICLES, AND THE AMU

The mass of a proton and the mass of a neutron are approximately equal, as shown in Table 4-1. Each has a mass that is slightly greater than 10^{-24} g, that is,

$$0.000000000000000000000001 \text{ g}$$

The electron is even lighter. Its mass is nearly 2000 times smaller than that of the proton or the neutron. The relative masses of the subatomic particles, given in the third column of Table 4-1, are found by taking the ratio of each mass to the mass of the proton. Relative mass shows how much greater one mass is than another.

TABLE 4-1 *Masses of the subatomic particles*

Particle	Mass (g)	Relative mass[a]	Mass (amu)
proton	1.673×10^{-24}	1	1.007
neutron	1.675×10^{-24}	1	1.008
electron	9.110×10^{-28}	1/1836	5.486×10^{-4}

[a]Compared to the mass of the proton.

Because electrons have so little mass compared to protons and neutrons, their mass is usually ignored in calculations of atomic mass. The mass of an atom is determined almost completely by the mass of the protons and neutrons concentrated in its nucleus. In terms of mass, we can think of the atom as an elephant with fleas: the elephant represents the nucleus and the fleas represent the electrons.

Chemists prefer to avoid using the small numbers that represent the actual values of mass in grams of the subatomic particles. Instead, they use a relative scale of masses for atomic measurements. This scale is based on comparisons with another small mass that is used as a standard. This reference mass is a particular atom (carbon) that has six protons and six neutrons in its nucleus. The **atomic mass unit** is defined as one-twelfth (1/12) the mass of this reference atom. The symbol for the atomic mass unit is **amu**. Thus, the mass of the reference atom (carbon) can be expressed as 12 amu. Since the actual measured mass of this atom is 1.993×10^{-23} g, one amu is equivalent to a mass of $(1/12) \times (1.993 \times 10^{-23}$ g$) = 1.661 \times 10^{-24}$ g. The proton and the neutron each have a mass that is approximately equal to one amu, as indicated in the final column of Table 4-1.

What we have done here is similar to what a baker might do in baking cupcakes. Cupcakes contain ingredients in definite proportions. To simplify recipe calculations, the baker can set up a relative scale using one of these cupcakes as a reference. Let us say that she assigns the cupcake a mass of 2 baking units because it contains 2 tablespoons of flour. (Thus, 1 baking unit corresponds to 1 tablespoon of flour.) She can then compare the mass of other baked goods to the mass of the cupcake. A pie may have a mass that is 25 times that of a cupcake. Its

FIGURE 4-2

Atomic mass unit (amu). The mass of an atom can be conveniently expressed using atomic mass units (amu). Here, atom X has a mass of 7 amu.

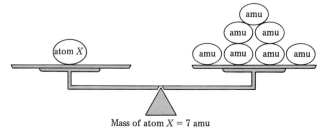

Mass of atom X = 7 amu

TABLE 4-2 *Summary of subatomic particles*

Particle	Symbol	Charge	Approximate mass (amu)
proton	p	+1	1
neutron	n	0	1
electron	e	−1	0

mass is then said to be 50 baking units. If two cookies together have the mass of one cupcake, each cookie is just 1 baking unit. The baker now knows the relative masses of the baked goods. Each pie has 25 times the mass of a cupcake, and each cookie has one-half the mass. In addition, because one tablespoon of flour has a mass of 1 baking unit, she knows the number of tablespoons of flour in each: 50 in the pie, and 1 in the cookie.

Just as the "baking unit" can be used to show the relative masses of baked goods, the atomic mass unit expresses the relative masses of atoms. An atom with a mass of 50 amu would have 25 times the mass of an atom with a mass of 2 amu, for example. Furthermore, since the proton and neutron each have masses of about one amu, the mass of an atom in amu indicates the total number of protons and neutrons in the nucleus of the atom. Figure 4-2 shows how the mass of an atom can be expressed using the amu. Table 4-2 summarizes the properties of the subatomic particles.

4.4
ATOMIC NUMBER
AND MASS
NUMBER

The number of protons in an atom's nucleus is called the **atomic number** of the atom. Since every atom has the same number of electrons as protons, the atomic number also equals the number of electrons outside the nucleus of the atom.

atomic number = number of protons

= number of electrons

Each of the 106 chemically different types of atoms has its own atomic number. Because the properties that enable us to distinguish one type of atom from another are related to the atomic number, the atomic number is the most important identification of an atom. It is the atom's signature.

The smallest atom possible has one proton and one electron, and therefore has an atomic number of 1. The next largest atom has an atomic number of 2 because it has two protons and two electrons. Every succeeding type of atom has one more proton in its nucleus, and one more electron outside its nucleus. The number of neutrons also increases, but does not affect the atomic number. The atomic numbers

TABLE 4-3 *Examples of atomic composition*

Atomic number	Mass number	Number of protons	Number of neutrons	Number of electrons
4	9	4	5	4
11	23	11	12	11
17	35	17	18	17
26	56	26	30	26
78	195	78	117	78

of the 106 different kinds of atoms range from 1 to 106. The largest known atom, with atomic number 106, has 106 protons and 106 electrons.

The **mass number** of an atom is found by adding the number of protons and neutrons in its nucleus, while ignoring the contribution of the electrons.

mass number = number of protons + number of neutrons

For example, the mass number of an atom with six protons and six neutrons equals 6 + 6, or 12. Because the proton and the neutron each have a mass of about one amu, the number 12 represents the atom's approximate mass in terms of atomic mass units.

We can also find the number of neutrons in an atom if we know its mass number and atomic number. We merely subtract the atomic number, which is the number of protons, from the mass number:

number of neutrons = mass number − atomic number
(number of protons)

Thus if we know that the mass of an atom is 12 and that its atomic number is 6, the number of neutrons is 12 − 6, or 6.

Table 4-3 lists examples of atoms. It gives their atomic number, mass number, and composition in terms of protons, neutrons, and electrons. Using the equations given in this section and the relationship:

atomic number = number of protons = number of electrons

you should be able to cover up one column of this table at a time and fill in the missing information.

Example   An atom has 12 protons and 13 neutrons. What is its mass number?

UNKNOWN	GIVEN	CONNECTION
mass number	12p 13n	mass number = number of protons + number of neutrons

mass number = 12 + 13 = 25

Example   An atom has an atomic number of 20 and a mass number of 42. Find its number of neutrons.

UNKNOWN	GIVEN	CONNECTION
number of neutrons	atomic number = 20 mass number = 42	number of neutrons = mass number − atomic number

number of neutrons = 42 − 20 = 22

Practice problem   Find the mass number of an atom that has an atomic number of 35 and 39 neutrons.

ANSWER   74

Practice problem   How many neutrons are present in an atom having a mass number of 125 and an atomic number of 61?

ANSWER   64

4.5
ELEMENTS

An **element** is a large collection of atoms that have the same atomic number (the same number of protons). For example, the element gold consists of many atoms each with an atomic number of 79. We can see elements like silver, iron, or gold, but we cannot see their individual atoms. An atom is the smallest part of an element that still has the properties of that element. For example, if we could take a piece of gold and cut it into smaller and smaller pieces, the tiniest possible piece that still behaved like gold would be a single atom of gold.

Since atoms exist that have atomic numbers from 1 to 106, there are 106 different elements. Of these, 90 elements can be found in nature. The others are made artificially as explained in Chapter Seventeen.

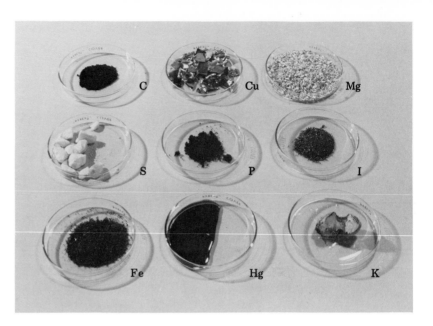

FIGURE 4-3

Elements. Shown here are samples of nine common elements. From left to right, top row: carbon (C), copper (Cu), magnesium (Mg); middle row: sulfur (S), phosphorous (P), iodine (I); bottom row: iron (Fe), mercury (Hg), potassium (K). (Photo by Al Green.)

Some examples of elements are shown in Figure 4-3. As more experiments are done, new elements with higher atomic numbers may be added to the present list of 106.

Each element can be represented by a symbol, a shorthand way of representing its name. Such a symbol consists of one to three letters. If the symbol has two or three letters, the first is always capitalized and the other letters are written in small letters. This distinction is very important as you can see by comparing the following symbols.

Co   Cobalt: an element

CO   Carbon monoxide:
        a combination of two elements, carbon and oxygen

Table 4-4 contains the names and symbols for the 40 most common elements. Since you will be coming across these elements often, you may find it helpful to memorize their names and symbols. (You may want to use flash cards for this task.) An easily confused symbol is P. P stands for phosphorus, *not* for potassium. The symbol for potassium is K. Another easily confused symbol is S. S stands for sulfur and *not* for sodium, whose symbol is Na.

Some of the stranger-looking symbols are based on the name of the element in a language other than English. Many symbols come from

TABLE 4-4 *Common elements*

Name	Symbol	Pronunciation
aluminum	Al	a-loo′mi-num
argon	Ar	ar′gon
arsenic	As	ars′nik
barium	Ba	bar′ee-um
beryllium	Be	be-ril′ee-um
boron	B	bore′on
bromine	Br	broh′meen
calcium	Ca	kal′see-um
carbon	C	kar′bun
chlorine	Cl	klor′een
chromium	Cr	kroh′mee-um
cobalt	Co	ko′ballt
copper	Cu	kopp′er
fluorine	F	flor′een
gold	Au	gohld
helium	He	hee′lee-um
hydrogen	H	hy′dro-jen
iodine	I	eye′uh-dine
iron	Fe	eye′urn
lead	Pb	led
lithium	Li	lith′ee-um
magnesium	Mg	mag-nee′zee-um
manganese	Mn	man′guh-neez
mercury	Hg	mer′kyuh-ree
neon	Ne	nee′on
nickel	Ni	nik′ell
nitrogen	N	ny′truh-jen
oxygen	O	ahk′sih-jen
phosphorus	P	fahs′fuh-rus
platinum	Pt	plat′num
plutonium	Pu	ploo-toh′nee-um
potassium	K	puh-tass′ee-um
silicon	Si	sill′ih-kun
silver	Ag	sill′ver
sodium	Na	sohd′ee-um
strontium	Sr	stront′chee-um
sulfur	S	sul′fer
tin	Sn	tinn
uranium	U	yoo-ray′nee-um
zinc	Zn	zink

Latin names: Au (gold) from *aurum*, Cu (copper) from *cuprum*, Hg (mercury) from *hydrargyrum*, Na (sodium) from *natrium*, Pb (lead) from *plumbum*, and Sn (tin) from *stannum*. Some come from other languages. The symbol K (potassium) comes from the Arabic word *kalium*, and the symbol W (tungsten) comes from the German word *wolfram*. Some of the earlier symbols that were used to represent elements are illustrated in Figure 4-4.

FIGURE 4-4

Early symbols for elements. The symbols we use today are very different from those used by the ancient Greeks, the alchemists, or John Dalton.

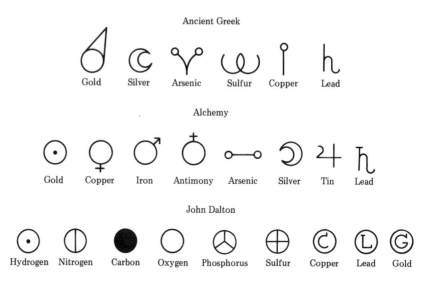

Ancient Greek

Gold   Silver   Arsenic   Sulfur   Copper   Lead

Alchemy

Gold   Copper   Iron   Antimony   Arsenic   Silver   Tin   Lead

John Dalton

Hydrogen   Nitrogen   Carbon   Oxygen   Phosphorus   Sulfur   Copper   Lead   Gold

4.6
ISOTOPES AND ATOMIC WEIGHT

All of the atoms of a particular element have the same atomic number, and therefore the same number of protons (and also the same number of electrons). However, they can have different numbers of neutrons. Hydrogen atoms, for example, all have one proton and one electron. But hydrogen atoms can have either 0, 1, or 2 neutrons. In other words, all hydrogen atoms have the same atomic number, 1, but can have a mass number of either 1, 2, or 3. Therefore, there are three forms of hydrogen, which differ from each other only in the number of neutrons in their nuclei (see Table 4-5). Thus, the atoms of a particular element can exist in several possible forms, each having different mass numbers. These forms are called **isotopes**. Isotopes are atoms of the same element that have different mass numbers.

TABLE 4-5   *Isotopes of hydrogen*

	Mass number of isotope		
	1	*2*	*3*
number of protons (atomic number)	1	1	1
number of neutrons	0	1	2
number of electrons	1	1	1
natural abundance	99.985%	0.015%	—[a]

[a]This isotope (tritium) does not occur naturally.

Because isotopes have the same atomic number (and, therefore, the same number of protons and of electrons), they are chemically alike. Their sole difference is in the number of neutrons, which contributes only to the mass of the atom. As we will see, the chemical identity and properties of an atom are determined by its atomic number and not by its mass number.

Of the 106 known elements, most exist in nature as mixtures of several isotopes. For example, 99.985% of all hydrogen atoms have a mass number of 1, while 0.015% have a mass number of 2. (This latter isotope is called deuterium or heavy hydrogen.) These percentages describe what we call the natural abundance of each isotope. A third isotope of hydrogen (called tritium) is not found in nature, but can be created artificially.

Because of the existence of isotopes, we cannot use a simple mass number to describe the mass of an element. Instead, since elements consist of isotopes that have various masses, we must use an average mass that takes into account the relative natural abundance of each isotope. The average of the atomic masses of the naturally-occurring isotopes of an element is called the **atomic weight** of the element.

Atomic weights are expressed in atomic mass units. They are based on comparison with the most common isotope of the element carbon (the one with six protons and six neutrons), called carbon-12, which is assigned a mass of exactly 12 amu. (As we explained in Chapter Two, the terms mass and weight have different meanings. The use of "weight" in atomic weight is not strictly correct, but is commonly used anyway. A more precise term, relative atomic mass, is not yet widely used.)

Consider the element chlorine. It consists of two isotopes: one with an atomic mass of 34.969 amu and a natural abundance of 75.53%, and the other with a mass of 36.947 amu and an abundance of 24.47%. The atomic weight of chlorine is found as follows:

$$\frac{75.53 \ (34.969 \ \text{amu}) \ + \ 24.47 \ (36.947 \ \text{amu})}{100} = 35.453 \ \text{amu}$$

This calculation involves forming what is called a "weighted average." The isotope of mass 34.969 amu is multiplied by 75.53 because 75.53%, or 75.53 out of every 100 atoms of chlorine, have this mass. Similarly, the isotope of greater mass (36.947 amu) is multiplied by the percentage of atoms that have this value. Dividing by 100 then gives the average atomic mass, which we call the atomic weight. Notice that the result, 35.453 amu, is closer in value to 34.969 than to 36.947 because a larger percentage of chlorine atoms have a mass of 34.969. In general, because most atoms have isotopes, the atomic weights of the elements are not simple whole numbers.

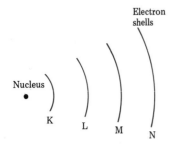

Electron shells

Nucleus

K L M N

FIGURE 4-5

Electron shells. Shells indicate the locations of electrons at certain average distances from the nucleus.

**4.7
SIMPLIFIED
ATOMIC
DIAGRAMS**

As we will see in Chapter Five, the chemical properties of an element depend on the number and arrangement of the electrons outside the nuclei of its atoms. Various types of diagrams can be used to show the electron arrangement of an atom. They are all unreal in the sense that they are simplified attempts to visualize the abstract mathematical equations that describe the behavior of electrons in atoms. Nevertheless, these diagrams are useful tools for describing atoms and their interactions.

The simplest diagram of the atom is shown in Figure 4-5. In this type of diagram, electrons are located in one or more **shells** outside the nucleus. These shells do not represent exact locations of the electrons. The shells merely provide a convenient way to identify electrons found at certain average distances from the nucleus.

Each shell in an atom can hold only a limited number of electrons. The closest shell to the nucleus, called the K shell, contains up to two electrons. The second closest shell, the L shell, holds up to eight electrons. The third closest shell, the M shell, can hold as many as 18 electrons. And the fourth shell, the N shell, can hold up to 32. In general, the maximum number of electrons in a shell is equal to $2n^2$, or $2 \times n \times n$, where n is the number of the shell (K = 1, L = 2, M = 3, N = 4, etc.). For example, the fourth shell ($n = 4$) can contain up to $2 \times (4)^2 = 2 \times (16) = 32$ electrons.

For atoms that have no more than 18 electrons, each inner shell must be completely filled before electrons can appear in the next shell. Thus the first (K) shell must contain its maximum of two electrons before an electron can appear in the second (L) shell. (We will see later that, in large atoms, inner shells are sometimes left only partially filled while electrons enter outer shells.)

According to this simplified way of representing electrons in shells, the major isotope of hydrogen (atomic number = 1; mass number = 1) can be represented as:

$$\left(\text{1p}\right) \text{1e}$$
K
hydrogen atom

The circle represents the nucleus, which has one proton and no neutrons. There is one electron outside the nucleus in the first (K) shell.

An even simpler way of representing the electrons in an atom is to write what we call its **Lewis symbol** (named after the American chemist, Gilbert Lewis). Also known as electron-dot notation, a Lewis symbol consists of the symbol for the element along with either dots (·) or crosses (×) that represent the outer shell electrons. For example, the Lewis symbol for hydrogen is

H·   or   H×

It does not matter on which side of the symbol the dot is placed. All of the following symbols have the same meaning:

H·   Ḣ   ·H   H̦

The next element, helium, with atomic number 2, consists of atoms that have two protons and two neutrons in the nucleus, and two electrons in the first shell.

He:

helium atom

The Lewis symbol now contains two dots, which represent the two electrons of helium.

Lithium, with atomic number 3, has three protons and three electrons. The most common isotope of lithium has four neutrons. The third electron must go into the second shell because the first shell can hold only two electrons.

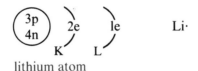

Li·

lithium atom

Because the Lewis symbol shows only the *outermost* shell of electrons, we draw only one dot for lithium. It represents the one electron in the L shell. The letters Li in the Lewis symbol now stand for both the nucleus and also the filled inner shell of electrons. Together, the nucleus and the filled inner shell are the kernel or core of the atom. The outermost electrons, which are represented by dots in the Lewis symbol, are often called *valence* electrons. Similarly, the outermost shell of the atom may be called the *valence* shell. The outermost electrons are the basis for chemical properties and changes, as we will see later.

Atomic number 4 corresponds to the element beryllium. Beryllium atoms have four protons, five neutrons, and four electrons. Two electrons are in the first shell and two electrons are in the second shell.

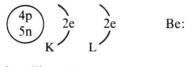

Be:

beryllium atom

Boron, with atomic number 5, has five protons and five electrons. The first shell contains two electrons, and the second shell contains three.

The major isotope of boron has six neutrons.

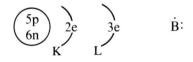

boron atom

Note that in writing Lewis symbols, we never place more than two dots on one side of the symbol.

Carbon, with atomic number 6, has six protons and six electrons. The most abundant isotope of carbon has six neutrons.

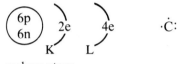

carbon atom

This atom, with mass number 12, was chosen as the standard for expressing atomic mass. The atomic mass unit is 1/12th the mass of this carbon atom. (Note that in boron and carbon atoms, the first two electrons may be written together as a pair while additional electrons are placed separately on the sides of the Lewis symbol. The reason for writing symbols in this way will become apparent in the following sections.)

Diagrams and Lewis symbols for atoms of the next four elements are as follows:

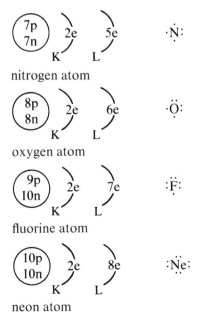

nitrogen atom

oxygen atom

fluorine atom

neon atom

With neon, atomic number 10, the second (L) shell is now complete. The Lewis symbol for neon contains eight dots, which represent the eight electrons in the filled outer shell. Neon is said to have a complete octet of electrons in its outermost shell.

The next element, sodium, has atomic number 11. The eleventh electron is in the third (M) shell, because the inner two shells are filled.

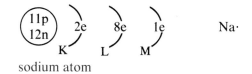

sodium atom

The letters Na in the Lewis symbol for sodium stand for a kernel consisting of the sodium nucleus and the two completely filled inner shells (the K shell with two electrons and the L shell with eight electrons). The dot represents the one outer shell electron.

The methods presented in this section for displaying the electronic arrangement within atoms are simple and useful, but they are oversimplified. A more accurate but complex picture of the atom is presented in the following sections.

4.8 QUANTUM THEORY

In the everyday world, objects that are acted upon by forces, like pushes or pulls, follow the laws of mechanics that were described by Isaac Newton in the late 17th century. However, experiments done in the early 20th century showed that Newtonian, or classical, mechanics does not apply to objects of atomic size. According to the contemporary physicist Richard Feynman, the atom "behaves like nothing you have every seen before." In the 1920s, a new branch of science, called quantum mechanics, was developed to describe the strange world of the atom. (See Figure 4-6.) The model of the atom that resulted from

"ACTUALLY I STARTED OUT IN QUANTUM MECHANICS, BUT SOMEWHERE ALONG THE WAY I TOOK A WRONG TURN."

FIGURE 4-6

(Copyright © 1976 by Sidney Harris/American Scientist Magazine.)

quantum mechanics is difficult to visualize because the behavior of atomic particles is so far removed from our own experience. Yet quantum mechanics provides the best theory we now have to explain the structure and behavior of atoms.

The **quantum theory** is based on the idea that *the energy of an object can have only certain fixed values,* which are determined by the nature of that object. The energies of the electrons in an atom are said to be quantized, or restricted to definite values. For example, imagine a car that can travel only at certain specific speeds, say 0, 8, 20, 35, 62, and 100 mi/hr but cannot travel at speeds between these values. This car's speed is said to be quantized.

Electrons that are restricted to the same allowed value of energy are said to occupy the same **energy level**. The electron shells (K, L, M, and so on) presented in the previous section correspond to these levels. Electrons in the first (K) shell occupy the first energy level. Electrons in the second (L) shell occupy the second energy level, and so on. The maximum number of electrons in an energy level is the same as the maximum number allowed in the shell. The energy of a particular level depends on the distance of the electrons from the nucleus of the atom. For example, those electrons in the shell closest to the nucleus (the K shell) have the lowest energy and are bound most strongly to the nucleus of the atom.

All energy levels except the first are divided into parts called **sublevels**. These sublevels are shown in Figure 4-10(a). There are four different sublevels, labelled s, p, d, and f. (The choice of these letters is of historical importance only and does not concern us here.) We will now use the term principal energy level to describe the energy level of which the sublevel is a part. The energy difference between sublevels is much smaller than the energy difference between principal energy levels.

Within a principal energy level, the energies of the sublevels follow the order (the symbol < means ''less than''):

$$s < p < d < f$$

The s sublevel has the lowest energy, followed by the p sublevel, then the d sublevel, and finally the f sublevel.

As we now see, electrons are arranged into sublevels of the principal energy levels of an atom. Just as the energy levels are divided into sublevels, the electron shells can be divided into corresponding subshells. The electrons in a subshell have the energy of a particular energy sublevel. As we will see in the next section, however, electrons are not really located in actual, distinct ''shells.'' For this reason, we will emphasize the terms energy level and sublevel in this book when describing electrons having the same energy and average distance from the nucleus.

**4.9
ATOMIC ORBITALS**

In the previous section, we saw that electrons in atoms are arranged in energy levels. But we have not yet answered the question: where *exactly* in the space around the nucleus are the electrons found? Electron shell diagrams do not indicate the true positions of electrons in atoms. They show only that electrons exist in a certain order outside the nucleus.

According to quantum mechanics, we can never know the *exact* positions of electrons in an atom at a given instant of time. This fundamental limitation is known as the Heisenberg uncertainty principle. However, we can predict the *probability* of finding electrons at certain locations. The electrons may be found in defined regions of space around the nucleus which are called **orbitals**. An orbital is a region in which the probability, or chance, of finding an electron is high. Each orbital can hold up to two electrons.

Figure 4-7 compares an orbital to a target on a shooting range. If you fire many shots at the target and your aim is reasonably good, the shots will distribute themselves around the center of the target, as shown in part (a). By looking at this target, someone else can then predict that there is a good chance that your next shot will land within the region containing many holes. The distribution of holes thus contains probability information about shots. Similarly, in part (b) of Figure 4-7, dots represent the probability of finding the electron in a region around the nucleus of an atom. An electron is most likely to be found in the central region of the orbital shown, because this is where the density of dots is greatest.

Such a representation of an orbital is called a probability cloud. The position of a single electron at any instant of time is unknown. It is also not known what path the electron follows from one moment to the next. The best we can do is to show with a probability cloud where the electron is most likely to be located. Another way of representing orbitals is to draw the outer boundaries within which there is a very high probability, say 99%, of finding the electron.

You can also think of an electron cloud in the following way. Imagine that you could take pictures at different times of an electron in an atom. Since the electron is moving, the location of the electron would be different in each picture. You could take a very large number of these pictures and put them together to make a movie. If you looked at this movie, you would see a blurry image of the electron. That image results from the electron being spread out over all its possible locations around the nucleus. This electron cloud is how we picture an orbital.

In the last section, we saw that electrons are arranged into sublevels of the principal energy levels of an atom. The electrons in a sublevel are located in one or more orbitals. The orbitals are labelled s, p, d, and f, like the sublevels. The s orbital, shown in Figure 4-8, is a spherical electron cloud. Also shown in Figure 4-8 are the three p orbitals. Each p electron cloud is similar in shape to a dumbbell. The three p

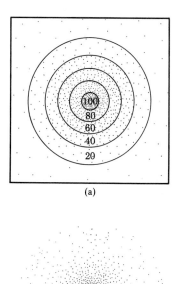

(a)

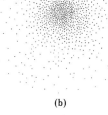

(b)

FIGURE 4-7

Electron cloud. The arrangement of holes in a target (a) provides information about the probability of shots landing at particular locations around the nucleus. Similarly, the electron cloud (b) shows the chance of finding an electron at particular locations around the nucleus.

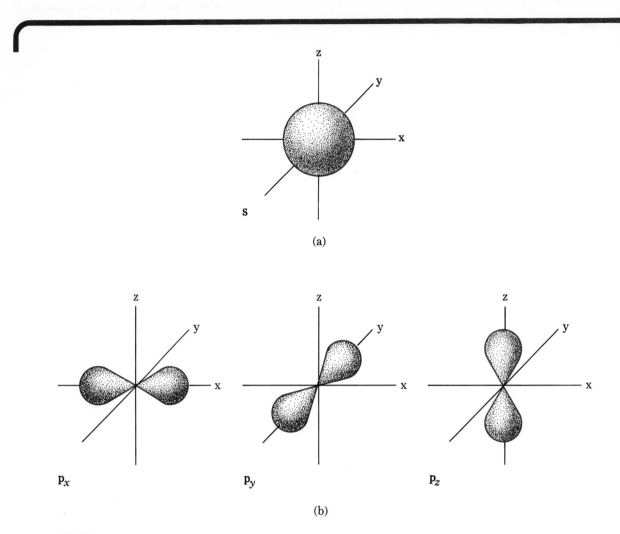

FIGURE 4-8

Atomic orbitals. The s orbital (a) is spherical, while the three p orbitals (b)
are dumbbell-shaped. Each orbital can hold up to two electrons.

orbitals differ only in direction. (The p_x, p_y, and p_z orbitals are 90°
apart from each other.) In addition, there are five d orbitals and seven
f orbitals. These orbitals have more complicated shapes and are not
illustrated here.

Each orbital can hold a maximum of two electrons. An s sublevel,
which contains one s orbital, can therefore hold two electrons. A p
sublevel contains the set of three p orbitals. Since each p orbital can
hold two electrons, a p sublevel can be occupied by a maximum of
3×2 electrons, or 6 electrons. Similarly, a d sublevel contains the
five d orbitals and can hold up to 5×2 electrons, or 10 electrons.
Finally, an f sublevel contains the seven f orbitals and can hold a

maximum of 7 × 2 electrons, or 14 electrons. The following list summarizes the relationship between sublevels, orbitals and electrons:

Sublevel	Number of orbitals	Maximum number of electrons
s	1	2
p	3	6
d	5	10
f	7	14

Let us now see how the orbitals are actually arranged in an atom. The first principal energy level (K shell) can hold two electrons. These electrons are located in the single s orbital of an s sublevel. This sublevel is designated 1s to show that it is part of the first principal energy level.

principal ↗ ^{1s} ↖ sublevel
energy level

The s orbital in the 1s sublevel is called a 1s orbital.

The second principal energy level (L shell) can hold a maximum of eight electrons. Two of these electrons occupy an s orbital in the s sublevel. This sublevel, and the corresponding orbital, are labelled 2s to indicate that they are part of the second principal energy level. (See Figure 4-9.) The remaining six electrons are located in the three p orbitals of the 2p sublevel.

The third principal energy level (M shell) can hold a maximum of 18 electrons. These electrons are organized in the following way:

s sublevel with one s orbital:	2 electrons
p sublevel with three p orbitals:	6 electrons
d sublevel with five d orbitals:	10 electrons
	18 electrons

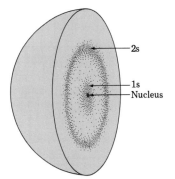

FIGURE 4-9

Orbitals in a lithium atom. The 1s orbital contains two electrons and the 2s orbital contains one electron.

105

TABLE 4-6 *Electrons in the first four principal energy levels*

Principal energy level	Shell	Sublevel	Number of orbitals	Maximum number of electrons
1	K	1s	1	2
2	L	2s	1	2
		2p	3	6
				8
3	M	3s	1	2
		3p	3	6
		3d	5	10
				18
4	N	4s	1	2
		4p	3	6
		4d	5	10
		4f	7	14
				32

The sublevels, and their corresponding orbitals, in the third principal energy level are labelled 3s, 3p, and 3d.

If we continued this process, we would find that the fourth type of orbital, the f orbital, appears in the fourth principal energy level. Table 4-6 summarizes the arrangement of electrons in the first four principal energy levels. Table 4-7 lists the electron sublevels that are occupied in atoms of known elements.

The shape of any particular type of orbital is the same regardless of what energy level it is in. For example, all s orbitals, the 1s, 2s, 3s, and so on, have spherical shapes. However, they differ in size. The 2s orbital is larger than the 1s, and the 3s orbital is larger than the 2s. This increase in orbital size corresponds to the increasing distance of electrons from the nucleus in each successive principal energy level.

Let us now summarize what we know about electrons in atoms. Electrons are tiny, negatively-charged particles that surround a dense, central nucleus. They are located in orbitals, regions having various shapes and sizes, in which the chance of finding an electron is great. Orbitals of the same type, whether s, p, d, or f, have the same energy and make up an energy sublevel of a principal energy level. The energy sublevels represent the allowed values of energy that electrons can have. Electrons in the 1s sublevel are lowest in energy and located in an orbital close to the nucleus. Additional electrons, which are in sublevels with higher energy, are located in orbitals further from the nucleus of the atom.

TABLE 4-7

Summary of sublevels in atoms of known elements

Principal energy level	Electron sublevels
1	1s
2	2s,2p
3	3s,3p,3d
4	4s,4p,4d,4f
5	5s,5p,5d,5f
6	6s,6p,6d
7	7s,7p

4-10
ELECTRON
CONFIGURATION

The arrangement of electrons in an atom is called the electron configuration. It can be described in terms of the orbitals of the energy sublevels that the electrons occupy. Electrons tend to enter those orbitals that are in the sublevels with the lowest energy. Figure 4-10 (a and b) shows two different ways of illustrating the relative energies of the sublevels. In both diagrams, the sublevels with lowest energies are at the bottom.

The 1s sublevel has the lowest energy, and therefore the 1s orbital is the first orbital to be filled with electrons. It is followed by 2s, 2p,

FIGURE 4-10

Energy level diagrams. In (a), the sublevels are indicated by horizontal lines. In (b), the orbitals within the sublevels are shown as circles. In both diagrams, the lowest energy levels are at the bottom.

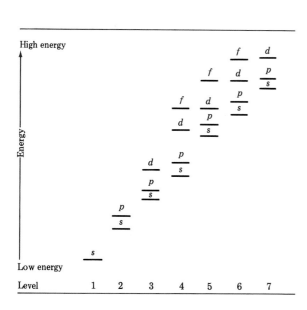

(a)

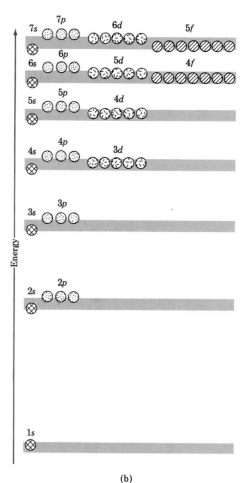

(b)

3s, 3p, 4s, 3d, 4p, 5s, and so on. (This order can be determined experimentally.) Within a principal energy level, the s sublevel is lower in energy than the p, the p sublevel is lower than the d, and the d sublevel is lower than the f, as we saw in the previous section. Notice, however, that a sublevel from one principal energy level can have a *lower* energy than a sublevel from a lower principal energy level. For example, the 4s sublevel is lower in energy than the 3d sublevel. Therefore the 4s orbital is filled by electrons before the 3d orbital. The following list is another way to represent the relative energies of the sublevels and their orbitals:

$$1s < 2s < 2p < 3s < 3p < 4s < 3d < 4p < 5s < 4d, \text{ and so on}$$

The arrangement shown in Figure 4-11 is an easy way to remember the order in which the orbitals are filled.

We can show the filling of the orbitals in atoms of all the elements by using the diagrams shown in Figure 4-10 and Figure 4-11. We call this procedure the *aufbau process*, which in German means to build up. The simplest element, hydrogen, has one electron in the lowest energy orbital, 1s. We can represent the 1s orbital with a circle, and draw an arrow within the circle to represent the one electron.

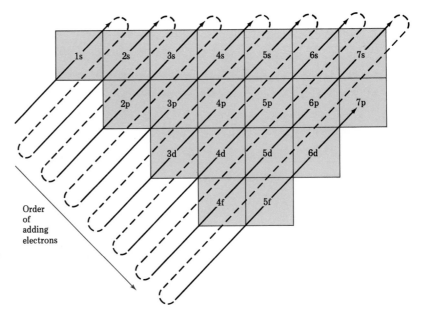

FIGURE 4-11

Order of filling sublevels. This diagram shows a way to remember the order in which electrons fill the sublevels in the aufbau *process. The first two electrons are placed in the 1s orbital, the next two in the 2s, the next six in the 2p, and so on.*

Or, we can write a superscript 1 next to the symbol to show that one electron is in this orbital.

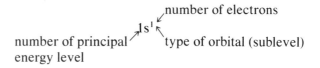

number of electrons

$1s^1$

number of principal energy level

type of orbital (sublevel)

The next element, helium, has two electrons. Its electron configuration is

He
$\;$ 1s
$\;\;\uparrow\downarrow\;$
or
$\;\;\; 1s^2$

When we represent an orbital with a circle, the second electron is shown by an arrow pointing down. Electrons are said to have **spin**, and the arrows show that the spin of the second electron in an orbital must be opposite in direction to that of the first electron. The rule that the two electrons in any orbital must have different spins, one "up" ($\uparrow$) and one "down" ($\downarrow$), is known as the Pauli principle (after Wolfgang Pauli, who won a Nobel prize in 1945). In the notation $1s^2$, the superscript 2 means that two electrons are in the 1s orbital.

The third element, lithium, has its three electrons arranged in the following way:

Li
$\;$ 1s $\;\;$ 2s
$\;\;\uparrow\downarrow\;\;\;\uparrow\;$
or
$\;\;\; 1s^22s^1$

The lowest energy orbital, the 1s, is completely filled. Therefore the third electron must enter the next lowest orbital, the 2s. This situation corresponds to the first principal energy level (K shell) filling up, and the next electron being placed in the second principal energy level (L shell), as described in Section 4.7.

In the next element, beryllium, the 2s orbital becomes filled.

Be
$\;$ 1s $\;\;$ 2s
$\;\;\uparrow\downarrow\;\;\;\uparrow\downarrow\;$
or
$\;\;\; 1s^22s^2$

The fifth electron of boron must enter the next lowest orbital, the 2p.

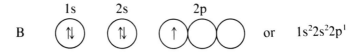

B
$\;$ 1s $\;\;$ 2s $\;\;$ 2p
$\;\;\uparrow\downarrow\;\;\;\uparrow\downarrow\;\;\;\uparrow\;\bigcirc\bigcirc\;$
or
$\;\;\; 1s^22s^22p^1$

The set of three circles under the symbol 2p represents the three p orbitals (p_x, p_y, p_z) of the 2p sublevel. The three p orbitals have equal energies and are called degenerate orbitals. According to Hund's rule (developed by Friedrich Hund in 1927), one electron is placed unpaired in each degenerate orbital before electrons are paired in the same orbital. For example, the first p orbital does not receive a second electron until after all three p orbitals have each received one electron. Hund's rule is based on the repulsion between the negative electrons. Because electrons repel each other, they enter different orbitals, rather than pair up, so that they can be as far apart from each other as possible. Thus, the electron configuration of the next element, carbon, is:

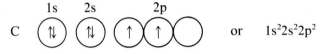

or $1s^2 2s^2 2p^2$

The sixth electron does not pair up with the electron in the first p orbital, but instead enters the second p orbital, according to Hund's rule. Carbon is said to have two *unpaired* electrons in its p orbitals. The placement of electron dots in the Lewis symbols for boron and carbon show the one unpaired electron of boron and the two unpaired electrons of carbon.

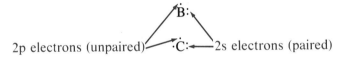

2p electrons (unpaired)   2s electrons (paired)

The 2 p electrons in the next four elements are placed as follows:

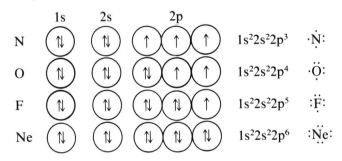

We can see that nitrogen has an unpaired electron in each of three p orbitals. In oxygen, fluorine, and neon, additional electrons pair up with the electrons already in these p orbitals. The eight dots in the Lewis symbol of neon correspond to the eight electrons in its outermost s and p orbitals.

	one 2s orbital	three 2p orbitals	total
maximum number of electrons	2	6	8

Thus an octet (eight electrons) represents the complete filling of the s and p orbitals within a principal energy level.

The next element, sodium, contains eleven electrons. The eleventh electron enters the next lowest orbital, the 3s. Since it becomes awkward to continue writing all the filled orbitals, a common shortcut is to use the following abbreviation:

Na $[Ne]3s^1$

Since neon has the first two principal energy levels completely filled, [Ne] stands for the kernel or core ($1s^2 2s^2 2p^6$) of the sodium atom.

We can find the electron configuration of elements with large atomic numbers by writing the orbitals in their order of filling (that is, from lower to higher energy) using the scheme shown in Figure 4-11. For example, to find the electron configuration of iron, which has atomic number 26, we must place 26 electrons in orbitals having the lowest energies. First we write down the orbitals in the order in which they are filled:

1s, 2s, 2p, 3s, 3p, 4s, 3d, 4p, 5s, and so on

We then add the maximum number of electrons to each orbital until we have added all 26 electrons.

Fe $\underbrace{1s^2 2s^2 2p^6 3s^2 3p^6 4s^2}_{20e}$ $\underbrace{3d^6}$

20e   +   6e = 26e

There are not enough electrons to completely fill the 3d orbitals, which can hold 10 electrons, but all the lower energy orbitals are completely filled. The final electron configuration is usually written in order of the principal energy levels, which is not necessarily the order in which the orbitals are filled:

Fe $1s^2 2s^2 2p^6 3s^2 3p^6 3d^6 4s^2$

This configuration can be abbreviated as

Fe $[Ar]3d^6 4s^2$

because the electron configuration of argon, Ar, is $1s^2 2s^2 2p^6 3s^2 3p^6$.

Table 4-8 presents the electron configuration of all the elements. You may notice that there are certain exceptions to the scheme that we have described here. These exceptions usually result from the special stability of a half-filled sublevel, such as d^5.

TABLE 4-8 *Electron configurations of the elements*

Atomic number	Element	Electron configuration[a]	Atomic number	Element	Electron configuration[a]
1	H	$1s^1$	28	Ni	$[Ar]\, 3d^8 4s^2$
2	He	$1s^2$	29	Cu	$*[Ar]\, 3d^{10} 4s^1$
3	Li	$[He]\, 2s^1$	30	Zn	$[Ar]\, 3d^{10} 4s^2$
4	Be	$[He]\, 2s^2$	31	Ga	$[Ar]\, 3d^{10} 4s^2 4p^1$
5	B	$[He]\, 2s^2 2p^1$	32	Ge	$[Ar]\, 3d^{10} 4s^2 4p^2$
6	C	$[He]\, 2s^2 2p^2$	33	As	$[Ar]\, 3d^{10} 4s^2 4p^3$
7	N	$[He]\, 2s^2 2p^3$	34	Se	$[Ar]\, 3d^{10} 4s^2 4p^4$
8	O	$[He]\, 2s^2 2p^4$	35	Br	$[Ar]\, 3d^{10} 4s^2 4p^5$
9	F	$[He]\, 2s^2 2p^5$	36	Kr	$[Ar]\, 3d^{10} 4s^2 4p^6$
10	Ne	$[He]\, 2s^2 2p^6$	37	Rb	$[Kr]\, 5s^1$
11	Na	$[Ne]\, 3s^1$	38	Sr	$[Kr]\, 5s^2$
12	Mg	$[Ne]\, 3s^2$	39	Y	$[Kr]\, 4d^1 5s^2$
13	Al	$[Ne]\, 3s^2 3p^1$	40	Zr	$[Kr]\, 4s^2 5s^2$
14	Si	$[Ne]\, 3s^2 3p^2$	41	Nb	$*[Kr]\, 4d^4 5s^1$
15	P	$[Ne]\, 3s^2 3p^3$	42	Mo	$*[Kr]\, 4d^5 5s^1$
16	S	$[Ne]\, 3s^2 3p^4$	43	Tc	$[Kr]\, 4d^5 5s^2$
17	Cl	$[Ne]\, 3s^2 3p^5$	44	Ru	$*[Kr]\, 4d^7 5s^1$
18	Ar	$[Ne]\, 3s^2 3p^6$	45	Rh	$*[Kr]\, 4d^8 5s^1$
19	K	$[Ar]\, 4s^1$	46	Pd	$*[Kr]\, 4d^{10}$
20	Ca	$[Ar]\, 4s^2$	47	Ag	$*[Kr]\, 4d^{10} 5s^1$
21	Sc	$[Ar]\, 3d^1 4s^2$	48	Cd	$[Kr]\, 4d^{10} 5s^2$
22	Ti	$[Ar]\, 3d^2 4s^2$	49	In	$[Kr]\, 4d^{10} 5s^2 5p^1$
23	V	$[Ar]\, 3d^3 4s^2$	50	Sn	$[Kr]\, 4d^{10} 5s^2 5p^2$
24	Cr	$*[Ar]\, 3d^5 4s^1$	51	Sb	$[Kr]\, 4d^{10} 5s^2 5p^3$
25	Mn	$[Ar]\, 3d^5 4s^2$	52	Te	$[Kr]\, 4d^{10} 5s^2 5p^4$
26	Fe	$[Ar]\, 3d^6 4s^2$	53	I	$[Kr]\, 4d^{10} 5s^2 5p^5$
27	Co	$[Ar]\, 3d^7 4s^2$			

[a]Asterisks indicate exceptions to the rule for adding electrons.
[b]Symbol proposed by the International Union of Pure and Applied Chemistry. (See Chapter Five for further discussion.)

4.11
ELECTRON TRANSITIONS AND RADIATION

The electron configuration predicted by the aufbau process gives us the lowest energy state of an atom. This state is called the electronic ground state. All of the electrons of an atom that is in its electronic ground state are in the lowest energy levels possible. If energy is added to an atom, its electrons may be made to jump to higher energy levels. The addition of energy is said to excite the electrons and the resulting

Atomic number	Element	Electron configuration[a]	Atomic number	Element	Electron configuration[a]
54	Xe	[Kr] $4d^{10}5s^25p^6$	81	Tl	[Xe] $4f^{14}5d^{10}6s^26p^1$
55	Cs	[Xe] $6s^1$	82	Pb	[Xe] $4f^{14}5d^{10}6s^26p^2$
56	Ba	[Xe] $6s^2$	83	Bi	[Xe] $4f^{14}5d^{10}6s^26p^3$
57	La	*[Xe] $5d^16s^2$	84	Po	[Xe] $4f^{14}5d^{10}6s^26p^4$
58	Ce	*[Xe] $4f^15d^16s^2$	85	At	[Xe] $4f^{14}5d^{10}6s^26p^5$
59	Pr	[Xe] $4f^36s^2$	86	Rn	[Xe] $4f^{14}5d^{10}6s^26p^6$
60	Nd	[Xe] $4f^46s^2$	87	Fr	[Rn] $7s^1$
61	Pm	[Xe] $4f^56s^2$	88	Ra	[Rn] $7s^2$
62	Sm	[Xe] $4f^66s^2$	89	Ac	*[Rn] $6d^17s^2$
63	Eu	[Xe] $4f^76s^2$	90	Th	*[Rn] $6d^27s^2$
64	Gd	*[Xe] $4f^75d^16s^2$	91	Pa	*[Rn] $5f^26d^17s^2$
65	Tb	[Xe] $4f^96s^2$	92	U	*[Rn] $5f^36d^17s^2$
66	Dy	[Xe] $4f^{10}6s^2$	93	Np	*[Rn] $5f^46d^17s^2$
67	Ho	[Xe] $4f^{11}6s^2$	94	Pu	[Rn] $5f^67s^2$
68	Er	[Xe] $4f^{12}6s^2$	95	Am	[Rn] $5f^77s^2$
69	Tm	[Xe] $4f^{13}6s^2$	96	Cm	*[Rn] $5f^76d^17s^2$
70	Yb	[Xe] $4f^{14}6s^2$	97	Bk	[Rn] $5f^97s^2$
71	Lu	*[Xe] $4f^{14}5d^16s^2$	98	Cf	[Rn] $5f^{10}7s^2$
72	Hf	[Xe] $4f^{14}5d^26s^2$	99	Es	[Rn] $5f^{11}7s^2$
73	Ta	[Xe] $4f^{14}5d^36s^2$	100	Fm	[Rn] $5f^{12}7s^2$
74	W	[Xe] $4f^{14}5d^46s^2$	101	Md	[Rn] $5f^{13}7s^2$
75	Re	[Xe] $4f^{14}5d^56s^2$	102	No	[Rn] $5f^{14}7s^2$
76	Os	[Xe] $4f^{14}5d^66s^2$	103	Lr	[Rn] $5f^{14}6d^17s^2$
77	Ir	[Xe] $4f^{14}5d^76s^2$	104	Unq[b]	[Rn] $5f^{14}6d^27s^2$
78	Pt	*[Xe] $4f^{14}5d^96s^1$	105	Unp[b]	[Rn] $5f^{14}6d^37s^2$
79	Au	*[Xe] $4f^{14}5d^{10}6s^1$	106	Unh[b]	[Rn] $5f^{14}6d^47s^2$
80	Hg	[Xe] $4f^{14}5d^{10}6s^2$			

atom is said to be in an excited state. This process is called the **absorption** of energy by an atom.

_____   _____ ↑   upper energy level

(↑ = electron)   absorption of energy

___↑_____   _____   lower energy level

Because the energy levels of the atom are fixed, only certain definite amounts of energy can be absorbed. These definite amounts of energy, must correspond to the difference in energy between two energy levels.

Electrons generally remain excited only for short periods, often a fraction of a second. They soon return to a more stable lower level, and release their extra energy. This process is called **emission**.

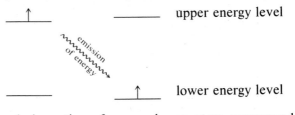

Both emission and absorption of energy by an atom correspond to electron transitions, the movement of an electron from one level to another.

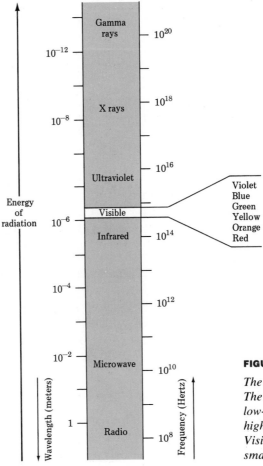

FIGURE 4-12

The electromagnetic spectrum. The spectrum ranges from low-energy radio waves to high-frequency gamma rays. Visible radiation is only a small part of the spectrum.

Radiation is one means by which energy is absorbed or emitted by an atom. (The type of radiation we are concerned with here is called electromagnetic radiation. Nuclear radiation is discussed in Chapter Seventeen.) The different forms of radiation are given in Figure 4-12. Our eyes see only visible radiation, a very narrow part of the overall range.

Box 4

Electromagnetic radiation

Electromagnetic radiation is a form of transmitted energy. It is called electromagnetic because it consists of electric and magnetic fields. These fields are at right angles (90°) to each other and to the direction in which the energy travels.

As shown in Figure 4-12, the various types of electromagnetic radiation differ greatly in their energy, wavelength, and frequency. All forms, however, travel at the speed of light which is 3.0×10^{10} m/sec in a vacuum.

Radio wave. This long wavelength, low energy radiation is emitted and received by large antennas. It includes the AM, FM, TV, and shortwave frequencies.

Microwave. This form of radiation is used in microwave ovens. It causes water molecules in food to rotate and vibrate, generating heat. Microwave radiation is also used in communications.

Infrared. Infrared, or IR radiation is produced by warm objects and is felt as heat. An electric heater, for example, emits infrared radiation.

Visible. The only type of electromagnetic radiation that we can see is called visible radiation. It contains the colors of the visible spectrum: red, orange, yellow, green, blue, and violet.

Ultraviolet. Known as UV radiation, ultraviolet radiation has more energy than visible radiation. It is the form of radiation from the sun that causes sunburn. The electron transitions described in Section 4.11 can be produced either by ultraviolet or visible radiation.

X ray. This high energy radiation can penetrate the body, and it is partially absorbed by the denser parts of the body such as bones or teeth. If a photographic film is placed behind a part of the body that is exposed to X rays, the denser portions of the body (such as bone) show up as lighter areas after the film is developed.

Gamma ray. Gamma rays are produced by nuclear reactions. They are an extremely high energy, short wavelength form of radiation.

All of these forms of electromagnetic radiation can be used to study atoms and molecules. For example, elements and compounds can be identified by the particular frequencies of radiation that they absorb or emit. In addition, the quantity or concentration of a substance often can be found from the intensity of radiation that is absorbed or emitted.

Radiation can be thought of as a travelling wave and described in terms of its wavelength λ (lambda). One **wavelength** is the distance between two wave peaks, as shown in Figure 4-13. One complete up and down pattern of the wave is called a cycle. The **frequency** of a wave ν (nu) is the number of cycles per second. The unit for frequency, one cycle per second (1/sec), is called the hertz, symbolized Hz. The two waves shown in Figure 4-13 have different frequencies. The frequency of wave A is low; the number of cycles per second is small. Notice that its wavelength is long. The frequency of wave B is high; the number of cycles per second is greater. The wavelength of the wave is short. This relationship between wavelength and frequency is summarized by the following equation:

$$\lambda \times \nu = c$$

The symbol c stands for the speed of light in a vacuum. All electromagnetic radiation travels at this speed, which is 3.00×10^8 m/sec. Note in this formula and in Figure 4-13 that wavelength and frequency are inversely related, that is, one increases when the other decreases.

Radiation can also be described in terms of particles of energy, which we call **photons**. The energy of a photon is given by the following equation, put forth by Einstein in 1905:

$$E = h \times \nu$$

According to this equation, the energy E of a photon equals the frequency ν of the radiation multiplied by a fundamental constant h, which has the value 6.63×10^{-34} J sec and is called Planck's constant. Thus the energy of radiation is proportional to the frequency. Radiation with high frequency (and therefore small wavelength), like X rays, has high energy.

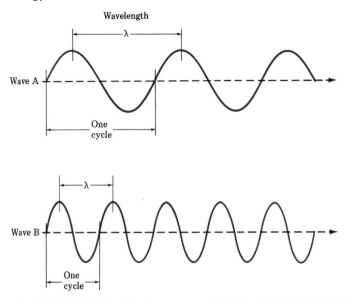

FIGURE 4-13

Wave nature of radiation. We can think of electromagnetic radiation traveling as a wave. The wavelength is the distance between adjacent peaks. Wave A has twice the wavelength of wave B.

FIGURE 4-14

Hydrogen absorption spectrum. The frequencies of radiation absorbed by hydrogen appear as lines in the spectrum (a). Each line results from the absorption of a specific amount of energy as an electron jumps from a lower energy level to a higher one (b). The absorptions are labelled to show which energy levels are involved.

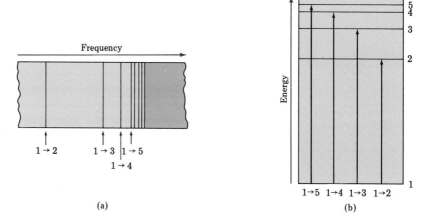

You may recall that only certain energies of radiation can be absorbed or emitted by atoms. And because frequency is directly related to energy, only certain frequencies of radiation can be absorbed or emitted. The pattern of the frequencies of radiation that can be absorbed or emitted by a substance is called the **spectrum** of that substance. Part of an absorption spectrum for the hydrogen atom is shown in Figure 4-14 (a). The lines represent those frequencies that are

---

Do-it-yourself 4:

Flaming colors

Artificial fireplace logs are made from sawdust and wax. They also contain substances that color the flame. The heat of the fire excites electrons in the atoms of these substances. When the excited electrons return to lower energy levels, the atoms emit visible radiation. You can create this effect yourself by using chemicals from around the house.

To do: To make colored flames, choose one of the substances listed below. Dissolve some in water, soak a wood log in the liquid, and then allow the log to dry before burning it.

	Color of flame
table salt	yellow
borax	green
salt substitute	lavender

As the log burns, the flame should appear yellow, green, or lavender, depending on which substance you used.

If you would rather not soak a log, you can instead carefully sprinkle some solid table salt, borax, or salt substitute directly onto the flames in the fireplace.

absorbed. They correspond to the hydrogen energy-level transitions that are shown in Figure 4-14 (b). Each line in the spectrum results from absorption of photons, whose energy is determined by the difference in energy of the two levels involved.

An absorption spectrum is made by exposing a substance to radiation that has a wide range of frequencies. We then use an instrument to record those frequencies that are absorbed by the substance. An emission spectrum, on the other hand, is made by first exciting the electrons in the atoms of a substance (by increasing the temperature, for example) and then recording the frequencies of radiation that are emitted as the excited electrons return to the lower energy levels. Both absorption and emission spectra provide important information in studying and identifying chemical substances.

"White" light is a mixture of all the frequencies present in visible radiation, and the frequencies correspond to all the visible colors. Substances appear colored when they absorb some of the visible radiation and reflect the rest of it. If a substance does not absorb any of the frequencies of white light, the substance will appear white or colorless. If a substance absorbs all of the frequencies, it will appear black. If on the other hand, a substance absorbs only some of the frequencies, then it will appear colored. The substance will have the color of the frequencies of light that are *not* absorbed. For example, plants appear green because they absorb radiation of the frequencies that we call red. The red frequencies of the visible region are removed from the white light, leaving a combination of frequencies that have the color green.

A brief history of the modern atom

When atomic spectra were first obtained in the late 1800s, they could not be fully explained. Atoms were still thought to be the smallest units of matter—tiny, hard, and indivisible. The breakthrough that led to our modern model of the atom was the discovery that the atom consists of even smaller pieces, now called the subatomic particles.

The first such particle to be discovered was the electron. In 1897, the English scientist J. J. Thomson (1856–1940) carried out experiments with evacuated (cathode ray) tubes in which he conclusively identified negatively charged particles. These particles were smaller than hydrogen, the lightest atom. Thomson pictured the atom as a sphere of positive charge in which negatively charged particles were uniformly distributed, like raisins in a pudding. These particles are now called electrons.

In 1908, Ernest Rutherford (1871–1937), a former student of Thomson at Cambridge University, performed a key experiment that disproved Thomson's model. Rutherford bombarded a very thin sheet of gold foil with positively charged particles (helium nuclei, or alpha particles), as illustrated in Figure 4-15. Most of the particles passed right through the foil, but much to Rutherford's surprise, some were greatly shifted in path and a few even reversed direction completely. As Rutherford stated, "It was almost as incredible as if you fired a 15-inch shell at a piece of tissue paper and it came back and hit you." To explain the results of his experiment, Rutherford proposed that the positive charge and almost all the mass of an atom must be concentrated in a very small nucleus. The extremely small electrons must be spread out in a large region surrounding the nucleus. Since most of the atom is therefore nearly empty space, the majority of bombarding particles would pass right through the atoms of the foil. But those few particles that came close to a dense nucleus would be deflected. A "direct hit" on a nucleus would cause the particle to return in the direction from which it came.

Another former student of Thomson, the Danish scientist Niels Bohr (1885–1962), combined Rutherford's model of the atom with the ideas of quantum mechanics developed by the German scientist Max Planck. He proposed that electrons orbited in circular paths at fixed distances from the nucleus. Each orbit corresponded to a certain energy. The transfer of an electron from one fixed orbit to another required an amount of energy equal to the difference in energy of the two orbits. Thus, Bohr's model could explain the definite frequencies observed in atomic spectra.

In 1919, shortly after Bohr presented his model of the atom, Rutherford showed experimentally that the nuclei of atoms contain positively charged particles. He called them protons. It was not until 1932, however, that James Chadwick (1891–1974), who had previously worked under Rutherford, discovered the neutron. (Nuclear chemistry, the study of atomic nuclei, is discussed in detail in Chapter Seventeen).

In 1923, the Frenchman Louis De Broglie (1892–   ) predicted that the electron, previously thought of only in terms of a particle, also exhibited the properties of a wave. He found that the wavelength of

FIGURE 4-15

The Rutherford scattering experiment. In the Thomson model of the atom (a), mass and charge are spread throughout the atom. A positively-charged beam of particles would experience little deflection. In the Rutherford model (b), the mass and positive charge are concentrated in a small nucleus. Most positively-charged particles would pass through without deflection, but the few that came close to a nucleus would be greatly deflected.

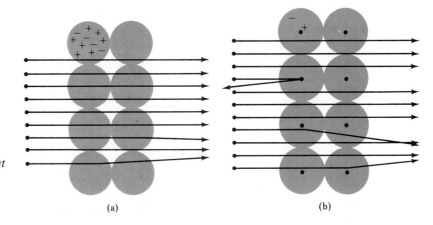

(a) (b)

a particle depended on its mass and velocity. This strange discovery helped support Einstein's view that matter (particles) and energy (waves) are different forms of the same thing.

An Austrian, Erwin Schrödinger (1887–1961), used De Broglie's concept as the basis for describing the electrons of an atom. In Schrödinger's model of the atom, electrons form a continuous cloud of negative charge around the nucleus. The electrons are described in terms of waves; the intensity of the wave at any point is proportional to the probability of finding an electron at that point. Peaks of the waves corresponded to the old Bohr orbits. Schrödinger developed a key relationship called the wave equation that gave quantum mechanics a firm mathematical basis.

At about the same time, the German scientist Werner Heisenberg (1901–1976) independently developed a different but equivalent mathematical basis for quantum mechanics. In addition, he introduced the uncertainty principle, which states that we cannot know the exact location of an electron in an atom at any instant of time. Because of this principle, we can only speak of an electron orbital and the probability of finding the electron within it, rather than the precise location of an electron.

The development of our current model of the atom is one of the greatest achievements of science. This model provides the theoretical basis for all of chemistry.

Summary **4.1** According to atomic theory, all matter is composed of particles called atoms. Atoms are the basic building blocks of nature. There are 106 chemically different kinds of atoms now known. Chemistry is the study of how atoms combine and how substances made of atoms are changed into other substances.

4.2 Charge is a basic property of nature. It exists in two possible forms: positive (+) and negative (−). Like charges repel each other, and unlike charges attract.

Atoms consist of even smaller parts called the subatomic particles; these are protons, neutrons and electrons. Protons have a positive charge, neutrons have no charge, and electrons have a negative charge. In a neutral atom, the number of protons exactly equals the number of electrons.

4.3 The mass of a proton is approximately equal to that of a neutron. The electron, however, has a mass that is nearly 2000 times smaller than that of a proton or neutron. The atomic mass unit (amu) is the basis for a scale of relative masses.

4.4 The atomic number equals the number of protons in an atom. It is the most important identification of an atom. The mass number is the sum of the numbers of protons and neutrons in an atom.

4.5 An element is a large collection of atoms having the same atomic number. There are 106 different elements known. Each element is represented by a symbol consisting of one, two, or three letters.

4.6 Isotopes are atoms of the same element that have different mass numbers. The atomic weight of an element is the average of the atomic masses of its isotopes as they are found in nature.

4.7 Electrons are arranged in shells that surround the nucleus. Each shell can hold only a certain number of electrons. For example, the first (K) shell can hold two electrons, the second (L) shell can hold eight electrons, and the third (M) shell can hold 18 electrons. The Lewis symbol for an atom uses dots or crosses to represent the number of outer electrons.

4.8 The quantum theory proposes that the energies of electrons in an atom are restricted to certain values. These values are given by the energy levels of the electrons. Electrons in the same shell occupy the same energy level. Electron energy levels actually consist of sublevels that differ slightly in energy. The electron shells are divided into corresponding subshells. The electrons in each subshell have the energy of a specific sublevel.

4.9 Electrons are located in orbitals. Orbitals are regions of space around the nucleus of an atom in which the chance of finding an electron is high. There are four different types of orbitals: s, p, d and f. Orbitals of the same type make up an electron sublevel.

4.10 The electron configuration of an atom is the arrangement of its electrons in orbitals. When we fill orbitals by the aufbau process, we place electrons in orbitals in order of increasing energy.

4.11 Electrons can be excited to a higher energy level when atoms absorb certain frequencies of radiation. Similarly, when excited electrons return to a lower level, radiation can be emitted. The characteristic pattern of frequencies absorbed or emitted by a substance is called the spectrum of that substance.

KEY-WORD MATCHING EXERCISE

For each term on the left, choose a phrase from the right-hand column that most closely matches its meaning.

1 atom	**a** outermost electron
2 proton	**b** central part of atom
3 electron	**c** amu
4 neutron	**d** protons + neutrons
5 nucleus	**e** collection of atoms
6 charge	**f** weighted mass of isotopes
7 atomic mass unit	**g** holds 2 to 36 electrons
8 atomic number	**h** charge of +1
9 mass number	**i** charge of −1
10 element	**j** either 0, +, or −
11 isotope	**k** neutral particle
12 atomic weight	**l** basic unit of matter
13 principal energy level	**m** shows outermost electrons
14 Lewis symbol	**n** energy values restricted
15 valence electron	**o** building-up procedure
16 quantum theory	**p** electron arrangement
17 orbital	**q** number of protons
18 electron configuration	**r** same element, different mass
19 aufbau process	**s** electron cloud
20 radiation	**t** pattern of frequencies
21 spectrum	**u** energy absorbed or emitted

EXERCISES BY TOPIC

Atoms (4.1)

22 What exactly is an atom?

23 Describe the subatomic particles.

24 Why is the atom hard to study?

25 What is the charge of the following: **a** electron? **b** neutron? **c** proton?

26 If atoms contain charged particles, how can atoms be neutral?

27 What is the relationship between the number of protons and the number of electrons in an atom?

Masses of subatomic particles (4.3)

28 Why are electrons ignored when we calculate the atomic mass?

29 What is an amu?

30 What is the reason for measuring mass in amu?

Atomic number and mass number (4.4)

31 Which one of the terms does not belong in the following: **a** number of protons? **b** atomic number? **c** number of neutrons? **d** number of electrons?

32 What is the difference between atomic number and mass number?

33 An atom has 15 protons and 17 neutrons. Find its atomic number and mass number.

34 If an atom has a mass number of 36 and an atomic number of 17, how many neutrons does it have?

35 Complete the following table:

	Atomic number	Mass number	Number of protons	Number of neutrons	Number of electrons
a	9	19	___	___	___
b	___	___	16	18	___
c	___	52	25	___	___
d	___	___	___	38	35
e	90	___	___	92	___
f	___	68	___	___	33
g	___	81	40	___	___
h	27	___	___	28	___

Elements (4.5)

36 What is an element?

37 What do all the atoms of an element have in common?

38 Write the symbol for: **a** aluminum **b** lead **c** silver **d** oxygen **e** potassium

39 Name these elements: **a** Ar **b** C **c** S **d** Hg **e** Cu

40 Match the name and symbol for these elements:

a phosphorus	**1** Mg
b iron	**2** N
c sodium	**3** Na
d magnesium	**4** P
e nitrogen	**5** Fe

Isotopes and atomic weight (4.6)

41 How are the isotopes of an element alike? How do they differ?

42 How does atomic weight differ from atomic mass?

43 Which pairs of atoms are isotopes?

	A	B	C	D
mass number	53	53	52	52
atomic number	25	24	24	25

44 Why are most atomic weights not whole numbers?

45 Lithium has two isotopes, one with a mass of 6.015 that is 7.42% abundant, and another with a mass of 7.016 that is 92.58% abundant. Find the atomic weight of lithium.

46 An atom has an atomic number of 18 and a mass number of 38. Give the number of protons and neutrons in a possible isotope of this atom.

Simplified atomic diagrams (4.7)

47 Draw an electron shell diagram for magnesium.

48 Write Lewis symbols for   **a** carbon   **b** bromine   **c** potassium **d** phosphorus   **e** argon

49 Draw an electron shell diagram for lithium and for sodium. How are these two diagrams similar?

50 What are valence electrons?

51 How many outermost electrons are in an atom of   **a** fluorine   **b** hydrogen **c** aluminum   **d** neon   **e** silicon

Quantum theory (4.8)

52 What is the basis for the quantum theory?

53 What does it mean for something to be quantized? Give an example.

54 What is the relationship between electron shells and energy levels?

55 What is the relationship between electron subshells and energy sublevels?

Atomic orbitals (4.9)

56 What is an orbital?

57 Why are orbitals represented by probability clouds?

58 Complete the following table:

	Sublevel	Number of orbitals	Number of electrons
a	s	_____	_____
b	p	_____	_____
c	d	_____	_____
d	f	_____	_____

59 Describe the meaning of these symbols: **a** 5s   **b** 3p   **c** 2s   **d** 4d

Electron configuration (4.10)

60 What is   **a** the Pauli principle?   **b** Hund's rule?

61 Write the electron configuration of argon.

62 Why does nitrogen have three unpaired electrons in its p orbitals?

63 Write the electron configuration for: **a** fluorine   **b** chlorine   **c** iodine. What do they have in common?

64 Write the electron configuration for: **a** sodium **b** phosphorus **c** copper **d** silver **e** neon

Electron transitions and radiation (4.11)

65 What is the **a** ground state of an atom? **b** excited state of an atom?

66 Compare the processes of absorption and emission.

67 What is radiation? Give examples of different types of radiation.

68 What is a spectrum? Why does it show distinct lines?

69 What is the difference between an absorption and emission spectrum?

70 Why do many objects appear colored?

ADDITIONAL EXERCISES

71 The following table contains information about five different atoms. Complete the table:

	Number of protons	Number of neutrons	Number of electrons	Atomic number	Mass number
a	32	___	___	___	73
b	___	14	14	___	___
c	___	___	___	28	59
d	48	64	___	___	___
e	___	115	___	77	___

72 Identify the elements in Exercise 71.

73 What number is most important for identifying an atom?

74 Calculate the actual mass in grams of an atom with a relative mass of 10 amu.

75 Match the name and symbol for these elements:

a Si	**1** plutonium
b Sn	**2** gold
c Au	**3** tin
d Pu	**4** zinc
e Zn	**5** silicon

76 Neon has three isotopes. One isotope has a mass of 19.992 and is 90.92% abundant, one has a mass of 20.994 amu and is 0.257% abundant, and one has a mass of 21.991 amu and is 8.82% abundant. Find the atomic weight of neon.

77 Given the following data, calculate the atomic weight of: **a** sulfur and **b** iron

a Mass	**Abundance**	**b Mass**	**Abundance**
31.972	95.00%	53.940	5.82%
32.971	0.76%	55.935	91.66%
33.968	4.22%	56.935	2.19%
		57.933	0.33%

78 What would be the maximum number of electrons in the fifth principal energy level?

79 Write Lewis symbols for **a** nitrogen **b** barium **c** iodine **d** sulfur **e** cesium

80 What are degenerate orbitals? Give an example.

81 Write the electron configuration for: **a** zinc **b** helium **c** lead **d** chromium **e** tin

82 Find the frequency of orange light. Its wavelength is 6.0×10^{-5} cm.

83 Find the energy of orange light (see Exercise 82).

84 An AM radio station broadcasts at 810 kHz. Find the **a** wavelength and **b** energy of its radio waves.

85 Which elements have the following electron configurations? **a** $1s^2 2s^2 2p^6 3s^2 3p^6$ **b** [Ne] $3s^2 3p^6 4s^2 3d^5$ **c** [Ne] $3s^2 3p^6 4s^1 3d^{10}$

86 The wavelength of violet light is 420 nm. Find its frequency and energy.

87 To which element do each of the following descriptions belong? **a** atomic number 54 **b** 50 electrons **c** 55 protons

88 Which of the following atoms are isotopes? **a** atomic number 34, mass number 79 **b** 34 protons, 44 neutrons **c** 35 protons, 44 neutrons **d** atomic number 44, 44 neutrons **e** 47 neutrons, 34 electrons

89 Write the atomic number and mass number for an atom with: **a** 28 protons, 30 neutrons **b** 33 electrons, 40 neutrons **c** 150 neutrons, 94 electrons

90 Explain the meaning of the following symbols: **a** $2s^1$ **b** $3p^4$ **c** $1s^2$ **d** $3d^{10}$ **e** $5p^3$

91 Arrange these sublevels from lowest to highest energy: 5s 3d 7s 4p 5f 4d

92 What is the reason that electrons are found *un*paired in degenerate orbitals whenever possible?

93 Cosmic rays from outer space are a form of radiation with extremely high energy. How would you expect the wavelength and frequency of cosmic rays to compare with the wavelength and frequency of visible radiation?

94 Why are only certain amounts of energy absorbed and emitted by atoms?

95 Predict the electron configuration for element number 107.

PROGRESS CHART Check off each item when completed.

_____ Previewed chapter

Studied each section:

_____ **4.1** Atoms

_____ **4.2** Charges of subatomic particles

_____ **4.3** Masses of subatomic particles, and the amu

_____ **4.4** Atomic number and mass number

_____ **4.5** Elements

_____ **4.6** Isotopes and atomic weight

_____ **4.7** Simplified atomic diagrams

_____ **4.8** Quantum theory

_____ **4.9** Atomic orbitals

_____ **4.10** Electron configuration

_____ **4.11** Electron transitions and radiation

_____ Reviewed chapter

_____ Answered assigned exercises:

#'s _____

FIVE

The periodic table

In Chapter Five we introduce a systematic arrangement of the elements called the periodic table. First, the organization of a standard form of the periodic table is described. The table then is used to reveal trends in the properties of elements. Based on their properties, elements are classified as metals, nonmetals, or semimetals. Forms of the periodic table other than the standard one also are presented. At the end of this chapter, the uses of important elements are summarized. To demonstrate an understanding of Chapter Five, you should be able to:

1 State the periodic law and explain its importance.

2 Describe the relationships between the periods of the periodic table and the filling of electron energy levels.

3 Identify groups in the periodic table and explain the relationship between group number and the number of outer electrons.

4 Compare the properties of metals, semimetals, and nonmetals.

5 Describe the reason for the periodic variation of atomic size, ionization energy, and electron affinity.

6 Explain the periodic trends in the metallic and nonmetallic properties of elements.

7 Compare the features of different forms of the periodic table.

8 Give examples of practical uses of an element from each main group.

**5.1
PERIODIC LAW**

By the beginning of the 19th century, a relatively large number of elements had been isolated and studied. Although each element had its individual characteristics, many clearly had certain properties in common. Scientists made numerous attempts to organize the known elements in a way that would show their relationship to one another. Figure 5-1 shows one of the earliest arrangements.

FIGURE 5-1

(Copyright © 1977 by Sidney Harris/American Scientist Magazine.)

"THE PERIODIC TABLE."

It was also at this time that a German chemist, Johann Wolfgang Dobereiner, organized the elements into groups of three, which he called triads. For example, the elements calcium, strontium, and barium formed one triad, and the elements lithium, sodium, and potassium formed another. The middle element in any of Dobereiner's triads had properties and atomic weight that were approximately the average of those of the other two members of the triad. An English chemist, John A. Newlands, later arranged the elements into groups of seven in such a way that any given element and the eighth element following it had similar properties. Because this scheme resembled the notes on a musical scale, it was referred to as Newlands' octaves.

The way that we arrange the elements today is based on the work of Dimitri Mendeléev. In 1869, Mendeléev and other independent workers discovered that when the 63 elements then known were listed by atomic weight, certain properties were repeated in a regular, or periodic, manner. Mendeléev's system of classifying elements is called a **periodic table**. A periodic table shows the relationships between the elements by revealing their periodicity, the tendency of their properties to repeat at regular intervals.

The periodic table is a way of organizing a large number of facts. The chemist J. F. Hyde once remarked that the table is "perhaps as close as we can come to putting all we know about the universe on one piece of paper." In addition to being a system of classification, the periodic table is also a powerful tool for predicting the relationships between elements. For example, Mendeléev's original table had several blank spaces in it. Mendeléev was able to predict accurately the properties of yet undiscovered elements, based on the positions of the spaces in the table.

Smithsonian Institution: Photo #58228

While he was a professor of chemistry at the University of St. Petersburg in Russia, Mendeléev prepared to write a chemistry textbook. He wanted to organize the 63 known elements in a way that would help students to understand chemistry. Mendeléev made one card for each element and tried arranging them in different ways. The inspiration to order them by increasing atomic weight supposedly came to him in his sleep.

Mendeléev's second inspiration was to place elements with similar chemical properties in the same vertical columns. This resulted in a table of short rows followed by longer rows. Previous less successful arrangements had the same number of elements in each row. Shown here is one of Mendeléev's early tables (1869). (Note that in this version, the elements with similar properties appear in the same row.)

There were empty spaces in Mendeléev's table into which no known element would fit. Mendeléev boldly claimed that these spaces (indicated by question marks in the symbol column) would be filled by new elements which were not yet discovered. By comparing the properties of elements above and below the gaps, he even predicted the properties of the missing elements.

Years later, between 1875 and 1885, the elements gallium, scandium, and germanium were discovered and were found to match Mendeléev's predictions! As a result, the Russian chemist became famous. In 1955, long after Mendeléev's death, element number 101 was named mendelevium (Md) in his honor.

```
                                    Ti = 50    Zr =  90    ? = 180
                                    V  = 51    Nb =  94    Ta = 182
                                    Cr = 52    Mo =  96    W  = 186
                                    Mn = 55    Rh = 104,4  Pt = 197,4
                                    Fe = 56    Ru = 104,4  Ir = 198
                            Ni = Co = 59       Pd = 106,6  Os = 199
  H = 1
                                    Cu = 63,4  Ag = 108    Hg = 200
          Be =  9,4   Mg = 24       Zn = 65,2  Cd = 112
          B  = 11     Al = 27,4     ?  = 68    Ur = 116    Au = 197 ?
          C  = 12     Si = 28       ?  = 70    Sn = 118
          N  = 14     P  = 31       As = 75    Sb = 122    Bi = 210 ?
          O  = 16     S  = 32       Se = 79,4  Te = 128 ?
          F  = 19     Cl = 35,5     Br = 80    J  = 127
  Li = 7  Na = 23     K  = 39       Rb = 85,4  Cs = 133    Tl = 204
                      Ca = 40       Sr = 87,6  Ba = 137    Pb = 207
                      ?  = 45       Ce = 92
                      ?Er = 56      La = 94
                      ?Yt = 60      [Di = 95
                      ?In = 75,6]   Th = 118 ?
```

If the elements are arranged by their atomic weights, as in Mendeléev's original periodic table, some of the elements seem to be in the wrong place. For example, the element potassium (atomic weight 39.098 amu) would come before argon (atomic weight 39.948 amu). Potassium would be grouped with a series of gases having properties very different from it.

This problem remained until the early 20th century when the English scientist Henry G. Moseley determined the atomic numbers of the elements. When the elements were arranged in the periodic table according to their atomic numbers, the problem was resolved. Argon, atomic number 18, now came before potassium, atomic number 19. This result is not surprising, since chemical properties depend on atomic number and not on atomic weight. In general, the ordering by atomic numbers is similar, but not identical, to the arrangement by atomic weights.

The modern periodic table is based on the following **periodic law**: *The properties of the elements repeat in a regular way when the elements are arranged by increasing atomic number.* As described in the previous chapter, the electronic structure of atoms varies in a regular way with atomic number. It is, therefore, not surprising that all those properties of the elements that depend on electronic structure also tend to vary periodically with atomic number.

5.2 PERIODS

A standard version of the periodic table is shown in Figure 5-2 (and on the inside front cover of this book). Each element has a separate box in the table. The symbol for the element, the atomic number, and the atomic weight are usually given in each box, as illustrated in Figure 5-3. Sometimes additional information, such as the element's name and the electron configuration, also appears.

The horizontal rows that run across the table are called **periods,** and are numbered 1 through 7. Each period represents the gradual filling of an energy level with electrons. A period begins when the first electron is added to a new principal energy level. Because the levels hold different numbers of electrons (see Table 4-6), the periods vary in length.

The first period corresponds to the filling of the 1s orbital of the first principal energy level. It consists of only two elements, hydrogen and helium. By comparing Figure 5-4 with the complete periodic table in Figure 5-2, we can see that the second period (beginning with lithium) and the third period (beginning with sodium) each consists of eight elements. The 2s and 2p orbitals are filled in the elements of the second period, and the 3s and 3p orbitals are filled in the elements of the third period.

The filling of the fourth principal energy level is interrupted after the 4s orbital of calcium by the filling of the 3d orbitals. Then, the 4p orbitals are filled. Thus, the fourth period has 18 elements, ending with

FIGURE 5-2

The periodic table of the elements. The table is a graphic way to show relationships between the elements.

1A	2A	3B	4B	5B	6B	7B	8B	8B	8B	1B	2B	3A	4A	5A	6A	7A	8A
1 H 1.0079																	2 He 4.0026
3 Li 6.941	4 Be 9.0122											5 B 10.81	6 C 12.011	7 N 14.0067	8 O 15.9994	9 F 18.9984	10 Ne 20.179
11 Na 22.9898	12 Mg 24.305											13 Al 26.9815	14 Si 28.0855	15 P 30.9738	16 S 32.06	17 Cl 35.453	18 Ar 39.948
19 K 39.098	20 Ca 40.08	21 Sc 44.9559	22 Ti 47.90	23 V 50.9415	24 Cr 51.996	25 Mn 54.9380	26 Fe 55.847	27 Co 58.9332	28 Ni 58.70	29 Cu 63.546	30 Zn 65.38	31 Ga 69.72	32 Ge 72.59	33 As 74.9216	34 Se 78.96	35 Br 79.904	36 Kr 83.80
37 Rb 85.4678	38 Sr 87.62	39 Y 88.9059	40 Zr 91.22	41 Nb 92.9064	42 Mo 95.94	43 Tc (98)	44 Ru 101.07	45 Rh 102.9055	46 Pd 106.4	47 Ag 107.868	48 Cd 112.41	49 In 114.82	50 Sn 118.69	51 Sb 121.75	52 Te 127.60	53 I 126.9045	54 Xe 131.30
55 Cs 132.9054	56 Ba 137.33	57 La* 138.905	72 Hf 178.49	73 Ta 180.9479	74 W 183.85	75 Re 186.207	76 Os 190.2	77 Ir 192.22	78 Pt 195.09	79 Au 196.9665	80 Hg 200.59	81 Tl 204.37	82 Pb 207.2	83 Bi 208.9804	84 Po (209)	85 At (210)	86 Rn (222)
87 Fr (223)	88 Ra 226.0254	89 Ac** (227.0278)	104 Unq (261)	105 Unp (262)	106 Unh (263)												

Transition elements

*Lanthanide series

58 Ce 140.12	59 Pr 140.9077	60 Nd 144.24	61 Pm (145)	62 Sm 150.4	63 Eu 151.96	64 Gd 157.25	65 Tb 158.9254	66 Dy 162.50	67 Ho 164.9304	68 Er 167.26	69 Tm 168.9342	70 Yb 173.04	71 Lu 174.67

**Actinide series

90 Th 232.0382	91 Pa 231.0359	92 U 238.029	93 Np 237.0482	94 Pu (244)	95 Am (243)	96 Cm (247)	97 Bk (247)	98 Cf (251)	99 Es (252)	100 Fm (257)	101 Md (258)	102 No (259)	103 Lr (260)

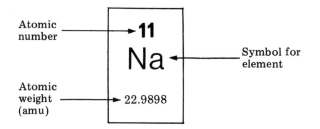

Atomic number → **11**

Na ← Symbol for element

Atomic weight (amu) → 22.9898

FIGURE 5-3

Information in the periodic table. Each box indicates the atomic number, symbol, and atomic weight (amu) of an element.

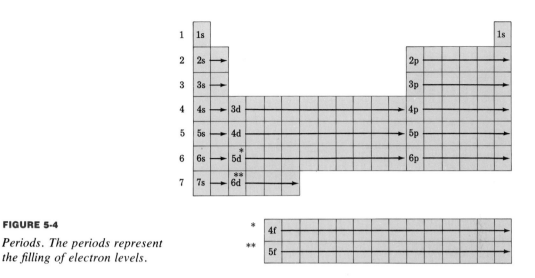

FIGURE 5-4

Periods. The periods represent the filling of electron levels.

krypton. The fifth period is completed in a similar way; after the 5s orbitals are filled, electrons enter the 4d orbitals. Then the 5p orbitals are filled. The fifth period also contains 18 elements.

Two interruptions take place in the sixth period. After the 6s orbital is filled (in cesium and barium) and one electron has been added to a 5d orbital (in lanthanum), the 4f orbitals are filled, as shown below the table. Then the remaining 5d orbitals are filled, followed by the filling of the 6p orbitals. Thus, the sixth period has 2 + 14 + 10 + 6 (s + f + d + p), or 32 elements. The seventh period should be filled in this manner also, but is presently incomplete.

**5.3
GROUPS**

The vertical columns of the periodic table are called **groups** or **families**. Each group is labeled with a number and letter. For example, the first group in the table is Group 1A. When the elements are arranged by atomic number into periods, as we just described, their properties repeat in a regular manner. *The elements in a group are those having similar properties.*

The number and arrangement of outermost electrons of the atoms of an element determine the properties of that element to a great extent. (The nature of this relationship will become clear later.) Elements in the same group are chemically similar because they have the same configuration of outermost electrons. Figure 5-5 shows the outermost electron configuration for each group in the periodic table. (There are some exceptions involving elements having partially filled d or f orbitals; see Table 4-8.)

The elements in Group 1A (except hydrogen) are called the **alkali metals**. The outer level of each of these elements has one electron in

FIGURE 5-5

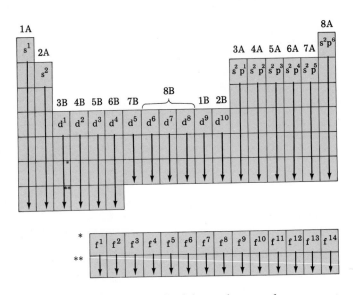

FIGURE 5-5

Groups. Elements in a group have the same number of electrons in the outermost level. (The outermost configuration for elements with d and f electrons may differ from that shown here; see Table 4-8 and Figure 5-10.)

an s orbital. Therefore, all of these elements have an outermost electron configuration of s^1, and a Lewis symbol with one dot.

	Lewis symbol	Electron configuration
H	H·	$1s^1$
Li	Li·	[He] $2s^1$
Na	Na·	[Ne] $3s^1$
K	K·	[Ar] $4s^1$
Rb	Rb·	[Kr] $5s^1$
Cs	Cs·	[Xe] $6s^1$
Fr	Fr·	[Rn] $7s^1$

The number of filled inner levels increases as we move down the group, but the common outermost s^1 orbital gives all of these elements (except hydrogen) similar properties. (The special properties of hydrogen are described in Section 5.6.)

The elements in Group 2A are called the **alkaline earth metals**. Every one of these elements has a filled outermost s orbital, s^2.

Moving to Group 3A, we see that these elements all have a filled outermost s orbital, s^2, plus one electron in a p orbital, p^1. The elements in Group 4A have an outermost electron configuration of s^2p^2, and the elements of Group 5A have an outermost electron configuration of s^2p^3. The Group 6A elements, called the **chalcogens**, have s^2p^4 as the outermost electron configuration. Group 7A, the **halogens**, has an s^2p^5 outermost electron arrangement. Finally, the **noble gases**, designated Group 8A (or Group 0), have an s^2p^6 configuration.

The following is a summary of the configurations and the total number of electrons (s + p) in the outermost level of each of these groups.

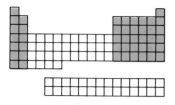

Main group elements

Group:		1A	2A	3A	4A	5A	6A	7A	8A
outermost electron configuration		s^1	s^2	s^2p^1	s^2p^2	s^2p^3	s^2p^4	s^2p^5	s^2p^6
number of electrons		1	2	3	4	5	6	7	8

These eight groups, in which the outermost s and p orbitals are being filled, are known as the **main group elements** or representative elements. For these main group elements, the group number is equal to the total number of outermost s and p electrons.

5.4 METALS AND NONMETALS

If you examine the periodic table in Figure 5-2, you will notice a zigzag line that looks like a staircase running diagonally across the right side of the table. This line separates the elements into two major categories. The elements on the left side of the line are **metals**, and the elements on the right side are **nonmetals**. The elements that touch the line on both sides are called the **semimetals** (or metalloids). Semimetals sometimes act as metals, sometimes act as nonmetals, or may have intermediate properties.

About three-fourths of the 106 known elements are metals. Metals are solids at room temperature, with the exception of mercury, which is a liquid. When polished, metals are shiny. Most metals appear gray to white, but the metal gold is yellow, and copper is reddish. Metals are malleable, which means they can be flattened by hammering or rolling. They are also ductile, which means they can be drawn into a wire. Metals are good conductors of both heat and electricity. They generally melt at high temperatures. For example, gold changes from a solid to a liquid at 1063°C.

The zigzag line separates metals (left) from nonmetals (right).

Do-it-yourself 5

A silver lining

Present-day coinage in the United States is based on the element copper. The penny is 95% copper and 5% zinc. A nickel is 75% copper and only 25% nickel. The dime, quarter, half-dollar and dollar coins consist of a core of pure copper covered by 75% copper and 25% nickel.

Not too long ago, the element silver was the basis for most U.S. coins. Until 1965 the dime, quarter, half dollar, and dollar were 90% silver. If you have any silver coins, they are worth more than their face value because of the high price of silver.

To do: Look up the value of silver in the financial section of your newspaper. The value is given in dollars per troy ounce of the element.

A silver dollar (dated 1935 or earlier) has a mass of 26.73 g and is 90.00% silver. If one troy ounce is 31.1035 g, calculate the value of a silver dollar based on its silver content.

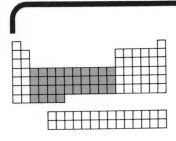

Transition elements

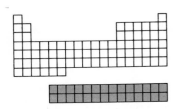

Inner transition elements

The metals that separate the two sections of the main group elements in the periodic table are called **transition elements**. These elements are in groups 3B, 4B, 5B, 6B, 7B, 8B (or 8), 1B and 2B. They form a transition between the filling of the s and p orbitals. It is in these elements that electrons fill the d orbitals. The d orbitals are not filled until the s orbital of the next higher principal energy level is filled. For example, the 3d orbitals are filled only *after* the 4s orbital is filled. Similarly, the 4d orbitals are filled after the outer 5s orbital. The orbitals being filled in the transition metals are *inner* d orbitals. Consequently, the transition metals differ from each other less than do the main group elements, in which the more exposed *outer* electron configuration (s and p) is changing.

The transition elements form four series: a series beginning with scandium (3d), a series beginning with yttrium (4d), a series beginning with lanthanum (5d), and a series beginning with actinium (6d). The series beginning with actinium is presently incomplete.

The final two series of transition elements are themselves interrupted by the filling of f orbitals before the d orbitals are completed. The elements in which the f orbitals are being filled are called **inner transition elements**. These elements are generally written separately, below the periodic table, to make the table more compact. The first row of inner transition elements is called the **lanthanides** (or lanthanoids). The lanthanides, plus the elements scandium and yttrium, are sometimes called the **rare earth elements**. The second row of inner transition elements consists of the **actinides** (or actinoids). Most of the synthetic elements, which were produced artificially after 1940, are actinides.

Nonmetals generally include all the noble gases (8A) and halogens (7A), along with some elements from groups 1A, 4A, 5A and 6A. There are only 17 nonmetals: hydrogen, carbon, nitrogen, phosphorus, oxygen, sulfur, selenium, fluorine, chlorine, bromine, iodine, helium, neon, argon, krypton, xenon, and radon. Although the noble gases are usually classified as nonmetals, their properties differ from those of typical nonmetals. These differences will be explained later in this chapter.

All elements that are gases at room temperature are nonmetals. Those nonmetals that are solids at room temperature are either hard and brittle like sulfur, or soft and crumbly like red phosphorus. One nonmetal, bromine, is a liquid. Nonmetals have a wide range of colors. They can be clear or colorless like the gas nitrogen, yellow like sulfur, or bluish-black like iodine. Nonmetals are generally poor conductors of heat and electricity.

The properties of the semimetals are intermediate between those of the metals and those of the nonmetals. For example, the semimetal silicon is neither an excellent conductor of electricity like a metal, nor an insulator like most nonmetals. Instead, it is a semiconductor. Be-

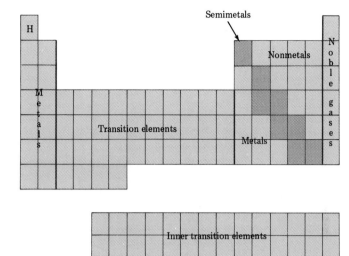

FIGURE 5-6

Types of elements. Elements of the same type are grouped together in the periodic table.

cause of this special property, silicon is used to make the microelectronic "chips" used in calculators, digital watches, and many other devices.

Figure 5-6 summarizes the locations in the periodic table of the different categories of elements.

5.5 PERIODIC PROPERTIES

The properties of the elements show certain trends that correspond to two progressive changes indicated in the periodic table:

1 The steady increase in atomic number by one, from each element to the next across a period.

2 The repeated filling of outermost electron levels, from one period to the next.

Consider the sizes of atoms of the elements shown in Figure 5-7. Notice that as we move across each period from left to right, the atomic size decreases. This decrease in atomic size results from change **1** in the list. As the atomic number increases, the positive charge of the nucleus increases (because protons are positively charged). This increase in charge causes electrons to be attracted more strongly to the nucleus. As a result, the atom "shrinks."

A second trend, resulting from change **2**, is the increase in atomic size down each group. In this case the atoms become larger because electrons are being added to levels that are farther away from the nucleus. These two trends, the decrease in atomic size across a period and the increase in atomic size down a group, are largest for the main group elements, smaller for the transition elements, and smallest for the inner transition elements. For example, the difference in atomic

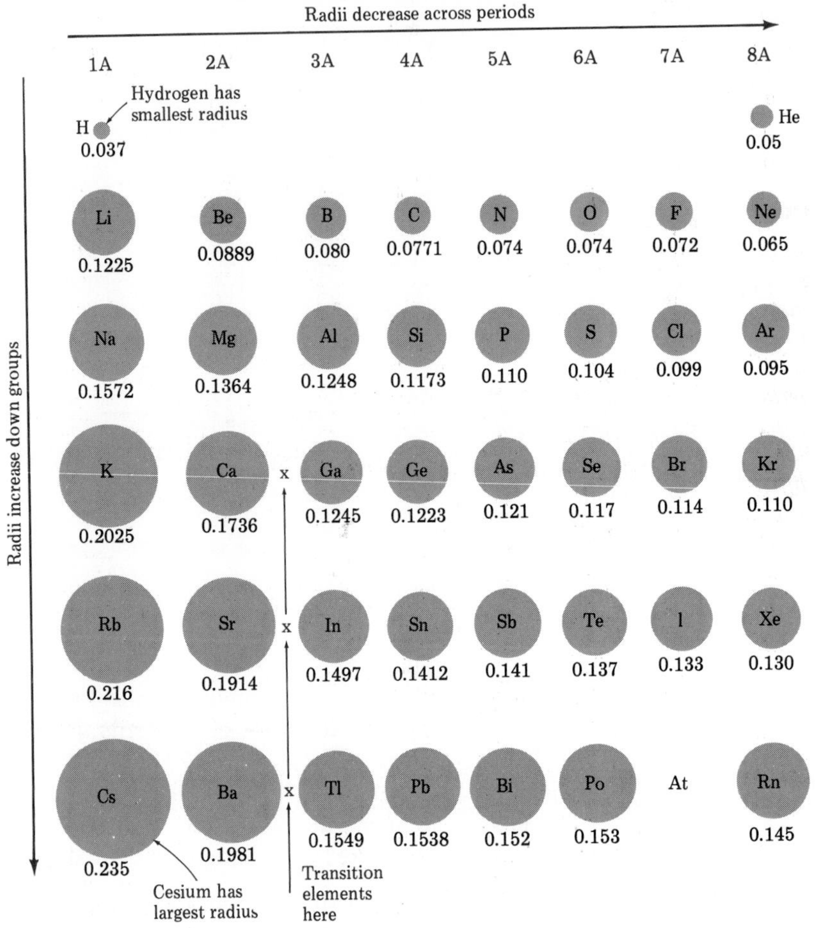

Radii decrease across periods →

| | 1A | 2A | 3A | 4A | 5A | 6A | 7A | 8A |

Hydrogen has smallest radius

H 0.037   He 0.05

Li 0.1225 | Be 0.0889 | B 0.080 | C 0.0771 | N 0.074 | O 0.074 | F 0.072 | Ne 0.065

Na 0.1572 | Mg 0.1364 | Al 0.1248 | Si 0.1173 | P 0.110 | S 0.104 | Cl 0.099 | Ar 0.095

K 0.2025 | Ca 0.1736 | x Ga 0.1245 | Ge 0.1223 | As 0.121 | Se 0.117 | Br 0.114 | Kr 0.110

Rb 0.216 | Sr 0.1914 | x In 0.1497 | Sn 0.1412 | Sb 0.141 | Te 0.137 | I 0.133 | Xe 0.130

Cs 0.235 | Ba 0.1981 | x Tl 0.1549 | Pb 0.1538 | Bi 0.152 | Po 0.153 | At | Rn 0.145

Radii increase down groups

Cesium has largest radius

Transition elements here

FIGURE 5-7

Sizes of atoms. Notice that the size increases down a group and decreases across a period (from left to right).

size between lithium (atomic number 3) and neon (10) is much greater than the size difference between scandium (21) and zinc (30).

Another property that changes in a systematic manner as we move through the periodic table is the **ionization energy**. The ionization energy of an element is the amount of energy needed to remove one electron from an atom of the element (in its gaseous state). A low ionization energy means that it is easy to remove an outermost electron from the atom. A high ionization energy means that it is difficult to remove an electron because the electrons are more strongly held.

Figure 5-8 presents a graph of ionization energy against atomic number. (For a review of how to read graphs, see Appendix C.4.). Note that as we move across a period, say from lithium to neon, the ionization energy increases from a relatively small value to a very high one. This increase in ionization energy results from the nuclei becoming more positive (change **1** in our list again). As the nuclei become more positive, they hold the electrons tighter and closer. This attraction makes the removal of electrons more difficult, and the ionization energy

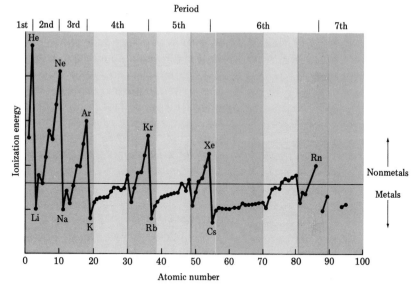

FIGURE 5-8

Ionization energies. This graph shows how ionization energy varies with atomic number. Notice the increase in ionization energy across a period [from lithium (Li) to neon (Ne), for example]. Shading is used here to differentiate main group, transition, and inner transition elements.

greater. Down a group, say from lithium to cesium, the ionization energies decrease. The outer electron of cesium is further away from the nucleus than the outer electron of lithium. Therefore, it is held less strongly, and is easier to remove.

Electron affinity is a measure of an atom's attraction for additional electrons. It is the amount of energy released on the addition of an electron to an atom (when the element is in the gaseous state). The more energy that is released by the addition of an electron, the greater the electron affinity of the atom. Across a period from left to right, the electron affinity *increases* until the last group, the noble gases, where the electron affinity is unusually low. (Noble gases will be discussed later.) The electron affinity increases across the period because the tendency of the atoms to gain an additional electron increases. Again, this trend arises from the increasing positive charges of the nuclei as they gain more protons (change **1**). Down a group, electron affinity *decreases* because of the increasing distance of the added electron from the nucleus of the atom (change **2**).

5.6 PERIODIC TRENDS IN METALS AND NONMETALS

The trends described in the preceding section are clearly related to each other because they are based on the same periodic patterns or changes. These trends help us to understand the differences between metals and nonmetals. *Metals have low ionization energies;* they have one or more electrons that are easily removed. This property is probably the most important characteristic of metals. Because metals have a strong tendency to *lose* their outer electrons, we would expect that their tendency to *gain* electrons is low. This tendency is, in fact, observed. Metals have low electron affinities.

With the exception of the noble gases, *nonmetals have high electron*

139

affinities. Their high electron affinities mean that these elements have a strong tendency to gain electrons. Therefore, we might expect non-metals to have strong tendencies to hold onto their electrons. And, in fact, they do. Nonmetals have high ionization energies.

The differences in properties between metals and nonmetals also are seen in atomic size. Atoms of nonmetals are generally smaller than atoms of metallic elements. Nonmetals are located in the upper right part of the periodic table, where atomic sizes are the smallest as a result of the trends described in the preceding section. To summarize, metals have low ionization energies, low electron affinities and relatively large atomic size. Nonmetals have high ionization energies, high electron affinities, and relatively small atomic size.

The most metallic elements are those having the greatest tendency to lose electrons. They are located in the bottom left corner of the periodic table. Cesium is probably the most metallic common element. (Francium is more metallic but is rare.) Cesium has a very low ionization energy, a small electron affinity, and a large size. In the top right corner of the table (excluding the noble gases) are the most nonmetallic elements. Fluorine is the most nonmetallic element. Its electron affinity is the greatest, and it has a high ionization energy and a small size. Figure 5-9 summarizes these trends.

The noble gas elements are exceptionally stable. They do not readily lose or gain electrons. Therefore, they have very high ionization energies and very low electron affinities. The stability of the noble gas elements results from their completely filled energy levels. The very large "gaps" in Figure 4-10 (b) help to explain this stability. They represent the differences in energy of the electrons between the energy levels. To add an electron to a noble gas atom would require a great deal of energy, since the added electron must have the energy of the next higher level. Removing an electron would also be difficult because

FIGURE 5-9

Periodic trends. The arrows show how various properties of elements change as we move across the periodic table. The most metallic elements are located at the bottom left-hand side of the table, and the most nonmetallic elements are at the top right-hand side.

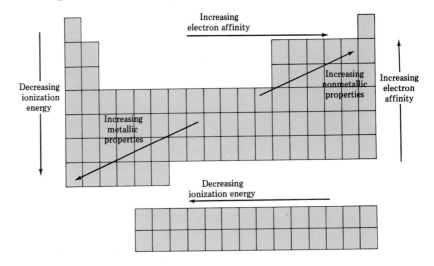

ASIDE . . .

Hydrogen and helium together make up 98% of the total number of atoms in the universe.

of the high positive charge of the nucleus, which exerts a "pull" on the outermost s and p electrons.

Hydrogen also has unique properties. Like the alkali metals of Group 1A, hydrogen has one electron in an s orbital (s^1). But, like the halogens of Group 7A, hydrogen is only one electron short of having a completely filled outer level. In addition, hydrogen is similar to the group 4A elements, because it has a half-filled outer level just as they do. To add to the confusion, hydrogen is usually included with the metals in the periodic table (above Group 1A), even though it is more properly classified as a nonmetal. The special nature of the hydrogen atom is due to its very small size. After all, hydrogen consists of only one proton and one electron.

5.7 OTHER PERIODIC TABLES

The standard form of the periodic table that is shown in Figure 5-2 and on the inside cover is not the only form possible. In fact, there are many different types of periodic tables, including rectangular, triangular, circular, spiral, and even three-dimensional tables. Each way of arranging the elements emphasizes certain aspects of the relationships between the elements. And each table has its own advantages and disadvantages.

Figure 5-10 shows one version of the long form of the periodic table.

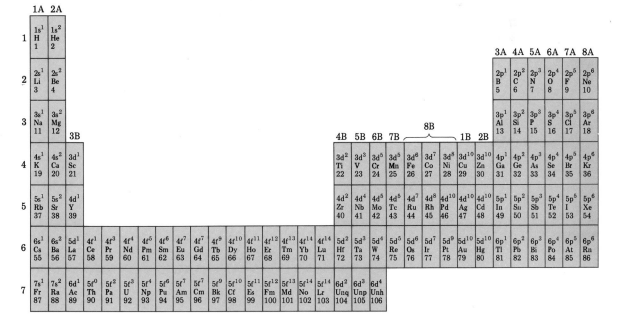

FIGURE 5-10

Long form of the periodic table. In this form, the lanthanides and actinides are included in the body of the table.

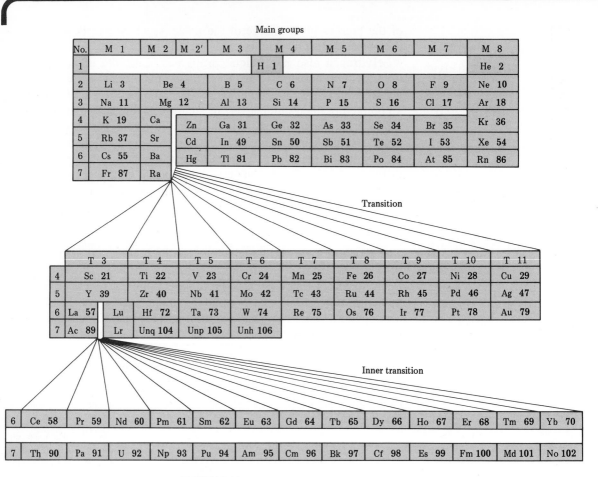

FIGURE 5-11

Separated form of the periodic table. The main group elements, transition elements, and inner transition elements appear separately.

The major advantage of this form is that the lanthanides and actinides are an integral part of the table; they are not separated from the rest of the elements as they are in the standard table. This particular table also shows which orbital is being filled in the atoms of each element.

The advantage of using the separated form of the periodic table, shown in Figure 5-11, is that the main group elements are written together, and are separate from both the transition and inner transition elements. This particular table (developed by R. T. Sanderson) uses simplified group numbers: M stands for main group and T stands for transition group. The number shows how many outermost electrons each atom has. Hydrogen is placed in a unique central position to emphasize its special properties.

A completely different form of the periodic table is shown in Figure 5-12. Here the elements are arranged on the basis of the aufbau

Period	Orbital	Main group elements							Transition elements										Inner transition elements														
		s		p						d										f														
		1	2	1	2	3	4	5	6	1	2	3	4	5	6	7	8	9	10	1	2	3	4	5	6	7	8	9	10	11	12	13	14
1	1 s	1 H	2 He																														
2	2 s	3 Li	4 Be																														
	2 p			5 B	6 C	7 N	8 O	9 F	10 Ne																								
3	3 s	11 Na	12 Mg																														
	3 p			13 Al	14 Si	15 P	16 S	17 Cl	18 Ar																								
4	4 s	19 K	20 Ca																														
	3 d									21 Sc	22 Ti	23 V	24 Cr	25 Mn	26 Fe	27 Co	28 Ni	29 Cu	30 Zn														
	4 p			31 Ga	32 Ge	33 As	34 Se	35 Br	36 Kr																								
5	5 s	37 Rb	38 Sr																														
	4 d									39 Y	40 Zr	41 Nb	42 Mo	43 Tc	44 Ru	45 Rh	46 Pd	47 Ag	48 Cd														
	5 p			49 In	50 Sn	51 Sb	52 Te	53 I	54 Xe																								
6	6 s	55 Cs	56 Ba																														
	4 f																			57 La	58 Ce	59 Pr	60 Nd	61 Pm	62 Sm	63 Eu	64 Gd	65 Tb	66 Dy	67 Ho	68 Er	69 Tm	70 Yb
	5 d									71 Lu	72 Hf	73 Ta	74 W	75 Re	76 Os	77 Ir	78 Pt	79 Au	80 Hg														
	6 p			81 Tl	82 Pb	83 Bi	84 Po	85 At	86 Rn																								
7	7 s	87 Fr	88 Ra																														
	5 f																			89 Ac	90 Th	91 Pa	92 U	93 Np	94 Pu	95 Am	96 Cm	97 Bk	98 Cf	99 Es	100 Fm	101 Md	102 No
	6 d			103 Lr	104 Unq	105 Unp	106 Unh																										

FIGURE 5-12

Periodic table based on the aufbau process. *The elements are arranged by energy sublevels.*

principle, in the order in which electrons enter the energy sublevels. The main group, transition, and inner transition elements appear in distinct areas of the table.

5.8
BRIEF SURVEY
OF THE
ELEMENTS

The 106 known elements shown in a periodic table are distributed unevenly in nature. Table 5-1 lists the percentage by weight of the most common elements present in the Earth's crust, atmosphere, and oceans, and in your own body. The Earth's crust consists mainly of oxygen and silicon, and the human body contains mostly atoms of oxygen, carbon, and hydrogen. Of the 106 elements, 90 elements occur naturally. The other elements can be produced artificially, using methods described in Chapter Seventeen, Nuclear Chemistry. Only 30 elements are found free and uncombined; the rest are found combined with other elements.

The following paragraphs summarize by group the important properties of many main group elements. These descriptions are meant to

TABLE 5-1 *Abundance of elements by mass*

Element	Percentage on Earth[a]	Percentage in human body
oxygen	49.5	65.0
silicon	25.7	trace
aluminum	7.5	—
iron	4.7	trace
calcium	3.4	1.5
sodium	2.6	0.15
potassium	2.4	0.35
magnesium	1.9	trace
hydrogen	0.9	9.5
titanium	0.6	—
chlorine	0.2	0.20
phosphorus	0.1	1.0
manganese	0.09	trace
carbon	0.08	18.5
sulfur	0.06	0.25
barium	0.04	—
nitrogen	0.03	3.3
fluorine	0.02	trace
nickel	0.02	trace
trace elements	0.16	0.25

[a]Includes Earth's crust, atmosphere, and oceans.

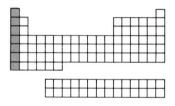

Group 1A

give you a better understanding of the elements and need not be memorized.

Of the *Group 1A* elements, hydrogen is clearly an exception. The other elements in Group 1A are soft metals with low melting points and low densities. Hydrogen, on the other hand, is a gas that acts as a nonmetal. Hydrogen is the lightest substance known. It is the most abundant element in the universe, making up 93% of the number of atoms and 75% of the mass. The sun, for example, consists mainly of

Box 5

Naming elements beyond 103

Elements heavier than plutonium (94) are not found at all in nature and must be produced artificially. Some of these artificially produced elements have been named after the place of their discovery. For example, element 98 is named californium because it was first produced in California. Elements are also named after famous scientists. Element 99, for example, einsteinium, was named after Albert Einstein.

In recent years, conflicts have arisen between scientists in the United States and scientists in the Soviet Union over who first discovered certain elements. For example, scientists in both countries claim credit for discovering element 104. Soviet scientists gave the element a Russian name, kurchatovium. American scientists called it rutherfordium, after James Rutherford.

To avoid this problem, a systematic method for naming elements beyond number 103 has been proposed recently. It involves naming the element according to its atomic number, using the following terms:

0 nil (n)	5 pent (p)
1 un (u)	6 hex (h)
2 bi (b)	7 sept (s)
3 tri (t)	8 oct (o)
4 quad (q)	9 en (e)

Names are made by combining the appropriate terms to represent the atomic number and then adding the ending "-ium." For example, element 104 would be called:

un	nil	quad	ium
1	0	4	ending

The symbol for unnilquadium is Unq. Element 105 is called unnilpentium using this system. Its symbol is Unp. Similarly, element 106 is unnilhexium, symbolized Unh.

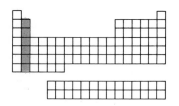

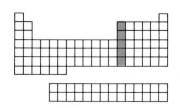

Group 2A

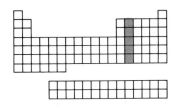

Group 3A

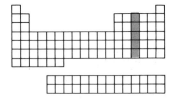

Group 4A

hydrogen, as do most stars. 63% of the atoms in your body are atoms of hydrogen. Hydrogen has many industrial applications. It is used in making ammonia, in refining petroleum, and in making margarine by "hardening" vegetable oils in a process called hydrogenation. Because hydrogen burns cleanly, it is being considered as a basic fuel for our future economy.

Of the other elements in Group 1A, the alkali metals sodium and potassium are the most abundant and are used most often. Liquid sodium, for example, is used as a heat exchanger in certain types of nuclear reactors. It conducts heat from the central core of the reactor to a boiler, which produces steam used to generate electrical power. The bright yellowish lamps seen on highways and streets are sodium vapor lamps.

In *Group 2A*, magnesium and calcium are the most abundant alkaline earth metals. Magnesium is a light structural metal (its density is 1.74 g/cm³) that is combined with other metals to build aircraft parts. Beryllium is hard, has a high melting point, and conducts heat well. For these reasons, it is used to make heat shields for space vehicles.

Group 3A includes one semimetal, boron, and four metals. Aluminum, the third most abundant element on earth, is a low density metal that is used in building and construction. It also is used as a foil for wrapping foods. Gallium, a rarer element, has an unusually low melting point, 30°C, which is just above normal room temperature.

Of the elements in *Group 4A,* carbon is a nonmetal, silicon and germanium are semimetals, and tin and lead are metals. Carbon is extremely important, because life is based on structures formed from its atoms. An entire branch of chemistry, called organic chemistry, deals with the study of carbon and how it combines with other elements. (Chapter Eighteen introduces organic chemistry.) Elemental carbon is found in several natural forms. Diamond is the rarest form. Graphite is the form of carbon that is used in pencils (referred to incorrectly as "lead"), and as a lubricant. In general, when an element can exist in different forms, these forms are called *allotropes.* Their composition is the same, but the internal arrangement of their atoms differs.

Silicon is the second most abundant element on Earth. In nature, silicon is found combined with oxygen in sand. Both silicon and germanium are semiconductors used in electronics. The "tin" cans in which food is stored consist of steel coated with a thin layer of tin, the fourth element in Group 4A. Lead, one of the earliest known metals, is very resistant to corrosion (the process of being "eaten away" by chemical action). Therefore, it is widely used in equipment for the chemical industry. Lead is poisonous because it causes damage to the nervous system.

In *Group 5A,* nitrogen and phosphorus are nonmetals, arsenic and antimony are semimetals, and bismuth is a metal. Nitrogen gas makes up about 80% of the air that we breathe. It is converted by special

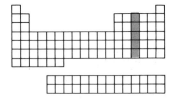

Group 5A

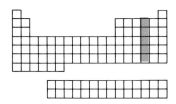

Group 6A

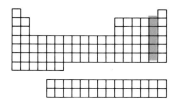

Group 7A

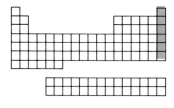

Noble gases

bacteria (called nitrogen-fixing bacteria) in plants to forms that are essential for life in all living cells. This process is part of the nitrogen cycle. Liquid nitrogen is used as a coolant because of its low temperature, −196°C. Nitrogen is used commercially to make ammonia. Phosphorus exists as several allotropes; the red form is mixed with powdered glass to make the coating on boxes of safety matches, and is also used in munitions. Arsenic is a toxic element used in gunshot and in semiconductors. Bismuth has a low melting point, and when combined with other metals serves as the "trigger" in automatic sprinkler systems that protect buildings against fires.

Of the elements in *Group 6A*, only the first two members, oxygen and sulfur, are common. Oxygen, the most abundant element on Earth, makes up about 20% of the air we breathe. It is the only element that can be used in its elemental form by the human body. (All the remaining necessary elements must be combined with others before they can be used.) Oxygen is essential for the chemical processes that take place in our cells. Oxygen also supports the process of combustion, commonly called burning. It is therefore used in welding, and in rocket propulsion as LOX, liquid oxygen. Ozone is an allotrope of oxygen. Present in the upper atmosphere, it helps prevent harmful ultraviolet radiation from reaching the surface of the Earth. Sulfur, called brimstone in the Bible, is a basic raw material in many industries. One important application of sulfur is in treating rubber to make it less brittle. This process is called vulcanization. The element selenium is a component of certain photoelectric cells, which are used to convert energy from sunlight into electricity.

The *Group 7A* elements are strongly nonmetallic. Except for astatine, the halogens are quite common. Fluorine, as we will see more clearly in Chapter Six, is the most chemically active element known. This gas readily combines with many other elements. Chlorine, a greenish-yellow poisonous gas, is used to disinfect drinking water and sewage by a process called chlorination. Bromine, unlike any other nonmetal, exists as a liquid at room temperature. It is red-brown and has an irritating vapor. Iodine, a shiny bluish-black solid, has the ability to sublime—to go directly from a solid to a gas when heated (skipping the liquid state). Tincture of iodine, which contains iodine in alcohol, is commonly used as a disinfectant.

The *noble gases* make up about 1% of the atmosphere by volume. Argon comprises 90% of the total amount of the noble gases. Because of their exceptional stability, these elements were once called the inert gases, but some of these gases now are known to combine with a few other elements. Helium is the second most common element in the universe, although it is rare on Earth. It is used to fill balloons and blimps because it is less dense than air, but does not burn. (Hydrogen, formerly used in blimps, *did* burn in the disastrous fire on the Hindenburg airship in 1937.) Helium is used to replace nitrogen in the breathing mixtures of deep sea divers to prevent nitrogen narcosis (a state of stupor, or altered consciousness). Neon is the gas that produces the red-orange glow of neon signs. Neon light fixtures are very efficient; about 25% of the electrical energy used is converted to light. (They are about four times more efficient than incandescent bulbs, which have tungsten filaments.) Argon is used in welding, and is the filler gas in incandescent light bulbs. Krypton is used, along with mercury vapor, in fluorescent lamps. Xenon is used as a filler gas in high intensity bulbs and flash lamps.

Summary

5.1 A periodic table is a systematic arrangement of the elements. It shows the relationships between the elements by revealing their periodicity—the tendency of their properties to repeat at regular intervals. The modern periodic table is based on the periodic law: The properties of elements repeat in a regular way when they are arranged by increasing atomic number.

5.2 The seven horizontal rows running across the standard periodic table are called periods. A period begins when the first electron is added to a new principal energy level.

5.3 The vertical columns of the periodic table are called groups (or families). The elements in a group have similar properties because they have the same outermost electron configuration.

5.4 About three-fourths of the elements are metals. Metals are generally solid, shiny, malleable, ductile, and are good conductors of heat and electricity. Only 17 elements are nonmetals. Many nonmetals are gases at room temperature. Nonmetals are poor conductors of heat and electricity.

5.5 The properties of the elements follow certain periodic trends. Atomic size decreases across a period and increases down a group. Both ionization energy and electron affinity increase across a period and decrease down a group.

5.6 Metals have low ionization energies (and low electron affinities), while nonmetals have high electron affinities (and high ionization energies). The noble gases are exceptionally stable, as a result of their filled energy levels. Hydrogen has unique properties because of its small size.

5.7 Many different forms of the periodic table exist. Each arrangement emphasizes different aspects of the relationships between the elements.

5.8 The Earth consists mainly of the elements oxygen and silicon, while the human body is made mostly of the elements oxygen, carbon, and hydrogen. Some properties and uses of the most important elements are described in the last section of this chapter.

Exercises

KEY-WORD MATCHING EXERCISE

For each term on the left, choose a phrase from the right-hand column that most closely matches its meaning.

1 periodic table   **a** energy to remove electrons

2 periodic law   **b** column in table

3 period   **c** row in table

4 group   **d** organization of elements

5 main group   **e** properties repeat regularly

6 metals   **f** tendency to gain electrons

7 nonmetals   **g** fill s and p orbitals

8 ionization energy   **h** most are solid and shiny

9 electron affinity   **i** only 17 elements

EXERCISES BY TOPIC

Periodic law (5.1)

10 State the modern periodic law.

11 Why is the periodic law important?

12 What was the significance of the empty spaces in Mendeléev's periodic table?

Periods (5.2)

13 What do periods represent?

14 Why do periods have different lengths?

15 To what does the number of elements in a period correspond?

16 What is the meaning of a period's number?

Groups (5.3)

17 What is a group?

18 What do the members of a group have in common? How do they differ?

19 Name the following groups: **a** 7A **b** 8A **c** 2A **d** 1A **e** 6A

20 Write the number of outermost electrons for each of the groups in Exercise 19.

21 Which groups have the following electron configurations? **a** s^2p^2 **b** s^2 **c** s^2p^6 **d** s^1 **e** s^2p^1

22 Which are the main group elements?

Metals and nonmetals (5.4)

23 Describe the relative locations of metals, nonmetals, and semimetals in the periodic table.

24 Summarize the properties of metals.

25 How do nonmetals differ from metals?

26 What are **a** transition elements? **b** actinides? **c** lanthanides?

27 Describe the properties of semimetals.

28 Identify each of the following elements as either a metal, nonmetal, or semimetal: **a** nitrogen **b** arsenic **c** argon **d** calcium **e** uranium

Periodic properties (5.5)

29 Why do trends exist in the periodic table?

30 Explain the variation of atomic size **a** across a period and **b** down a group.

31 What is ionization energy? What does a high ionization energy mean?

32 Explain the periodic trends in ionization energies of the elements.

33 Define electron affinity. Which elements have the greatest electron affinities?

34 Explain the trend in electron affinity **a** down a group and **b** across a period.

35 In each pair, which has the higher ionization energy? **a** Be or Ra **b** K or Kr **c** O or Hg

Periodic trends in metals and nonmetals (5.6)

36 In terms of ionization energies, how do metals differ from nonmetals?

37 How do metals and nonmetals differ in terms of electron affinities?

38 Why are noble gases relatively stable?

39 What makes hydrogen unique?

40 In general, which are larger in size, atoms of metals or atoms of nonmetals? Explain.

Other periodic tables (5.7)

41 How does the long form of the periodic table differ from the standard form?

42 What are the advantages of the separated form of the periodic table?

43 Describe the organization of the periodic table shown in Figure 5-10.

Brief survey of the elements (5.8)

44 Which is the most abundant element by mass **a** in your body? **b** on earth? **c** in the universe? **d** in the atmosphere?

45 Why is carbon an important element?

46 What is an allotrope? Give one example.

47 Which is the only element that can be used by your body in its elemental form?

48 Which nonmetal exists as a liquid at room temperature?

ADDITIONAL EXERCISES

49 State the number of outermost electrons in **a** Si **b** P **c** C **d** K **e** Ne

50 Why do transition elements differ less from each other than do main group elements?

51 Fill in each of the blanks with either "increases," "decreases," or "remains the same":

	Down a group	Across a period (left to right)
electron affinity	_____	_____
ionization energy	_____	_____
atomic size	_____	_____

52 Why should elements with low electron affinities have low ionization energies?

53 Why should elements with high ionization energies also have high electron affinities?

54 Describe a use for one element from each of the main groups.

55 If the periodic table were arranged by atomic weight instead of by atomic number, which elements would be in different locations?

56 If someone handed you an element, how could you tell if it were a metal or nonmetal?

57 Why are semiconductors manufactured from semimetals?

58 Write the outermost s and p electron configuration for: **a** P **b** I **c** Ba **d** Xe **e** Al

59 Name an element that **a** supports combustion **b** is found on match boxes **c** "hardens" vegetable oil **d** is a blue-black solid that readily sublimes **e** is used in liquid form for rocket propulsion

60 How do the noble gases differ from typical nonmetals?

61 Why does the fifth period have 18 elements?

62 For each pair of elements, state which is larger in size: **a** K or Br **b** Li or K **c** O or Na

63 What is similar about the electron configuration for the elements in Group 6A?

64 Name the element in which an electron first appears in a **a** d orbital **b** p orbital **c** f orbital

65 How does knowing the group number help you draw the Lewis symbol for an element?

66 Using each element only once, pick from the following list to match the descriptions: Ne, O, K, Ra, W, Al.
a element with low ionization energy **b** transition element **c** noble gas
d element with large radius **e** element with high electron affinity

67 Which of the following electron configurations belong to elements in the same group? **a** $[Kr]4d^{10}5s^25p^2$ **b** $[Ne]3s^23p^1$ **c** $[Rn]7s^1$ **d** $[He]2s^22p^1$
e $[Xe]4f^{14}5d^{10}6s^26p^1$

PROGRESS CHART Check off each item when completed.

_____ Previewed chapter

Studied each section:

_____ **5.1** Periodic law

_____ **5.2** Periods

_____ **5.3** Groups

_____ **5.4** Metals and nonmetals

_____ **5.5** Periodic properties

_____ **5.6** Periodic trends in metals and nonmetals

_____ **5.7** Other periodic tables

_____ **5.8** Brief survey of the elements

_____ Reviewed chapter

_____ Answered assigned exercises:

#'s _____

Review Exercises for Chapters Two–Five

1 Which states of matter **a** do not have a definite shape? **b** can expand without limit? **c** can be compressed?

2 Give an example of energy changing from one form into another.

3 A measurement is usually expressed in two parts. What are they?

4 Write the following numbers in exponential notation: **a** 29.1 **b** 0.073 **c** 10,541

5 Identify each of the following as a unit of mass, volume, or length: **a** mL **b** cg **c** μm **d** dm³ **e** Gm

6 Using exponential notation, express the size of the units in Exercise 5 in terms of the base units.

7 Arrange in order of increasing size: 1 cm³ 10 mL 0.10 dm³ 1 L

8 Perform the following calculation, expressing the result with the proper number of significant figures:

$$\frac{0.983 - 0.79}{11.5 + 6.124}$$

9 Estimate the following calculation:

$$\frac{(8.71 \times 10^{-3})}{(3.1 \times 10^{6}) \times (9.73 \times 10^{-10})}$$

10 Carry out the calculation given in Exercise 9.

11 Express the distance from the Earth to the sun, 150. gm, in **a** kilometers and **b** feet.

12 A newborn baby weighed 7 lb 11 oz. Convert the weight to **a** kilograms **b** grams

13 If there are 20 drops of water in a milliliter, how many drops are there in **a** one cubic decimeter? **b** one quart?

14 What is the mass in grams of a hamburger called the "quarter-pounder?"

15 How many cubic centimeters are in a half-liter bottle of wine?

16 Ether has a boiling point of 34.5°C. The temperature in a warm room is 85°F. Will the ether boil?

17 Find the normal (oral) body temperature, 98.6°F, in kelvins.

18 A heavy meal may contain 1.50×10^{3} kcal. Express this energy in kilojoules.

19 If it takes 3.0 pounds of plutonium to make a nuclear weapon, and the density of plutonium is 19.8 g/cm³, what volume of the metal is needed?

20 What is the mass in grams of a uranium atom with mass number 238?

21 How many protons, neutrons, and electrons are in an atom with mass number 53 and atomic number 24?

22 Write the symbol for: **a** copper **b** iron **c** manganese **d** potassium **e** mercury

23 Name these elements: **a** Sn **b** Ag **c** N **d** Cl **e** Pb

24 An element with atomic weight 10.811 amu has two isotopes, one with a mass of 10.0129 amu and one whose mass is 11.00931 amu. Which isotope is more abundant? Explain.

25 Draw an electron shell diagram for **a** sulfur **b** argon

26 Write the Lewis structures for the atoms in Exercise 25.

27 What do the following symbols mean? **a** $3p^4$ **b** $5d^7$ **c** $4s^1$

28 Write the electron configuration for zinc.

29 Why is it more dangerous for you to be exposed to ultraviolet radiation than to the same amount of infrared radiation.

30 Describe the filling of orbitals in the fourth period of the periodic table.

31 Why do the halogens have similar chemical properties?

32 Explain the trend in metallic properties for elements **a** down a group and **b** across a period from left to right.

33 Pick from the following elements: K, O, Ni, Si, Ca, N, Lu, Xe, the one that is **a** most nonmetallic **b** a transition element **c** an alkaline earth element **d** most metallic **e** a semimetal

SIX

Chemical bonding

In Chapter Six we describe how atoms can combine with other atoms. These combinations make possible the tremendous variety of substances that now exist or are being created in the laboratory. Various types of combinations of atoms are presented and compared in this chapter. To demonstrate an understanding of Chapter Six, you should be able to:

1 Describe the nature of a chemical bond.

2 Describe the formation of a covalent bond.

3 Identify and draw Lewis structures for the diatomic elements.

4 Define electronegativity and describe its effect on chemical bonds.

5 Describe the difference between polar covalent bonding and nonpolar covalent bonding.

6 Give the common combining capacities of the important nonmetals.

7 Draw Lewis structures for the molecules of covalent compounds.

8 Describe the formation of simple cations and anions, and predict their charges based on group numbers.

9 Using diagrams, show the formation of ionic bonds from single atoms.

10 Identify and draw Lewis structures of polyatomic ions.

11 Summarize the major features of chemical bonding.

6.1
THE CHEMICAL BOND

When two atoms combine, they are joined by a **chemical bond**. This bond is a force that holds atoms together. An early notion about chemical bonding was that atoms had hooks on them that could attach to other atoms. We now know that electrons, rather than actual hooks, play the key role in chemical bonding.

To form a chemical bond, atoms must come near each other. Their closest contact is made by the outermost orbitals. It is for this reason that *the outermost electrons determine the chemical behavior of an atom.*

Chemical bonds are based on the attraction between opposite charges. In particular, a bond results from the attraction of two atomic nuclei for the same electrons. The attraction produces an arrangement that is more stable and has lower energy than the two atoms by themselves. For this reason, energy is needed to pull atoms apart after they form a bond.

As we will see later in this chapter, atoms either share electrons, lose electrons, or gain electrons to form chemical bonds. The nature of the atoms involved determines which of these three possibilities will occur. In most cases, the result of sharing, losing, or gaining electrons is to produce atoms that have filled outer energy levels like the noble gases. A filled outer energy level generally means eight electrons in the outermost s and p orbitals. Thus, these eight electrons would have an s^2p^6 configuration. This end result is the basis for the **octet rule**. The octet rule states that atoms tend to combine in such a way that each obtains a filled outermost level of eight electrons upon forming a chemical bond.

Do-it-yourself 6

Sticking together

An adhesive or glue is a substance that holds materials together. It does *not* necessarily attract other substances by chemical bonding. Instead, the attraction may be created by *van der Waals forces.* These weak forces exist between atoms that are very close together.

When you break something, you know that the two pieces won't stick back together by themselves. The rough surfaces prevent the atoms in the separate pieces from coming close enough to form van der Waals bonds. An adhesive, however, can flow between the rough surfaces, make close contact, and form van der Waals bonds with each surface. The adhesive then hardens, either by chemical action or the evaporation of a liquid. Having become a solid, it holds together the surfaces of the materials.

You can make a simple adhesive at home. It is called paste, which means a mixture of flour and water. (A similar word, pasta, is the name of a familiar Italian food which is also a mixture of flour and water.)

To do: Blend 4 tablespoons of wheat flour into 6 tablespoons of water. Boil 1½ cups of water, and add the flour mixture. Continue boiling for about 5 minutes, while stirring to remove lumps.

This simple but not very strong adhesive is useful for children's paper work and for household repairs.

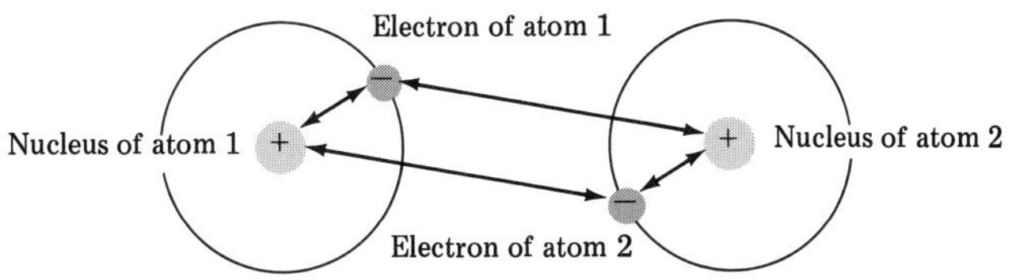

FIGURE 6-1

The basis for chemical bonding. Two different atomic nuclei attract the same electrons.

Electron of atom 1

Nucleus of atom 1 +   + Nucleus of atom 2

Electron of atom 2

6.2 COVALENT BOND FORMATION

The simplest chemical bond is formed by two hydrogen atoms. Each H atom consists of one proton and one electron. When two such atoms approach each other, the negative electrons are attracted by the positive nuclei of *both* atoms, as shown in Figure 6-1. *This mutual attraction of two nuclei for the same pair of electrons is the basis for the formation of a chemical bond between atoms.* For two atoms that form a chemical bond, the combined attraction of the first electron for the second nucleus and the second electron for the first nucleus is greater than the repulsion between the two electrons and between the two nuclei. In other words, the attractive forces between opposite charges overcome the repulsive forces between like charges.

Figure 6-2 shows the formation of a bond between two hydrogen atoms through the overlap of their 1s orbitals (each of which contains one electron). This type of interaction, in which atoms share electrons, is called a **covalent bond**. A covalent bond is the most common kind of chemical bond. It generally forms only between atoms of *nonmetals*, as we will see later on.

Through the covalent bond, each hydrogen atom gets to ''use'' an electron from the other atom as well as its own, thus obtaining a filled outer level. Both hydrogen atoms now have the electron arrangement of the noble gas helium. The shared pair of electrons in the bond between two hydrogen atoms can be illustrated using Lewis symbols:

H:H   or   H×H

Crosses (×) and dots (·) are often used together to indicate that the electrons in the bond came from different atoms, although in reality all electrons are alike.

The following analogy may help you to understand the formation of a covalent bond. Suppose that two people are reading books while sitting next to lamps at opposite ends of a room. If these readers move close together and put the two lamps between them, they both will receive more light than if they stay apart. The light from the two lamps will overlap, and each reader will receive light from both lamps.

In a covalent bond, two nuclei share a pair of electrons just as the two readers share a pair of lamps. And, just as the readers cannot tell which light comes from which lamp because the light from both lamps appears the same, the electrons become equivalent in a chemical bond.

157

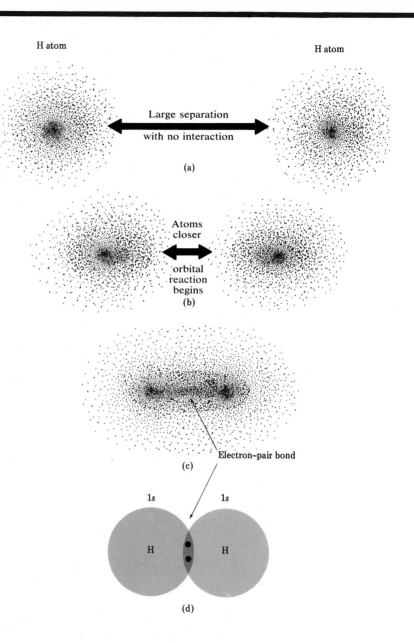

FIGURE 6-2

Formation of a covalent bond. In (a), the hydrogen atoms are too far apart to interact. In (b), the 1s orbitals begin to interact. In (c), they overlap and a chemical bond forms. In (d), another representation of the covalent bond of H₂ is shown.

All electrons are alike, even though they may originally have come from different atoms.

As shown in Figure 6-2, the overlap of hydrogen orbitals increases the electron density between the nuclei (in a way similar to the increase of light "density" when the two readers move the lamps together). The electrons now have a high probability of being located somewhere between the two positive nuclei. This arrangement has a lower energy and is more stable than two separate hydrogen atoms. Figure 6-3

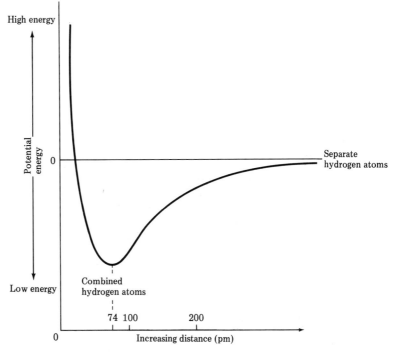

FIGURE 6-3

H_2 potential energy curve. At the far right, the two H atoms are separate. As we follow the curve from right to left, the distance between the two atoms decreases. Because of the electron–nucleus attractions, the energy also decreases. The minimum energy corresponds to the bond distance in H_2, 74 pm. The curve rises steeply on the left, showing that great energy is needed to overcome nucleus–nucleus repulsion and bring the atoms still closer.

illustrates this point. The right side of the curve represents the energy of the two atoms when they are far apart. No significant force exists between these two separated atoms. As they come closer together (moving along the curve to the left), however, the energy decreases because of the attraction between oppositely charged protons and electrons. The bond forms at the lowest point on the curve, when the atoms are 74 pm (0.74 Å) apart. This distance between the centers of the atoms is called the bond distance or bond length. To move the two atoms still closer would require tremendous energy due to repulsions between the positively-charged nuclei. This repulsion is shown by the very steep portion of the curve at the extreme left.

6.3 MOLECULES

The unit formed by the covalent bonding of two or more atoms is called a **molecule**. The properties of a molecule are different from the properties of the individual atoms that compose the molecule. Molecules range in size from a combination of just two atoms to tens of thousands of atoms. (Very large molecules are called macromolecules, since "macro" means large or great.) Chemistry has been described as the science of molecules because so many substances are composed of these units.

The number and kind of atoms in a molecule are given by th

chemical **formula** of the molecule. A hydrogen molecule is represented by the formula H_2 (read H–two).

$$H_2$$

↗ ↖

kind of atom   number of atoms

The subscript 2, written at the lower right side of the symbol for the element, shows that two hydrogen atoms are present in the molecule. We do *not* write the formula as 2H, because this notation means two *separate* atoms of hydrogen. We always use a subscript on the right when more than one atom of a given element is part of the molecule.

This simplest type of molecule is called **diatomic** because it consists of only two atoms ("di" means two) bonded together. The element hydrogen is normally found in nature as the diatomic molecule H_2, and not as a single atom. As shown in Table 6-1, there are six other non-metallic elements that also exist normally as diatomic molecules. Note that these covalently bonded pairs of atoms all have formulas that consist of the symbol for the element with the subscript 2 at the lower right-hand side.

Look at the Lewis structures of the diatomic elements shown in Table 6-1. Like hydrogen, the halogens obtain the noble gas electron arrangement by sharing a pair of electrons, one from each atom. Each of these covalent bonds is called a **single bond** because it consists of a single pair of electrons being shared between the nuclei of the two atoms. The single bond in these molecules can be represented by a dash (—).

H—H   F—F   Cl—Cl   Br—Br   I—I
single bonds

In the hydrogen molecule, electrons from an s orbital are shared to

TABLE 6-1   *The diatomic elements*

Element	Formula	Structure	Type of bond
hydrogen	H_2	H:H	single
fluorine	F_2	:F:F:	single
chlorine	Cl_2	:Cl:Cl:	single
bromine	Br_2	:Br:Br:	single
iodine	I_2	:I:I:	single
oxygen	O_2	:O::O:	double
nitrogen	N_2	:N:::N:	triple

form a single bond. Each hydrogen atom obtains a complete outer level of two electrons. In the halogens one p electron from each atom enters the bond. Each halogen atom in a diatomic molecule obtains a complete octet, an outermost s^2p^6 configuration.

Oxygen, in Group 6A, has six outer electrons ($2s^22p^4$) per atom and needs *two* more electrons to complete the outermost level. To obtain these electrons, oxygen can share *two* pairs of electrons with a second oxygen atom. Two p electrons are shared by each oxygen atom, as shown in Table 6-1. Check the Lewis structure yourself to see that each oxygen atom originally had six outermost electrons, but now has a total of eight. (Note that when drawing Lewis structures, the location of the electron dots can be shifted around the atomic symbol but their total number cannot be changed.) The covalent bond formed by the sharing of two pairs of electrons is called a **double bond**. A double bond can be represented by a double dash (=).

O=O   oxygen molecule
double bond

Nitrogen, in Group 5A, has five outermost electrons ($2s^22p^3$) per atom. Therefore, it requires *three* more electrons to fill the outermost level. The nitrogen atom obtains these three electrons through a **triple**

Box 6

The triple bond in N_2

The element nitrogen is essential for life. We obtain it in the form of protein in our diet. Of the three major categories of food—carbohydrate, fat, and protein—protein is generally the scarcest in nature.

Four-fifths of the atmosphere consists of nitrogen gas. However, this nitrogen is in the form N_2. The triple bond between the atoms in a nitrogen molecule is very strong. Because our bodies cannot break down the strong triple bond of a nitrogen molecule, we cannot use the nitrogen gas in the atmosphere. Instead we must obtain nitrogen atoms indirectly, through what is called the nitrogen cycle.

Certain plants have special microorganisms attached to their roots called nitrogen-fixing bacteria. These bacteria are able to break the triple bond in N_2 and convert the molecule into ammonia, NH_3. In addition, some N_2 in the atmosphere is converted into NO_2 by lightning. Some of the NO_2 enters the soil, and water in the soil further converts the NO_2 into HNO_3 and NO_3^- ions.

Plants are able to use the ammonia from nitrogen-fixing bacteria, and the nitrate ions from lightning, to make proteins. We then eat the plants, or eat the animals that ate the plants, and thus obtain nitrogen for ourselves. Nitrogen-containing wastes from animals and humans eventually return to the soil, completing the nitrogen cycle.

bond to another nitrogen atom. A triple bond consists of three pairs of electrons shared between the nuclei of two atoms. Again, look at Table 6-1 to check that each nitrogen atom in a nitrogen molecule has eight electrons (a complete octet). Three dashes can be used to represent a triple bond.

$$N\equiv N \qquad \text{nitrogen molecule}$$
triple bond

We refer to double and triple bonds as **multiple bonds**.

6.4
ELECTRONEGATIVITY

Electronegativity is a measure of the ability of an atom to attract to itself the electrons in a chemical bond. It is the amount of pull experienced by a negative outer electron toward the positively charged nucleus. Electronegativity depends both on the number of protons in the nucleus and the size of the atom. Figure 6-4 presents values for the electronegativities of the elements.

Electronegativity varies in a way similar to electron affinity (the energy released when an atom gains an electron). In the periodic table, across a period, from left to right, electronegativity increases because the nucleus becomes more and more positively charged. Down a group, it decreases because the electrons in the outermost orbitals are farther away from the nucleus of the atom. Thus the most electronegative elements are the nonmetals at the upper right of the periodic table. Fluorine is the most electronegative element, followed by oxygen and chlorine.

The electronegativity of the atoms involved in a chemical bond influences the strength of the bond. This result is not surprising, since electronegativity deals with an atom's attraction for electrons in a chemical bond. **Bond energy**, the amount of energy required to pull the bonded atoms apart, is a measure of the strength of a bond. A high

FIGURE 6-4

Electronegativities of the elements. Electronegativity is a measure of the ability of an ... m to attract electrons in a ... l bond. The higher the ... greater the ... ity.

																		He
Li 1.0	Be 1.5												B 2.0	C 2.5	N 3.1	O 3.5	F 4.1	Ne
Na 1.0	Mg 1.3												Al 1.5	Si 1.8	P 2.1	S 2.4	Cl 2.9	Ar
K 0.9	Ca 1.1	Sc 1.2	Ti 1.3	V 1.5	Cr 1.6	Mn 1.6	Fe 1.7	Co 1.7	Ni 1.8	Cu 1.8	Zn 1.7	Ga 1.8	Ge 2.0	As 2.2	Se 2.5	Br 2.8	Kr
Rb 0.9	Sr 1.0	Y 1.1	Zr 1.2	Nb 1.3	Mo 1.3	Tc 1.4	Ru 1.4	Rh 1.5	Pd 1.4	Ag 1.4	Cd 1.5	In 1.5	Sn 1.7	Sb 1.8	Te 2.0	I 2.2	Xe
Cs 0.9	Ba 0.9	La 1.1	Hf 1.2	Ta 1.4	W 1.4	Re 1.5	Os 1.5	Ir 1.6	Pt 1.5	Au 1.4	Hg 1.5	Tl 1.5	Pb 1.6	Bi 1.7	Po 1.8	At 2.0	Rn
Fr 0.9	Ra 0.9	Ac 1.0	Lanthanides: 1.0–1.2														
			Actinides: 1.0–1.2														

bond energy means that the atoms are strongly attracted to each other. This mutual attraction depends in part on the electronegativity of the atoms in the bond, which are joined through their attraction for the same pair or pairs of electrons.

Bond energy also depends on the number of electrons involved in the bond. The more electrons shared between two atoms, the greater the number of orbitals that overlap, and the greater the mutual attraction of the atoms. In other words, when more than one pair of electrons is involved in a chemical bond, there is more chemical "glue" holding the atoms together. For the same two atoms, a triple bond would be stronger than a double bond, which in turn would be stronger than a single bond.

Bond length, the distance between the atoms in a molecule, is also related to electronegativity and the number of bonding electrons. If the attraction between atoms is great, due either to high electronegativity or to a multiple bond, the bond length is usually relatively short. Thus, for the same two atoms, a triple bond would be shorter than a double bond, which would be shorter than a single bond.

**6.5
POLAR COVALENT
BONDING**

As we have seen, identical atoms of certain elements can combine to form molecules of that element. For example, in Section 6.3 we saw elements that exist as diatomic molecules. However, when atoms of two or more *different* elements combine by forming chemical bonds, the result is molecules of a **compound**. Chemical compounds have unique properties that differ from those of the elements whose atoms they contain. All matter consists of either elements (which contain one kind of atom), or compounds (which contain combinations of different kinds of atoms), or of mixtures of elements and compounds.

In a molecule of a diatomic element, electrons are shared *equally* between identical atoms. This type of bond is known as a **nonpolar covalent bond**. In the molecules of a compound, however, the bonding is different because electrons are *not* shared equally. For example, hydrogen and chlorine atoms combine to form a molecule of hydrogen chloride, HCl (read H–C–L).

$$\overset{\text{xx}}{\underset{\text{xx}}{\text{H}_{\text{x}}\text{Cl}_{\text{x}}^{\text{x}}}} \quad \text{hydrogen chloride}$$

Chlorine is more electronegative than hydrogen. Thus chlorine has a greater attraction for the pair of electrons in the covalent bond to hydrogen. The nucleus of the chlorine atom exerts a greater attraction for the bonding electrons than does the hydrogen atom nucleus. It is as if two people in a bed are sharing the same blanket; the stronger person is able to pull more of the blanket closer to one side of the bed more of the time.

This type of *unequal* sharing of electrons is called a **polar covalent bond**. The electrons spend over half of the time more closely associated with the atom that has greater electronegativity. Thus one part or pole of the molecule has an average excess of electrons and becomes slightly negative in charge compared to the other pole. Since chlorine is more electronegative than hydrogen, the chlorine side of the HCl molecule has more of the electrons and is partially negative. We use the symbol $\delta-$ to show partial negativity (the Greek letter δ, delta, means partial.) The hydrogen end of the HCl molecule then becomes partially positive, $\delta+$. Its negative electron is being pulled away by the chlorine atom, leaving the hydrogen atom with a net excess of positive charge.

$$\overset{\delta+}{H}\text{——}\overset{\delta-}{Cl}$$

polar covalent bond

The greater the difference in electronegativity between the atoms in a bond, the more polar the bond. A polar covalent molecule stays neutral as a whole, however, no matter how unequally charge is distributed within it.

The polar covalent bond in a hydrogen chloride molecule creates what we call a **dipole**. A dipole is a pair of equal but opposite charges that are separated by a small distance. We represent a dipole by drawing an arrow pointing from the positive to the negative end of the dipole.

$$H\overset{\longmapsto}{\text{——}}Cl$$

Hydrogen chloride is called a **polar molecule** because it contains a dipole. Molecules are called **nonpolar** if they contain no dipole. Nonpolar molecules contain either nonpolar covalent bonds, or polar covalent bonds whose dipoles cancel. For example, the carbon dioxide molecule contains two polar covalent bonds:

$$O{=}\overset{\longleftrightarrow}{C}{=}O$$

The two dipoles of this molecule point in opposite directions. Therefore, the dipoles cancel and the molecule is nonpolar overall.

6.6
COMBINING CAPACITY

Most molecules consist of more than two atoms each. We have seen that a nitrogen atom forms one triple bond to another nitrogen atom in the molecule N_2. But a nitrogen atom can also form single bonds to three other atoms. For example, in ammonia, the nitrogen atom forms single bonds to three hydrogen atoms. We write the formula of this molecule NH_3 (read N–H–three). This formula shows that the molecule contains four atoms: one nitrogen atom and three hydrogen atoms. NH_3 is a shorthand way of writing N_1H_3. The subscript 1 is understood when no subscript is written at the right side of the symbol for an atom. The Lewis structure for ammonia is:

$$H \overset{xx}{\underset{\overset{\cdot x}{H}}{\overset{x}{N}}} H \qquad ammonia$$

Each hydrogen atom contributes its 1s electron to a bond, and the nitrogen atom ($1s^22s^22p^3$) contributes one of its three p electrons to each of the three bonds. The hydrogen atoms now have filled outer levels of two electrons each, like helium (s^2), and the nitrogen atom has a complete octet, like neon (s^2p^6).

Carbon has the electron configuration $1s^22s^22p^2$. It has four electrons in its outermost level ($2s^22p^2$), and therefore needs four more electrons to complete this level. To obtain these four additional electrons, carbon can form four covalent bonds. For example, methane, CH_4 (read C–H–four), consists of a carbon atom with single bonds to four hydrogen atoms.

$$\overset{\overset{H}{x\cdot}}{H \overset{x}{\underset{\overset{\cdot x}{H}}{C}} H} \qquad methane$$

The electron configuration of oxygen is $1s^22s^22p^4$. Its outermost level contains six electrons ($2s^22p^4$), and therefore oxygen needs only two electrons to complete the level. To obtain two more electrons, oxygen can form two single bonds with other atoms, as it does in water, H_2O (read H-two-O.) Oxygen can also obtain two electrons by forming one double bond as in carbon dioxide, CO_2 (read C–O–two.)

$$H \overset{xx}{\underset{\overset{\cdot x}{H}}{\overset{x}{O}}} \qquad water \qquad \overset{xx}{\underset{xx}{O}} \overset{}{x} \overset{}{x} C \overset{}{x} \overset{}{x} \overset{xx}{\underset{xx}{O}} \qquad carbon\ dioxide$$

TABLE 6-2

Combining capacities of common atoms

1	2	3	4
H	O	N	C
F	S	P	
Cl			
Br			
I			

Note that in carbon dioxide, the carbon atom completes its outer level of electrons by forming two double bonds instead of four single bonds.

Depending on their outer electron configurations, atoms form certain numbers of covalent bonds. The number of covalent bonds that an atom can form is called its **combining capacity** (also called **valence**). Since carbon must form four bonds to obtain a complete outer level, its combining capacity is four. Similarly, nitrogen has a combining capacity of three, and oxygen a combining capacity of two. Fluorine, along with the other halogens and hydrogen, has a combining capacity of one. Table 6-2 summarizes this information. (A combining capacity greater than four may be possible when d orbitals, as well as s and p orbitals, are involved in bond formation.)

Note that with the exception of hydrogen, the combining capacity of these common nonmetal atoms is equal to eight minus the group number. (For example, the combining capacity of nitrogen in Group 5A, is 8 − 5 or 3.) This difference represents the number of electrons the nonmetal atom must gain to reach a filled outer level. Thus the combining capacity of an atom is determined by the number of outermost electrons, which is reflected in the group number.

6.7

DRAWING LEWIS STRUCTURES

In order to draw the Lewis structure of a molecule, we must know the combining capacities of the atoms involved. As explained in the previous section, the combining capacity of a nonmetal atom depends on the number of electrons needed to fill its outermost level.

Many covalent molecules have a central atom to which others are bonded. For example, the central atom in methane is carbon, the central atom in ammonia is nitrogen, and the central atom in water is oxygen. The central atom is easy to identify because it appears only once in the chemical formula: CH_4, NH_3, H_2O. Also, the central atom is generally the atom having the highest combining capacity. Hydrogen, and the halogens, generally cannot serve as central atoms because they form only one bond in filling their outer levels.

Once we have identified the central atom of the molecule, we draw the Lewis structure by placing the other atoms around it to complete the central atom's octet, as well as the outer levels of the surrounding atoms. For example, in phosphorus trichloride, PCl_3, phosphorus is the central atom. It is in Group 5A and has five outermost electrons:

·P: phosphorus

Phosphorus has a combining capacity of three because it needs three more electrons to fill its outermost level. It therefore forms three bonds. Each chlorine atom must form a single bond to the central phosphorus atom.

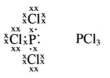 PCl_3

Note that by forming these bonds the chlorine atoms have gained one electron each. Their outermost levels are now filled.

Some molecules require the formation of double or triple bonds to complete the octets of all their atoms. For example, carbon dioxide, CO_2, contains two double bonds. Here carbon, the central atom, has four outermost electrons and must gain four additional electrons through bonding. To gain them, carbon must share two electrons from each of the two oxygen atoms. Therefore, carbon forms one double bond with each oxygen atom.

$$\overset{\text{xx}}{\underset{\text{xx}}{O}} \text{ⅹⅹ} C \text{ⅹⅹ} \overset{\text{xx}}{\underset{\text{xx}}{O}} \qquad CO_2$$

Oxygen atoms have six outermost electrons. By forming a double bond with carbon, each oxygen atom gains two electrons and completes its outer level.

Example   Draw the Lewis structure of carbon tetrachloride, CCl_4.

$$\overset{\text{xx}}{\underset{\text{xx}}{\text{ⅹ}Cl\text{ⅹ}}}$$
$$\overset{\text{xx}}{\underset{\text{xx}}{\text{ⅹ}Cl\text{ⅹ}}} \overset{\cdot\text{x}}{\underset{\text{x}\cdot}{C}} \overset{\text{xx}}{\underset{\text{xx}}{\text{ⅹ}C\text{ⅹ}}}$$
$$\overset{\text{xx}}{\underset{\text{xx}}{\text{ⅹ}Cl\text{ⅹ}}}$$

As shown in the following examples, it is possible for a pair of atoms to serve together as central atoms.

Example   Draw the Lewis structure of hydrogen peroxide, H_2O_2.

$$H\text{ⅹ}\overset{\text{xx}}{\underset{\text{xx}}{O}}\text{ⅹ}\overset{..}{\underset{..}{O}}\text{ⅹ}H$$

Example   Draw the Lewis structure of ethylene, C_2H_4.

$$\begin{matrix} H & H \\ \cdot\text{x} & \text{x}\cdot \\ C & \text{xx} & C \\ \cdot\text{x} & & \cdot\text{x} \\ H & H \end{matrix}$$

It is important to realize that although Lewis structures show how atoms are bonded, they do not indicate the actual geometry or shape of the molecule. Figure 6-5 shows the shapes of molecules based on the number of electron pairs surrounding the central atom. Notice that in ammonia (NH_3) and water (H_2O), the unshared pairs of electrons, called *lone pairs,* help to determine the molecular shape even though no atoms are bonded to these positions.

The shapes of simple covalent molecules are the result of repulsion

Number of electron pairs	Geometry of electron pairs	Number of atoms bonded	Example	Structure[a]
2	linear	2	BeF_2	F—Be—F
3	triangular	3	BF_3	(see figure)
4	tetrahedral	4	CH_4[b]	(see figure)
		3	NH_3[c]	(see figure)
		2	H_2O[c]	(see figure)

FIGURE 6-5

The geometry of molecules. The shape of a covalent molecule depends on the number of electron pairs surrounding the nucleus.

[a]The dashed lines indicate bonds extending behind the page, and the shaded line shows a bond coming out of the page.
[b]See Figure 18-1 for a three-dimensional model of CH_4 and the tetrahedron.
[c]Note that both NH_3 and H_2O have four electron pairs like CH_4, but that one or two "corners" of the tetrahedron, respectively, are not occupied by atoms.

between the outer electron pairs. These electrons pairs tend to distribute themselves around the central atom so that they are as far apart as possible. For example, two electron pairs move to opposite sides of the central atom, forming a linear ("straight line") molecule. In a molecule with three electron pairs, the pairs form a triangle around the central atom. Four electron pairs arrange themselves into a tetrahedron [also shown in Figure 18-1(c)]; if some of these electron pairs are lone pairs, the actual shape of the molecule is not tetrahedral (like CH_4), but may be pyramidal (as in NH_3) or bent (as in H_2O).

6.8 IONS

The molecules considered so far all consist of *non*metallic atoms covalently bonded to each other. But when *metals* form bonds with nonmetals, the difference in electronegativity between the atoms is so great that it becomes difficult to describe the bonding as simply the sharing of electrons. Metals have very low electronegativities and low ionization potentials. They have little attraction for electrons, and in fact they readily lose their outermost electrons. Nonmetals, on the other hand, have much higher electronegativities; they tend to attract electrons to themselves. When a metal interacts with a nonmetal, this large difference in electronegativity results in the *transfer of outermost electrons* from the metal atom's orbitals to those of the nonmetal atom. We can think of this situation as an extreme case of polar covalent bonding.

Consider the following example. Lithium, a Group 1A metal, has one outermost electron ($1s^2 2\underline{s^1}$.) Fluorine, a Group 7A nonmetal, has seven outermost electrons ($1s^2 2s^2 2p^5$.) If atoms of these two elements interact, lithium will transfer its one outermost electron to fluorine, which has a much greater electronegativity. By losing its one outer electron, lithium attains the completed energy level arrangement of helium ($1s^2$). Fluorine, by gaining one electron, attains the completed energy level arrangement of neon ($1s^2 2s^2 2p^6$). The number of electrons in each atom has changed, but the number of protons remains the same. Therefore the number of positive protons in each atom no longer equals the number of negative electrons, and the two atoms are no longer neutral. These electrically charged atoms are called **ions**. Using Lewis diagrams, we can show the formation of a lithium ion and a fluoride ion as follows:

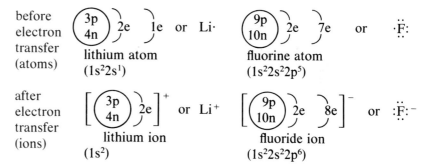

before electron transfer (atoms)

lithium atom ($1s^2 2s^1$)

fluorine atom ($1s^2 2s^2 2p^5$)

after electron transfer (ions)

lithium ion ($1s^2$)

fluoride ion ($1s^2 2s^2 2p^6$)

Lithium ion has a charge of 1+ because it has 3 protons but only 2 electrons:

protons:	3+	
electrons:	2−	Li^+, lithium ion
total charge:	1+	

Fluoride ion has a charge of 1− because it has 9 protons and 10 electrons:

protons:	9+	
electrons:	10−	F^-, fluoride ion
total charge:	1−	

Note that the name of a simple negative ion, like fluoride ion, is formed by changing the ending of the element name to "ide." (Chapter Seven will provide further information on naming ions.)

A positively charged ion, like Li^+, is called a **cation** (pronounced kat' eye-on). A cation consists of a positive nucleus surrounded by electrons that are too few to balance the charges of all its protons. Cations are formed by a process called **oxidation**, which means the loss of one or more electrons. The elements most easily oxidized are the metals. The more metallic an element (the lower its ionization energy), the easier its atoms are oxidized to become cations.

Metals in Group 1A, such as lithium, have only one outermost electron to lose. Therefore, they form cations with a charge of $1+$. Similarly, metals in Group 2A form ions with charges of $2+$ by losing their two outermost electrons. Metals in Group 3A form ions with charges of $3+$ by losing their three outermost electrons.

Group number	Outer electron configuration	Number of outer electrons	Charge of cation	Example
1A	s^1	1	$1+$	Na^+
2A	s^2	2	$2+$	Ca^{2+}
3A	s^2p^1	3	$3+$	Al^{3+}

Negatively charged ions, like F^-, are called **anions**. An anion consists of a positive nucleus surrounded by more than enough electrons to balance the charge of its protons. Anions are formed by **reduction**, which means the gain of one or more electrons. The elements reduced most easily are the nonmetals. The more nonmetallic an element (the greater its electronegativity), the easier it is reduced to form an anion.

The halogens of Group 7A have seven outermost electrons. They complete their outer levels by gaining one electron, and therefore form $1-$ anions. In the same way, nonmetals in Group 6A, which have six outer electrons, gain two electrons ($6 + 2 = 8$) to form ions with charges of $2-$. Nonmetals in Group 5A, which have five outer electrons, gain three more electrons ($5 + 3 = 8$) to form ions with charges of $3-$.

Group number	Outer electron configuration	Number of outer electrons	Electrons needed to complete level	Charge of anion	Example
5A	s^2p^3	5	3	$3-$	N^{3-}
6A	s^2p^4	6	2	$2-$	O^{2-}
7A	s^2p^5	7	1	$1-$	Cl^-

Anions, as well as cations, have the electron configurations of noble gas elements. They are charged, however, because they do not have the same number of protons as the noble gas atoms. They have unequal numbers of protons and electrons.

Ions differ in size from their original atoms. A cation is smaller than the neutral metal atom because an electron has been lost and a smaller number of electrons is now attracted by the same positive nuclear charge. Anions are larger than neutral nonmetal atoms because they have extra electrons, which are held by a less than equal number of protons in the nucleus. Figure 6-6 compares the sizes of atoms and ions for the alkali metals and the halogens.

6.9 IONIC BONDING

An **ionic bond** results from the attraction between oppositely charged ions, namely cations and anions. As illustrated here, the formation of an ionic bond can be thought of as taking place in two steps:

1 electron transfer   Li^{x} ⟶ $\ddot{\text{F}}$:

2 attraction of ions   Li^{+} $^{x}\ddot{\text{F}}$:$^{-}$
                              →   ←

It is important to realize that even though an electron is transferred from a metal orbital to a nonmetal orbital, it remains relatively close to the nucleus of the metal atom. In bond formation, electrons redistribute themselves so that they are closer to the more electronegative atom. However, because the oppositely charged ions attract each other, the electrons remain near the nuclei of *both* atoms involved in the bond.

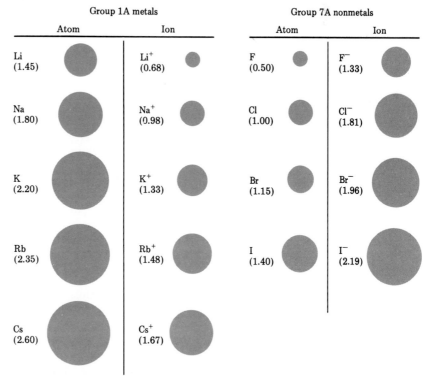

Group 1A metals		Group 7A nonmetals	
Atom	Ion	Atom	Ion
Li (1.45)	Li^{+} (0.68)	F (0.50)	F^{-} (1.33)
Na (1.80)	Na^{+} (0.98)	Cl (1.00)	Cl^{-} (1.81)
K (2.20)	K^{+} (1.33)	Br (1.15)	Br^{-} (1.96)
Rb (2.35)	Rb^{+} (1.48)	I (1.40)	I^{-} (2.19)
Cs (2.60)	Cs^{+} (1.67)		

FIGURE 6-6

Relative sizes of atoms and ions. As shown for Group 1A, a cation is smaller than a neutral atom of the same element. As shown for Group 7A, an anion is larger than a neutral atom of the same element.

The substance formed by lithium and fluorine is an ionic compound containing $1+$ cations (Li^+), and $1-$ anions (F^-) in equal numbers. Its formula is LiF (read L-I-F). The compound as a whole is neutral because the positive charges of the cations exactly balance the negative charges of the anions. Whether ionic or covalent, *all compounds, like all atoms, are electrically neutral.*

Let's look at another example. The formation of an ionic bond between sodium and chlorine also begins with an electron transfer from the metal to the more electronegative nonmetal:

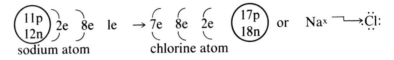

Sodium is oxidized, losing its one outermost electron to form sodium ion.

$1s^2 2s^2 2p^6 3s^1$ $1s^2 2s^2 2p^6$
sodium atom, Na   sodium ion, Na^+

Chlorine is reduced, gaining the electron from sodium to form chloride ion.

$1s^2 2s^2 2p^6 3s^2 3p^5$ $1s^2 2s^2 2p^6 3s^2 3p^6$
chlorine atom, Cl   chloride ion, Cl^-

The sodium cation and the chloride anion have opposite charges and therefore attract each other.

$$\left[\begin{array}{c} 11p \\ 12n \end{array} \; 2e \; 8e \right]^+ \qquad {}^-\left[8e \; 8e \; 2e \begin{array}{c} 17p \\ 18n \end{array} \right] \qquad \text{or} \qquad Na^{+ \; -x}\ddot{\underset{..}{Cl}}{:}$$

The ionic compound sodium chloride, NaCl (read N-A-C-L), is the result.

Part of the structure of sodium chloride is shown in Figure 6-7. It is an orderly pattern of Na^+ cations and Cl^- anions in which each ion is surrounded by six other ions of opposite charge. A regular repeating three-dimensional arrangement of ions or atoms is called a **crystal lattice**. In a crystal lattice, no one positive ion is paired with any one negative ion. Instead, each positive ion is surrounded by several negative ions, and each negative ion is surrounded by several positive ions. This arrangement allows ions with like charges to be relatively far apart. By increasing attractive forces and decreasing repulsive forces between ions, the crystal lattice increases the stability of ionic compounds.

As we can see from Figure 6-7, there are no distinct pairs of ions, no molecules of sodium chloride. In contrast to covalent compounds,

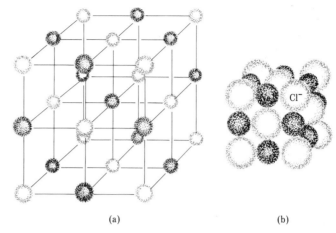

FIGURE 6-7

Two views of sodium chloride, NaCl. In this ordered arrangement of Na⁺ and Cl⁻ ions, each ion is surrounded by six ions of opposite charge.

(a)   (b)

ionic compounds do not exist as molecules. One sodium cation and one chloride anion, however, together make up a **formula unit** of sodium chloride. For every one sodium cation there is one chloride anion. A formula unit gives the relation between the number of cations and the number of anions in the crystal lattice of an ionic compound.

The common name for sodium chloride is table salt. The properties of this compound differ greatly from the properties of the two elements that make it up. Sodium is a shiny metal that burns spontaneously when dropped into water (see Figure 6-8). Chlorine is a poisonous gas that was used for chemical warfare during World War I. Yet we sprinkle sodium chloride on food! It is clear that when atoms lose or gain electrons to form ions, a great change takes place in their properties. Therefore, it is extremely important to distinguish between a neutral atom and an ion. When dealing with an ion, always be sure to specify its charge.

FIGURE 6-8

Sodium metal and sodium chloride. Sodium metal, Na (left), and sodium chloride, NaCl (right), are very different even though Na and Na⁺ (in NaCl) differ by only one electron. (Photo by Al Green.)

In the examples of ionic compound formation discussed so far, only one electron was transferred from one atom to another. But ionic compounds also can be formed by the transfer of more than one electron. For example, the elements magnesium and chlorine can form an ionic compound. Magnesium ($1s^2 2s^2 2p^6 3s^2$) must lose two outermost electrons, and chlorine ($1s^2 2s^2 2p^6 3s^2 3p^5$) can accept only one. Therefore,

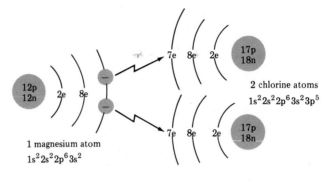

or

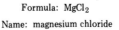

forming

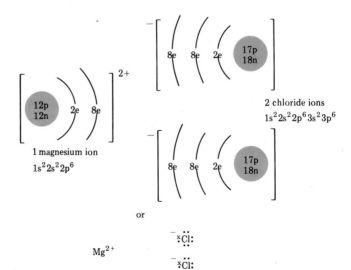

FIGURE 6-9

Formation of magnesium chloride, MgCl₂. One magnesium atom transfers its outer electrons to two chlorine atoms. The resulting ionic compound contains one magnesium ion (Mg²⁺) and two chloride ions (Cl⁻) in the formula unit.

two chlorine atoms are needed for each magnesium atom. The magnesium atom donates one electron to each chlorine atom, as shown in Figure 6-9. The resulting compound is magnesium chloride, $MgCl_2$ (read M-G-C-L-two).

Using the same reasoning, you should be able to understand how sodium and oxygen form the compound sodium oxide, Na_2O (read N-A-two-O), shown in Figure 6-10.

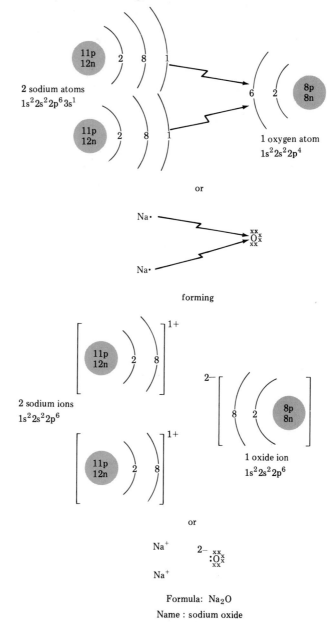

FIGURE 6-10

Formation of sodium oxide, Na_2O. Two sodium atoms each transfer one outer electron to an oxygen atom. The resulting ionic compound contains two sodium ions (Na^+) and one oxide ion (O^{2-}) in the formula unit.

6.10
POLYATOMIC
IONS

So far, we have considered only ions derived from single atoms. For example, the sodium ion Na^+ is derived from the sodium atom Na. But groups of atoms that are covalently bonded also can have a positive or negative charge and act as ions. Because they consist of several atoms that together carry a charge, they are called **polyatomic ions** ("poly" means many). For example, the hydroxide ion, OH^-, is an ion composed of one hydrogen atom and one oxygen atom.

$$\left[\overset{xx}{\underset{xx}{\overset{x}{O}}} \overset{}{:} H \right]^-$$ hydroxide ion, OH^-

There is a single covalent bond between the hydrogen and oxygen atoms, consisting of one electron from each atom. This single covalent bond completes the hydrogen atom's outer level, but the oxygen atom still needs one more electron. This electron (shown by an open dot in the Lewis structure) comes from the oxidation of a metal atom during the formation of a compound. For example, in the formation of the compound sodium hydroxide, NaOH, the one outermost electron of a sodium atom is transferred to the outer level of an oxygen atom.

$$Na° \longrightarrow \overset{xx}{\underset{\text{,x}}{\overset{x}{O}}} : H$$

Because this combination of atoms of oxygen and hydrogen now has one more electron than it has protons, it has a charge of $1-$ and is written OH^-.

One of the few common polyatomic cations is the ammonium ion, NH_4^+. It consists of an ammonia molecule, NH_3, which has formed a covalent bond to a hydrogen ion, H^+. (The H^+ ion is formed by the loss of one electron from a hydrogen atom.) The covalent bond is called a **coordinate covalent bond**. It is a bond in which one of the atoms provides both of the electrons being shared. In this case, the nitrogen atom makes its unshared pair of electrons available to a hydrogen ion, which has no electrons.

$$\overset{..}{H \overset{x}{N} \overset{}{:} H} \atop \underset{H}{\overset{x\cdot}{}} \qquad \left[\overset{H}{\underset{H}{\overset{..}{H \overset{x}{N} \overset{}{:} H}}} \right]^+$$

ammonia, NH_3 ammonium ion, NH_4^+

Since the hydrogen ion has a charge of $1+$, the neutral ammonia molecule becomes a positively charged ion when it is bonded to a hydrogen ion. An example of an ionic compound containing this polyatomic ion is ammonium chloride, NH_4Cl.

Other examples of polyatomic ions are shown in Table 6-3. In each case, the bonds within the group of atoms are covalent, but the unit

TABLE 6-3 *Examples of polyatomic ions*

Structure	Formula	Name
$\left[\begin{array}{c} \text{O:N:O} \\ \text{O} \end{array} \right]^{-}$	NO_3^-	nitrate ion
$\left[\begin{array}{c} \text{O} \\ \text{O:S:O} \\ \text{O} \end{array} \right]^{2-}$	SO_4^{2-}	sulfate ion
$\left[\begin{array}{c} \text{O} \\ \text{O:P:O} \\ \text{O} \end{array} \right]^{3-}$	PO_4^{3-}	phosphate ion

as a whole is an ion. More examples of polyatomic ions are presented in Chapter Seven.

6.11 OVERVIEW OF BONDING

There are several major factors that determine the type of bonding in a compound. One factor is the *number of outermost electrons* in the atoms involved. The number of outermost electrons ranges from one in the Group 1A elements to eight in the noble gases. Another way to think of this factor is as the number of vacancies, or the number of electrons that can be added to the outer level. The number of vacancies in the outer level ranges from seven in the Group 1A elements to zero in the noble gases. (These outermost vacancies are actually regions of space around the nucleus of an atom where electrons from another atom could be attracted by the nuclear charge of the first atom.) An atom with few outer electrons (a metal atom) tends to lose these electrons to an atom having a few vacancies (a nonmetal atom). In this case, an ionic bond results. When both atoms have more than a few outer electrons, as do two nonmetal atoms, they share electrons in a covalent bond.

Another factor that determines the type of bonding in a compound is the *electronegativity* of the atoms involved in the bond. The greater the difference in electronegativity between the two atoms, the more polar, and generally stronger the bond. When the electronegativity difference is large, as it is between a metal atom and a nonmetal atom, an ionic bond forms. When the electronegativity difference is small, as it is between two nonmetals, covalent bonding takes place.

We have studied two extreme types of bonding: "pure" nonpolar

Photo courtesy of Linus Pauling Institute.

Linus Pauling, an American chemist, is the only person ever to receive both a Nobel Prize in science and a Nobel Peace Prize. The science prize was given to him in 1954 for his work in chemistry. He won the Peace Prize for his efforts to promote nuclear disarmament, which led to the signing of an international test ban treaty in 1962. This work disturbed many colleagues, who considered his activities un-American.

Pauling's work in chemistry is remarkable because of the broad range and importance of his contributions. Pauling was introduced early to quantum mechanics when he studied abroad under the physicists Niels Bohr and Erwin Schrödinger. Pauling combined quantum mechanics with the notion of electron-pair bonding, which was developed by his friend Gilbert N. Lewis. In this way, he created a modern theory of chemical bonding based on electron orbitals. Quantum mechanics, which had been developed by physicists, thus became available to chemists.

Pauling showed that chemical bonds are partly covalent and partly ionic. In addition, he explained the bonding in ionic crystals. These concepts are presented in Pauling's classic book, *The Nature of the Chemical Bond*, which he dedicated to Lewis.

In the field of biochemistry, Pauling identified a key structural feature of protein molecules—the alpha helix. He also proposed that the genetic disease sickle cell anemia is caused by a defect in the structure of the hemoglobin molecule. Currently, he is exploring the effects of vitamins on disease and the possibility that some kinds of mental illness may have a chemical basis.

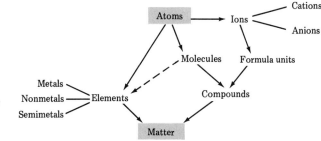

FIGURE 6-11

The composition of matter. This chart summarizes the relationship between matter and atoms. Matter consists of elements and compounds, which are made up of atoms.

covalent bonding, with equal sharing of electrons, and "pure" ionic bonding, with complete transfer of electrons. Most bonding in compounds, however, is somewhere between these two extremes. Polar covalent bonds, for example, can be thought of as a blend of covalent and ionic bonding. The exact nature of the bonding that exists between the atoms of a compound determines to a large degree the properties of the compound.

Figure 6-11 presents a summary of the composition of matter based on the concepts presented in this and previous chapters. All matter is composed of atoms, either in the form of an element or a compound. Elements are composed of atoms having the same atomic number. Compounds are composed of two or more different elements. Compounds can consist of molecules or of regular arrangements of oppositely charged ions. All the matter that surrounds us, as well as the matter in our own bodies, is composed of elements and compounds.

Summary

6.1 The number of outermost electrons in an atom determines the chemical behavior of that atom. According to the octet rule, elements tend to combine so that each atom obtains a filled outer electron configuration (s^2p^6) like a noble gas. The force that holds atoms together when they combine is called a chemical bond.

6.2 A covalent chemical bond results from the sharing of electrons between two atoms. It is based on the mutual attraction of two nuclei for the same pair of electrons.

6.3 A molecule is the unit formed by the covalent bonding of two or more atoms. The simplest type of covalent molecule is called diatomic because it consists of only two atoms. Seven nonmetallic elements exist in nature as diatomic molecules (H_2, N_2, O_2, F_2, Cl_2, Br_2, I_2). Multiple bonds consist of two or more electron pairs shared between two atoms.

6.4 Electronegativity is a measure of the attraction of an atom for the electrons in a chemical bond. The nonmetals fluorine, oxygen, and chlorine are the most electronegative elements.

6.5 A compound results from the bonding of the atoms of two or more different elements. Because of the difference in electronegativity of the atoms in a molecule, polar covalent bonds form. Polar covalent bonds involve unequal sharing of electrons.

6.6 The number of covalent bonds an atom can form is called its combining capacity. The combining capacity of most common non-metals is equal to eight minus the group number. Some of these combining capacities are: C,4; N,3; O,2; H and halogens, 1.

6.7 It is important to know the combining capacity of an atom in order to draw Lewis structures of molecules. The central atom in the Lewis structure is drawn first, and then the remaining atoms are placed around it so that all the atoms in the molecule have complete outer levels.

6.8 An ion is an electrically charged atom. Ions are formed by the transfer of outermost electrons from the orbitals of metal atoms to the orbitals of nonmetal atoms. The metal atoms lose electrons (undergo oxidation) to form positive ions, called cations. The nonmetal atoms gain electrons (undergo reduction) to form negative ions, called anions.

6.9 Ionic bonds result from the attraction between oppositely charged cations and anions. An ionic compound is neutral because the positive charge of the cation exactly balances the negative charge of the anion. A formula unit is the smallest repeating unit in an ionic compound.

6.10 A polyatomic ion is a group of covalently bonded atoms that has a positive or negative charge. Most polyatomic ions are anions.

6.11 The major factors that affect bonding are the number of outermost electrons and the electronegativities of the atoms involved. Most chemical bonds are somewhere between the two extremes of "pure" covalent or "pure" ionic bonds.

Exercises

KEY-WORD MATCHING EXERCISE For each term on the left, choose a phrase from the right-hand column that most closely matches its meaning.

1 chemical bond   **a** sharing of electrons

2 octet rule   **b** formed from different elements

3 covalent bond   **c** indicates atoms in molecule

4 molecule   **d** uneven electron distribution

5 formula	**e** charged atom
6 electronegativity	**f** atomic "glue"
7 compound	**g** gain of electrons
8 polar	**h** combination of atoms
9 combining capacity	**i** negative ion
10 ion	**j** attraction of ions
11 cation	**k** eight electrons in outer level
12 anion	**l** relates anions and cations
13 oxidation	**m** group of atoms with charge
14 reduction	**n** attraction for electrons
15 ionic bond	**o** positive ion
16 formula unit	**p** loss of electrons
17 polyatomic ion	**q** number of possible bonds

EXERCISES BY TOPIC

The chemical bond (6.1)

18 What force is the basis for chemical bonding?

19 What determines the chemical behavior of an atom?

20 What is the octet rule?

Covalent bond formation (6.2)

21 Describe the formation of a covalent bond.

22 How does the electron density around a hydrogen atom change as it forms a covalent bond to another hydrogen atom?

Molecules (6.3)

23 What is the difference in meaning between the symbols 2N and N_2?

24 What is a molecule? a diatomic molecule?

25 Which elements normally exist as diatomic molecules?

26 Why does a double bond exist in O_2, but only a single bond in F_2?

27 What information is given by the formula of a molecule?

Electronegativity (6.4)

28 What determines the electronegativity of an atom?

29 What does bond energy measure?

30 Explain how electronegativity varies across a period and down a group in the periodic table.

31 Which elements are most electronegative?

32 How is bond length related to bond energy?

Polar covalent bonding (6.5)

33 How does a compound differ from an element?

34 How does a nonpolar covalent bond differ from a polar covalent bond?

35 Explain how a molecule can have polar covalent bonds yet be nonpolar overall.

36 Describe the bonding in HCl.

Combining capacity (6.6)

37 What does the combining capacity of an atom represent?

38 How is combining capacity related to group number?

39 Why does nitrogen have a combining capacity of three?

40 What combining capacity would you expect for each of the following? **a** phosphorus **b** iodine **c** silicon **d** sulfur **e** boron

41 What is the maximum possible combining capacity of an atom if only s and p orbitals are involved? Explain your answer.

Drawing Lewis structures (6.7)

42 Why is knowing the combining capacities of atoms useful for drawing Lewis structures?

43 Draw the Lewis structure of chloroform, $CHCl_3$.

44 Draw the Lewis structure of ethane, C_2H_6.

45 Hydrogen cyanide, HCN, is a poisonous gas. Draw its Lewis structure.

46 Hydrogen sulfide, H_2S, has the odor of rotten eggs. Draw its Lewis structure.

Ions (6.8)

47 How does an ion differ from an atom?

48 Explain how a metal reacts with a nonmetal.

49 What is the difference between a cation and anion?

50 An ion has 12 protons, 13 neutrons, and 10 electrons. What is its charge?

51 Which is larger: **a** a cation or the neutral atom from which it forms? **b** an anion or the neutral atom from which it forms? Explain.

52 Compare the electron configurations of: **a** Na and Na^+ **b** Cl and Cl^- **c** Mg and Mg^{2+} **d** O and O^{2-} **e** Al and Al^{3+}

Ionic bonding (6.9)

53 What is an ionic bond? How does it form?

54 How is an ionic bond similar to and how does it differ from a covalent bond?

55 Why are ionic compounds neutral even though they contain charged particles?

56 How do ionic compounds exist in the solid state?

57 What is a formula unit?

58 Use diagrams to show the formation of: **a** calcium fluoride, CaF_2, from calcium and fluorine **b** lithium nitride, Li_3N, from lithium and nitrogen **c** magnesium oxide, MgO, from magnesium and oxygen

Polyatomic ions (6.10)

59 What is a polyatomic ion? How does it differ from a simple ion?

60 Draw the Lewis structure of the carbonate ion, CO_3^{2-}.

61 Draw the Lewis structure of the perchlorate ion, ClO_4^-.

Overview of bonding (6.11)

62 What factors affect the type of bonding between two atoms?

63 Summarize the differences between ionic, polar covalent, and nonpolar covalent bonding.

64 Predict the type of bond formed between each of the following pairs of elements: **a** sodium and sulfur **b** nitrogen and bromine **c** calcium and oxygen **d** phosphorus and iodine **e** carbon and oxygen

65 Classify each of the following as an element, compound, or mixture: **a** zinc sulfide **b** blood **c** selenium **d** sea water **e** sodium fluoride

ADDITIONAL EXERCISES

66 Why are the noble gases relatively stable?

67 Describe the bonding in N_2.

68 What factors determine the strength of a bond?

69 Explain why carbon has a combining capacity of four.

70 Acetylene, C_2H_2, is a gas used as a fuel. Draw its Lewis structure.

71 Why do metals form cations and nonmetals form anions?

72 Compare the processes of oxidation and reduction. Why must they take place at the same time?

73 Many cations and anions have noble gas electronic structures. Why are they reactive when noble gases are not?

74 Predict the ion formed from the following elements: **a** sulfur **b** cesium **c** bromine **d** strontium **e** aluminum

75 Why can't we speak about a sodium chloride molecule?

76 Draw the Lewis structure of the hydrogen sulfide ion, HS^-.

77 What factors affect the shape of a molecule?

78 What is a lone pair?

79 Why do elements in Group 6A form ions with a charge of $2-$?

80 Which are the four most electronegative elements?

81 Use a diagram to show the formation of Na_2S from sodium and sulfur.

82 What is a coordinate covalent bond?

83 Draw Lewis structures for the following: **a** F_2O **b** BeH_2 **c** SeF_4 **d** H_2SO_4 **e** $AsCl_5$

84 How are **a** bond energy and **b** bond length related to electronegativity?

85 Classify the bonds that would form between the following pairs of elements as mainly nonpolar covalent, polar covalent, or ionic: **a** Na and O **b** O and O **c** H and O **d** S and O **e** Sr and Br

86 Identify each of the following as an atom or ion: **a** atomic number 9, $1s^22s^22p^6$ **b** atomic number 55, $[Xe]\,6s^1$ **c** atomic number 16, $[Ne]\,3s^23p^4$ **d** atomic number 19, $[Ne]\,s^23p^6$

87 If a covalent bond forms between atoms of two different elements, why is this bond generally polar?

88 Draw the Lewis structure for each of these ions: **a** O^{2-} **b** BF_4^- **c** HSO_4^- **d** I^- **e** ClO^-

89 An ion has 30 protons, 35 neutrons, and 28 electrons. What is its **a** atomic number? **b** mass number? **c** charge?

90 Write the electron configuration for **a** Na^+ **b** Cl^- **c** O^{2-} **d** Ca^{2+} **e** Al^{3+}

91 Silicon and fluorine form a compound. By drawing a Lewis structure, determine the formula of the compound.

PROGRESS CHART   Check off each item when completed.

_____ Previewed chapter

Studied each section:

_____ **6.1** The chemical bond

_____ **6.2** Covalent bond formation

_____ **6.3** Molecules

_____ **6.4** Electronegativity

_____ **6.5** Polar covalent bonding

_____ **6.6** Combining capacity

_____ **6.7** Drawing Lewis structures

_____ **6.8** Ions

_____ **6.9** Ionic bonding

_____ **6.10** Polyatomic ions

_____ **6.11** Overview of bonding

_____ Reviewed chapter

_____ Answered assigned exercises:

#'s _____

SEVEN

Writing names and formulas of compounds

In Chapter Seven we describe how the composition of a chemical compound can be represented in two ways: by its formula and by its name. Systematic methods are presented for naming covalent and ionic compounds based on the number and type of atoms or ions in the molecule or formula unit. A method for predicting the formulas of ionic compounds is illustrated. Examples of compounds and their uses are given. To demonstrate an understanding of Chapter Seven you should be able to:

1 Write the names and formulas of covalent compounds.

2 Write the names and formulas of simple ionic compounds.

3 Write the names and formulas of compounds involving polyatomic ions.

4 Predict the formulas of ionic compounds, given the elements or ions from which they form.

5 Write the names and formulas of common acids.

6 Give an example of an important chemical compound for each main group.

7.1 COVALENT COMPOUNDS

Covalent compounds containing only two elements are called **binary compounds**. When we write the name or formula of a binary covalent compound we place the two elements in order of increasing electronegativity. The name or symbol of the more electronegative element is written last. The order of the elements is shown by the following list:

B Si C Sb As P N H Se S I Br Cl O F →

order of elements in binary covalent compounds

Smithsonian Institution: Photo # 46834

It was Berzelius, a prominent 19th century Swedish chemist, who proposed our modern system of chemical symbols and formulas. He suggested using the first letter, or the first two letters, of the Latin name of each element as the element's symbol. These symbols, when accompanied by subscripts, provide a shorthand way of expressing the composition of a compound. This notation is now the international language of chemistry. When Berzelius first suggested this system, however, it was opposed by John Dalton, who preferred his own circular symbols for the elements (see Figure 4-4).

Berzelius improved on Dalton in another way. He prepared a much more accurate list of atomic weights than Dalton had. In addition, he carefully analyzed the composition of several thousand compounds. This work provided strong support for the law of definite proportions, which had been proposed by the French chemist, Joseph-Louis Proust. The law states that a compound contains elements in a fixed proportion by mass, regardless of where it is found or how it is made. This law helped Dalton establish his atomic theory.

Berzelius also discovered the elements selenium, thorium, and cerium. And for nearly 30 years he published highly respected reviews of the annual progress of chemistry.

At the age of 56, Berzelius married a 24-year-old woman. As a wedding gift from the Swedish King Charles XIV, he was made a baron.

TABLE 7-1

Roots of common elements

Element	Root
hydrogen	hydr-
carbon	carb-
nitrogen	nitr-
phosphorus	phosph-
oxygen	ox-
sulfur	sulf-
fluorine	fluor-
chlorine	chlor-
bromine	brom-
iodine	iod-

The first (less electronegative) element of a binary covalent compound is named in full. Then, the root of the second element is written as a separate word with the ending *ide*. (Common roots are given in Table 7-1.) In addition to writing the element names, a prefix that shows the number of atoms is placed before each part of the name:

Prefix	Meaning
mono-	one
di-	two
tri-	three
tetra-	four
penta-	five
hexa-	six

Generally, the prefix *mono-* is omitted.

For example, consider CO_2, which consists of one carbon atom and two oxygen atoms. Carbon is less electronegative (it appears before oxygen in the list) and is written out in full. The root of oxygen is *ox-*. The prefix *di-* is used to indicate two atoms of oxygen, and the ending *-ide* is added. The complete name is *carbon dioxide*.

carbon	*di*	*ox*	*ide*
first element	two atoms	root of second element	ending

Other examples are:

Formula	Name
CO	carbon monoxide
NO_2	nitrogen dioxide
SO_3	sulfur trioxide
CCl_4	carbon tetrachloride
PBr_5	phosphorus pentabromide
XeF_6	xenon hexafluoride

(Note that NO_2 in this list is a molecule of nitrogen dioxide; it is not the same as NO_2^-, the nitrite ion, described in Chapter Six.)

A few binary covalent compounds have common names that are nearly always used instead of their formal names. The most important examples of such common names are water (H_2O) and ammonia (NH_3).

Molecules containing more than two elements are discussed later in this chapter.

Example   Write the formula and name of the covalent molecule with the following composition.

Composition	Formula	Name
1 sulfur atom 2 chlorine atoms	SCl_2	sulfur dichloride
1 boron atom 1 nitrogen atom	BN	boron nitride
1 phosphorus atom 3 iodine atoms	PI_3	phosphorus triiodide
1 selenium atom 2 oxygen atoms	SeO_2	selenium dioxide
1 carbon atom 4 iodine atoms	CI_4	carbon tetraiodide
2 phosphorus atoms 5 oxygen atoms	P_2O_5	diphosphorus pentoxide
1 nitrogen atom 3 bromine atoms	NBr_3	nitrogen tribromide
2 boron atoms 3 sulfur atoms	B_2S_3	diboron trisulfide
3 phosphorus atoms 5 nitrogen atoms	P_3N_5	triphosphorus pentanitride

7.2
IONIC
COMPOUNDS

The symbol for a cation or an anion is simply the symbol for the element with the appropriate charge written as a superscript at the right. Cations formed from a single atom have the same name as the atoms from which they form. For example, Na^+ is called sodium ion or sodium cation. Anions formed from a single atom are named by replacing the ending of the element name with -*ide*. Thus Cl^-, which is formed from the element chlorine, is called chloride ion or chloride anion. Table 7-2 lists the most common simple cations and anions. You can use this table to name many ionic compounds.

As with covalent compounds, the formulas and names of ionic compounds are written with the most electronegative element last. Sodium chloride, NaCl, is a typical example.

sodium	*chlor*	*ide*
name of cation	root of anion	ending of anion

TABLE 7-2 *Names of simple ions*

Cation		Anion	
Symbol	*Name*	*Symbol*	*Name*
H^+	hydrogen ion	F^-	fluoride ion
Li^+	lithium ion	Cl^-	chloride ion
Na^+	sodium ion	Br^-	bromide ion
K^+	potassium ion	I^-	iodide ion
Mg^{2+}	magnesium ion	O^{2-}	oxide ion
Ca^{2+}	calcium ion	S^{2-}	sulfide ion
Al^{3+}	aluminum ion	N^{3-}	nitride ion

Other examples are: magnesium oxide, MgO; potassium sulfide, K_2S; calcium bromide, $CaBr_2$; aluminum chloride, $AlCl_3$.

As will be described in detail in Chapter Fifteen, atoms can be assigned **oxidation numbers**. These numbers reflect the way charges are distributed among the atoms in a molecule or formula unit. In the case of simple metal cations, the oxidation number of the metal is the same as the charge of the ion. For example, the oxidation number of sodium in Na^+ is $+1$. This number is said to represent the **oxidation state** of the element.

Most transition metals have more than one oxidation state. To properly identify the compound, the oxidation number must be written as a Roman numeral after the name of the metal. (The first six Roman numerals are: I, II, III, IV, V, VI.) For example, the name iron chloride is incomplete. It could mean either $FeCl_2$, iron(II) chloride, or $FeCl_3$, iron(III) chloride.

Certain common elements, like iron, have only two major oxidation states, and an older system of naming used the endings *-ous* and *-ic* to distinguish these states. In that system, iron(II), Fe^{2+}, was *ferrous* ion, and iron(III), Fe^{3+} was *ferric* ion. In general, the lower state ends in *-ous* and the higher state in *-ic*. Other examples of this system are given in Table 7-3. Names and formulas of typical compounds involving elements that have more than one oxidation state are:

Formula	*Name*
SnF_2	tin(II) fluoride [stannous fluoride]
SnF_4	tin(IV) fluoride [stannic fluoride]
Hg_2O	mercury(I) oxide [mercurous oxide]
HgO	mercury(II) oxide [mercuric oxide]
$CuBr$	copper(I) bromide [cuprous bromide]
$CuBr_2$	copper(II) bromide [cupric bromide]

TABLE 7-3 *Atoms that form several ions*

Element	Ion	Formal name	Common name
iron	Fe^{2+}	iron(II) ion	ferrous ion
	Fe^{3+}	iron(III) ion	ferric ion
copper	Cu^+	copper(I) ion	cuprous ion
	Cu^{2+}	copper(II) ion	cupric ion
tin	Sn^{2+}	tin(II) ion	stannous ion
	Sn^{4+}	tin(IV) ion	stannic ion
mercury	Hg^{+a}	mercury(I) ion	mercurous ion
	Hg^{2+}	mercury(II) ion	mercuric ion

[a]Occurs as Hg_2^{2+}

Example Write the formula and name of the compound formed from the following ions.

Composition	Formula	Name
1 sodium ion 1 iodide ion	NaI	sodium iodide
3 potassium ions 1 nitride ion	K_3N	potassium nitride
2 cesium ions 1 oxide ion	Cs_2O	cesium oxide
1 calcium ion 2 fluoride ions	CaF_2	calcium fluoride
1 aluminum ion 3 bromide ions	$AlBr_3$	aluminum bromide
1 copper(II) ion 1 sulfide ion	CuS	copper(II) sulfide (cupric sulfide)
1 iron(III) ion 3 iodide ions	FeI_3	iron(III) iodide (ferric iodide)
1 tin(IV) ion 4 chloride ions	$SnCl_4$	tin(IV) chloride (stannic chloride)
1 platinum(IV) ion 4 bromide ions	$PtBr_4$	platinum(IV) bromide

7.3
COMPOUNDS WITH POLYATOMIC IONS

Polyatomic ions, such as those listed in Table 7-4, generally have the ending *-ate*. Many polyatomic ions consist of oxygen atoms bonded to a central atom. The central atom is usually a nonmetal, such as nitrogen in nitrate ion (NO_3^-), or sulfur in sulfate ion (SO_4^{2-}). Often, another polyatomic ion exists having one less oxygen atom. The names of these ions end in *-ite,* as in nitrite and sulfite.

sulfate ion	SO_4^{2-}	nitrate ion	NO_3^-
sulfite ion	SO_3^{2-}	nitrite ion	NO_2^-

TABLE 7-4 *Common polyatomic ions*

Charge	Name	Formula
1−	acetate ion	$C_2H_3O_2^-$
	cyanide ion	CN^-
	hydrogencarbonate (bicarbonate) ion	HCO_3^-
	hydroxide ion	OH^-
	nitrate ion	NO_3^-
	nitrite ion	NO_2^-
	permanganate ion	MnO_4^-
2−	carbonate ion	CO_3^{2-}
	peroxide ion	O_2^{2-}
	sulfate ion	SO_4^{2-}
	sulfite ion	SO_3^{2-}
3−	phosphate ion	PO_4^{3-}
1+	ammonium ion	NH_4^+

Sometimes additional forms are possible. The prefix *per-* is placed in front of the *-ate* form if *more* oxygen atoms are attached. The prefix *hypo-* is placed in front of the *-ite* form if there are *fewer* oxygen atoms. These forms are illustrated in the following series:

Ion	Name
ClO_4^-	*per*chlor*ate* ion
ClO_3^-	chlor*ate* ion
ClO_2^-	chlor*ite* ion
ClO^-	*hypo*chlor*ite* ion

A few polyatomic ions have names ending in *-ide*, such as hydroxide (OH^-) and peroxide (O_2^{2-}). Polyatomic cations like ammonium (NH_4^+) end in *-onium*. Note that the formulas of polyatomic ions are usually written by placing the atoms in order of increasing electronegativity.

Certain related polyatomic ions differ from each other only in their charge and number of hydrogens. Examples are the carbonate ion CO_3^{2-}, and the hydrogencarbonate ion, HCO_3^-, commonly called bicarbonate ion. The hydrogencarbonate ion has one less negative charge than the carbonate ion because it contains an additional positive charge, a hydrogen ion, H^+, as shown:

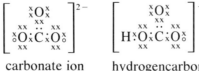

carbonate ion
CO_3^{2-}

hydrogencarbonate (bicarbonate) ion
HCO_3^-

Another such group of related polyatomic ions and their names are:

PO_4^{3-} phosphate ion
HPO_4^{2-} hydrogenphosphate ion
$H_2PO_4^-$ dihydrogenphosphate ion

Names and formulas of compounds containing polyatomic ions are written in the same way as the names and formulas of simple ionic compounds. We write the name of the cation, followed by the name of the anion. To prevent confusion when compounds contain more than one of the same polyatomic ion in the formula unit, we place the ion in parentheses before writing its subscript. Thus calcium hydroxide is written

$Ca(OH)_2 \leftarrow$ two hydroxide ions

to show that it contains one calcium ion and two hydroxide ions (read C–A–O–H taken twice). If instead we wrote $CaOH_2$, the subscript 2 would apply only to the nearest atom, hydrogen, and not to the oxygen. By placing the parentheses around the hydroxide ion, we make the subscript apply to the entire group, OH^-, and not just to part of it.

Sometimes there are subscripts within the parentheses. For example, the formula for aluminum nitrate is $Al(NO_3)_3$. The formula unit of this compound consists of one aluminum ion, Al^{3+} and three nitrate ions, NO_3^-.

The following are examples of ionic compounds, and their formulas and composition.

Name	Formula	Composition
ammonium carbonate	$(NH_4)_2CO_3$	2 ammonium ions 1 carbonate ion
potassium permanganate	$KMnO_4$	1 potassium ion 1 permanganate ion
aluminum sulfate	$Al_2(SO_4)_3$	2 aluminum ions 3 sulfate ions
magnesium acetate	$Mg(C_2H_3O_2)_2$	1 magnesium ion 2 acetate ions

Name	Formula	Composition
sodium peroxide	Na_2O_2	2 sodium ions 1 peroxide ion
calcium phosphate	$Ca_3(PO_4)_2$	3 calcium ions 2 phosphate ions
copper(II) nitrate	$Cu(NO_3)_2$	1 copper(II) ion 2 nitrate ions
colbalt(II) hydroxide	$Co(OH)_2$	1 cobalt(II) ion 2 hydroxide ions

**7.4
WRITING
FORMULAS
OF IONIC
COMPOUNDS**

An easy way to write formulas of ionic compounds is based on the charges of their ions. Because all compounds are neutral, one formula unit must contain equal numbers of positive and negative charges. For example, consider again the formation of a compound from sodium and chlorine. The cation from sodium is Na^+ and the anion from chlorine is Cl^-. To be neutral, the compound sodium chloride must contain one Na^+ for every Cl^-. Therefore the formula is NaCl.

We can determine the formula of the compound formed from magnesium and chlorine in a similar way. Magnesium is in Group 2A and can lose two electrons to form the ion Mg^{2+}. Chlorine forms the ion Cl^-. Two negative charges are needed to balance the two positive charges of each magnesium ion. Thus the formula unit must contain one Mg^{2+} ion and two Cl^- ions. So we write the formula of magnesium chloride as $MgCl_2$.

In general, the procedure for writing the formula of an ionic compound consists of the following three steps:

1 Write the symbols for the ions involved.

2 Multiply one or more of the ions by the lowest possible whole numbers to get the same number of positive and negative charges.

3 Write these numbers (in lowest terms) as subscripts in the formula.

Example

Write the formula of the compound formed from aluminum and iodine.

1 Al^{3+} (aluminum is in Group 3A) I^- (iodine is in Group 7A)

2 $1(Al^{3+})$, one aluminum ion $3(I^-)$, three iodide ions
The three positive charges balance the three negative charges.

3 AI_3, aluminum iodide

Example   Write the formula of the compound formed from calcium and nitrogen.

1 Ca^{2+} (calcium is in Group 2A) N^{3-} (nitrogen is in Group 5A)
2 $3(Ca^{2+})$, three calcium ions $2(N^{3-})$, two nitride ions
Note: You must multiply both ions to balance these charges; the lowest common multiple of 2 and 3 is 6. The six positive charges balance the six negative charges.
3 Ca_3N_2, calcium nitride

By balancing the positive and negative charges, we make sure that the number of electrons lost by one atom is equal to the number gained by the other atom. This principle is the basis for all processes involving oxidation and reduction, which are discussed further in Chapter Fifteen.

Example   Write the formula and name of the ionic compound formed from the following atoms:

a potassium and sulfur
1 K^+ S^{2-}
2 $2(K^+)$ $1(S^{2-})$
   two + charges   balance   two − charges
3 K_2S, potassium sulfide

b strontium and chlorine
1 Sr^{2+} Cl^-
2 $1(Sr^{2+})$ $2(Cl^-)$
   two + charges   balance   two − charges
3 $SrCl_2$, strontium chloride

c aluminum and sulfur
1 Al^{3+} S^{2-}
2 $2(Al^{3+})$ $3(S^{2-})$
   six + charges   balance   six − charges
3 Al_2S_3, aluminum sulfide

d lithium and nitrogen
1 Li^+ N^{3-}
2 $3(Li^+)$ $1(N^{3-})$
   three + charges   balance   three − charges
3 Li_3N, lithium nitride

Example   Write the formula and name of the ionic compound formed from the following ions:

a ammonium ion and bromide ion
1 NH_4^+ Br^-
2 $1(NH_4^+)$ 1 Br^-
   one + charge   balances   one − charge
3 NH_4Br, ammonium bromide

b magnesium ion and nitrate ion
1 Mg^{2+} NO_3^-
2 $1(Mg^{2+})$ $2(NO_3^-)$
  two + charges   balance   two − charges
3 $Mg(NO_3)_2$, magnesium nitrate

c molybdenum(VI) ion and fluoride ion
1 Mo^{6+} F^-
2 $1(Mo^{6+})$ $6F^-$
  six + charges   balance   six − charges
3 MoF_6, molybdenum(VI) fluoride

d colbalt(II) ion and phosphate ion
1 Co^{2+} PO_4^{3-}
2 $3(Co^{2+})$ $2(PO_4^{3-})$
  six + charges   balance   six − charges
3 $Co_3(PO_4)_2$, cobalt(II) phosphate

7.5 NAMING ACIDS

Acids are a special class of compounds which we will discuss in Chapter Fourteen. For now, we will simply define acids as substances whose molecules can release hydrogen ions, H^+. Upon release of the hydrogen ion, an anion remains. For example, HCl is a typical acid. It can lose its H^+ ion, leaving the anion Cl^-. Notice that in the formula of an acid, the H is always written first.

Acids are named on the basis of the resulting anion. For acids whose anions end in *-ide* (like chloride), the acid is named as a binary compound. Thus HCl is called hydrogen chloride. When binary acids are dissolved in water, another name is usually used. This name is formed by adding the prefix *hydro-* to the root of the anion, followed by the ending *-ic acid*. In the case of HCl, the name is hydrochloric acid.

hydro	*chlor*	*ic acid*
prefix	root	ending

As described in Section 7.2, the names of oxygen-containing anions generally end in *-ate* or *-ite*. For ions with names ending in *-ate*, like nitrate ion, the acid is named by adding *-ic acid* to the root.

HNO_3

nitr	*ic acid*
root	ending for *-ate* anion

If the anion ends in *-ite* like nitrite ion, the acid ending becomes *-ous acid*.

HNO_2

nitr	*ous acid*
root	ending for *-ite* anion

Additional examples of naming acids are given in Table 7-5.

TABLE 7-5 *Names of common acids*

I. *Acids whose anions end in -ide*

anion		acid	
chloride ion	Cl⁻	hydrogen chloride (hydrochloric acid)	HCl
cyanide ion	CN⁻	hydrogen cyanide (hydrocyanic acid)	HCN
fluoride ion	F⁻	hydrogen fluoride (hydrofluoric acid)	HF
sulfide ion	S²⁻	hydrogen sulfide (hydrosulfuric acid)	H₂S

II. *Acids whose anions end in -ate*

anion		acid	
acetate ion	$C_2H_3O_2{}^-$	acetic acid	$HC_2H_3O_2$
carbonate ion	$CO_3{}^{2-}$	carbonic acid	H_2CO_3
nitrate ion	$NO_3{}^-$	nitric acid	HNO_3
phosphate ion	$PO_4{}^{3-}$	phosphoric acid	H_3PO_4
sulfate ion	$SO_4{}^{2-}$	sulfuric acid	H_2SO_4

III. *Acids whose anions end in -ite*

anion		acid	
nitrite	$NO_2{}^-$	nitrous acid	HNO_2
sulfite	$SO_3{}^{2-}$	sulfurous acid	H_2SO_3

7.6
BRIEF SURVEY
OF CHEMICAL
COMPOUNDS

The compounds described in this section are presented here primarily to give you practice in reading their names and formulas. This section also provides a sampling of the applications of some common chemical compounds. Keep in mind that this large amount of information is not meant to be memorized.

Hydrogen is present in a tremendous number of compounds. Most organic compounds (compounds based on carbon) contain hydrogen atoms covalently bonded to carbon atoms. As mentioned in the previous section, the class of compounds called acids have special properties because they contain hydrogen. Compounds called hydrides consist of hydrogen bonded to less electronegative elements, particularly metals. An example of a hydride is NaH, sodium hydride.

The Group 1A metals serve as cations in large numbers of ionic compounds. Lithium carbonate, Li_2CO_3, has recently been used by psychiatrists to treat manic-depressives. Sodium chloride, table salt, NaCl, has over 14,000 separate uses (see Figure 7-1). Sodium hydrox-

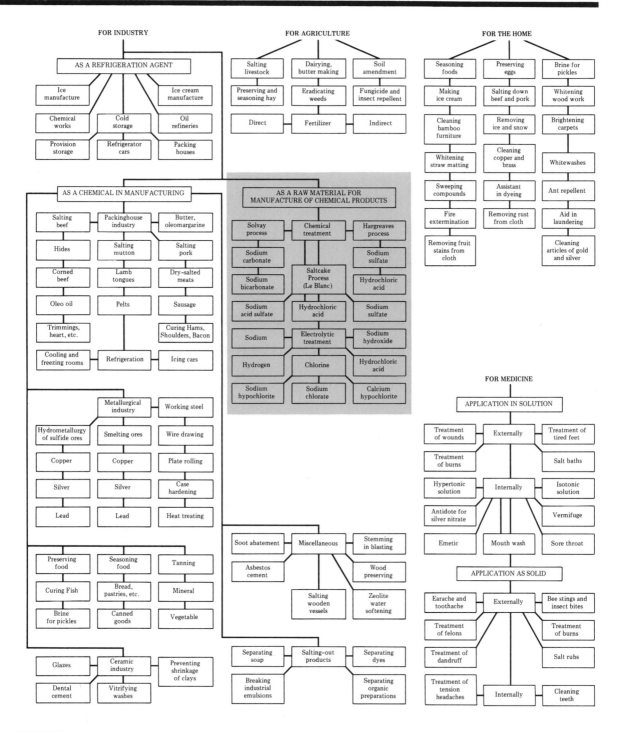

FIGURE 7-1

Uses of sodium chloride, NaCl. In addition to its use as table salt, sodium chloride has a tremendous range of applications.

ide, NaOH, also called lye and caustic soda, is a major industrial chemical. It is also present in solid drain-cleaners like Drano. Sodium carbonate, Na_2CO_3, called washing soda or soda ash, is used in manufacturing glass, soap, paper, and paints. Sodium hydrogencarbonate, $NaHCO_3$, commonly called sodium bicarbonate, is also known as baking soda. Baking power is baking soda with other ingredients added. When dough containing baking powder is heated, the sodium hydrogencarbonate releases gaseous carbon dioxide, CO_2, which causes the dough to rise.

Because plants need K^+ ions to grow, potassium oxide, K_2O, is used in fertilizers. Potassium chloride, KCl, is present in salt substitutes (see Figure 7-2). Salt substitutes are used by people with high blood pressure, who must restrict their intake of Na^+ ions. Potassium nitrate, KNO_3, known as saltpeter, is a component of gunpowder and is used to cure meats.

Box 7

Lithium in psychiatry

Lithium carbonate, Li_2CO_3, is a relatively simple compound. However, its behavior in the human body is complex and still somewhat of a mystery. Li_2CO_3 is considered the most effective drug for treating a type of mental illness called manic-depressive illness.

People who suffer from this illness tend to become highly overactive or deeply depressed for weeks or months at a time. Mania is characterized by excitement, elation, overactivity and talkativeness. On the other hand, a person feels downhearted during periods of depression, and everything seems to require tremendous effort. A very depressed person might even attempt suicide.

Other drugs previously used for treating manic depression are complex organic compounds. (Organic compounds are introduced in Chapter Eighteen.) Lithium carbonate is not only a much simpler compound than those drugs, but works better and has fewer side effects. It calms the patient during the manic stage, and appears to help control mood swings.

The way that this drug works in the body is not yet well understood. We know that signals are transmitted within the nervous system through a process that depends on a balance between sodium and potassium ions. Lithium ions, from lithium carbonate, may also effect the way that nerve cells communicate.

Lithium therapy helps 70% of those treated. It represents one of the highpoints in the use of drugs to treat mental illness. Over the past 10 years, lithium therapy has saved approximately $4 billion, based on the cost of lost work time and the cost of previous methods of treatment.

FIGURE 7-2

Salt substitute. This substitute contains potassium chloride (KCl) instead of sodium chloride (NaCl). It is used by people who must restrict their intake of sodium ion. (Photo courtesy of Diamond Crystal Salt Company.)

The Group 2A elements also serve as cations in a large number of ionic compounds. Magnesium oxide or magnesia, MgO, is used on tops of electric kitchen ranges because it conducts heat well, but conducts electricity poorly. This unusual property protects the users of electric stoves from the danger of electric shock. Magnesium hydroxide, $Mg(OH)_2$, suspended in water, is known as milk of magnesia and is an antacid. Magnesium sulfate, or epsom salts, $MgSO_4$, is used in tanning leather (converting hides into leather) and fixing the colors of dyes. It is also used as a laxative.

Calcium oxide, or lime, CaO, is the second most produced industrial chemical (after sulfuric acid, H_2SO_4). It is used to make other chemicals and to treat soil. At one time, lime was used to produce light on theatrical stages. The expression "in the limelight" refers to the use of lime in the theater, and has now come to mean "at the center of attention." Calcium hydroxide, or slaked lime, $Ca(OH)_2$, is present in mortar, which is used to hold together bricks or stones. Calcium carbonate, $CaCO_3$, is found as limestone, marble, and chalk. Its presence makes water "hard" (unable to form suds with soap). It is used to make cement. Calcium chloride, $CaCl_2$, is a drying agent, since it absorbs water from liquids and gases. It is spread on roads in the winter to prevent ice formation, because it lowers the temperature at which ice freezes. Calcium sulfate, $CaSO_4$, is found in plaster of paris (see Section

12.8). Barium sulfate, $BaSO_4$, is used as an opaque medium for gastrointestinal tract X rays, as shown in Figure 7-3.

Most compounds of the Group 3A elements contain boron or aluminum. Borax, or sodium borate, $Na_2B_4O_7$, is used to clean metal surfaces in soldering and welding, and to make glass and enamels. It is present as an additive in some laundry detergents. Boric oxide, B_2O_3, is used with silicon dioxide, SiO_2, in making heat resistant borosilicate glass like Pyrex and Kimax. Boron nitride, BN, known also as borazon, is one of the hardest substances known. It is used in cutting tools, and as an abrasive.

Aluminum forms "double salts" known as alums, which contain Al^{3+} and one of four other cations (Li^+, Na^+, K^+, or NH_4^+). The most common example of an alum is aluminum potassium sulfate, $AlK(SO_4)_2$, which is used to purify water. Aluminum oxide or alumina, Al_2O_3, is a raw material for making furnace linings, abrasives, artificial gems, cement, and ceramics. Aluminum hydroxide, $Al(OH)_3$, is an antacid. It is combined with magnesium hydroxide in such products as Di-Gel and Maalox.

Carbon, in Group 4A, forms 96% of all known compounds. Organic chemistry, which encompasses most of these compounds, is introduced

Do-it-yourself 7

Baking soda

Baking soda is the common name for sodium hydrogencarbonate (sodium bicarbonate), $NaHCO_3$. It is a leavening agent, which is a substance that causes a dough or batter to rise. Baking soda works by releasing the gas carbon dioxide, CO_2. The gas makes the dough or batter expand and become less dense.

An acid is generally needed to activate baking soda. In foods that are acidic, like those containing sour milk, baking soda can be used by itself as a leavening agent. In foods that are not acidic, baking *powder* must be used for leavening. Baking powder contains baking soda plus a solid acid such as cream of tartar. (Cream of tartar, which is potassium hydrogentartrate, $KHC_4H_4O_6$, is obtained from grapes after they have been used for making wine.)

To do: Set up three small glasses. Add one teaspoon of baking soda to each. Observe what happens when you do the following:

First glass—Add a small amount of water.
Second glass—Add a small amount of vinegar (acetic acid).
Third glass—Add two teaspoons of cream of tartar and then add a small amount of water.

Can you explain what happens?

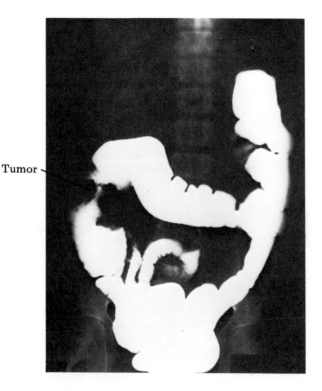

Tumor

FIGURE 7-3

X-ray photograph of colon. Barium sulfate ($BaSO_4$) permits X-ray examination of internal organs. Because X rays cannot penetrate the barium sulfate, the coated organs show up as light areas on the film. Here, the arrow points to a tumor that is blocking the colon. (Photo courtesy of American Cancer Society.)

in Chapter Eighteen. A few simple carbon compounds are mentioned here. Carbon monoxide, CO, is a poisonous gas used industrially in making other substances and in purifying metals. It is a major air pollutant and is present in car exhaust fumes and cigarette smoke. Carbon monoxide forms when carbon or its compounds do not burn completely to form carbon dioxide, CO_2. Carbon dioxide is the gas we exhale. It is a waste product of the breakdown of food in our bodies. Plants use CO_2 from the air as a starting material for their growth. CO_2 is also a major industrial gas and is used to make carbonated drinks. The solid form of CO_2, at $-79°C$, is called dry ice, and is shown in Figure 7-4. Dry ice evaporates and becomes a gas; it does not melt and leave a liquid puddle. (This process is called sublimation.) Hydrogen cyanide, HCN, is a very poisonous gas which has been used in the gas chambers of prisons. Carbides contain carbon bonded to a less electronegative element. Calcium carbide, CaC_2, was once a source of acetylene when mixed with water in miners' lamps. Silicon carbide, SiC, or carborundum, and boron carbide, B_4C, are used as abrasives because of their hardness.

Silicon dioxide, or silica, SiO_2, is found in many forms, such as quartz, and is the main component in sand. It provides a basic raw material for glass, ceramics, and enamels. (Ordinary glass consists of

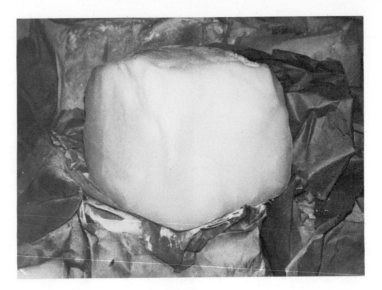

FIGURE 7-4

Dry ice. This solid is frozen carbon dioxide (CO_2). (Photo by Christine Osinski.)

FIGURE 7-5

Glass manufacturing. Silica (SiO_2) is the primary raw material used in making glass. (Photo courtesy of Corning Glass.)

70–74% SiO_2, 10–13% CaO, and 13–16% Na_2O.) Figure 7-5 illustrates part of the glass manufacturing process. Tin(II) fluoride, or stannous fluoride, SnF_2, is a toothpaste additive that serves as a source of fluoride ions. Lead(II) oxide, also called lead monoxide, PbO, is used as plates in electric storage batteries.

Because of their compounds, nitrogen and phosphorous are the most prominent elements in Group 5A. Ammonia, NH_3, is the third most important industrial chemical (after sulfuric acid and lime). It is used in making other nitrogen compounds and synthetic fibers like nylon, in dyeing textiles, and in producing fertilizer. (See Figure 7-6.) It also serves as a refrigerant. "Laughing gas," or nitrous oxide, N_2O, is an anesthetic and is used as an aerosol propellant in whipped cream. Nitrogen monoxide (nitric oxide), NO, present in automobile exhausts, is converted in the air to NO_2, nitrogen dioxide. This brown gas forms organic nitrates (from unburned gasoline fumes) that irritate the lungs and burn the eyes. Nitric acid, HNO_3, is used to make explosives like TNT, fertilizers, and other nitrogen compounds. Phosphoric acid, H_3PO_4, is used in manufacturing sugar, soft drinks, gelatin, drugs, and cements.

Of the Group 6A elements, oxygen and sulfur are the most important because they form compounds with nearly all other elements. General

FIGURE 7-6

Fertilizer. Nitrogen in the form of ammonia (NH_3) or its derivatives is an important fertilizer. (Photo courtesy of USDA-SCS; photo by M.G. Hassler.)

formulas for the oxides of main group elements are given in the following table, where E stands for a main group element:

Group number	1A	2A	3A	4A	5A	6A	7A
Formula of oxide	E_2O	EO	E_2O_3	EO_2	E_2O_5	EO_3	E_2O_7
Example	Na_2O sodium oxide	MgO magnesium oxide	Al_2O_3 aluminum oxide	CO_2 carbon dioxide	N_2O_5 dinitrogen pentoxide	SO_3 sulfur trioxide	Cl_2O_7 dichlorine heptaoxide

The greatest amount of combined oxygen exists as silicon compounds (such as sand, silicon dioxide) in the Earth's crust, and as water, H_2O, in the oceans. In addition to oxides, oxygen also forms peroxides. Hydrogen peroxide, H_2O_2, is a bleach and a disinfectant.

Sulfur dioxide, SO_2, is an irritating air pollutant released by burning coal and oil that contain sulfur. This gas is used industrially for refrigeration and to preserve foods. Hydrogen sulfide, H_2S, is a poisonous gas that has an odor like that of rotten eggs. Sulfuric acid, sometimes called oil of vitriol, H_2SO_4, is the workhorse of the chemical industry because over 70 billion pounds per year are used in manufacturing fertilizer, detergents, plastics, fibers, paints and other materials. Sodium thiosulfate, $Na_2S_2O_3$, known as fixer or hypo, is used in photography to prevent film from remaining light-sensitive after it has been exposed and developed (see Figure 7-7).

FIGURE 7-7

Fixer or hypo. Sodium thiosulfate ($Na_2S_2O_3$) is used by photographers to make exposed film insensitive to further light by removing unreacted silver. (Courtesy of Eastman Kodak Company. Photo by Christine Osinski.)

Noble gas compounds did not exist until 1962. In that year, a Canadian chemist, Neil Bartlett, prepared $XePtF_6$.

The halogens, Group 7A, have already been mentioned, since they serve as anions in many of the ionic compounds discussed previously. (The halogen anions are known as halides.) A few other examples of halogen compounds are given here. Fluorocarbons like Freon-12 (dichlorodifluoromethane), CF_2Cl_2, are used as refrigerants and aerosol propellants. They may have harmful effects on the protective ozone (O_3) layer in our atmosphere. Hydrogen fluoride, HF, is used for etching glass. For example, it is used to frost the insides of light bulbs. Fluoride ion is often added to water supplies in the form of sodium fluoride, NaF, and other fluoride compounds, to inhibit tooth decay.

Hydrogen chloride, HCl, is used in manufacturing glue, soap, and dyes. In water, HCl is hydrochloric (or muriatic) acid. The most common use of HCl is in dissolving the rock formations in which petroleum is found. Sodium hypochlorite, NaClO, is the active ingredient in household bleach. Sodium iodide, NaI, is present in "iodized" salt. Adding iodized salt to your food helps to prevent iodine deficiencies.

Until relatively recently, no noble gas compounds had ever been made because of the stability of these elements. However, we now know that noble gases can be made to combine with other elements. Xenon is the noble gas that forms the largest number of known compounds. Examples of xenon compounds are xenon difluoride, XeF_2, xenon tetrafluoride, XeF_4, xenon hexafluoride, XeF_6, and xenon trioxide, XeO_3. Krypton compounds have also been made.

Summary

7.1 Covalent compounds that contain two elements are called binary compounds. When writing either the name or the formula of a binary compound, place the two elements in order of increasing electronegativity. Prefixes are used to indicate the number of atoms, and the ending is *-ide*.

7.2 For ionic compounds, the name or formula of the cation is written first, followed by the name or formula of the anion. For transition metal compounds, the oxidation number of the metal should be indicated.

7.3 The names of polyatomic ions generally end in *-ate*. Names and formulas of compounds containing polyatomic ions are written in the same way as the names and formulas of simple ionic compounds.

7.4 Formulas of ionic compounds can be written based on the charges of the ions present. The formula unit must contain equal numbers of positive and negative charges.

7.5 Acids are named for the anions that remain after the release of hydrogen ions by the molecules. If the name of the anion ends in *-ide*, the acid is named either as a binary compound, or in the form: *hydro* (root of anion) *ic acid*. Oxygen-containing acids end either in *-ic acid* or *-ous acid*.

7.6 A number of representative compounds and their practical applications are described. The industrial chemicals produced in the largest amounts are: sulfuric acid (H_2SO_4), calcium oxide or lime (CaO), and ammonia (NH_3).

Exercises

EXERCISES BY TOPIC **Covalent compounds (7.1)**

1 Write the name and formula of the compound formed from: **a** one P atom and five Cl atoms   **b** one Si atom and four I atoms   **c** two I atoms and five O atoms

2 Name the following compounds: **a** H_2S **b** SO_2 **c** BF_3 **d** XeO_3 **e** P_3N_5

3 Write the formula of: **a** phosphorus tribromide   **b** diarsenic trioxide   **c** diboron hexachloride   **d** silicon dioxide   **e** carbon monoxide

4 What is the name of each of the following compounds?   **a** P_2S_5 **b** S_2Cl_2 **c** BP   **d** CS_2 **e** NI_3

Ionic compounds (7.2)

5 Write the name and formula for the compound formed from: **a** one Al^{3+} ion and three Cl^- ions   **b** one Ca^{2+} ion and two F^- ions   **c** three Mg^{2+} ions and two N^{3-} ions   **d** two Li^+ ions and one S^{2-} ion

6 Name the following compounds: **a** CaO   **b** KI   **c** AlF_3 **d** BaS   **e** Li_3N

7 Write the name and formula for the compound formed from: **a** one Pb(II) ion and two Cl^- ions   **b** two Cu(I) ions and one O^{2-} ion   **c** one Mo(VI) ion and six Br^- ions   **d** one Zn(II) ion and two F^- ions

8 Name the following: **a** FeI_2 **b** AgBr   **c** $HgCl_2$ **d** CoF_3 **e** PbO

Compounds with polyatomic ions (7.3)

9 Write the name and formula of the compound formed from: **a** one Ba^{2+} ion and one CO_3^{2-} ion   **b** two Cs^+ ions and one SO_4^{2-} ion   **c** three Li^+ ions and one PO_4^{3-} ion   **d** one Hg(II) ion and two CN^- ions

10 Name the following compounds: **a** Ag_2CO_3 **b** $(NH_4)_2SO_4$ **c** $Zn(NO_3)_2$ **d** $AlPO_4$ **e** $Fe(HCO_3)_3$

11 For each compound in Exercise 10, identify the number of ions of each type that are present.

12 Name the following compounds and describe their composition:
a $NaHSO_4$ **b** $Cu_3(PO_4)_2$ **c** $LiC_2H_3O_2$ **d** NH_4NO_2 **e** $Pt(CN)_4$

Writing formulas of ionic compounds (7.4)

13 Write the formula and name of the compound formed from: **a** sodium ion and oxalate ion ($C_2O_4^{2-}$) **b** magnesium ion and iodide ion **c** Al^{3+} ion and Cl^- ion

14 Write the formula and name of the compound formed from: **a** Fe(III) ion and Br^- ion **b** Cr(VI) ion and F^- ion **c** mercuric ion and oxide ion

15 Write the formula and name of the compound formed from: **a** lithium ion and acetate ion **b** Ca^{2+} ion and $H_2PO_4^-$ ion **c** ammonium ion and sulfate ion

16 Write the formula and name of the compound formed from: **a** Ti(IV) ion and OH^- ion **b** Ag(I) ion and CO_3^{2-} ion **c** Zn(II) ion and PO_4^{3-} ion

Naming acids (7.5)

17 Name the following acids: **a** $HC_2H_3O_2$ **b** HNO_2 **c** $HClO_4$

18 Write the formula of: **a** hydrosulfuric acid **b** phosphoric acid **c** hypochlorous acid

19 Write the formula and name of the acid that contains the oxalate ion, $C_2O_4^{2-}$.

20 Write the name and formula of the acid whose anion is borate, BO_3^{3-}.

21 Name the following: **a** HIO_4 **b** H_2Se **c** HBr

Brief survey of chemical compounds (7.6)

22 What are hydrides?

23 What compounds are called by these names: **a** washing soda **b** lye **c** saltpeter **d** borax **e** lime **f** hypo

24 Give an example of a double salt (other than the one given in the textbook).

25 From what is limestone made?

26 Name several important industrial compounds and their applications.

ADDITIONAL EXERCISES

27 Ammonium carbonate is used in some smelling salts. Write the formula of this compound.

28 The compound ClO_2 is used for bleaching. What is its name?

29 Write the formula of iron(III) sulfate, a compound used as a pigment (it colors other materials).

30 Name the compound with the formula H_3PO_2.

31 Tetraphosphorus hexasulfide is used in organic chemistry to make other compounds. Write its formula.

32 Write the formula of zinc(II) oxide, a substance found in calamine lotion.

33 In each box, write the formula of the compound formed by the corresponding ions:

	chloride ion	phosphate ion	carbonate ion	acetate ion
aluminum ion				
calcium ion				
ammonium ion				
vanadium(IV) ion				

34 Name the following compounds: **a** $KClO_3$ **b** P_2S_5 **c** XeO_3 **d** $Cu(CN)_2$ **e** SF_6 **f** $KHSO_4$ **g** BPO_4 **h** $Ca(C_2H_3O_2)_2$ **i** $AgNO_2$ **j** UCl_4

35 Write the formula of: **a** hydrofluoric acid **b** oxygen difluoride **c** copper phosphide **d** zinc(II) cyanide **e** hypochlorous acid **f** silver(I) nitride **g** potassium hydrogencarbonate **h** sulfurous acid **i** calcium hydroxide **j** bromine pentafluoride

36 Name the following compounds: **a** Na_3N **b** HF **c** Li_2O **d** I_2O_5 **e** BCl_3 **f** $Ca(NO_2)_2$

37 Write the formula for: **a** iodine monochloride **b** sodium periodate **c** ammonium nitride **d** mercurous sulfate **e** bismuth(III) fluoride

38 The active ingredient in liquid bleach is $NaClO$. What is the name of this compound?

39 Write the name of: **a** $Ca(C_2H_3O_2)_2$ **b** $(NH_4)_2SO_3$ **c** $CsNO_3$ **d** $Al(HCO_3)_3$ **e** $NaMnO_4$

40 As shown in this table, certain compounds have common names that differ from their chemical names. In each case, write the chemical name:

	Common name	Formula
a	galena	PbS
b	potash	K_2CO_3
c	slaked lime	$Ca(OH)_2$
d	sal ammoniac	NH_4Cl
e	laughing gas	N_2O

41 Write formulas for the following compounds: **a** ruthenium(III) chloride **b** platinum(II) nitrate **c** molybdenum(IV) acetate **d** chromium(III) carbonate **e** manganese(V) sulfate

42 The head of a safety match contains $KClO_3$ and Sb_2S_3. Write the names of these two compounds.

43 Complete the following table with the appropriate formulas in each row:

	S^{2-}	CN^-	PO_4^{3-}	SO_4^{2-}	NO_3^-
a Ba^{2+}					
b Fe^{3+}					
c NH_4^+					
d Pb^{4+}					
e Co^{6+}					

44 Name the compounds in Exercise **43a**.

45 Name the compounds in Exercise **43b**.

46 Name the compounds in Exercise **43c**.

47 Name the compounds in Exercise **43d**.

48 Name the compounds in Exercise **43e**.

49 Solid toilet-bowl cleaners contain the compound sodium hydrogensulfate. Write the formula of this compound.

50 Aqua regia, which means "royal water," is the name given to a mixture of HCl and HNO_3 that dissolves gold. What are the names of the two acids in aqua regia?

51 Lead(II) oxide, tetraphosphorus trisulfide, and sulfur are found on the heads of strike-anywhere matches. Write the formulas of these three substances.

PROGRESS CHART   Check off each item when completed.

———— Previewed chapter

Studied each section:

———— **7.1** Covalent compounds

———— **7.2** Ionic compounds

———— **7.3** Compounds with polyatomic ions

———— **7.4** Writing formulas of ionic compounds

———— **7.5** Naming acids

———— **7.6** Brief survey of chemical compounds

———— Reviewed chapter

———— Answered assigned exercises:

#'s ————————————————

Chemical formulas and the mole

Chemical formulas are the words in the language of chemistry. (The alphabet consists of symbols for the elements.) In the previous chapter we concentrated on the basic rules for spelling the words—how to write formulas. In this chapter, we introduce calculations that use formulas or that determine formulas of compounds.

Chapter Eight is an introduction to chemical arithmetic. The calculations in this chapter provide the foundation for concepts developed later in the book. To demonstrate an understanding of Chapter Eight, you should be able to:

1 Find the molecular or formula weight of a compound.

2 Explain the unit mole in terms of mass and the number of particles.

3 Convert a given amount of substance in moles to its mass in grams.

4 Convert the mass of a substance in grams to the amount in moles.

5 Perform conversions using Avogadro's number.

6 Calculate the percentage composition of a compound.

7 Find the chemical formula of a compound, using its percentage composition.

8.1
FORMULA OR MOLECULAR WEIGHT OF COMPOUNDS

The smallest part of a compound that still has the properties of that compound is the molecule (for covalent compounds) or formula unit (for ionic compounds). It is very useful to be able to calculate the mass of these smallest units of compounds. The mass of a molecule is called the **molecular weight**. The mass of a formula unit is called the **formula weight**. Just as in the term atomic weight, we use the word *weight* when we really mean *mass*. (The more precise term *relative molecular mass*, which is defined as the average mass per formula of a compound, is rarely used.) Even though all compounds do not exist as molecules,

TABLE 8-1 *Composition of compounds based on formulas*

Formula	Composition
H_2O	2 hydrogen atoms, 1 oxygen atom
$C_6H_{12}O_6$	6 carbon atoms, 12 hydrogen atoms, 6 oxygen atoms
Na_3PO_4	3 sodium atoms, 1 phosphorus atom, 4 oxygen atoms
$Al_2(SO_4)_3$	2 aluminum atoms, 3 sulfur atoms, 12 oxygen atoms
$(NH_4)_2CO_3$	2 nitrogen atoms, 8 hydrogen atoms, 1 carbon atom, 3 oxygen atoms

the term molecular weight is sometimes used for both molecules and formula units.

When measuring something, it is often helpful to know the mass of the smallest unit. For example, if someone asked you for 1000 marbles, you could sit down and count them out one by one. But it would be much simpler to calculate the mass of 1000 marbles and then pour marbles into a container on a balance until the balance indicates that mass. And in order to calculate the mass of 1000 marbles, you would first need to know the mass of a single marble, the smallest unit. The same principle holds true when we measure the mass of chemical compounds.

The molecular weight of a compound is the *sum of the atomic weights of all the atoms present in the molecule*. Similarly, the formula weight of a compound is the sum of the atomic weights of all the ions present in the formula unit. The ability to read a chemical formula correctly is the most important part of finding these weights. Table 8-1 illustrates how to interpret the formula of a compound in terms of atoms. As you know from preceding chapters, the subscript at the lower right side of the symbol for each element gives the number of that type of atom present.

Remember that when a subscript is written next to a group of atoms in parentheses, all atoms inside the parentheses are multiplied by that subscript. Thus, in aluminum sulfate, $Al_2(SO_4)_3$, there are 3×1 sulfur, or 3 sulfur atoms, and 3×4 oxygens, or 12 oxygen atoms present, in addition to the 2 aluminum atoms. When finding formula weights, we do not have to distinguish between atoms and ions because the mass of an electron (the difference between an atom and an ion) is so small.

Example Describe the composition of a urea molecule, $CO(NH_2)_2$. (Urea is a waste product of the body, excreted in urine.)

$CO(NH_2)_2$ contains:

1 carbon atom	2 nitrogen atoms
1 oxygen atom	4 hydrogen atoms

Once we know how to read a chemical formula, it is easy to find the molecular or formula weight. We simply add up the atomic weights of each of the atoms present in one unit of the compound. These values can be taken from either the periodic table or a table of atomic weights.

Atomic weights generally should be rounded-off before calculating a molecular or formula weight. For example, potassium has an atomic weight of 39.0983 amu. (The amu, atomic mass unit, was described in Section 4.3.) This value may be rounded-off to 39.1 amu, using the method described in Section 2.9. Working with atomic weights written to the nearest tenth is generally sufficient for calculations in this book. Other examples of rounding-off atomic weights are listed in Table 8-2.

Except for the first four elements, rounding-off atomic weights in this way provides at least three significant figures, enough for most basic calculations. For the lightest elements, a second decimal place may be used. For example, the atomic weight of hydrogen, 1.0079 amu, can be rounded-off to 1.01 amu.

Using rounded-off atomic weights, we can find the molecular weight of water, H_2O, by simply adding the atomic weights of 2 hydrogen atoms and 1 oxygen atom:

$$2 \text{ atoms H} \times \frac{1.01 \text{ amu}}{1 \text{ atom H}} = 2.02 \text{ amu}$$

$$1 \text{ atom O} \times \frac{16.0 \text{ amu}}{1 \text{ atom O}} = 16.0 \text{ amu}$$

molecular weight of H_2O = 18.0 amu

Remember that when adding numbers, the answer cannot have more decimal places than the number with the *least* decimal places, which is 16.0 in this example. The same procedure can be followed for a more

TABLE 8-2 *Rounding-off atomic weights*

Element	Atomic weight from table	Rounded-off atomic weight[a]
calcium	40.08	40.1
carbon	12.011	12.0
chlorine	35.453	35.5
fluorine	18.99840	19.0
nitrogen	14.0067	14.0
sodium	22.98977	23.0
sulfur	32.06	32.1

[a]Rounded-off to one decimal place.

complicated compound, aluminum acetate, $Al(C_2H_3O_2)_3$. (This compound is found in Burrow's solution, an astringent that contracts tissue.)

$Al(C_2H_3O_2)_3$	Number of atoms		Atomic weight (amu)	Total mass (amu)
1 aluminum atom	1 atom Al	$\times$	$\dfrac{27.0 \text{ amu}}{1 \text{ atom Al}}$	$= 27.0$ amu
6 carbon atoms	6 atoms C	$\times$	$\dfrac{12.0 \text{ amu}}{1 \text{ atom C}}$	$= 72.0$ amu
9 hydrogen atoms	9 atoms H	$\times$	$\dfrac{1.01 \text{ amu}}{1 \text{ atom H}}$	$= 9.09$ amu
6 oxygen atoms	6 atoms O	$\times$	$\dfrac{16.0 \text{ amu}}{1 \text{ atom O}}$	$= \underline{96.0}$ amu

formula weight of $Al(C_2H_3O_2)_3 = 204.1$ amu

Example Find the molecular or formula weight of the following compounds:

a C_5H_5N, pyridine (used in making vitamins and drugs)

$$5 \text{ atoms C} \times \frac{12.0 \text{ amu}}{1 \text{ atom C}} = 60.0 \text{ amu}$$

$$5 \text{ atoms H} \times \frac{1.01 \text{ amu}}{1 \text{ atom H}} = 5.05 \text{ amu}$$

$$1 \text{ atom N} \times \frac{14.0 \text{ amu}}{1 \text{ atom N}} = \underline{14.0 \text{ amu}}$$

molecular weight of $C_5H_5N = 79.1$ amu

b K_2SO_4, potassium sulfate (used in fertilizers)

$$2 \text{ atoms K} \times \frac{39.1 \text{ amu}}{1 \text{ atom K}} = 78.2 \text{ amu}$$

$$1 \text{ atom S} \times \frac{32.0 \text{ amu}}{1 \text{ atom S}} = 32.0 \text{ amu}$$

$$4 \text{ atoms O} \times \frac{16.0 \text{ amu}}{1 \text{ atom O}} = \underline{64.0 \text{ amu}}$$

formula weight of $K_2SO_4 = 174.2$ amu

c $Mg_3(PO_4)_2$, magnesium phosphate (used as an antacid and dentifrice polishing agent)

$$3 \text{ atoms Mg} \times \frac{24.3 \text{ amu}}{1 \text{ atom Mg}} = 72.9 \text{ amu}$$

$$2 \text{ atoms P} \times \frac{31.0 \text{ amu}}{1 \text{ atom P}} = 62.0 \text{ amu}$$

$$8 \text{ atoms O} \times \frac{16.0 \text{ amu}}{1 \text{ atom O}} = \underline{128 \text{ amu}}$$

formula weight of $Mg_3(PO_4)_2$ = 263 amu

d NH_4NO_3, ammonium nitrate (used in fertilizer and explosives)

$$2 \text{ atoms N} \times \frac{14.0 \text{ amu}}{1 \text{ atom N}} = 28.0 \text{ amu}$$

$$4 \text{ atoms H} \times \frac{1.01 \text{ amu}}{1 \text{ atom H}} = 4.04 \text{ amu}$$

$$3 \text{ atoms O} \times \frac{16.0 \text{ amu}}{1 \text{ atom O}} = \underline{48.0 \text{ amu}}$$

formula weight of NH_4NO_3 = 80.0 amu

e $Pb(C_2H_3O_2)_2$, lead(II) acetate (used for dyeing textiles)

$$1 \text{ atom Pb} \times \frac{207.2 \text{ amu}}{1 \text{ atom Pb}} = 207.2 \text{ amu}$$

$$4 \text{ atoms C} \times \frac{12.0 \text{ amu}}{1 \text{ atom C}} = 48.0 \text{ amu}$$

$$6 \text{ atoms H} \times \frac{1.01 \text{ amu}}{1 \text{ atom H}} = 6.06 \text{ amu}$$

$$4 \text{ atoms O} \times \frac{16.0 \text{ amu}}{1 \text{ atom O}} = \underline{64.0 \text{ amu}}$$

formula weight of $Pb(C_2H_3O_2)_2$ = 325.3 amu

Practice problems   Find the atomic, molecular, or formula weights of the following substances.

a $K_4P_2O_7$ **d** $Na_3Fe(C_2O_4)_3$

b U   **e** $B_{10}H_9BrC_2H_2$

c C_6H_6O

ANSWERS   **a** 330.4 amu   **b** 238.0 amu   **c** 94.1 amu   **d** 388.8 amu
e 223.0 amu

8.2
THE MOLE

The mass of a single molecule or formula unit cannot be measured on a balance, because the mass of each atom is so small. The smallest mass an ordinary laboratory balance can measure is about a milligram. The heaviest atom weighs only about 10^{-19} mg. To weigh atoms or groups of atoms, we must therefore take an extremely large number of them and measure them together. It is as if we were weighing feathers; they are so light that very many would be needed before an ordinary balance would begin to register. Chemists therefore work with large collections of atoms or molecules.

The **mole** is a special unit that describes such a big collection. A mole of any substance contains *a certain number of particles* which is always the same, just as a dozen always means 12 things. A mole of any substance contains the following number of particles of that substance: 602,000,000,000,000,000,000,000 or 6.02×10^{23} particles. This huge number of particles is called **Avogadro's number,** after Amadeo Avogadro, an Italian chemist. To give yourself an idea of how large this number is, imagine that you could stack a mole of the thinnest sheets of paper on top of each other. The pile of papers would reach from the Earth to a nearby star! (Also see Figure 8-1.)

One mole of any substance always contains Avogadro's number of units of that substance, but the mass or weight of one mole varies,

FIGURE 8-1

A mole of feathers. This drawing represents the tremendous number of particles in a mole. (Based on an original drawing by Wendy Lindboe.)

Photo courtesy of the Institute of Physics, Niels Bohr Library

Although he was a physics professor at the University of Turin in Italy, Avogadro's greatest contributions to science were in the field of chemistry. He was the first person to distinguish clearly between an atom and a molecule. Avogadro based this distinction between atoms and molecules on the work of the French chemist Joseph Louis Gay-Lussac, who showed that every gas expands by the same amount for a given increase in temperature. Gay-Lussac also found that when gases combine, they do so in simple whole-number ratios (at constant temperature and pressure). For example, one volume of hydrogen gas combines with an equal volume of chlorine gas to form two volumes of hydrogen chloride gas.

Avogadro suggested that equal volumes of gases (under identical conditions of temperature and pressure) must therefore contain equal numbers of particles. This statement is now known as Avogadro's hypothesis. Furthermore, the particles are not necessarily single atoms. They can be atoms joined together into groups, which Avogadro called molecules.

For example, in the hydrogen chloride synthesis mentioned above, the equal volumes of hydrogen and chlorine gas must contain the same numbers of molecules. Furthermore, these molecules must be diatomic (H_2 and Cl_2) in order to produce twice the volume, and therefore twice the number of molecules, of hydrogen chloride (2HCl).

Unfortunately, Avogadro's ideas were ignored for nearly 50 years. They did not begin to catch on until his countryman Stanislao Cannizzaro presented them at the First International Chemical Congress at Karlsruhe, Germany, in 1860.

Today, the number of particles in a mole, 6.02×10^{23}, is called Avogadro's number in his honor.

*If you stacked
Avogadro's number of
sheets of paper on top
of each other, they
would reach from the
Earth to a nearby star.*

depending on the mass of the individual unit. Think of a mole of feathers and a mole of marbles. The number of each is the same (Avogadro's number), but the mole of marbles weighs much more because each marble is heavier than each feather. In the same way, a mole of oxygen atoms and a mole of hydrogen atoms each contains Avogadro's number of atoms, but the mole of oxygen atoms weighs 16 times as much as the mole of hydrogen atoms. The reason is, of course, that each atom of oxygen (16.0 amu) weighs 16 times as much as each hydrogen atom (1.0 amu) because of the greater number of protons and neutrons in the oxygen nucleus.

Avogadro's number is the basis for the mole because it makes calculating the mass or weight of one mole very easy. Consider the hydrogen atom, whose mass is 1.01 amu. Since the mass corresponding to one amu is 1.66×10^{-24} g, one mole of hydrogen atoms must have the following mass:

$$\begin{matrix} \text{mass of} \\ \text{one mole} \\ \text{of H atoms} \end{matrix} = \frac{6.02 \times 10^{23}\ \cancel{\text{H atoms}}}{1\ \text{mole H atoms}} \times \frac{1.01\ \cancel{\text{amu}}}{1\ \cancel{\text{H atom}}} \times \frac{1.66 \times 10^{-24}\ \text{g}}{1\ \cancel{\text{amu}}}$$

$$= 1.01\ \text{g/mole of H atoms}$$

Because one hydrogen atom has a mass of 1.01 amu, one mole of hydrogen atoms (containing 6.02×10^{23} H atoms) has a mass of 1.01 g. This relationship is true in general; Avogadro's number of atoms of an element has a mass in grams that is exactly equal to the atomic weight of the element (in amu). For example, the atomic weight of oxygen is 16.0 amu. One mole of oxygen atoms therefore has a mass of 16.0 g (see Figure 8-2).

Individual atoms or molecules, whose mass is given in amu, cannot be measured in the laboratory. Because a mole of a substance contains so many atoms or molecules, however, its mass in grams is easy to measure in the laboratory. Figure 8-3 shows one mole of carbon (C)

FIGURE 8-2

*A mole of H atoms and a
mole of O atoms. Both contain
the same number of atoms,
but the mole of O atoms has a
mass 16 times greater than the
mole of H atoms.*

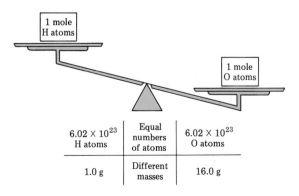

217

Box 8

Measuring Avogadro's number

Avogadro's number can be determined experimentally in several ways. One method involves using the compound oleic acid, which can spread out on a water surface to form a film that is only one molecule thick. From the known volume of oleic acid used, and the measured area of the film, we can determine Avogadro's number.

The first step is to find the *volume of oleic acid* present in one drop of a solution of oleic acid in alcohol and water. We do this by counting the number of drops needed to make one milliliter in a graduated cylinder. Then we correct for the dilution of the oleic acid by the alcohol and water in which it is dissolved.

Next we fill a tray with water and lightly dust the water surface with a powder that will help make the film of oleic acid visible. We then add one drop of the oleic acid solution to the center of the powder on the water. The oleic acid pushes back the powder as it flattens out and forms a circular film.

We measure the *diameter of the film* in several places and calculate an average value. We then use the average diameter to find the area of the film.

$$\text{area (cm}^2) = 3.14 \times \left(\frac{\text{average diameter (cm)}}{2}\right)^2$$

Next, we find the *height or thickness of the film* from the film's volume and area. The volumes of the film and the drop are equal.

$$\text{height of film (cm)} = \frac{\text{volume (cm}^3)}{\text{area (cm}^2)}$$

Assuming that an oleic acid molecule has the shape approximately of a cube, we can find the *volume of one molecule* from the film's height.

$$\text{volume of one molecule (cm}^3) = [\text{height (cm)}]^3$$

Finally, we can calculate Avogadro's number from the ratio of the number of grams in one mole (molar mass) to the mass of one molecule (density × volume):

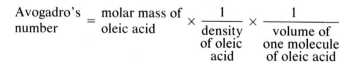

$$\begin{matrix}\text{Avogadro's} \\ \text{number}\end{matrix} = \begin{matrix}\text{molar mass of} \\ \text{oleic acid}\end{matrix} \times \frac{1}{\begin{matrix}\text{density} \\ \text{of oleic} \\ \text{acid}\end{matrix}} \times \frac{1}{\begin{matrix}\text{volume of} \\ \text{one molecule} \\ \text{of oleic acid}\end{matrix}}$$

Since the molar mass (282.5 g) and the density (0.895 g/mL) of oleic acid are already known, the only missing information in this equation is the volume of one molecule, which we have just calculated. The answer, Avogadro's number, is given in terms of the number of oleic acid molecules per mole of oleic acid.

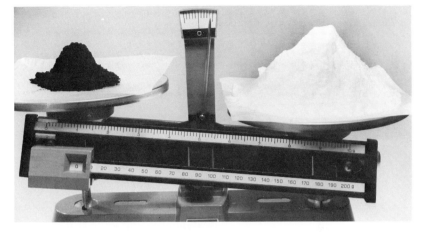

FIGURE 8-3

A mole of carbon atoms and a mole of glucose molecules. One mole of carbon atoms (C), on the left, has a mass of 12 g. One mole of glucose molecules ($C_6H_{12}O_6$), on the right, has a mass of 180 g. The number of carbon atoms, however, is exactly the same as the number of glucose molecules: 6.02×10^{23}. (Photo by Al Green.)

and one mole of glucose ($C_6H_{12}O_6$). Each contains the same huge number of particles, Avogadro's number. The glucose has a much greater mass because each glucose particle (glucose molecule) has a larger mass than each carbon particle (carbon atom).

The mass of one mole of atoms is the atomic weight written in grams. Similarly, the mass of one mole of molecules is the molecular weight written in grams. And the mass of one mole of formula units is the formula weight written in grams. (These amounts used to be called the gram-atomic weight, gram-molecular weight, and gram-formula weight, respectively. Now the general term mole replaces these more limited expressions.)

Table 8-3 compares the mole for different substances. The last

TABLE 8-3 *Mass of 1 mole of various substances*

Atom	Atomic weight	Mass of 1 mole
C	12.0 amu	12.0 g
Fe	55.8 amu	55.8 g
Hg	200.6 amu	200.6 g

Molecule	Molecular weight	Mass of 1 mole
NH_3	17.0 amu	17.0 g
CO_2	44.0 amu	44.0 g
$C_6H_{12}O_6$	180.1 amu	180.1 g

Formula unit	Formula weight	Mass of 1 mole
NaCl	58.5 amu	58.5 g
$KMnO_4$	158.0 amu	158.0 g
$AgNO_3$	169.9 amu	169.9 g

column gives the mass in grams of one mole of each substance. This quantity is called the **molar mass** or molar weight of that substance. It can be defined as:

$$\text{molar mass} = \frac{\text{mass (g) of 1 mole of substance}}{\text{one mole}}$$

Thus, the molar mass of carbon is 12.0 g/mole, and the molar mass of sodium chloride is 58.5 g/mole. (A pinch of salt, which has a mass smaller than 1 g, contains about 10^{-3} mole, that is, 6×10^{20} NaCl formula units.) The molar mass is simply the atomic, molecular, or formula weight, in grams.

In summary, *the mole is a measure of the amount of a substance.*

1 A mole contains Avogadro's number (6.02×10^{23}) of atoms, formula units, or molecules, and

2 has a mass in grams equal to the atomic, formula, or molecular weight of the substance.

Small quantities of a substance can be described in terms of the mole if we add appropriate metric prefixes. Millimole for example, means an amount that is 1/1000th of a mole (10^{-3} mole).

8.3
CONVERTING
MOLES TO GRAMS

The mole is very useful in calculations involving quantities of chemical substances. But to carry out laboratory procedures, we must often first convert moles to grams. Chemists seldom use exactly one mole of a substance. Instead, they must be able to measure some multiple or fraction of a mole and convert this quantity into grams. For example, since one mole of sodium chloride weighs 58.5 g, two moles weigh 2.00 moles $\times$ 58.5 g/mole = 117 g, three moles weigh 3.00 moles $\times$ 58.5 g/mole = 176 g, and so on. Similarly, one-half mole weighs 0.500 mole $\times$ 58.5 g/mole = 29.3 g, one-quarter mole weighs 0.250 mole $\times$ 58.5 g/mole = 14.6 g, and so on. In each case, to find the number of grams, we simply multiply the number of moles by the molar mass, which is the mass of 1 mole:

$$\frac{\text{number}}{\text{of grams}} = \text{number of moles} \times \frac{\text{mass (g) of 1 mole of substance}}{\text{1 mole}}$$

The molar mass is thus used as a conversion factor in the following examples:

Example   Find the mass in grams of 0.100 mole of glucose, $C_6H_{12}O_6$, the sugar in your blood.

$$6 \text{ atoms C} \times \frac{12.0 \text{ amu}}{1 \text{ atom C}} = 72.0 \text{ amu}$$

$$12 \text{ atoms H} \times \frac{1.01 \text{ amu}}{1 \text{ atom H}} = 12.1 \text{ amu}$$

$$6 \text{ atoms O} \times \frac{16.0 \text{ amu}}{1 \text{ atom O}} = \underline{96.0 \text{ amu}}$$

molecular weight of glucose = 180.1 amu

mass of 1 mole of glucose   = 180.1 g

UNKNOWN	GIVEN	CONNECTION	
grams glucose	0.100 mole glucose	$\dfrac{1 \text{ mole glucose}}{180.1 \text{ g glucose}}$ or	$\dfrac{180.1 \text{ g glucose}}{1 \text{ mole glucose}}$

$$\text{grams} = 0.100 \text{ mole glucose} \times \frac{180.1 \text{ g glucose}}{1 \text{ mole glucose}}$$

ESTIMATE $10^{-1} \text{ mole glucose} \times \dfrac{2 \times 10^2 \text{ g glucose}}{1 \text{ mole glucose}} = 2 \times 10^1 \text{ g glucose}$

CALCULATION   18.0 g glucose

In these problems, the first step always involves finding the molar mass by calculating the molecular or formula weight and expressing it in grams. We then use the molar mass as a conversion factor, making sure that units cancel properly.

   The ability to convert from moles to grams is extremely important in chemistry, especially in laboratory work. Therefore, you should try to master this skill. Study the following example.

Example   You need 3.50 moles of ammonium chloride, NH_4Cl to prepare a battery. How many grams do you need?

$$1 \text{ atom N} \times \frac{14.0 \text{ amu}}{1 \text{ atom N}} = 14.0 \text{ amu}$$

$$4 \text{ atoms H} \times \frac{1.01 \text{ amu}}{1 \text{ atom H}} = 4.04 \text{ amu}$$

$$1 \text{ atom } Cl \times \frac{35.5 \text{ amu}}{1 \text{ atom } Cl} = \underline{35.5 \text{ amu}}$$

formula weight of NH_4Cl = 53.5 amu

mass of 1 mole of NH_4Cl = 53.5 g

UNKNOWN	GIVEN	CONNECTION	
grams NH_4Cl	3.50 moles NH_4Cl	$\dfrac{1 \text{ mole } NH_4Cl}{53.5 \text{ g } NH_4Cl}$ or	$\dfrac{53.5 \text{ g } NH_4Cl}{1 \text{ mole } NH_4Cl}$

$$\text{grams} = 3.50 \text{ moles } NH_4Cl \times \frac{53.5 \text{ g } NH_4Cl}{1 \text{ mole } NH_4Cl}$$

ESTIMATE $\quad 3 \text{ moles } NH_4Cl \times \dfrac{5 \times 10^1 \text{ g } NH_4Cl}{1 \text{ mole } NH_4Cl} = 2 \times 10^2 \text{ g } NH_4Cl$

CALCULATION $\quad$ 187 g NH_4Cl

Always check your calculations to be sure that they make sense. In the first example, you were asked to find the mass in grams of 0.100 mole of glucose. The required amount is much less than 1 mole. Your answer, then, must be much *less* than 180 g, the mass of 1 mole. In the second example, you were asked to find the mass in grams of 3.50 moles of ammonium chloride. Because this amount is more than 1 mole, your answer must be *more* than 53.5 g, the mass of 1 mole. Checking your answers by this method, and making sure that units cancel properly, will help you to catch mistakes.

Examples Convert each of the following quantities into grams.

a 1.80 moles H_3PO_4,

$$3 \text{ atoms } H \times \frac{1.01 \text{ amu}}{1 \text{ atom } H} = 3.03 \text{ amu}$$

$$1 \text{ atom } P \times \frac{31.0 \text{ amu}}{1 \text{ atom } P} = 31.0 \text{ amu}$$

$$4 \text{ atoms } O \times \frac{16.0 \text{ amu}}{1 \text{ atom } O} = \underline{64.0 \text{ amu}}$$

molecular weight of H_3PO_4 = 98.0 amu

mass of 1 mole of H_3PO_4 $\quad$ = 98.0 g

UNKNOWN	GIVEN	CONNECTION
grams H_3PO_4	1.80 moles	$\dfrac{1 \text{ mole } H_3PO_4}{98.0 \text{ g } H_3PO_4}$ or $\dfrac{98.0 \text{ g } H_3PO_4}{1 \text{ mole } H_3PO_4}$

$$\text{grams} = 1.80 \text{ moles } H_3PO_4 \times \frac{98.0 \text{ g } H_3PO_4}{1 \text{ mole } H_3PO_4}$$

ESTIMATE $\quad 2 \text{ moles } H_3PO_4 \times \dfrac{1 \times 10^2 \text{ g } H_3PO_4}{1 \text{ mole } H_3PO_4}$

$$= 2 \times 10^2 \text{ g } H_3PO_4$$

$$= 200 \text{ g } H_3PO_4$$

CALCULATION $\quad 176 \text{ g } H_3PO_4$

b 5.10×10^{-4} moles $HgCl_2$

$1 \text{ atom Hg} \times \dfrac{200.6 \text{ amu}}{1 \text{ atom Hg}} = 200.6 \text{ amu}$

$2 \text{ atoms Cl} \times \dfrac{35.5 \text{ amu}}{1 \text{ atom Cl}} = \underline{\quad 71.0 \text{ amu}}$

formula weight of $HgCl_2 = 271.6$ amu

mass of 1 mole $HgCl_2 \quad = 271.6$ g

UNKNOWN	GIVEN	CONNECTION
grams $HgCl_2$	5.10×10^{-4} mole $HgCl_2$	$\dfrac{1 \text{ mole } HgCl_2}{271.6 \text{ g } HgCl_2}$ or $\dfrac{271.6 \text{ g } HgCl_2}{1 \text{ mole } HgCl_2}$

$$\text{grams } HgCl_2 = 5.10 \times 10^{-4} \text{ mole } HgCl_2 \times \frac{271.6 \text{ g } HgCl_2}{1 \text{ mole } HgCl_2}$$

ESTIMATE $\quad 5 \times 10^{-4} \text{ mole } HgCl_2 \times \dfrac{3 \times 10^2 \text{ g } HgCl_2}{1 \text{ mole } HgCl_2}$

$$= 2 \times 10^{-1} \text{ g} \quad HgCl_2 = 0.2 \text{ g } HgCl_2$$

CALCULATION $\quad 0.139 \text{ g } HgCl_2$

c 0.335 mole $Zn(NO_3)_2$

$1 \text{ atom Zn} \times \dfrac{65.4 \text{ amu}}{1 \text{ atom Zn}} \quad = 65.4 \text{ amu}$

$$2 \text{ atoms N} \times \frac{14.0 \text{ amu}}{1 \text{ atom N}} = 28.0 \text{ amu}$$

$$6 \text{ atoms O} \times \frac{16.0}{1 \text{ atom O}} = \underline{96.0 \text{ amu}}$$

formula weight of $Zn(NO_3)_2 = 189.4 \text{ amu}$

mass of 1 mole of $Zn(NO_3)_2 = 189.4 \text{ g}$

UNKNOWN	GIVEN	CONNECTION
grams $Zn(NO_3)_2$	0.335 mole $Zn(NO_3)_2$	$\dfrac{1 \text{ mole } Zn(NO_3)_2}{189.4 \text{ g } Zn(NO_3)_2}$ or $\dfrac{189.4 \; Zn(NO_3)_2}{1 \text{ mole } Zn(NO_3)_2}$

$$\text{grams } Zn(NO_3)_2 = 0.335 \text{ mole } Zn(NO_3)_2 \times \frac{189.4 \text{ g } Zn(NO_3)_2}{1 \text{ mole } Zn(NO_3)_2}$$

ESTIMATE $\quad 3 \times 10^{-1} \text{ mole } Zn(NO_3)_2 \times \dfrac{2 \times 10^2 \text{ g } Zn(NO_3)_2}{1 \text{ mole } Zn(NO_3)_2}$

$$= 6 \times 10^1 \text{ g } Zn(NO_3)_2 = 60 \text{ g } Zn(NO_3)_2$$

CALCULATION $\quad 63.4 \text{ g } Zn(NO_3)_2$

Practice problem Find the mass in grams of 2.15 moles of potassium perchlorate, $KClO_4$.

ANSWER 298 g

Practice problem What is the mass in grams of 0.867 mole of methylmercury(II) chloride, CH_3HgCl?

ANSWER 218 g

8.4
CONVERTING
GRAMS TO MOLES

So far, we have seen how to convert moles into grams. The reverse calculation, converting the mass of a substance in grams into the corresponding quantity in moles, is equally important. At times, a chemist may prepare a certain mass of a compound, and then need to know the amount in moles. The procedure is similar to the one we carried out in the preceding section. Again, we must find the molecular or formula weight of the substance, which in turn gives us the molar mass, the mass in grams of 1 mole. In this case, however, we then multiply the given number of *grams* by a conversion factor relating grams and moles.

$$\text{number of moles} = \text{number of grams} \times \frac{1 \text{ mole}}{\text{mass (g) of 1 mole}}$$

Example How many moles of SO_3 are present in a 20.0 g sample?

$$1 \text{ atom S} \times \frac{32.0 \text{ amu}}{1 \text{ atom S}} = 32.0 \text{ amu}$$

$$3 \text{ atoms O} \times \frac{16.0 \text{ amu}}{1 \text{ atom O}} = \underline{48.0 \text{ amu}}$$

molecular weight of SO_3 = 80.0 amu

mass of 1 mole SO_3 = 80.0 g

UNKNOWN	GIVEN	CONNECTION	
moles SO_3	20.0 g SO_3	$\dfrac{1 \text{ mole } SO_3}{80.0 \text{ g } SO_3}$ or	$\dfrac{80.0 \text{ g } SO_3}{1 \text{ mole } SO_3}$

$$\text{moles } SO_3 = 20.0 \text{ g } SO_3 \times \frac{1 \text{ mole } SO_3}{80.0 \text{ g } SO_3}$$

ESTIMATE $\quad 2 \times 10^1 \text{ g } SO_3 \times \dfrac{1 \text{ mole } SO_3}{8 \times 10^1 \text{ g } SO_3} = 0.3 \text{ mole } SO_3$

CALCULATION $\quad$ 0.250 mole SO_3

When setting up the calculation we must be careful to use the correct conversion factor. In this case, 1 mole is in the numerator, and the mass in grams of 1 mole is in the denominator. We can check our conversion factor by making sure that the unit grams cancels out, leaving moles as the units of the answer. As always, we check to see that the result makes sense. In this example, we had 20.0 g of SO_3. Because the molar mass of SO_3 is 80.0 g, we know that the amount is less than one mole.

Example A 1.00 mL blood sample contains 2.10×10^{-3} g cholesterol, $C_{27}H_{46}O$. How many moles of cholesterol are present in this sample?

$$27 \text{ atoms C} \times \frac{12.0 \text{ amu}}{1 \text{ atom C}} = 324 \text{ amu}$$

$$46 \text{ atoms H} \times \frac{1.01 \text{ amu}}{1 \text{ atom H}} = 46.5 \text{ amu}$$

$$1 \text{ atom O} \times \frac{16.0 \text{ amu}}{1 \text{ atom O}} \qquad = \underline{16.0 \text{ amu}}$$

molecular weight of cholesterol = 387 amu

mass of 1 mole of cholesterol   = 387 g

UNKNOWN	GIVEN	CONNECTION
moles cholesterol	2.10×10^{-3} g cholesterol	$\dfrac{1 \text{ mole cholesterol}}{387 \text{ g cholesterol}}$ or $\dfrac{387 \text{ g cholesterol}}{1 \text{ mole cholesterol}}$

$$\text{moles cholesterol} = 2.10 \times 10^{-3} \text{ g cholesterol} \times \frac{1 \text{ mole cholesterol}}{387 \text{ g cholesterol}}$$

ESTIMATE 2×10^{-3} g cholesterol $\times \dfrac{1 \text{ mole cholesterol}}{4 \times 10^2 \text{ g cholesterol}}$

$= 5 \times 10^{-6}$ mole cholesterol

CALCULATION 5.43×10^{-6} mole cholesterol

As you can see, both calculations—converting moles to grams and converting grams to moles—are straightforward. Just be sure that you set up the problem correctly. In one case you perform a multiplication, and in the reverse case you perform a division, using the molar mass. If you pay attention to the units, you can avoid making mistakes.

Example   Perform the following conversions.

a 200. g of H_2SO_4 to moles

$$2 \text{ atoms H} \times \frac{1.01 \text{ amu}}{1 \text{ atom H}} \quad = \quad 2.02 \text{ amu}$$

$$1 \text{ atom S} \times \frac{32.0 \text{ amu}}{1 \text{ atom S}} \quad = 32.0 \quad \text{amu}$$

$$4 \text{ atoms O} \times \frac{16.0 \text{ amu}}{1 \text{ atom O}} \quad = \underline{64.0 \quad \text{amu}}$$

molecular weight of H_2SO_4 = 98.0   amu

mass of 1 mole H_2SO_4 = 98.0 g

UNKNOWN	GIVEN	CONNECTION	
moles H_2SO_4	200. g H_2SO_4	$\dfrac{1 \text{ mole } H_2SO_4}{98.0 \text{ g } H_2SO_4}$	or $\dfrac{98.0 \text{ g } H_2SO_4}{1 \text{ mole } H_2SO_4}$

$$\text{moles } H_2SO_4 = 200. \text{ g } \cancel{H_2SO_4} \times \frac{1 \text{ mole } H_2SO_4}{98.0 \text{ g } \cancel{H_2SO_4}}$$

ESTIMATE $\quad 2 \times 10^2 \text{ g } \cancel{H_2SO_4} \times \dfrac{1 \text{ mole } H_2SO_4}{1 \times 10^2 \text{ g } \cancel{H_2SO_4}} = 2 \text{ moles } H_2SO_4$

CALCULATION $\quad 2.04 \text{ moles } H_2SO_4$

b 200. moles of H_2SO_4 to grams
mass of 1 mole $H_2SO_4 = 98.0$ g (same as in **a**)

UNKNOWN	GIVEN	CONNECTION	
grams H_2SO_4	200. moles H_2SO_4	$\dfrac{1 \text{ mole } H_2SO_4}{98.0 \text{ g } H_2SO_4}$	or $\dfrac{98.0 \text{ g } H_2SO_4}{1 \text{ mole } H_2SO_4}$

$$\text{grams } H_2SO_4 = 200. \cancel{\text{moles } H_2SO_4} \times \frac{98.0 \text{ g } H_2SO_4}{1 \cancel{\text{ mole } H_2SO_4}}$$

ESTIMATE $\quad 2 \times 10^2 \cancel{\text{moles } H_2SO_4} \times \dfrac{1 \times 10^2 \text{ g } H_2SO_4}{1 \cancel{\text{ mole } H_2SO_4}}$
$\quad\quad\quad = 2 \times 10^4 \text{ g } H_2SO_4$

CALCULATION $\quad 1.96 \times 10^4 \text{ g } H_2SO_4$

c 185 g of $NaHCO_3$ to moles

$1 \cancel{\text{ atom Na}} \times \dfrac{23.0 \text{ amu}}{1 \cancel{\text{ atom Na}}} \quad = 23.0 \text{ amu}$

$1 \cancel{\text{ atom H}} \times \dfrac{1.01 \text{ amu}}{1 \cancel{\text{ atom H}}} \quad = 1.01 \text{ amu}$

$1 \cancel{\text{ atom C}} \times \dfrac{12.0 \text{ amu}}{1 \cancel{\text{ atom C}}} \quad = 12.0 \text{ amu}$

$3 \cancel{\text{ atoms O}} \times \dfrac{16.0 \text{ amu}}{1 \cancel{\text{ atom O}}} \quad = \underline{48.0 \text{ amu}}$

formula weight of $NaHCO_3 = 84.0$ amu
mass of 1 mole $NaHCO_3 \quad = 84.0$ g

UNKNOWN	GIVEN	CONNECTION
moles NaHCO$_3$	185 g NaHCO$_3$	$\dfrac{1 \text{ mole NaHCO}_3}{84.0 \text{ g NaHCO}_3}$ or $\dfrac{84.0 \text{ g NaHCO}_3}{1 \text{ mole NaHCO}_3}$

$$\text{moles NaHCO}_3 = 185 \text{ g NaHCO}_3 \times \frac{1 \text{ mole NaHCO}_3}{84.0 \text{ g NaHCO}_3}$$

ESTIMATE $\quad 2 \times 10^2 \text{ g NaHCO}_3 \times \dfrac{1 \text{ mole NaHCO}_3}{8 \times 10^1 \text{ g NaHCO}_3}$

$\qquad = 3$ moles NaHCO$_3$

CALCULATION $\quad$ 2.20 moles NaHCO$_3$

Practice problem What amount in moles of strontium carbonate, SrCO$_3$, is present in 480. g?

ANSWER $\quad$ 3.25 moles

Practice problem Convert 135 g of mercury(II) cyanide, Hg(CN)$_2$, into moles of this compound.

ANSWER $\quad$ 0.534 mole

8.5 CONVERSIONS BASED ON AVOGADRO'S NUMBER You should now be able to do conversions between the number of moles of a substance and its mass in grams. It is also important to understand the relation between the quantity of matter in moles or grams and the number of particles (atoms, ions, or molecules) that it contains. One mole of any substance always contains Avogadro's number of particles, 6.02×10^{23}. So, one mole of iron (55.8 g) contains 6.02×10^{23} iron atoms. One mole of magnesium oxide (36.3 g) contains 6.02×10^{23} formula units of MgO, or 6.02×10^{23} Mg^{2+} ions and 6.02×10^{23} O^{2-} ions. One mole of ammonia (17.0 g) contains 6.02×10^{23} molecules of NH$_3$, or 6.02×10^{23} nitrogen atoms and $3 \times (6.02 \times 10^{23})$ or 1.81×10^{24} hydrogen atoms.

Example Describe the number and type of particles present in one mole of rubbing alcohol, C$_3$H$_8$O.

One mole of alcohol contains 6.02×10^{23} alcohol molecules, or a total of:

C: $3 \times (6.02 \times 10^{23}) = 1.81 \times 10^{24}$ carbon atoms
H: $8 \times (6.02 \times 10^{23}) = 4.82 \times 10^{24}$ hydrogen atoms
O: $1 \times (6.02 \times 10^{23}) = 6.02 \times 10^{23}$ oxygen atoms

If we want to know the number of particles in an amount of substance that is greater or less than one mole, we can perform the calculation by using the conversion factor 6.02×10^{23} particles/1 mole. In general,

$$\frac{\text{number of}}{\text{particles}} = \frac{\text{number of}}{\text{moles}} \times \frac{6.02 \times 10^{23} \text{ particles}}{1 \text{ mole}}$$

For example, the number of mercury atoms in 2.50 moles of mercury is:

$$\text{number of atoms} = 2.50 \text{ moles} \times \frac{6.02 \times 10^{23} \text{ atoms}}{1 \text{ mole}}$$
$$= 1.51 \times 10^{24} \text{ atoms}$$

Example   How many molecules of sulfur trioxide, SO_3, an air pollutant, are present in 0.650 mole of SO_3?

UNKNOWN	GIVEN	CONNECTION
molecules SO_3	0.650 mole SO_3	$\dfrac{1 \text{ mole } SO_3}{6.02 \times 10^{23} \text{ molecules } SO_3}$ or $\dfrac{6.02 \times 10^{23} \text{ molecules } SO_3}{1 \text{ mole } SO_3}$

$$\text{molecules } SO_3 = 0.650 \text{ mole } SO_3 \times \frac{6.02 \times 10^{23} \text{ molecules } SO_3}{1 \text{ mole } SO_3}$$

ESTIMATE $$7 \times 10^{-1} \text{ mole } SO_3 \times \frac{6 \times 10^{23} \text{ molecules } SO_3}{1 \text{ mole } SO_3}$$
$$= 4 \times 10^{23} \text{ molecules } SO_3$$

CALCULATION 3.91×10^{23} molecules SO_3

As always, we check to see that our answer makes sense. In this example, the answer should be smaller than Avogadro's number because 0.650 mole is less than 1 mole. We also make sure that units cancel correctly.

Sometimes, instead of an amount in moles, we have to deal with the mass of a substance in grams. The calculation is essentially the same, only now we need to know the atomic, molecular, or formula weight. The preceding example could have been stated, "How many molecules of sulfur trioxide are present in 52.0 g of SO_3?" If so, the answer would be the same, 3.91×10^{23}, since 52.0 g of SO_3 contains 0.650 mole:

$$\text{moles } SO_3 = 52.0 \text{ g } SO_3 \times \frac{1 \text{ mole } SO_3}{80.0 \text{ g } SO_3}$$

$$= 0.650 \text{ mole } SO_3$$

The additional information needed is the molecular weight of SO_3, which is 80.0 amu.

Example How many formula units of potassium sulfate, K_2SO_4, are present in 200. g of K_2SO_4?

$$2 \text{ atoms K} \times \frac{39.1 \text{ amu}}{1 \text{ atom K}} = 78.2 \text{ amu}$$

$$1 \text{ atom S} \times \frac{32.0 \text{ amu}}{1 \text{ atom S}} = 32.0 \text{ amu}$$

$$4 \text{ atoms O} \times \frac{64.0 \text{ amu}}{1 \text{ atom O}} = \underline{64.0 \text{ amu}}$$

formula weight of K_2SO_4 = 174.2 amu

mass of 1 mole of K_2SO_4 = 174.2 g

UNKNOWN	GIVEN	CONNECTION
formula units K_2SO_4	200. g K_2SO_4	$\dfrac{1 \text{ mole } K_2SO_4}{174.2 \text{ g } K_2SO_4}$ or $\dfrac{174.2 \text{ g } K_2SO_4}{1 \text{ mole } K_2SO_4}$
		$\dfrac{1 \text{ mole } K_2SO_4}{6.02 \times 10^{23} \text{ formula units } K_2SO_4}$
		or $\dfrac{6.02 \times 10^{23} \text{ formula units } K_2SO_4}{1 \text{ mole } K_2SO_4}$

$$\text{formula units } K_2SO_4 = 200. \text{ g } K_2SO_4 \times \frac{1 \text{ mole } K_2SO_4}{174.2 \text{ g } K_2SO_4} \times$$

$$\frac{6.02 \times 10^{23} \text{ formula units } K_2SO_4}{1 \text{ mole } K_2SO_4}$$

ESTIMATE $\quad 2 \times 10^2$ g̶ ̶K̶₂̶S̶O̶₄̶ $\times \dfrac{1 \text{ mole K₂SO₄}}{1 \times 10^2 \text{ g K₂SO₄}} \times$

$$\dfrac{6 \times 10^{23} \text{ formula units K}_2\text{SO}_4}{1 \text{ mole K}_2\text{SO}_4}$$

$$= 1 \times 10^{24} \text{ formula units K}_2\text{SO}_4$$

CALCULATION $\quad 6.91 \times 10^{23}$ formula units K_2SO_4

Notice how easy the calculation becomes when we use the conversion factor method. The additional step, changing grams to moles, involves only multiplying by another conversion factor in the computation. Of course we must be certain that the units are properly arranged to cancel out.

Example Find the number of silver atoms in 1.00 g of silver metal.

UNKNOWN	GIVEN	CONNECTION
atoms Ag	1.00 g Ag	$\dfrac{1 \text{ mole Ag}}{107.9 \text{ g Ag}}$ or $\dfrac{107.9 \text{ g Ag}}{1 \text{ mole Ag}}$
		$\dfrac{1 \text{ mole Ag}}{6.02 \times 10^{23} \text{ atoms Ag}}$ or $\dfrac{6.02 \times 10^{23} \text{ atoms Ag}}{1 \text{ mole Ag}}$

$$\text{atoms Ag} = 1.00 \text{ g Ag} \times \dfrac{1 \text{ mole Ag}}{107.9 \text{ g Ag}} \times \dfrac{6.02 \times 10^{23} \text{ atoms Ag}}{1 \text{ mole Ag}}$$

ESTIMATE $\quad 1$ g Ag $\times \dfrac{1 \text{ mole Ag}}{1 \times 10^2 \text{ g Ag}} \times \dfrac{6 \times 10^{23} \text{ atoms Ag}}{1 \text{ mole Ag}}$

$$= 6 \times 10^{21} \text{ atoms Ag}$$

CALCULATION $\quad 5.58 \times 10^{21}$ atoms Ag

Example A sample of hydrogen sulfide, H_2S, contains 8.35×10^{24} molecules. How many moles of H_2S are present?

UNKNOWN	GIVEN	CONNECTION
moles H₂S	8.35×10^{24} molecules H₂S	$\dfrac{1 \text{ mole H}_2\text{S}}{6.02 \times 10^{23} \text{ molecules H}_2\text{S}}$
		or $\dfrac{6.02 \times 10^{23} \text{ molecules H}_2\text{S}}{1 \text{ mole H}_2\text{S}}$

$$\text{moles} \quad H_2S \quad = \quad 8.35 \quad \times \quad 10^{24} \quad \frac{\text{molecules} \quad H_2S}{} \quad \times$$

$$\frac{1 \text{ mole } H_2S}{6.02 \times 10^{23} \text{ molecules } H_2S}$$

ESTIMATE $\quad 8 \times 10^{24} \text{ molecules } H_2S \times \dfrac{1 \text{ mole } H_2S}{6 \times 10^{23} \text{ molecules } H_2S}$

$$= 1 \times 10^1 \text{ moles } H_2S = 10 \text{ moles } H_2S$$

CALCULATION $\quad 1.39 \times 10^1 \text{ moles } H_2S = 13.9 \text{ moles } H_2S$

Example What is the mass in grams of 2.10×10^{22} atoms of lead?

UNKNOWN	GIVEN	CONNECTION
g Pb	2.10×10^{22} atoms Pb	$\dfrac{1 \text{ mole Pb}}{6.02 \times 10^{23} \text{ atoms Pb}}$ or $\dfrac{6.02 \times 10^{23} \text{ atoms Pb}}{1 \text{ mole Pb}}$ $\dfrac{1 \text{ mole Pb}}{207.2 \text{ g Pb}}$ or $\dfrac{207.2 \text{ g Pb}}{1 \text{ mole Pb}}$

$$\text{g Pb} = 2.10 \times 10^{22} \text{ atoms Pb} \times \frac{1 \text{ mole Pb}}{6.02 \times 10^{23} \text{ atoms Pb}} \times \frac{207.2 \text{ g Pb}}{1 \text{ mole Pb}}$$

ESTIMATE $\quad 2 \times 10^{22} \text{ atoms Pb} \times \dfrac{1 \text{ mole Pb}}{6 \times 10^{23} \text{ atoms Pb}} \times \dfrac{2 \times 10^2 \text{ g Pb}}{1 \text{ mole Pb}}$

$$= 7 \text{ g Pb}$$

CALCULATION $\quad 7.23 \text{ g Pb}$

Example How many atoms of each element are present in 2.5 moles of sulfuric acid, H_2SO_4?

$$2.5 \text{ moles } H_2SO_4 \times \frac{6.02 \times 10^{23} \text{ molecules } H_2SO_4}{1 \text{ mole } H_2SO_4} = 1.5 \times 10^{24} \text{ molecules } H_2SO_4$$

$$1.5 \times 10^{24} \text{ molecules } H_2SO_4 \times \frac{2 \text{ H atoms}}{1 \text{ molecule } H_2SO_4} = 3.0 \times 10^{24} \text{ H atoms}$$

$$1.5 \times 10^{24} \text{ molecules } H_2SO_4 \times \frac{1 \text{ S atom}}{1 \text{ molecule } H_2SO_4} = 1.5 \times 10^{24} \text{ S atoms}$$

$$1.5 \times 10^{24} \text{ molecules } H_2SO_4 \times \frac{4 \text{ O atoms}}{1 \text{ molecule } H_2SO_4} = 6.0 \times 10^{24} \text{ O atoms}$$

Practice problem   How many formula units of potassium acetate, $KC_2H_3O_2$, are present in 325 g of this compound?

ANSWER 1.99×10^{24} formula units

Practice problem   Find the mass in grams of 2.15×10^{21} molecules of boron phosphate, BPO_4. (This compound is used in ceramics.)

ANSWER   0.378 g

**8.6
PERCENTAGE
COMPOSITION**   One of the most important properties of a compound is that it has a *definite composition,* which is described by its formula. For example, everywhere that it is found, the compound water has exactly the same makeup. Each water molecule contains two atoms of hydrogen and one atom of oxygen. The mass of the water molecule has a fixed value (18.0 amu). The contributions of the two hydrogen atoms (2.0 amu) and

---

Do-it-yourself 8

**Determining
composition**

One way to determine the percentage composition of a compound containing carbon and hydrogen is to burn it completely in oxygen. When such a compound is burned in oxygen, carbon is converted to carbon dioxide, CO_2, and hydrogen is converted to water, H_2O. Both the CO_2 and H_2O can be trapped by special compounds, which are weighed before and after the experiment. From the gain in weight of the trapping compound, the amount of CO_2 and H_2O produced, and thus the percentage of C and H in the original compound, can be calculated.

You can demonstrate the fact that burning a substance converts carbon to CO_2 and hydrogen to H_2O by doing the following.

To do: Place a candle in a holder and light the candle. Briefly hold a cold glass upside down above the flame to collect some of the vapor. Rub your finger along the inside of the glass. What do you see?

Hold a glass or metal funnel upside down above the flame. Place a lighted match above the narrow end of the funnel. What happens? What is in the gas that makes the match go out?

the one oxygen atom (16.0 amu) to that mass are also fixed. In other words, the percentage of the total mass of water that is hydrogen and the percentage that is oxygen have definite values.

Percentage is just a special type of fraction, meaning parts per hundred. For example, fifty percent, 50%, means 50 parts out of 100 equal parts, or 50/100. We can express a percentage like 50% as a decimal, 0.50 or fifty-hundredths, by moving the decimal point two places to the left (or dividing by 100).

To convert any fraction to a percent, we must first write the fraction as a decimal and then multiply by 100%. Multiplying by 100% or 100/100, is the same as multiplying by one; it does not change the value of the number being multiplied. (See Appendix B.3 for conversions of fractions to decimals.)

The percentage by weight of each element in a compound is called the compound's **percentage composition**. To find the percentage composition of a compound, we determine the fraction of the compound's total mass that is contributed by each element. To find this fraction, we *divide the mass of all the atoms of each element by the formula or molecular weight of the compound.* For example, carbon monoxide, CO, has a molecular weight of 28.0 amu (12.0 amu + 16.0 amu). The fraction of its mass contributed by carbon is: 12.0 amu divided by 28.0 amu, or 0.428. We multiply by 100% to convert the decimal to a percent: 0.428 × 100% = 42.8%. In carbon monoxide, then, 42.8% of the total mass is due to the mass of the carbon atom. The rest of the compound's mass, 16.0 amu/28.0 amu = 0.572 × 100% = 57.2%, is contributed by the oxygen atom.

We can check a calculation of percentage composition by adding up the percentage of each element to make sure that the total is 100%. (The total may be slightly more or slightly less than 100% due to rounding-off.) If the sum is not approximately 100%, we know that we probably made a mistake, and should repeat the calculations. When we check our calculation of the percentage composition of carbon monoxide, we get:

$$42.8\% + 57.2\% = 100.0\%$$

This result means that our calculations were probably correct.

Now let's determine the percentage composition of two compounds whose formulas are closely related: water, H_2O, and hydrogen peroxide, H_2O_2.

water, H_2O

$$\text{mass of 2 H atoms} = 2 \text{ atoms H} \times \frac{1.01 \text{ amu}}{1 \text{ atom H}} = 2.02 \text{ amu}$$

$$\underline{\text{mass of 1 O atom} = 1 \text{ atom O} \times \frac{16.0 \text{ amu}}{1 \text{ atom O}} \quad = \underline{16.0 \text{ amu}}}$$

$$\text{mass of } H_2O \text{ molecule} \qquad\qquad\qquad = 18.0 \quad \text{amu}$$

$$\% \text{ H} = \frac{2.02 \text{ amu}}{18.0 \text{ amu}} \times 100\% = 0.112 \times 100\% = 11.2\%$$

$$\% \text{ O} = \frac{16.0 \text{ amu}}{18.0 \text{ amu}} \times 100\% = 0.889 \times 100\% = 88.9\%$$

Check: $\quad$ 11.2%
$$\underline{88.9\%}$$
$$100.1\%$$

hydrogen peroxide, H_2O_2

$$\text{mass of 2 H atoms} \qquad = 2 \text{ atoms H} \times \frac{1.01 \text{ amu}}{1 \text{ atom H}} = 2.02 \text{ amu}$$

$$\underline{\text{mass of 2 O atoms} \qquad = 2 \text{ atoms O} \times \frac{16.0 \text{ amu}}{1 \text{ atom O}} = 32.0 \quad \text{amu}}$$

$$\text{mass of } H_2O_2 \text{ molecule} \qquad\qquad\qquad\qquad = 34.0 \quad \text{amu}$$

$$\% \text{ H} = \frac{2.02 \text{ amu}}{34.0 \text{ amu}} \times 100\% = 0.0594 \times 100\% = 5.94\%$$

$$\% \text{ O} = \frac{32.0 \text{ amu}}{34.0 \text{ amu}} \times 100\% = 0.941 \times 100\% = 94.1\%$$

Check: $\quad$ 5.94%
$$\underline{94.1\%}$$
$$100.0\%$$

Therefore a molecule of water consists of two parts of hydrogen, or about 11% of the mass, to one part of oxygen, 89% of the mass. Hydrogen peroxide, a poisonous compound used in disinfectants and hair bleach, has two parts hydrogen, about 6% of the mass, to two parts oxygen, 94% of the mass. Thus, we have two ways to describe the makeup of a molecule or formula unit. We can speak of either the number of atoms of each element that are present, or of the percentage composition based on the molecular or formula weight.

Percentage composition calculations can also be carried out by using the molar masses of the elements instead of their atomic masses. That is, the unit *g/mole* can be substituted for *amu/atom* without affecting the results. The formula of a compound shows the relationship between the elements both in terms of atoms and in terms of moles.

Example Find the percentage composition of the feldspar mineral $CaAl_2Si_2O_8$.

$$\text{mass of 1 Ca atom} = 1 \text{ atom Ca} \times \frac{40.1 \text{ amu}}{1 \text{ atom Ca}} = 40.1 \text{ amu}$$

$$\text{mass of 2 Al atoms} = 2 \text{ atoms Al} \times \frac{27.0 \text{ amu}}{1 \text{ atom Al}} = 54.0 \text{ amu}$$

$$\text{mass of 2 Si atoms} = 2 \text{ atoms Si} \times \frac{28.1 \text{ amu}}{1 \text{ atom Si}} = 56.2 \text{ amu}$$

$$\text{mass of 8 O atoms} = 8 \text{ atoms O} \times \frac{16.0 \text{ amu}}{1 \text{ atom O}} = \underline{128 \text{ amu}}$$

$$\text{mass of } CaAl_2Si_2O_8 \text{ formula unit} = 278 \text{ amu}$$

$$\% \text{ Ca} = \frac{40.1 \text{ amu}}{278 \text{ amu}} \times 100\% = 0.144 \times 100\% = 14.4\%$$

$$\% \text{ Al} = \frac{54.0 \text{ amu}}{278 \text{ amu}} \times 100\% = 0.194 \times 100\% = 19.4\%$$

$$\% \text{ Si} = \frac{56.2 \text{ amu}}{278 \text{ amu}} \times 100\% = 0.202 \times 100\% = 20.2\%$$

$$\% \text{ O} = \frac{128 \text{ amu}}{278 \text{ amu}} \times 100\% = 0.460 \times 100\% = \underline{46.0\%}$$

Check: 100.0%

Example What is the percentage composition of table sugar, sucrose, $C_{12}H_{22}O_{11}$?

$$\text{mass of 12 C atoms} = 12 \text{ atoms C} \times \frac{12.0 \text{ amu}}{1 \text{ atom C}} = 144 \text{ amu}$$

$$\text{mass of 22 H atoms} = 22 \text{ atoms H} \times \frac{1.01 \text{ amu}}{1 \text{ atom H}} = 22.2 \text{ amu}$$

$$\text{mass of 11 O atoms} = 11 \text{ atoms O} \times \frac{16.0 \text{ amu}}{1 \text{ atom O}} = \underline{176 \text{ amu}}$$

$$\text{mass of } C_{12}H_{22}O_{11} \text{ molecule} = 342 \text{ amu}$$

$$\% \text{ C} = \frac{144.0 \text{ amu}}{342 \text{ amu}} \times 100\% = 0.421 \times 100\% = 42.1\%$$

$$\% \text{ H} = \frac{22.2 \text{ amu}}{342 \text{ amu}} \times 100\% = 0.0649 \times 100\% = 6.49\%$$

$$\% \text{ O} = \frac{176 \text{ amu}}{342 \text{ amu}} \times 100\% = 0.514 \times 100\% = \underline{51.4\%}$$

Check: 100.0%

Example Determine the percentage composition of potassium alum, $KAl(SO_4)_2$.

$$\text{mass of 1 K atom} = 1 \text{ atom K} \times \frac{39.1 \text{ amu}}{1 \text{ atom K}} = 39.1 \text{ amu}$$

$$\text{mass of 1 Al atom} = 1 \text{ atom Al} \times \frac{27.0 \text{ amu}}{1 \text{ atom Al}} = 27.0 \text{ amu}$$

$$\text{mass of 2 S atoms} = 2 \text{ atoms S} \times \frac{32.0 \text{ amu}}{1 \text{ atom S}} = 64.0 \text{ amu}$$

$$\text{mass of 8 O atoms} = 8 \text{ atoms O} \times \frac{16.0 \text{ amu}}{1 \text{ atom O}} = \underline{128 \text{ amu}}$$

$$\text{mass of } KAl(SO_4)_2 \text{ formula unit} = 258 \text{ amu}$$

$$\% \text{ K} = \frac{39.1 \text{ amu}}{258 \text{ amu}} \times 100\% = 0.152 \times 100\% = 15.2\%$$

$$\% \text{ Al} = \frac{27.0 \text{ amu}}{258 \text{ amu}} \times 100\% = 0.105 \times 100\% = 10.5\%$$

$$\% \text{ S} = \frac{64.0 \text{ amu}}{258 \text{ amu}} \times 100\% = 0.248 \times 100\% = 24.8\%$$

$$\% \text{ O} = \frac{128 \text{ amu}}{258 \text{ amu}} \times 100\% = 0.496 \times 100\% = \underline{49.6\%}$$

$$\text{Check:} \quad 100.1\%$$

Practice problem Calculate the percentage composition of tin(II) sulfate, $SnSO_4$, used in dyeing and tinplating.

ANSWERS 55.3% Sn, 14.9% S, 29.8% O

Practice problem Find the percentage composition of calcium gluconate, $CaC_{12}H_{22}O_{14}$, a food additive and source of calcium.

ANSWERS 9.3% Ca, 33.5% C, 5.2% H, 52.0% O

**8.7
DETERMINING
CHEMICAL FORMULAS** In the preceding section, we saw how to find the percentage composition of a compound based on its formula. Using our knowledge of moles, we can now carry out the reverse procedure—finding the formula of a compound from its percentage composition. This procedure is very useful to chemists, who often make new compounds and then need to know their formulas. A chemist often analyzes a new compound

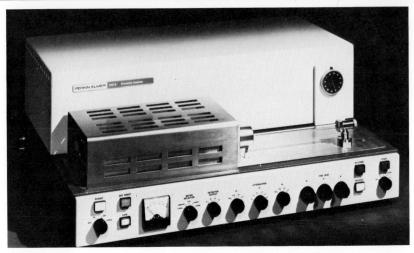

to determine its percentage composition. Figure 8-4 shows a commercial elemental analyzer that measures the percentages of carbon, hydrogen, and nitrogen in a compound. Once the percentage composition of the compound is known, the chemist can determine its chemical formula.

For example, suppose that analysis of a compound reveals its composition to be 78.3% boron and 21.7% hydrogen. No matter how much of the compound we consider, these percentages always describe its makeup. To make our calculations simple, we will assume that 100 g of compound are present. This 100 g of compound must contain 78.3 g of boron (78.3% B × 100 g) and 21.7 g of hydrogen (21.7% H × 100 g.) To find the formula, we first convert these weights to moles, using the appropriate molar masses.

$$78.3 \ \cancel{g \ B} \times \frac{1 \ \text{mole B}}{10.8 \ \cancel{g \ B}} = 7.25 \ \text{moles B}$$

$$21.7 \ \cancel{g \ H} \times \frac{1 \ \text{mole H}}{1.01 \ \cancel{g \ H}} = 21.5 \ \text{moles H}$$

So far our calculation tells us how many moles of each different type of atom are present in 100 g of compound. The compound's formula is based on the *relative* numbers of moles of each atom, that is, the number of moles of one atom compared to the number of moles of another atom. Therefore, we must now find the *ratio* between the number of moles of boron and the number of moles of hydrogen. To find this ratio, we divide both numbers by the *smaller* number of moles: 7.25 moles of boron.

$$B: \quad \frac{7.25 \ \cancel{\text{moles}}}{7.25 \ \cancel{\text{moles}}} = 1.00$$

$$H: \quad \frac{21.5 \ \cancel{\text{moles}}}{7.25 \ \cancel{\text{moles}}} = 2.97$$

TABLE 8-4

Molecular and empirical formulas

Molecular formula	Empirical formula
H_2O_2	HO
N_2H_4	NH_2
P_4O_{10}	P_2O_5
C_2H_2	CH
C_6H_6	CH
$C_6H_{12}O_6$	CH_2O

The number 2.97 can be rounded off to 3. Then the ratio becomes: 1 mole of boron atoms to 3 moles of hydrogen atoms. This result, the relative number of moles of boron to hydrogen in 100 g of the compound, must be the same as the relative numbers of each kind of atom in one molecule. It does not matter how many molecules we examine; the relation between the atoms is the same because compounds have a definite, constant composition. So instead of choosing 100 g to begin with, we could have chosen any amount. Taking 100 g, however, makes the computation simplest.

The result of the calculation is that the formula of the compound seems to be BH_3, in which one boron atom is present for every three hydrogen atoms. But this result is not necessarily the true molecular formula because it only represents the *simplest* whole number ratios of the atoms. This kind of result is known as the **empirical formula.** The *molecular* formula may be a multiple of the empirical formula, such as B_2H_6, B_3H_9, B_4H_{12}, and so on. Every one of these possibilities has three hydrogen atoms for every one boron atom. The molecular formula states the *actual* numbers of each type of atom in a molecule. The empirical formula only states the *relation* between the atoms. Table 8-4 compares other examples of molecular and empirical formulas.

Note that the empirical formula does not contain as much information as the molecular formula. For example, acetylene, C_2H_2, contains two carbon atoms and two hydrogen atoms per molecule. Its empirical formula, CH, shows only that there are equal numbers of carbon and hydrogen atoms, but does not show how many of each. It is also possible for different compounds to have the same empirical formulas. As shown in the table, benzene (C_6H_6) has the same empirical formula, CH, as acetylene (C_2H_2).

Once we have found the empirical formula, we need an additional piece of information to obtain the molecular formula, namely, the molecular weight of the compound. Molecular weights can be determined experimentally. Suppose that the molecular weight of our compound of boron and hydrogen is 27.6 amu. The mass based on the empirical formula BH_3 is

$$1 \text{ atom B} \times \frac{10.8 \text{ amu}}{1 \text{ atom B}} = 10.8 \text{ amu}$$

$$3 \text{ atoms H} \times \frac{1.01 \text{ amu}}{1 \text{ atom H}} = \underline{ 3.03 \text{ amu}}$$

$$13.8 \quad \text{amu}$$

Since the molecular weight, 27.6 amu is twice the mass based on the empirical formula $\left(\dfrac{27.6 \text{ amu}}{13.8 \text{ amu}} = 2.00\right)$, each molecule must contain twice the number of atoms as BH_3. Thus the molecular formula is B_2H_6.

We can think of the molecular formula as a whole number multiple of the empirical formula. In the last example, the molecular formula was $(BH_3)_n$, where n is some whole number greater than or equal to 1. To find the value of n, we divide the molecular or formula weight by the mass based on the empirical formula.

$$n = \frac{\text{molecular or formula weight}}{\text{mass of empirical formula}}$$

$$= \frac{27.6 \, \text{amu}}{13.8 \, \text{amu}} = 2$$

The molecular formula $(BH_3)_n$ is therefore $(BH_3)_2$ or B_2H_6.

Example Upon analysis, a compound is found to contain 68.9% C, 4.9% H, and 26.2% O. Its molecular weight is 122.1. Find its formula.

In 100. g of compound, there exists:

68.9% C × 100. g = 68.9 g C
 4.9% H × 100. g = 4.9 g H
26.2% O × 100. g = 26.2 g O

Converting to moles:

$$68.9 \, \text{g C} \times \frac{1 \, \text{mole C}}{12.0 \, \text{g C}} = 5.74 \text{ moles C}$$

$$4.9 \, \text{g H} \times \frac{1 \, \text{mole H}}{1.01 \, \text{g H}} = 4.9 \text{ moles H}$$

$$26.2 \, \text{g O} \times \frac{1 \, \text{mole O}}{16.0 \, \text{g O}} = 1.64 \text{ moles O}$$

Dividing by the smallest amount, 1.64 moles:

$$\text{C:} \quad \frac{5.74 \, \text{moles}}{1.64 \, \text{moles}} = 3.50$$

$$\text{H:} \quad \frac{4.9 \, \text{moles}}{1.64 \, \text{moles}} = 3.0$$

$$\text{O:} \quad \frac{1.64 \, \text{moles}}{1.64 \, \text{moles}} = 1.00$$

empirical formula: $C_{3.5}H_3O_1$. Converting to whole numbers: $C_7H_6O_2$. (*Note:* Multiplying by 2 is necessary here to convert 3.5 to a whole number. In a calculation like this, 3.5 is too far from a whole number to be rounded off.)

$$7 \text{ atoms C} \times \frac{12.0 \text{ amu}}{1 \text{ atom C}} \quad = \quad 84.0 \text{ amu}$$

$$6 \text{ atoms H} \times \frac{1.01 \text{ amu}}{1 \text{ atom H}} \quad = \quad 6.06 \text{ amu}$$

$$2 \text{ atoms O} \times \frac{16.0 \text{ amu}}{1 \text{ atom O}} \quad = \quad \underline{32.0 \text{ amu}}$$

mass of empirical formula = 122.1 amu = molecular weight (given),

so empirical formula   = molecular formula = $C_7H_6O_2$

Example A white powder has the composition 4.8% hydrogen, 17.5% boron, and 77.7% oxygen. Find its empirical formula.

In 100. g of compound, there exists:
  4.8% H × 100. g =   4.8 g H
 17.5% B × 100. g = 17.5 g B
 77.7% O × 100. g = 77.7 g O

Converting to moles:

$$4.8 \text{ g H} \times \frac{1 \text{ mole H}}{1.01 \text{ g H}} = 4.8 \text{ moles H}$$

$$17.5 \text{ g B} \times \frac{1 \text{ mole B}}{10.8 \text{ g B}} = 1.62 \text{ moles B}$$

$$77.7 \text{ g O} \times \frac{1 \text{ mole O}}{16.0 \text{ g O}} = 4.86 \text{ moles O}$$

Dividing by the smallest amount (1.62 moles):

$$\text{H:} \quad \frac{4.8 \text{ moles}}{1.62 \text{ moles}} = 3.0$$

$$\text{B:} \quad \frac{1.62 \text{ moles}}{1.62 \text{ moles}} = 1.00$$

$$\text{O:} \quad \frac{4.86 \text{ moles}}{1.62 \text{ moles}} = 3.00$$

empirical formula: H_3BO_3

Example Another white powder has been analyzed, and its composition is 27.4% sodium, 1.2% hydrogen, 14.3% carbon, and 57.1% oxygen. Find its empirical formula.

In 100. g of compound there exists:

27.4% Na × 100. g = 27.4 g Na
1.2% H × 100. g = 1.2 g H
14.3% C × 100. g = 14.3 g C
57.1% O × 100. g = 57.1 g O

Converting to moles:

$$27.4 \ \cancel{g \ Na} \times \frac{1 \ \text{mole Na}}{23.0 \ \cancel{g \ Na}} = 1.19 \ \text{moles Na}$$

$$1.2 \ \cancel{g \ H} \times \frac{1 \ \text{mole H}}{1.01 \ \cancel{g \ H}} = 1.2 \ \text{moles H}$$

$$14.3 \ \cancel{g \ C} \times \frac{1 \ \text{mole C}}{12.0 \ \cancel{g \ C}} = 1.19 \ \text{moles C}$$

$$57.1 \ \cancel{g \ O} \times \frac{1 \ \text{mole O}}{16.0 \ \cancel{g \ O}} = 3.57 \ \text{moles O}$$

Dividing by the smallest amount (1.19 moles):

$$\text{Na:} \quad \frac{1.19 \ \cancel{\text{moles}}}{1.19 \ \cancel{\text{moles}}} = 1.00$$

$$\text{H:} \quad \frac{1.2 \ \cancel{\text{moles}}}{1.19 \ \cancel{\text{moles}}} = 1.0$$

$$\text{C:} \quad \frac{1.19 \ \cancel{\text{moles}}}{1.19 \ \cancel{\text{moles}}} = 1.00$$

$$\text{O:} \quad \frac{3.57 \ \cancel{\text{moles}}}{1.19 \ \cancel{\text{moles}}} = 3.00$$

empirical formula: $NaHCO_3$

Example The formula weight of a compound is 612 amu. Its empirical formula is $NaPO_3$. Find the actual formula.

$$1 \ \cancel{\text{atom Na}} \times \frac{23.0 \ \text{amu}}{1 \ \cancel{\text{atom Na}}} = 23.0 \ \text{amu}$$

$$1 \ \cancel{\text{atom P}} \times \frac{31.0 \ \text{amu}}{1 \ \cancel{\text{atom P}}} = 31.0 \ \text{amu}$$

$$3 \ \cancel{\text{atoms O}} \times \frac{16.0 \ \text{amu}}{1 \ \cancel{\text{atom O}}} = \underline{48.0 \ \text{amu}}$$

mass of empirical formula, $NaPO_3$ = 102.0 amu

$$n = \frac{\text{formula weight}}{\text{mass of empirical formula}}$$

$$= \frac{612 \text{ amu}}{102.0 \text{ amu}} = 6.00$$

actual formula is $Na_6P_6O_{18}$.

Practice problem Potassium bisulfite has the following composition: 35.2% K, 28.8% S, and 36.0% O. Find its empirical formula.

ANSWER $K_2S_2O_5$

Practice problem The percentage composition of sodium saccharin, found in artificial sweetners, is 11.2% Na, 41.0% C, 2.0% H, 6.8% N, 23.4% O, and 15.6% S. What is its empirical formula?

ANSWER $NaC_7H_4NO_3S$

Summary **8.1** Molecular weight is the mass of a molecule. Formula weight is the mass of a formula unit. Therefore, the molecular or formula weight is the sum of the atomic weights of all the atoms (or ions) present in a molecule or formula unit.

8.2 The mole is a unit that measures the amount of a substance. One mole contains Avogadro's number (6.02×10^{23}) of atoms, formula units, or molecules. It has a mass equal to the atomic, formula, or molecular weight of the substance.

8.3–8.5 To convert moles to grams, we multiply the given number of moles by the molar mass (the mass of one mole of the substance). To convert grams to moles, we multiply the given number of grams by a conversion factor that relates moles to grams. These conversions are basic to most chemical calculations and laboratory work. Using Avogadro's number, we can also relate the number of moles (or grams) of a substance to the number of atoms or molecules present.

8.6 Compounds have a definite composition that is reflected by their percentage composition. The percentage by weight of each element in a compound is the compound's percentage composition. We can find the percentage composition of a compound by dividing the mass of all the atoms of each element in the compound by the formula or molecular weight.

8.7 Finding the formula of a compound from its percentage composition is also necessary at times. It is based on finding the relative numbers of moles of each atom in the compound. The simplest whole number ratios of the atoms is expressed as the empirical formula. If the molecular weight is known, the actual molecular formula can be found from the empirical formula.

For each term on the left, choose a phrase from the right-hand column that most closely matches its meaning.

1 molecular weight	**a** mass of one mole
2 formula weight	**b** mass of molecule
3 mole	**c** simplest whole number formula
4 Avogadro's number	**d** mass of formula unit
5 molar mass	**e** percentages of elements in compound
6 percentage composition	**f** unit for amount of substance
7 empirical formula	**g** number of particles in mole

EXERCISES BY TOPIC

Formula or molecular weight of compounds (8.1)

8 List the number of atoms of each type in **a** $K_2Cr_2O_7$ **b** $Fe(NH_4)(SO_4)_2$ **c** $N(CH_3)_3$ **d** $Zn(C_2H_3O_2)_2$

9 Round-off each of the following atomic weights to the nearest tenth: **a** 186.207 **b** 54.9380 **c** 102.9055 **d** 200.59 **e** 26.98154

10 What is the molecular weight of DDT, whose formula is $C_{14}H_9Cl_5$?

11 Find the molecular weight of the antibiotic penicillin G, whose formula is $C_{16}H_{18}N_2O_4S$.

12 Calculate the molecular or formula weights of the compounds in Exercise 8.

13 Determine the molecular or formula weight of **a** Li_3PO_4 **b** $Al_2(SO_4)_3$ **c** C_2H_5OH **d** $Ca(NO_3)_2$ **e** $NaHCO_3$.

The mole (8.2)

14 What does the unit *mole* measure?

15 What does Avogadro's number represent?

16 Why should a mole of one substance have a different mass than a mole of another substance?

17 Find the molar mass of **a** gold, Au **b** sulfur dioxide, SO_2 **c** calcium nitrate, $Ca(NO_3)_2$

18 What is the molar mass of aspirin, $C_9H_8O_4$?

19 Calculate the mass of one mole of: **a** Na_3PO_4 **b** C_5H_5N **c** $Sr(C_2H_3O_2)_2$

Converting moles to grams (8.3)

20 What mass of nickel contains 0.0138 mole of the metal?

21 Convert to grams: **a**   5.5 moles $Pb(NO_3)_2$ **b**   0.236 mole KH_2PO_4 **c** 1.48 moles $NaClO_4$

22 A bottle contains 0.890 mole of rubbing alcohol, C_3H_8O. What mass of alcohol does it hold?

23 Find the mass in grams of the following amounts of potassium permanganate, $KMnO_4$: **a** 10.0 mole **b** 0.450 moles **c** 1.55×10^{-3} mole

24 Determine the mass in grams of **a** 9.75 moles $C_8H_{16}O_2$ **b** 5.4×10^{-3} mole $HgCl_2$ **c** 0.153 mole $Ca(C_2H_3O_2)_2$

25 Nitroglycerin, used medically for heart patients and in the manufacture of explosives, has the formula $C_3H_5N_3O_9$. Find the mass of 5.22 moles of this compound.

Converting grams to moles (8.4)

26 Find the number of moles of ammonium chloride present in the following masses of NH_4Cl: **a** 50.0 g **b** 5.00 g **c** 294 g **d** 1.00×10^{-2} g

27 How many moles of nicotine, $C_{10}H_{14}N_2$, are present in 10.4 g of this poisonous compound?

28 If you weigh out 75.0 g of hypo (sodium thiosulfate, $Na_2S_2O_3$) for a photographic solution, what is the amount in moles?

29 Convert to moles: **a** 20.0 g of $Mg(OH)_2$ **b** 3.80 g of C_6H_6 **c** 410. g of $Al_2(SO_4)_3$

30 Find the number of moles in: **a** 17.0 g of H_3PO_4 **b** 0.983 g of $KAl(SO_4)_2$ **c** 7.31×10^{-2} g of CH_4O

31 Find the number of moles present in **a** 13.5 g I_2 **b** 0.995 g of phosphoric acid, H_3PO_4

Conversions based on Avogadro's number (8.5)

32 How many oxygen atoms are present in one mole of sodium carbonate, Na_2CO_3?

33 What is the number of K_3PO_4 formula units present in 3.20 moles of potassium phosphate?

34 Find the number of lithium ions in 0.820 mole of lithium borate, Li_3BO_3.

35 A sample of iron oxide, Fe_2O_3, has a mass of 250. g. How many formula units are present in this sample?

36 How many molecules are present in 18.0 g of glucose, $C_6H_{12}O_6$?

37 What is the mass in grams of 3.12×10^{24} uranium atoms?

Percentage composition (8.6)

38 What does the percentage composition of a compound represent?

39 Find the percentage composition of the male hormone testosterone, $C_{19}H_{28}O_2$.

40 Determine the percentage by weight of each element in mercury(II) thiocyanate, $Hg(SCN)_2$.

41 A form of Vitamin D has the formula $C_{28}H_{44}O$. What is its percentage composition?

42 Chlorophyll in plants has the formula $C_{55}H_{72}MgN_4O_5$. Find its percentage composition.

Determining chemical formulas (8.7)

43 Why is it useful to know the percentage composition of a compound?

44 How does an empirical formula differ from a molecular formula?

45 Find the empirical formula of a compound containing 83.23% C and 16.77% H.

46 A compound has the composition 11.62% Na, 64.13% I, and 24.25% O. Find its empirical formula.

47 A yellow dye contains 72.28% C, 3.64% H, and 24.07% O. Find its empirical formula.

48 A compound with a molecular weight of 180. amu has the composition 79.98% C, 4.48% H, and 15.55% N. Find its molecular formula.

49 A bleach contains 28.93% K, 23.72% S, and 47.35% O and has a formula weight of 270. amu. Find its formula.

ADDITIONAL EXERCISES

50 What is the formula or molecular weight of: **a** $C_{21}H_{30}O_4$ **b** K_2SiF_6 **c** $V(CO)_6$ **d** $UO_2(C_2H_3O_3)_2$ **e** $C_{16}H_{21}NO_3$ **f** $Cu(C_3H_5O_3)_2$ **g** $Ba(ClO_3)_2$ **h** $CH_3CH_2CO_2H$ **i** $Ag_2S_2O_3$ **j** $Ti_2(C_2O_4)_3$

51 Calculate the mass in grams of: **a** 2.98 moles $PbSO_4$ **b** 0.312 mole K_2PtCl_6 **c** 1.00×10^{-3} mole PH_3 **d** 15.2 moles $Hg(CN)_2$ **e** 0.0159 mole $C_8H_{11}NO_2$ **f** 2.215 moles $Bi_2(C_2O_4)_3$ **g** 2.02×10^2 moles NaH_2PO_2 **h** 0.98 mole C_8H_9Br **i** 18.2 moles KH_2AsO_4 **j** 7.9 moles $Mg(NH_2)_2$

52 Determine the number of moles of each substance in: **a** 152 g of $C_{18}H_{34}O_2$ **b** 51.3 g of $KNaHPO_4$ **c** 1.99 g of $Pb(NO_2)_2$ **d** 532 g of $C_9H_{13}N_3O$ **e** 19.2 g of $AgMnO_4$ **f** 210. g of $C_9H_{10}O_4$ **g** 6.12 moles of $KSCN$ **h** 1.83×10^{-3} g of $C_6H_{12}O_6$ **i** 17.3 g of $Ba(H_2PO_2)_2$ **j** 6.2×10^4 g of $Al_2(S_2O_3)_3$

53 Find the mass of one carbon atom.

54 Which substance is present in greater amount (more formula units), 100 g of KCl or 100 g of NaI? Explain.

55 Find the mass in grams of 5.0×10^{22} formula units of CuS.

56 Can an empirical formula and a molecular formula be the same? Give an example.

57 A compound contains calcium, carbon, hydrogen and oxygen. An analysis shows that it is 25.34% C, 3.82% H, and 40.46% O. Find its empirical formula.

58 A certain balance registers only masses that are greater than 0.001 g. What is the smallest number of gold atoms that it can measure?

59 Determine the percentage composition of the compounds listed in Exercise 50.

60 Determine the empirical formula of compounds having the following composition: **a** 67.57% C, 9.93% H, 22.49% O   **b** 32.87% Fe, 28.27% C, 1.19% H, 37.67% O   **c** 65.44% C, 5.49% H, 29.06% O   **d** 55.39% Pb, 18.95% Cl, 25.66% O   **e** 56.97% C, 3.94% H, 7.82% N, 22.32% O, 8.95% S

61 Find the actual formulas of the compounds in Exercise 60 if their molecular or formula weights are respectively: **a** 142.2 amu   **b** 169.9 amu   **c** 110.1 amu   **d** 374.1 amu   **e** 358.4 amu

62 Arrange the following quantities of gold in order of increasing mass: 200. mg Au   0.100 mole Au 2.50×10^{20} atoms Au

63 Determine the percent composition of KClO, $KClO_2$, $KClO_3$, and $KClO_4$.

64 Calculate the molar mass of: **a** H_3PO_4 **b** Fe_3O_4 **c** $HgBr_2$ **d** $Ba(OH)_2$ **e** XeF_6 **f** $C_{12}H_{22}O_{11}$ **g** Mn^{2+} **h** N **i** $H_2PO_4^-$

65 Talc has the following composition by weight: 19.22% Mg, 29.63% Si, 50.62% O, and 0.53% H. Find its empirical formula.

66 Which is the larger amount of substance in moles, 3.5 g of CO_2 or 3.5 g of NaCl?

67 The recommended nutritional daily allowance for iron is 18 mg for an adult. How many iron atoms are needed each day?

68 Which iron oxide has the highest percentage of iron: FeO, Fe_2O_3, or Fe_3O_4?

69 Find the mass in grams that corresponds to **a** 1.01 moles of quartz, SiO_2 **b** 0.42 mole of saccharin, $C_7H_5O_3SN$ **c** 1.0×10^{40} moles of HCl

70 How many atoms are present in each example in Exercise 69?

71 Calculate the percentage composition of **a** carborundum, SiC **b** LSD, $C_{20}H_{25}N_3O$ **c** ether, C_4H_6O **d** riboflavin, $C_{17}H_{20}N_4O_6$ **e** potassium pyrophosphate, $K_4P_2O_7$

72 An organic compound has the composition 52.18% C, 13.04% H, and 34.78% O, and a molecular weight of 92.1. Find its molecular formula.

73 What mass of barium contains the same number of atoms as 10.0 g of calcium?

74 The hormone epinephrine, also called adrenaline, is secreted by the body during periods of stress. Its composition is 59.00% C, 7.15% H, 7.65% N, and 26.60% O. Find its empirical formula.

75 Vitamin B_{12} has the formula $C_{63}H_{88}CoN_{14}O_{14}P$. Find **a** its formula weight **b** the number of millimoles in 25 μg of vitamin B_{12} **c** its percentage composition

PROGRESS CHART Check off each item when completed.

_____ Previewed chapter

Studied each section:

_____ **8.1** Formula or molecular weight of compounds

_____ **8.2** The mole

_____ **8.3** Converting moles to grams

_____ **8.4** Converting grams to moles

_____ **8.5** Conversions based on Avogadro's number

_____ **8.6** Percentage composition

_____ **8.7** Determining chemical formulas

_____ Reviewed chapter

_____ Answered assigned exercises:

    #'s _____

Review Exercises for Chapters Six–Eight

1 Draw the Lewis structure of: **a** PH_3 **b** H_2SO_3

2 What are the major differences between ionic, nonpolar covalent, and polar covalent bonding?

3 Nitrogen atoms are less electronegative than oxygen atoms. Explain this statement.

4 Explain why carbon atoms have a combining capacity of 4.

5 Write the formula of the ion that can be formed by **a** cesium   **b** sulfur   **c** bromine   **d** strontium

6 Write the name and formula of the ionic compound formed from: **a** calcium and iodine   **b** potassium and nitrogen

7 What is the difference between ammonia and ammonium ion?

8 Name the following: **a** BI_3 **b** MgF_2 **c** Cs_2S **d** P_2S_5 **e** CuI_2

9 Write the formula of: **a** nitric acid   **b** silicon tetraiodide   **c** lead(II) nitrate   **d** acetic acid   **e** aluminum sulfide

10 Name the following compounds: **a** H_2SO_4 **b** $LiOH$ **c** H_2SO_3 **d** NH_3

11 For each of the following pairs of elements, predict whether an ionic or covalent compound can be formed: **a** sulfur and oxygen   **b** barium and bromine   **c** phosphorus and bromine   **d** lithium and sulfur

12 Write the formulas of compounds formed from the pairs of elements in Exercise 11.

13 Find the formula weight of **a** $(NH_4)_2CO_3$ **b** $K_2Cr_2O_7$ **c** $CaHPO_4$

14 What is the mass in grams of 0.731 mole of each compound in Exercise 13?

15 Find the number of moles in 58.1 g of each compound in Exercise 13.

16 How many formula units are present in 1.000 kg of each compound in Exercise 13?

17 The hormone hydrocortisone has the formula $C_{21}H_{30}O_5$. Calculate its percentage composition by mass.

18 A local anesthetic has the composition: 65.43% C, 6.71% H, 8.48% N, 19.37% O. What is its empirical formula?

19 A component of cigarettes has the composition 74.03% C, 8.70% H, 17.27% N and has a molecular weight of 162.23 amu. Find its molecular formula.

NINE

Chemical reactions

In Chapter One we defined chemistry as the study of the composition, structure, and properties of substances, and the changes that they undergo. So far, by studying the nature of elements and compounds, we have concentrated on the first part of the definition of chemistry. Chapter Nine deals with the second part of our definition: the study of chemical change. This chapter describes how we represent these changes, gives examples of different types of changes, and relates energy to chemical change. To demonstrate an understanding of Chapter Nine, you should be able to:

1 Describe the relationship between compound formation and chemical change.

2 Identify the symbols in chemical equations.

3 State the law of conservation of mass and describe its relationship to chemical equations.

4 Balance simple chemical equations by trial and error.

5 Interpret balanced chemical equations.

6 Classify chemical reactions according to the type of reaction taking place.

7 Write chemical equations given the reactions in words or given the reactants and type of reaction.

8 Describe the relationship between energy changes and chemical reactions.

9 Describe the major sources and uses of energy.

10 Summarize the chemical process by which a major industrial product is manufactured.

FIGURE 9-1

Iron and sulfur. In (a), these two elements exist as a mixture, and the magnet can separate the iron. In (b), the elements have been combined into a compound, iron sulfide. The iron can no longer be separated with a magnet. (Photo by Al Green.)

(a)   (b)

9.1
CHEMICAL AND PHYSICAL CHANGE

When iron filings are added to powdered sulfur at room temperature, the result is a mixture of iron and sulfur. Once mixed, the iron and sulfur can be physically separated. For example, as shown in Figure 9-1(a), iron can be removed from the mixture with a magnet, leaving only the sulfur behind. In a **mixture**, the properties of each substance remain the same; iron still acts like iron, and sulfur still acts like sulfur.

However, if we mix iron with sulfur and then *heat* the mixture, a **chemical change** takes place. A new substance forms that has different properties from either iron or sulfur. The magnet no longer can separate the iron and sulfur (Figure 9-1(b)). Chemical bonds have formed between the atoms of these two elements. A single new substance is now present instead of two. This new substance is a compound, iron(II) sulfide.

Chemical changes generally result in **physical changes** as well. Physical changes involve changes in the state, color, odor, density, and other characteristics of a substance. The occurence of physical change in a substance, however, does not guarantee that a chemical change has taken place. For example, if we boil water, we see a physical change—the liquid bubbles and steam forms. However, the chemical composition and identity of the water remains the same, and no chemical change takes place.

A **chemical reaction**, the process of chemical change, is caused by the making or breaking of chemical bonds. Reactions occur when atoms or molecules bump into each other in just the right way, and also with enough energy to cause existing chemical bonds to break or new bonds to form. During a chemical reaction, atoms rearrange themselves into new combinations. You are already familiar with some chemical reactions. The rusting of iron, the burning of wood, and the digesting of food all involve chemical change.

All compounds are formed by chemical reactions. The same compound can be formed by several different chemical reactions. But regardless of how a particular compound is made, its composition is always the same. This observation is summarized by the **law of definite**

proportions: *a given compound always contains its component elements in fixed proportions by weight.* The law, illustrated in the discussion of percentage composition in Section 8.6, follows from the atomic theory. A compound consists of molecules or formula units, which are fixed combinations of atoms or ions, joined by bonds formed through a chemical reaction. For example, the compound water consists of molecules of H_2O. No matter how water is made or where it is found, it always contains two hydrogen atoms for every one oxygen atom.

Mixtures, on the other hand, consist of substances that have not reacted with each other. There are no chemical bonds formed between the components in a mixture, and the mixture's composition therefore is not fixed.

9.2
CHEMICAL
EQUATIONS

A **chemical equation** is a shorthand representation of a chemical reaction. Using the formulas of the substances involved in the reaction, the chemical equation summarizes the chemical change that takes place. We can think of chemical equations as chemical sentences. We write them by following certain rules, just as we do when we write sentences in English (see Figure 9-2).

The equation for the reaction of iron and sulfur is:

$$\underbrace{Fe + S}_{\text{reactants}} \rightarrow \underbrace{FeS}_{\text{product}}$$

The starting substances, called **reactants**, are separated from the new substance, called the **product**, by an arrow. The arrow shows the direction of chemical change by pointing from the reactants to the products. An equation is read in the direction of the arrow, which is usually from left to right. When more than one substance appears on the same side of the arrow, they are separated by a plus sign ($+$). For example, in our sample equation, the two reactants, Fe and S, are separated by

FIGURE 9-2

(Peanuts by Charles M. Schultz. Copyright © 1980 by United Feature Syndicate, Inc.)

TABLE 9-1 *Symbols used in chemical equations*

Symbol	Meaning
$+$	plus; separates substances
$\rightarrow$	forms, produces, yields; points from reactants to products
$\leftrightarrows$	reversible reaction; proceeds in both forward and reverse direction
$\uparrow$	gas formed, as in $H_2\uparrow$
$\downarrow$	solid formed, as in $AgCl\downarrow$
(s)	exists as solid, as in $Zn(s)$
(l)	exists as liquid, as in $Br_2(l)$
(g)	exists as gas, as in $CO_2(g)$
(aq)	aqueous, dissolved in water, as in $Na^+(aq)$
Δ	heat

a plus sign. We can read the sample question in any of the following ways:

Iron and sulfur react to form iron(II) sulfide.
Iron plus sulfur produces iron(II) sulfide.
Iron plus sulfur yields iron(II) sulfide.

Table 9-1 presents other symbols used in chemical equations. A double arrow $\rightleftarrows$ indicates that the reaction is reversible. (Reversible reactions are discussed in Chapter Sixteen.) Sometimes it is useful to show that a particular product separates physically from the other substances in the reaction. An upward arrow $\uparrow$ shows that a product escapes as a gas. A downward arrow $\downarrow$ means that a product deposits as a solid, called a **precipitate**. For example, in the reaction

$$AgNO_3 + HCl \rightarrow AgCl\downarrow + HNO_3$$

silver   hydrochloric   silver   nitric
nitrate   acid   chloride   acid

silver chloride settles to the bottom of the reaction flask as a white powder, while the other substances remain dissolved.

It is sometimes necessary to specify the physical states of the substances involved in a reaction. The symbols used are (s) for solid, (l) for liquid, (g) for gas, and (aq) for aqueous (meaning dissolved in water). These symbols are placed in parentheses on the right sides of the formulas. For example, examine the following equation:

$$Zn(s) + H_2SO_4(aq) \rightarrow ZnSO_4(aq) + H_2(g)$$

zinc   sulfuric   zinc   hydrogen
acid   sulfate

Here, solid zinc reacts with aqueous sulfuric acid to form aqueous zinc sulfate and gaseous hydrogen.

Reaction conditions, such as the temperature needed for the reaction to occur, may appear above the arrow. For example:

$$CaCO_3 \xrightarrow{1000°C} CaO + CO_2$$

calcium   calcium   carbon
carbonate   oxide   dioxide

This equation indicates that a temperature of 1000°C is required to convert calcium carbonate to calcium oxide and carbon dioxide. If we want to show that an unspecified amount of heat is needed, we can simple use the Greek symbol Δ (delta), which stands for heat in general. For example:

$$Fe + S \xrightarrow{\Delta} FeS$$

iron   sulfur   iron(II)
               sulfide

Substances called **catalysts** are often added to reactants to make the reaction take place faster. Because catalysts themselves are unchanged at the end of the reaction, small quantities can be used over and over again. The catalytic converters in automobile speed up reactions that make exhaust fumes less harmful. For example, they hasten the conversion of carbon monoxide to carbon dioxide.

$$2CO + O_2 \xrightarrow{catalyst} 2CO_2$$

carbon   oxygen   carbon
monoxide   dioxide

Catalysts are discussed further in Chapter Sixteen.

Example   Translate the following equations into words.

a $FeS + Cu_2O \rightarrow FeO + Cu_2S$
Iron(II) sulfide reacts with copper(I) oxide to form iron(II) oxide and copper(I) sulfide.

b $C(s) + O_2(g) \rightarrow CO_2(g)$
Solid carbon reacts with gaseous oxygen to form gaseous carbon dioxide.

c $H_2CO_3 \xrightarrow{\Delta} H_2O + CO_2\uparrow$
Upon heating, hydrogen carbonate (or carbonic acid) forms water and carbon dioxide. The carbon dioxide escapes as a gas.

Smithsonian Institution: Photo #46-835D

The French chemist Lavoisier is known as "the father of modern chemistry." One of his greatest contributions was his study of the process of combustion, in which a substance is burned in air. In one experiment, Lavoisier even burned a diamond to show that it consisted only of the element carbon. Lavoisier demonstrated that in combustion a substance combines with an element in the air. He learned of that element, oxygen, from Joseph Priestley, although Lavoisier did not give Priestley credit. Apparently Lavoisier wanted others to think that he had discovered it himself.

Lavoisier helped to develop a system of naming compounds that is similar to the one that we use today. He replaced names like "butter of antimony" with names based on the elements present in the compound. In addition, he applied careful measurement to the study of chemical reactions. Using a balance, he established the law of conservation of mass, which became the foundation of 19th century chemistry.

Lavoisier held various government positions. While he was director of the gunpowder commission, one of his associates was Éleuthère Irénée du Pont de Nemours, who later built his own powder mill in the U.S. That mill in Delaware was the start of the Du Pont corporation.

To help finance his research, Lavoisier worked for a private firm that collected taxes for the French government. When the Revolution broke out, he was arrested and killed by the guillotine. Upon learning of Lavoisier's death, the great mathematician Lagrange mourned: "A moment was all that was necessary to strike off his head, and probably a hundred years will not be sufficient to produce another like it."

**9.3
CONSERVATION
OF MASS**

The **law of conservation of mass**, demonstrated by Antoine Lavoisier in the late 18th century, is a basic principle of chemistry. It states that in any chemical reaction, the total mass of the reactants must equal the total mass of the products. Therefore, in the reaction of iron with sulfur, the mass of the iron, plus the mass of the sulfur with which it reacts, is equal to the mass of iron sulfide produced. (See Figure 9-3.) A chemical reaction took place which reshuffled the atoms, but left the total amount of mass unchanged. In other words, matter was neither created nor destroyed. In Chapter Two we saw that under ordinary conditions energy is conserved. Now we see that the same holds true for mass.

In terms of atoms, the law of conservation of mass means that *the total number of atoms of each kind in the reactants must equal the total number of atoms of each kind in the products.* In other words, the number of atoms of each element must be the same before and after the reaction. An atom of one element cannot be changed to an atom of another element by a chemical reaction because the nuclei are not affected. Chemical reactions involve only the electrons of atoms.

In the iron-sulfur equation

$$Fe + S \rightarrow FeS$$

Reactants	*Products*
1 iron atom	1 iron sulfide formula
plus	unit, containing 1 iron
1 sulfur atom	atom and 1 sulfur atom

the number of iron atoms is the same on both sides of the equation. One iron atom appears on the left, and one appears on the right,

FIGURE 9-3

Conservation of mass. In a chemical reaction, the mass of the reactants equals the mass of the products. Here, the mass of the reactants iron and sulfur (left side) is shown to be equal to the mass of the product, iron sulfide (right side). (Photo by Al Green.)

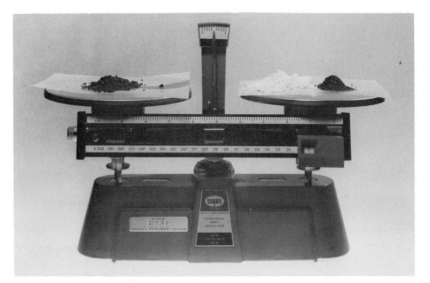

combined with sulfur. (For the purpose of counting, no distinction is made between atoms and ions.) Similarly, one sulfur atom is present on each side of the equation. This equation is said to be **balanced**—it is written in such a way that it obeys the law of conservation of mass. The following equation is *not* balanced.

$$H_2 + O_2 \rightarrow H_2O \qquad \text{not balanced}$$

One oxygen atom appears to have disappeared in the reaction as it is written here. Since an atom cannot be destroyed in a chemical reaction, the equation must be modified by balancing, as described in the following section.

9.4
BALANCING
CHEMICAL
EQUATIONS

Every chemical equation must be balanced. That is, the number of atoms of each kind of element must be the same on both sides of the equation. To balance an equation, we place numbers called **coefficients** in front of the reactants and products. In the iron and sulfur equation that we considered in the last section, coefficients of 1 are understood to exist in front of each substance and do not need to be written.

$$1\,Fe + 1\,S \rightarrow 1\,FeS$$

However, coefficients other than than 1 must always be written. There are no definite rules for balancing most equations. Usually, we must use trial and error to determine which numbers to use as coefficients in an equation. However, the following guidelines are helpful:

1 Start with the atoms of the element that appears in the fewest number of formulas on both sides of the equation. If each element appears in the same number of formulas, then it does not matter where you start. For example, in the equation:

$$C_6H_{12} \rightarrow C_6H_6 + H_2$$

Start by balancing carbon atoms. Carbon appears in only one formula on each side of the equation.

2 Place coefficients in front of the reactants and products containing the chosen element to make the total number of its atoms the same on both sides of the equation. For example, the equation

$$H_2O_2 \rightarrow O_2 + H_2O \qquad \text{not balanced}$$

is balanced by placing a 2 in front of H_2O_2 (hydrogen peroxide) and a 2 in front of H_2O (water).

$$2H_2O_2 \rightarrow O_2 + 2H_2O \qquad \text{balanced}$$

There are now four hydrogen atoms and four oxygen atoms on each side of the equation. Note that placing a coefficient in front of a formula multiplies each atom in that formula by the value of the coefficient.

3 If the equation is still not balanced, continue the process by choosing atoms of another element and adjusting their coefficients. Generally, when atoms of many elements must be balanced, it is best first to balance all the atoms except hydrogen and oxygen. Then balance hydrogen atoms, and finally, oxygen atoms.

Consider the following unbalanced equation, which represents the formation of water from hydrogen and oxygen molecules:

$$H_2 + O_2 \rightarrow H_2O \qquad \text{not balanced}$$

To a beginning student, it might make sense to balance the equation by changing the formula of water:

$$H_2 + O_2 \rightarrow H_2O_2 \qquad \text{incorrectly balanced}$$

But this equation no longer represents the same reaction! The product of the reaction is now hydrogen peroxide, rather than water. Therefore, the following rule is very important: When you are given an equation to balance, *never change any formulas*. While you are learning how to balance equations, you might want to place each formula of an unbalanced equation in a box to help you to remember this rule. Then balance the equation by placing the appropriate coefficients *in front* of each box:

$$2 \boxed{H_2} + \boxed{O_2} \rightarrow 2 \boxed{H_2O} \qquad \text{balanced}$$

We will now use our three guidelines to balance the following equation. Glucose (known as blood sugar), $C_6H_{12}O_6$, reacts with oxygen (in the breakdown of carbohydrates in the body), to form carbon dioxide and water.

$$\underset{\text{glucose}}{C_6H_{12}O_6} + \underset{\text{oxygen}}{O_2} \rightarrow \underset{\substack{\text{carbon} \\ \text{dioxide}}}{CO_2} + \underset{\text{water}}{H_2O} \qquad \text{not balanced}$$

This equation is not balanced because there are different numbers of atoms of carbon, hydrogen, and oxygen on each side of the equation.

Reactants	*Products*
6 carbon atoms	1 carbon atom
12 hydrogen atoms	2 hydrogen atoms
8 oxygen atoms	3 oxygen atoms

Because this equation violates the law of conservation of mass, the equation is incorrect as written. Therefore, we must balance it.

The atoms that appear in the fewest formulas are carbon and hydrogen. They are each in only two compounds in the equation. Oxygen, on the other hand, appears in all four compounds in the equation and should be left for last. We will begin with carbon. There are six carbon atoms on the left side of the equation and only one on the right side. Therefore, we place the coefficient 6 in front of CO_2 on the right side of the equation:

$$C_6H_{12}O_6 + O_2 \rightarrow 6CO_2 + H_2O \qquad \text{not balanced}$$

Reactants
 6 carbon atoms
 12 hydrogen atoms
 8 oxygen atoms

Products
 6 carbon atoms
 2 hydrogen atoms
 13 oxygen atoms

There are now equal numbers of carbon atoms on each side of the equation: six in the glucose molecule and six in the six molecules of carbon dioxide. Notice that the coefficient 6 multiplies the entire formula next to it. Thus, $6CO_2$ means 6 carbon atoms and 6×2 or 12 oxygen atoms. However, the coefficient 6 does not apply to H_2O because it is a different compound, separated from carbon dioxide by a plus (+) sign.

Next we balance the hydrogen atoms. There are 12 hydrogen atoms on the left side of the equation and only 2 on the right side. We balance the hydrogen atoms by placing the coefficient 6 in front of the H_2O on the right side of the equation.

$$C_6H_{12}O_6 + O_2 \rightarrow 6CO_2 + 6H_2O \qquad \text{not balanced}$$

Reactants
 6 carbon atoms
 12 hydrogen atoms
 8 oxygen atoms

Products
 6 carbon atoms
 12 hydrogen atoms
 18 oxygen atoms

There are now equal numbers of hydrogen atoms on each side: 12 in glucose, and 12 in the six molecules of H_2O (because $6 \times 2 = 12$).

Finally, we balance the oxygen atoms by placing the coefficient 6 in front of O_2:

$$C_6H_{12}O_6 + 6O_2 \rightarrow 6CO_2 + 6H_2O \qquad \text{balanced}$$

Reactants
 6 carbon atoms
 12 hydrogen atoms
 18 oxygen atoms

Products
 6 carbon atoms
 12 hydrogen atoms
 18 oxygen atoms

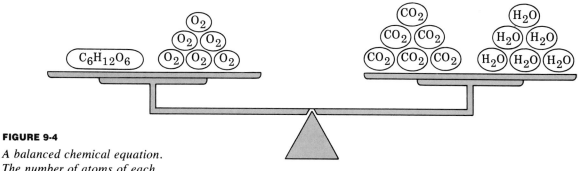

FIGURE 9-4

A balanced chemical equation. The number of atoms of each kind (C, H, O) is the same on both sides: 6 C atoms, 12 H atoms, 18 O atoms.

$$C_6H_{12}O_6 + 6O_2 \longrightarrow 6CO_2 + 6H_2O$$

The 18 oxygen atoms on the left (6 in glucose and 12 in the six O_2 molecules) balance the 18 oxygen atoms on the right (12 in the six CO_2 molecules and 6 in the six H_2O molecules). As illustrated by Figure 9-4, the equation is now completely balanced. There are the same number of atoms of each kind (carbon, hydrogen, and oxygen) in the reactants and in the products.

To balance equations involving compounds containing polyatomic ions, we can often simplify the process by treating the polyatomic ion as a single unit (as long as the polyatomic ion is not changed in the reaction). Consider the reaction between silver nitrate and sodium sulfate to form silver sulfate and sodium nitrate:

$$AgNO_3 + Na_2SO_4 \rightarrow Ag_2SO_4 + NaNO_3 \qquad \text{not balanced}$$

silver nitrate   sodium sulfate   silver sulfate   sodium nitrate

Reactants	*Products*
1 silver atom	2 silver atoms
1 nitrate group	1 nitrate group
2 sodium atoms	1 sodium atom
1 sulfate group	1 sulfate group

Both the nitrate ion, NO_3^-, and the sulfate ion, SO_4^{2-}, can be balanced as single units rather than as separate nitrogen, oxygen, and sulfur atoms. In the first step, the coefficient 2 is placed in front of silver nitrate to balance the silver atoms.

$$2AgNO_3 + Na_2SO_4 \rightarrow Ag_2SO_4 + NaNO_3 \qquad \text{not balanced}$$

Reactants	*Products*
2 silver atoms	2 silver atoms
2 nitrate groups	1 nitrate group
2 sodium atoms	1 sodium atom
1 sulfate group	1 sulfate group

Next, the nitrate groups (as well as sodium atoms) are balanced by placing the coefficient 2 in front of sodium nitrate.

$$2AgNO_3 + Na_2SO_4 \rightarrow Ag_2SO_4 + 2NaNO_3 \qquad \text{balanced}$$

Reactants	*Products*
2 silver atoms	2 silver atoms
2 nitrate groups	2 nitrate groups
2 sodium atoms	2 sodium atoms
1 sulfate group	1 sulfate group

The equation is now balanced. You may have noticed that in the first step of this procedure, the nitrate groups, which initially were balanced, became *un*balanced when the coefficient of silver nitrate was changed. Do not let this side effect disturb you. It was a necessary setback that was easily corrected in a later step.

The coefficients of a balanced equation should be in the lowest whole number terms. For example, each of these equations is balanced.

$$2H_2 + O_2 \rightarrow 2H_2O$$

$$4H_2 + 2O_2 \rightarrow 4H_2O$$

$$18H_2 + 9O_2 \rightarrow 18H_2O$$

However, only the first equation contains coefficients in their lowest whole number ratios. Balanced equations should be written in the simplest and clearest way possible, and therefore the *first* equation is best.

To summarize, we balance equations by adjusting the coefficients of the reactants and products. We adjust only the coefficients, and never change the formulas of elements or compounds in the equation. The coefficients should be expressed in the lowest whole number terms.

Example Balance the following equations.

a HF $\;\;\;+$ SiO$_2$ $\rightarrow$ SiF$_4$ $\;\;+$ H$_2$O
  hydrogen   silicon   silicon   water
  fluoride   dioxide   fluoride

Balance F: $4HF + SiO_2 \rightarrow SiF_4 + H_2O$
Balance H: $4HF + SiO_2 \rightarrow SiF_4 + 2H_2O$ \qquad balanced

b Al $\;\;\;+$ H$_2$O $\rightarrow$ Al$_2$O$_3$ $\;\;+$ H$_2$
  aluminum   water   aluminum   hydrogen
                     oxide

Balance Al: $2Al + H_2O \rightarrow Al_2O_3 + H_2$

Balance O: $2Al + 3H_2O \rightarrow Al_2O_3 + H_2$

Balance H: $2Al + 3H_2O \rightarrow Al_2O_3 + 3H_2$ balanced

c $KNO_3 + C \rightarrow K_2CO_3 + CO + N_2$

   potassium carbon potassium carbon   nitrogen
   nitrate   carbonate monoxide

Balance K: $2KNO_3 + C \rightarrow K_2CO_3 + CO + N_2$
  (and N)

Balance O: $2KNO_3 + C \rightarrow K_2CO_3 + 3CO + N_2$

Balance C: $2KNO_3 + 4C \rightarrow K_2CO_3 + 3CO + N_2$ balanced

d $Al_2(SO_4)_3 + Na_3PO_4 \rightarrow AlPO_4 + Na_2SO_4$

  aluminum   sodium   aluminum sodium
  sulfate   phosphate phosphate sulfate

Balance Al: $Al_2(SO_4)_3 + Na_3PO_4 \rightarrow 2AlPO_4 + Na_2SO_4$

Balance SO_4^{2-}: $Al_2(SO_4)_3 + Na_3PO_4 \rightarrow 2AlPO_4 + 3Na_2SO_4$

Balance PO_4^{3-}: $Al_2(SO_4)_3 + 2Na_3PO_4 \rightarrow 2AlPO_4 + 3Na_2SO_4$
  (and Na)   balanced

e $Ca(OH)_2 + K_3PO_4 \rightarrow Ca_3(PO_4)_2 + KOH$

  calcium   potassium calcium   potassium
  hydroxide phosphate phosphate hydroxide

Balance Ca: $3Ca(OH)_2 + K_3PO_4 \rightarrow Ca_3(PO_4)_2 + KOH$

Balance OH^-: $3Ca(OH)_2 + K_3PO_4 \rightarrow Ca_3(PO_4)_2 + 6KOH$

Balance PO_4^{3-}: $3Ca(OH)_2 + 2K_3PO_4 \rightarrow Ca_3(PO_4)_2 + 6KOH$
  (and K)   balanced

9.5 INTERPRETING BALANCED EQUATIONS

A balanced equation provides a useful summary of a chemical reaction. It not only states which elements or compounds are used up and which are formed, but also shows the relative amounts of each. Equations can be read or interpreted in various ways. The most obvious way to interpret an equation is in terms of atoms, molecules or formula units. For example:

$C_6H_{12}O_6 + 6O_2 \rightarrow 6CO_2 + 6H_2O$
One glucose molecule plus six oxygen molecules form six carbon dioxide molecules plus six water molecules.

$Fe_2O_3 + 3C \rightarrow 2Fe + 3CO$
One iron(III) oxide formula unit plus three carbon atoms form two iron atoms plus three carbon monoxide molecules.

Anything that we can say about the relative numbers of atoms, molecules, or formula units in a reaction can also be said about the relative numbers of moles. This principle is extremely important because moles are practical amounts that can be easily measured in the laboratory. The principle is based on the fact that a mole is just a large collection of atoms, molecules, or formula units, and that one mole of any substance contains the same number of particles. Thus, if one molecule of hydrogen reacts with one molecule of oxygen in a particular reaction, it must also be true that one *mole* of hydrogen molecules reacts with one *mole* of oxygen molecules. The relative amounts of hydrogen and oxygen that react are the same (one to one in this case) whether we speak of molecules or of moles. Only the total number of molecules being considered has increased. As you will see in Chapter Ten, we interpret equations in terms of moles whenever we perform calculations based on reactions.

When we interpret an equation in moles, we must be careful to say exactly what we mean. For example, "one mole of hydrogen" can mean two different things. It can mean one mole of H atoms or one mole of H_2 molecules. To avoid this problem, we must always specify the type of particle that is involved, as shown in these examples:

$N_2 + 3H_2 \rightarrow 2NH_3$
One mole of nitrogen molecules plus three moles of hydrogen molecules form two moles of ammonia molecules.

$2K + 2H_2O \rightarrow 2KOH + H_2$
Two moles of potassium atoms plus two moles of water molecules form two moles of potassium hydroxide formula units plus one mole of hydrogen molecules.

Example   Interpret the following reactions in terms of moles.

a $2ZnS + 3O_2 \rightarrow 2ZnO + 2SO_2$
Two moles of zinc sulfide formula units plus three moles of oxygen molecules form two moles of zinc oxide formula units plus two moles of sulfur dioxide molecules.

b $3H_2 + CO \rightarrow CH_4 + H_2O$
Three moles of hydrogen molecules plus one mole of carbon monoxide molecules form one mole of methane molecules plus one mole of water molecules.

c $V_2O_5 + 5Ca \rightarrow 2V + 5CaO$
One mole of vanadium(V) oxide formula units plus five moles of calcium atoms form two moles of vanadium atoms plus five moles of calcium oxide formula units.

d $2NH_3 + CO_2 + H_2O + 2NaCl \rightarrow Na_2CO_3 + 2NH_4Cl$
Two moles of ammonia molecules plus one mole of carbon dioxide molecules plus one mole of water molecules plus two moles of sodium chloride formula units form one mole of sodium carbonate formula units plus two moles of ammonium chloride formula units.

e $2Fe_2O_3 + 6C + 3O_2 \rightarrow 4Fe + 6CO_2$

Two moles of iron(III) oxide formula units plus six moles of carbon atoms plus three moles of oxygen molecules form four moles of iron atoms plus six moles of carbon dioxide molecules.

These equations give no information about how the reaction actually takes place. For example, several steps may be involved in the conversion of reactants to products. The equations do, however, show the net change that occurs through the reaction.

9.6

TYPES OF CHEMICAL REACTIONS

Chemical reactions can be classified into four major reaction types.

1 Combination, or **synthesis**, **reactions** are reactions in which two or more reactants combine to form one product. The formation of iron(III) oxide (rust) from iron and oxygen is an example of a combination reaction. Here two elements, a metal and a nonmetal, combine to form an ionic compound.

$$4Fe + 3O_2 \rightarrow 2Fe_2O_3$$
iron   oxygen   iron(III) oxide

Since oxygen is the nonmetal, the product is called an oxide. The synthesis of ammonia is another example of a combination reaction. In this case, two nonmetallic elements combine to form a covalent compound.

$$N_2 + 3H_2 \rightarrow 2NH_3$$
nitrogen   hydrogen   ammonia

If one of the nonmetals is oxygen, a covalent oxide results, as in the formation of sulfur dioxide.

$$S + O_2 \rightarrow SO_2$$
sulfur   oxygen   sulfur dioxide

Compounds can themselves combine. For example, the compounds ammonia and hydrogen chloride can react to form ammonium chloride, shown in Figure 9-5(a).

$$NH_3 + HCl \rightarrow NH_4Cl$$
ammonia   hydrogen   ammonium
         chloride   chloride

Polymerization is a special type of combination reaction in which many small molecules, either of the same or different kinds, join together to make a huge macromolecule called a polymer. Plastics, nylon, starch, proteins, and the genetic material, DNA, are all polymers. Polymerization is discussed in Chapter Eighteen.

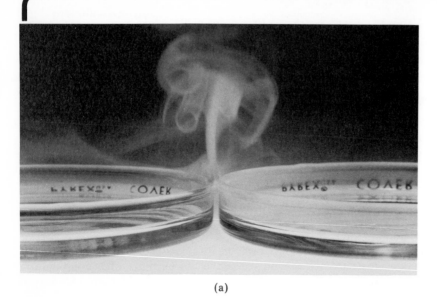

(a)

(c)

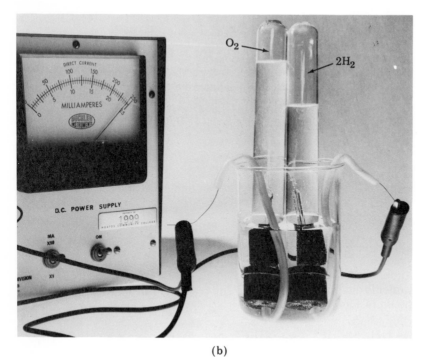

(b)

(d)

FIGURE 9-5

Examples of chemical reactions. (a) synthesis: $NH_3(g) + HCl(g) \rightarrow$
$NH_4Cl(s)$; (b) decomposition: $2H_2O(l) \rightarrow 2H_2(g) + O_2(g)$; (c) replacement or
substitution: $Fe(s) + 2HCl(aq) \rightarrow FeCl_2(aq) + H_2(g)$; (d) double
replacement or metathesis: $BaCl_2(aq) + Na_2SO_4(aq) \rightarrow BaSO_4(s) +$
$2NaCl(aq)$. (Photos by Al Green.)

2 Decomposition reactions involve the breakdown of a compound into simpler compounds or elements. Simple binary compounds (composed of two elements) can be decomposed into their elements. Water, for example, is broken down into hydrogen and oxygen molecules by an electric current. This process, called electrolysis, is shown in Figure 9-5(b).

$$2H_2O \xrightarrow[\text{current}]{\text{electric}} 2H_2 \ + \ O_2$$

$$\text{water} \qquad\qquad \text{hydrogen} \quad \text{oxygen}$$

In 1774, Joseph Priestley used a decomposition reaction to produce a new gas, which was later called oxygen. He decomposed mercury(II) oxide by heating it with sunlight through a magnifying glass.

$$2HgO \ \rightarrow \ 2Hg + O_2$$

$$\text{mercury(II)} \quad \text{mercury oxygen}$$
$$\text{oxide}$$

Some compounds decompose to form both an element and a simpler compound. For example, potassium chlorate can be broken down to the element oxygen and the compound potassium chloride.

$$2KClO_3 \rightarrow 2KCl \ + \ 3O_2$$

$$\text{potassium} \quad \text{potassium oxygen}$$
$$\text{chlorate} \qquad \text{chloride}$$

This reaction can be used in the laboratory to generate oxygen. Other compounds decompose to give several simpler compounds. Calcium carbonate, for example, produces calcium oxide and carbon dioxide.

$$CaCO_3 \xrightarrow{\Delta} CaO + CO_2$$

$$\text{calcium} \quad \text{calcium carbon}$$
$$\text{carbonate} \quad \text{oxide} \quad \text{dioxide}$$

3 Replacement or **substitution reactions** are reactions in which one element takes the place of another element in a compound. For example, iron can replace hydrogen in the reaction given by the following equation:

$$Fe \ + \ 2HCl \ \rightarrow \ FeCl_2 + H_2$$

$$\text{iron} \quad \text{hydrochloric} \quad \text{iron(II)} \quad \text{hydrogen}$$
$$\text{acid} \qquad \text{chloride}$$

The iron combines with the chlorine of HCl, replacing the hydrogen, which is released as H_2. (See Figure 9-5(c))

It is possible to predict which elements can replace others in a compound. To make this prediction, we use a list of elements called an **activity series**. An element can replace any element that appears after it in this series. The activity series of the metals (and hydrogen) is:

>—Li—K—Ba—Sr—Ca—Na—Mg—Al—Mn—Zn—Cr—Fe—
Cd—Co—Ni—Sn—Pb—H—Cu—Ag—Pd—Hg—Pt—Au→
Activity series

Because hydrogen appears after iron, it can be replaced by iron, as we have already seen. Similarly, if we drop a copper penny into water containing silver nitrate ($AgNO_3$), the penny becomes coated with shiny silver. Copper replaces silver in the reaction:

$$Cu \ + \ AgNO_3 \rightarrow Ag \ + \ CuNO_3$$
copper   silver   silver   copper
nitrate   nitrate

The halogens also have an activity series:

Halogen activity series   >—F_2—Cl_2—Br_2—I_2→

From this series, we can predict that bromine can be replaced by chlorine. For example, chlorine combines with the sodium of NaBr, and releases bromine as Br_2, in the following reaction.

$$2NaBr \ + \ Cl_2 \ \rightarrow \ 2NaCl \ + \ Br_2$$
sodium   chlorine   sodium   bromine
bromide   chloride

Do-it-yourself 9

Removing tarnish

Many metals tarnish. They lose their bright finish when surface atoms react with other substances to form a dull coating. For example, silver tarnishes by reacting with sulfur to form a dark coating of silver sulfide, Ag_2S. The sulfur comes from certain foods, such as eggs, or from traces of hydrogen sulfide, H_2S, in the air.

Tarnish can be scraped away by rubbing it with an abrasive (a rough material). But chemistry provides gentler methods for tarnish removal. One approach is to use a more active metal to react with the tarnish in a replacement reaction. The metal substitutes for the substance that formed the tarnish. For example, silver tarnish can be removed by a reaction with aluminum:

$$3Ag_2S(s) \ + \ 2Al(s) \rightarrow Al_2S_3(s) \ + \ 6Ag(s)$$
silver   aluminum   aluminum   silver
sulfide   sulfide
(tarnish)

To do: Place tarnished silverware in an old aluminum pan so that each piece touches the aluminum. Cover the silver with hot water and add one tablespoon of detergent for each quart of water.

After several minutes, the tarnish should be gone.

In general, a more active element replaces a less active element in a compound. The basis for these activity series is discussed in Chapter Fifteen.

4 Double replacement or **metathesis reactions** are reactions in which two reactants exchange atoms to form two new compounds. These reactions often go to completion, that is, they continue until one or more of the reactants is entirely used up. In order for a reaction to go to completion, one of the products must remove itself from the reaction, as illustrated in the following examples. In Figure 9-5(d), sodium sulfate reacts with barium chloride in water to form sodium chloride and a white precipitate of barium sulfate. The precipitate is the solid settling to the bottom of the test tube in the photograph.

$$Na_2SO_4 + BaCl_2 \rightarrow BaSO_4\downarrow + 2NaCl$$

sodium sulfate barium chloride barium sulfate sodium chloride

The sulfate ion, originally part of sodium sulfate, is now present as the precipitate barium sulfate, which does not dissolve in water. Another example of double replacement is the reaction of potassium sulfide with hydrogen bromide:

$$K_2S + 2HBr \rightarrow 2KBr + H_2S\uparrow$$

potassium sulfide hydrobromic acid potassium bromide hydrogen sulfide

Here, one of the compounds formed by "switching partners" is hydrogen sulfide. Because hydrogen sulfide is a gas, it can escape from the system and the reaction can go to completion. Still another example is the reaction of sodium hydroxide with sulfuric acid to form sodium sulfate and water.

$$2NaOH + H_2SO_4 \rightarrow Na_2SO_4 + 2H_2O$$

sodium hydroxide sulfuric acid sodium sulfate water

The exchange that takes place here becomes clear if we think of the formula of water as HOH. Since water is not an ionic compound, its formation as a product effectively removes hydrogen ions and hydroxide ions from the reaction. (This kind of reaction, between a base and an acid, is called neutralization and is discussed in Chapter Fourteen.) In each of these three examples of double replacement reactions, a set of ions was eliminated from the reaction, either as a precipitate, a gas, or a covalent compound.

Table 9-2 summarizes the four major reaction types just discussed. The letters A, B, C, and D are general symbols for the elements involved.

Another major category of reaction, discussed in detail in Chapter Fifteen, involves the transfer of electrons. These reactions are called

TABLE 9-2 *Summary of reaction types*

Type	Reaction
combination (synthesis)	$A + B \rightarrow AB$
decomposition	$AB \rightarrow A + B$
replacement (substitution) [a]	$AB + C \rightarrow AC + B$
double replacement (metathesis)	$AB + CD \rightarrow AD + CB$

[a]Another possible reaction is $AB + D \rightarrow DB + A$

oxidation-reduction, or **redox reactions** because they involve the simultaneous processes of oxidation and reduction. One reactant loses electrons, becoming oxidized. Another reactant gains these electrons, becoming reduced. As explained in Sections 6.7 and 6.8, the formation of an ionic compound is an example of an oxidation-reduction reaction. We can separate an oxidation-reduction reaction into two *half-reactions* that clearly show the loss and gain of electrons:

$$
\begin{aligned}
2Na &\rightarrow 2\,Na^+ + 2e^- \qquad \text{oxidation} \\
Cl_2 + 2e^- &\rightarrow 2Cl^- \qquad\qquad \text{reduction} \\
\hline
2Na + Cl_2 &\rightarrow 2NaCl
\end{aligned}
$$

The cation is formed by oxidation and the anion is formed by reduction. Notice that in this case, and in all oxidation-reduction reactions, the number of electrons lost by one reactant equals the number of electrons gained by the other reactant.

**9.7
WRITING CHEMICAL
EQUATIONS**

We are now able to write simple balanced chemical equations. Given a reaction in words, we can translate it into a chemical equation. Here are some examples:

Example

The following reaction takes place in the storage batteries of automobiles: Lead(II) oxide plus lead plus sulfuric acid form lead(II) sulfate and water. Write the equation for this reaction.

$$PbO_2 + Pb + H_2SO_4 \rightarrow PbSO_4 + H_2O \qquad \text{not balanced}$$
$$PbO_2 + Pb + 2H_2SO_4 \rightarrow 2PbSO_4 + 2H_2O \qquad \text{balanced}$$

First write the correct formulas of the reactants, separated from each other by plus signs on the left side of a reaction arrow. Then write the correct formulas for the products on the right side of the arrow. Finally, balance the equation using the method described in Section 9.4.

Based on the four reaction types described in the previous section, we also can write equations given only the reactants and the type of reaction, as shown in the following example.

Example Hydrogen fluoride reacts with silicon dioxide by a double replacement reaction. Write the balanced chemical equation.

$$HF + SiO_2 \rightarrow SiF_4 + H_2O \qquad \text{not balanced}$$
$$4HF + SiO_2 \rightarrow SiF_4 + 2H_2O \qquad \text{balanced}$$

First write the correct formulas of the reactants on the left side of the reaction arrow. Then determine the correct formulas of the products. Since this example is a double replacement reaction, you know that hydrogen must combine with oxygen and that silicon must combine with fluorine. And, because hydrogen has a combining capacity of one, it forms one bond. Oxygen has a combining capacity of two and forms two bonds. Therefore, the product of this combination is H_2O. Similarly, silicon has a combining capacity of four and must combine with four fluorine atoms, each of which has a combining capacity of one. The product is SiF_4, silicon tetrafluoride. The equation is then balanced by the usual procedure.

It is sometimes possible to predict both the type of reaction and the products of the reaction, given only the reactants. Consider the following example.

Example Magnesium reacts with oxygen, producing a brilliant flame and a white powder. Write a balanced chemical equation for this reaction.

$$Mg + O_2 \rightarrow MgO \qquad \text{not balanced}$$
$$2Mg + O_2 \rightarrow 2MgO \qquad \text{balanced}$$

Given magnesium and oxygen as reactants, only one product is possible: magnesium oxide, the oxide of the metal formed by the combination of magnesium with oxygen.

Since it is not always possible to predict the products of a new chemical reaction, the products generally must be analyzed in the laboratory and identified completely before a chemical equation can be written.

Practice problem Sodium reacts violently with water by a substitution reaction. (Hint: Think of H_2O as HOH.) Write the balanced chemical equation.

ANSWER $Na + H_2O \rightarrow NaOH + H_2$ not balanced
$2Na + 2H_2O \rightarrow 2NaOH + H_2$ balanced

Chemical reactions involve energy changes. Chemical energy is a form of internal energy. It is the energy stored in the bonds of molecules and in ionic lattices (described in Section 6.9).

When bonds break and bonds form in chemical reactions, energy can be released or absorbed as light, electricity, or heat. For example, some reactions produce light, such as those that take place in fireflies or in flashbulbs. Other reactions absorb light energy. For example, plants absorb light energy during photosynthesis. Chemical reactions that create electricity are used in dry cells and in car batteries. On the other hand, some chemical reactions require electricity. Electrolysis reactions, such as the breakdown of water molecules described previously, require electrical energy.

Most chemical reactions involve changes in thermal energy. Thermal energy is the form of energy that is related to temperature and is transferred as heat. If heat is given off during a reaction, the reaction is called **exothermic.** If heat is absorbed, the reaction is called **endothermic.** The amount of energy that is transferred as heat during a reaction is called the **heat of reaction** or the enthalpy change.

The heat of reaction can be measured by using a calorimeter, illustrated in Figure 9-6. The reaction that we want to measure is allowed to take place in the inner chamber of the calorimeter. The heat that is given off or absorbed by the reaction changes the temperature of the surrounding water. The temperature change is then used to calculate the heat of reaction. Heats of reaction, symbolized ΔH (read delta-H) are expressed in kilocalories (kcal) or kilojoules (kJ). One kilocalorie

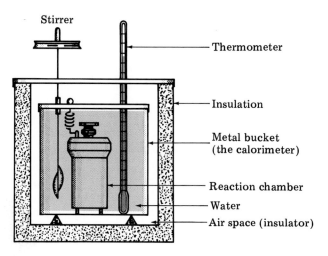

FIGURE 9-6

A calorimeter. The heat change in the reaction chamber causes a temperature change in the surrounding water. This change is measured with a thermometer.

is the amount of heat energy needed to raise the temperature of 1 kg of water by 1°C. In terms of SI units, one kcal is equal to 4.184 kJ.

The heat that is given off or absorbed by a reaction depends on the bond energies of the reactants and the products. According to the law of conservation of energy, the total amount of energy must be the same before and after a reaction takes place. Thus, the change in chemical energy must be equal and opposite to the heat change in the reaction. The initial chemical energy is found by adding the bond energies of all the bonds in the reactant molecules. The final chemical energy is found by adding the bond energies of all the bonds in the product molecules. The difference between the final energy and the initial energy equals the amount of energy given off or absorbed during the reaction.

Consider the reaction of hydrogen and fluorine to form hydrogen fluoride:

$$H_2(g) + F_2(g) \rightarrow 2HF(g)$$

hydrogen   fluorine   hydrogen
                       fluoride

The initial energy equals the hydrogen bond energy (104 kcal/mole) plus the fluorine bond energy (37 kcal/mole). The sum equals 141 kcal/mole. Since two HF molecules form, the final energy is found by multiplying the hydrogen-fluorine bond energy (135 kcal/mole) by two, which gives us 270 kcal/mole. The energy difference, 270 kcal/mole − 141 kcal/mole = 129 kcal/mole, is the energy that is released as heat by this exothermic reaction.

The chemical reactions most likely to take place are those that are exothermic and that produce smaller molecules. Reactions are generally exothermic when the products contain stronger bonds than the reactants. (Stronger bonds form when a greater electronegativity difference exists between the atoms involved.) The stronger bonds make the products lower in energy, and more stable than the reactants.

When more stable (lower energy) compounds are produced in a chemical reaction, the excess energy is usually given off as heat, as we have seen. Combustion, for instance, is an important chemical process that involves the oxidation of a substance, such as the burning of a fuel. Combustion reactions are extremely exothermic, producing large amounts of heat. For example, the combustion of octane (a component of gasoline) releases 1302 kcal/mole (or 5452 kJ/mole).

$$2C_8H_{18} + 25O_2 \rightarrow 16CO_2 + 18H_2O + heat$$

octane   oxygen   carbon   water
                        dioxide

combustion of octane

Such combustion reactions are not only exothermic, but they usually produce carbon dioxide and water molecules, which are smaller molecules than the starting fuel.

Reactions that are endothermic and produce more complex molecules do not occur on their own. The can be forced to take place, however, if an outside energy source is available. For example, during photosynthesis in green plants, light energy from the sun forces the formation of glucose from carbon dioxide and water.

Box 9

Fuel value

A fuel is a substance that produces energy in a controlled reaction. The most common type of energy-producing reaction is combustion. In combustion, heat is produced by the exothermic oxidation of a fuel. For example, the most common fuels, which contain carbon and hydrogen atoms, form carbon dioxide (CO_2) and water (H_2O) when completely burned in air.

Fossil fuels—petroleum, coal, and natural gas—are formed from the decay of animal and vegetable remains over millions of years. Artificial fuels can be made from these natural ones. Gasoline is made from petroleum, and synthetic natural gas can be made from coal.

The energy given off when a fuel is burned is known as the **fuel value.** It is expressed as the number of kilocalories (or kilojoules) released per gram of the fuel. For octane, a component of gasoline, the fuel value is

$$\frac{1303 \text{ kcal}}{1 \text{ mole octane}} \times \frac{1 \text{ mole octane}}{114 \text{ g octane}} = \frac{11.4 \text{ kcal}}{\text{g octane}}$$

Some representative fuel values are given in the table below.

Food is the fuel for our bodies. Carbohydrates and proteins have a fuel value of about 4 kcal/g, while fats provide 9 kcal/g.

Representative fuel values

	Fuel value	
Fuel	*kcal/g*	*kJ/g*
coal	7.6	32
gasoline	11.5	48
hydrogen	33.9	142
methyl alcohol	5.3	22
natural gas	11.7	49
petroleum	10.8	45
wood	4.3	18

$$6CO_2 + 6H_2O \xrightarrow{\text{energy}} C_6H_{12}O_6 + 6O_2$$

carbon dioxide · water · glucose · oxygen

Here, 673 kcal/mole (2816 kJ/mole) of solar energy is stored in plants in the newly formed bonds of glucose molecules. We depend on light energy from the sun because we must eat either plants, or animals that eat plants, to obtain our own supply of energy.

9.9 SOURCES AND USES OF ENERGY

The major energy sources for our industrial society are three fuels: petroleum, natural gas, and coal. Their energy is released by exothermic combustion reactions. For example:

$$C_7H_{16} + 11O_2 \rightarrow 7CO_2 + 8H_2O \qquad \text{petroleum}$$

heptane · oxygen · carbon dioxide · water

$$CH_4 + 2O_2 \rightarrow CO_2 + 2H_2O \qquad \text{natural gas}$$

methane · oxygen · carbon dioxide · water

(Both heptane and methane are hydrocarbons, which we describe in Chapter Eighteen.) Figure 9-7 shows how these fuels can be used to generate electricity in a power plant. The burning fuel produces heat, which converts water to steam. The steam turns the rotors of a turbine to generate electricity. (The steam is then condensed to water by cooling and the process is repeated.) The efficiency of this process is only about 40%. In general, some energy is always wasted when it is converted from one form to another. In this example, chemical energy is converted to thermal energy, then to mechanical energy, and finally to electrical energy.

Other energy sources are hydropower, produced by water falling down a dam, and nuclear power (described in Chapter Seventeen). The sun is a potential source of enormous amounts of energy. Unlike

FIGURE 9-7

Generation of electricity. Heat is produced by burning fuel such as coal or petroleum. The heat converts water into steam in a boiler. The steam turns the blades of a turbine, which is connected to an electrical generator.

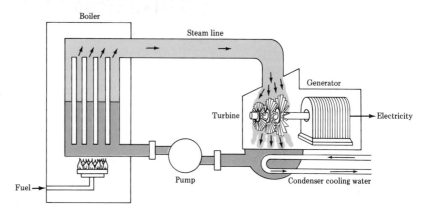

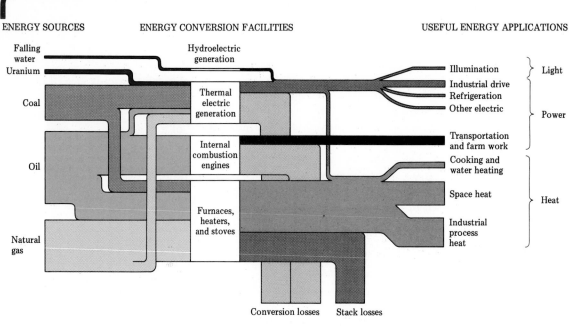

Falling water
Uranium
Coal
Oil
Natural gas

Hydroelectric generation
Thermal electric generation
Internal combustion engines
Furnaces, heaters, and stoves

Illumination — Light
Industrial drive
Refrigeration
Other electric — Power
Transportation and farm work
Cooking and water heating
Space heat — Heat
Industrial process heat

Conversion losses   Stack losses

UNAVAILABLE ENERGY

FIGURE 9-8

Energy flow. This chart shows the major sources (left) and uses (right) of energy in the United States. The width of a channel is proportional to the amount of energy.

petroleum, natural gas, and coal, which are being used up, solar energy is nearly unlimited in supply.

Figure 9-8 summarizes our sources and uses of energy. The major uses of energy are: industrial, transportation, residential, and commercial. About 40% of all the energy used in one year is consumed by industry, largely to make steam, provide direct heat, and generate electricity. The greatest *single* use of energy, however, is as fuel for transportation.

9.10 IMPORTANT INDUSTRIAL REACTIONS

The reactions described in this section are of major economic importance. They are presented here as further examples of chemical reactions and their equations.

More *steel* is produced each year than any other chemical product. Over 300 billion pounds of steel are made each year in the United States alone. Steel is prepared from iron ores, which contain the iron oxides Fe_3O_4 (magnetite) and Fe_2O_3 (hematite). The first step in steel production involves conversion of these iron oxides to iron metal in a blast furnace, such as the structure, almost 200 ft tall, shown in Figure 9-9(a). The furnace is loaded with iron ore, coke (the carbon residue that remains after coal is heated in the absence of air), and

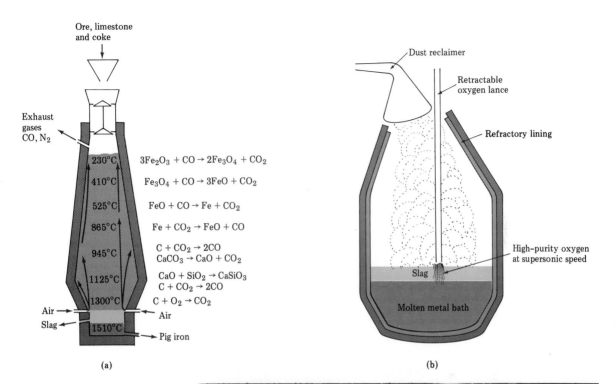

Ore, limestone and coke

Exhaust gases CO, N_2

230°C	$3Fe_2O_3 + CO \rightarrow 2Fe_3O_4 + CO_2$
410°C	$Fe_3O_4 + CO \rightarrow 3FeO + CO_2$
525°C	$FeO + CO \rightarrow Fe + CO_2$
865°C	$Fe + CO_2 \rightarrow FeO + CO$
945°C	$C + CO_2 \rightarrow 2CO$ $CaCO_3 \rightarrow CaO + CO_2$
1125°C	$CaO + SiO_2 \rightarrow CaSiO_3$ $C + CO_2 \rightarrow 2CO$
1300°C	$C + O_2 \rightarrow CO_2$
1510°C	

Air
Slag
Air
Pig iron

(a)

Dust reclaimer

Retractable oxygen lance

Refractory lining

High-purity oxygen at supersonic speed

Slag

Molten metal bath

(b)

FIGURE 9-9

Steel production: The blast furnace (a) converts iron ore into pig iron. In the basic oxygen furnace (b), impurities in the pig iron are oxidized and steel is produced. A blast furnace in operation is shown in (c). (Photo courtesy of American Iron and Steel Institute.)

(c)

limestone ($CaCO_3$). Preheated air is then blown into the bottom of the furnace. In the lower part of the furnace shaft, some of the coke burns, forming carbon monoxide, which reacts with iron ore (represented here as Fe_2O_3):

$$Fe_2O_3(s) + 3CO(g) \rightarrow 2Fe(l) + 3CO_2(g)$$

hematite   carbon   iron   carbon
[iron(III) oxide]   monoxide   dioxide

In the upper part of the shaft, coke directly reduces the ore:

$$Fe_2O_3(s) + 3C(s) \rightarrow 2Fe(l) + 3CO(g)$$

hematite   carbon   iron   carbon monoxide
[iron(III) oxide]   (coke)

The molten iron produced in these two oxidation-reduction reactions flows to the bottom of the furnace.

The limestone ($CaCO_3$) removes impurities, which consist mostly of silicon dioxide, SiO_2. The limestone first decomposes:

$$CaCO_3(s) \rightarrow CaO(s) + CO_2(g)$$

calcium   calcium   carbon
carbonate   oxide   dioxide

The resulting CaO then reacts with the silicon dioxide impurities:

$$CaO(s) + SiO_2(s) \rightarrow CaSiO_3(l)$$

calcium   silicon   calcium
oxide   dioxide   silicate

At the high temperature of the blast furnace, the calcium silicate forms a molten layer called slag. Since it is less dense than iron, slag floats on top and can be separated from the iron.

Both molten iron and slag are removed periodically. The molten iron is not pure, and is called pig iron. A blast furnace can produce about 4000 tons of pig iron per day. Pig iron is 85–95% iron, combined with carbon and small amounts of manganese, sulfur, phosphorus, and silicon.

The second step in steel production removes most of these non-iron components from pig iron. A basic oxygen furnace, illustrated in Figure 9-9(b), is typically fed 110 tons of the molten pig iron, 45 tons of scrap iron, and 10 tons of limestone. Oxygen, blown into the top of the furnace, reacts with the carbon present to produce carbon monoxide, carbon dioxide, and heat:

$$2C(s) + O_2(g) \rightarrow 2CO(g)$$

carbon   oxygen   carbon
                   monoxide

$$C(s) + O_2(g) \rightarrow CO_2(g)$$

carbon   oxygen   carbon
                   dioxide

Carbon thus escapes from the mixture as CO and CO_2 gases. The other impurities in pig iron also form oxides and react further with the limestone to make slag, which is removed. The final product, steel, consists of iron with 0.2–1.5% carbon, depending on the type of steel desired.

Sulfuric acid, H_2SO_4, is the chemical compound that is produced in the largest amounts. Over 60 billion pounds of it are produced in the United States alone every year. Sulfuric acid is made by the contact process from sulfur dioxide. Sulfur dioxide can be formed by reacting sulfur or hydrogen sulfide with oxygen:

$$S(g) + O_2(g) \rightarrow SO_2(g)$$
sulfur   oxygen   sulfur
                     dioxide

$$2H_2S(g) + 3O_2(g) \rightarrow 2SO_2(g) + 2H_2O(g)$$
hydrogen   oxygen   sulfur   water
sulfide   dioxide

Sulfur dioxide is oxidized in air to sulfur trioxide at temperatures between 400° and 600°C.

$$2SO_2(g) + O_2(g) \rightarrow 2SO_3(g)$$
sulfur   oxygen   sulfur
dioxide   trioxide

Sulfur trioxide then reacts with water to form sulfuric acid:

$$SO_3(g) + H_2O(l) \rightarrow H_2SO_4(l)$$
sulfur   water   sulfuric
trioxide   acid

Many industrial plants make their own sulfuric acid by this method, rather than purchase it (see Figure 9-10).

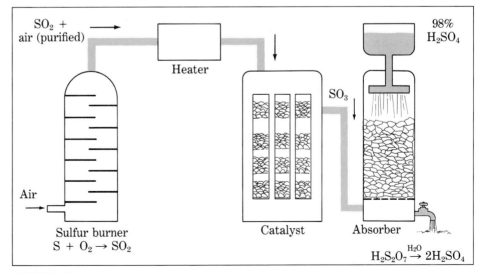

FIGURE 9-10

Sulfuric acid production. Sulfuric acid, H_2SO_4, is manufactured in greater amounts than any other chemical. The contact process for manufacturing sulfuric acid is illustrated here.

Ammonia, another major industrial chemical, is made by the Haber process. This process combines nitrogen and hydrogen:

$$N_2(g) + 3H_2(g) \rightarrow 2NH_3(g)$$
nitrogen   hydrogen   ammonia

A high temperature and pressure as well as a catalyst are needed for this combination reaction. Thus the Haber process consumes large amounts of energy and is expensive. Chemists today are trying to develop less expensive ways to make ammonia in order to reduce the cost of nitrogen fertilizer.

Ammonia is used to make *nitric acid* by the Ostwald process. Ammonia is first oxidized by preheated air in the presence of a catalyst:

$$4NH_3(g) + 5O_2(g) \xrightarrow[\text{oxygen}]{930°C} 4NO(g) + 6H_2O(g)$$
ammonia   nitric   water
                                  oxide

The nitric oxide that forms is then further oxidized to nitrogen dioxide,

$$2NO(g) + O_2(g) \rightarrow 2NO_2(g)$$
nitric   oxygen   nitrogen
oxide   dioxide

which reacts with water to form nitric acid.

$$3NO_2(g) + H_2O(l) \rightarrow 2HNO_3(l) + NO(g)$$
nitrogen   water   nitric   nitric
dioxide   acid   oxide

Air is used to continue the process, converting the nitric oxide from this last step to nitrogen dioxide, which then reacts with more water. The primary use of nitric acid is in the synthesis of ammonium nitrate:

$$NH_3(aq) + HNO_3(aq) \rightarrow NH_4NO_3(aq)$$
ammonia   nitric   ammonium
                acid   nitrate

Ammonium nitrate, like ammonia, is used as a fertilizer.

Another interesting series of chemical reactions used in industry is called the Solvay process. This process is used to manufacture *sodium carbonate* (soda ash), a main ingredient of glass. The raw materials are table salt (NaCl), limestone, and ammonia. First carbon dioxide is made from the limestone by a decomposition reaction.

$$CaCO_3(s) \xrightarrow{\Delta} CaO(s) + CO_2(g)$$
calcium   calcium   carbon
carbonate   oxide   dioxide

The CO_2 then reacts with ammonia in water to produce ammonium hydrogencarbonate (ammonium bicarbonate) by a combination reaction.

$$NH_3(g) + CO_2(g) + H_2O(l) \rightarrow NH_4HCO_3(aq)$$

ammonia   carbon   water   ammonium
dioxide   hydrogencarbonate

Ammonium hydrogencarbonate is converted to sodium hydrogencarbonate (sodium bicarbonate) by a double replacement reaction with salt.

$$NH_4HCO_3(aq) + NaCl(aq) \rightarrow NaHCO_3(s) + NH_4Cl(aq)$$

ammonium   sodium   sodium   ammonium
hydrogencarbonate chloride   hydrogencarbonate chloride

The sodium hydrogencarbonate is then filtered off and heated to form sodium carbonate by decomposition.

$$2NaHCO_3(s) \xrightarrow{\Delta} Na_2CO_3(s) + CO_2(g) + H_2O(g)$$

sodium   sodium   carbon   water
hydrogencarbonate carbonate dioxide

Ammonia can then be recovered from this process and used again. It is recovered by reacting the ammonium chloride with calcium hydroxide (sometimes called slaked lime).

$$2NH_4Cl(s) + Ca(OH)_2(aq) \rightarrow 2NH_3(g) + CaCl_2(aq) + 2H_2O(l)$$

ammonium   calcium   ammonia   calcium   water
chloride   hydroxide   chloride

Unfortunately, the large amount of calcium chloride formed as a waste product can become a serious pollutant when it is dumped in lakes and streams. The difficulty involved in disposing of calcium chloride limits the usefulness of the Solvay process.

Summary **9.1** A chemical change results in the formation of a new substance called a compound. A chemical reaction is a process of chemical change in which chemical bonds are broken and new bonds are formed. Reactions occur when atoms or molecules collide, causing atoms to rearrange into new combinations.

9.2 A chemical equation is a shorthand representation of a chemical reaction. In a chemical equation, the starting substances, called reactants, are separated from the new substances, called products, by an arrow. The arrow shows the direction of the chemical change.

9.3 The law of conservation of mass states that in any chemical reaction, the total mass of the reactants must equal the total mass of the products. Matter is neither created nor destroyed in a chemical reaction.

9.4 Every chemical equation must be balanced. In a balanced equation, the number of atoms of each kind of element is the same on both sides of the equation. To balance an equation, we place coefficients in front of the reactants and products.

9.5 A balanced equation provides a useful summary of a chemical reaction. It not only states what elements or compounds are used up and formed, but also shows the relative amounts of each. Equations can be interpreted either in terms of numbers of atoms, molecules, and formula units, or in terms of numbers of moles.

9.6 Chemical reactions can be classified in various ways. There are four major reaction types: combination (synthesis), decomposition, replacement (substitution), and double replacement (metathesis). Another major category, oxidation-reduction or redox reactions, involves the transfer of electrons.

9.7 It is important to be able to write simple balanced chemical equations by translating from words into chemical language. You should be able to write equations when you are given only the reactants and the type of reaction.

9.8 Chemical reactions involve energy changes. Exothermic reactions give off heat. Endothermic reactions absorb heat. The reactions most likely to take place are those that release heat and produce simpler molecules.

9.9 Today's major energy sources are the fuels petroleum, natural gas, and coal. The major categories of energy use are industrial, transportation, residential, and commercial.

9.10 Several chemical reactions used in industry are described in this chapter. They are the reactions used in the production of steel, sulfuric acid, ammonia, nitric acid, and sodium carbonate.

For each term on the left, choose a phrase from the right-hand column that most closely matches its meaning.

1 chemical reaction	**a** speeds up reaction
2 law of definite proportions	**b** starting substance
3 chemical equation	**c** substance produced
4 reactant	**d** process of chemical change
5 product	**e** matter not created or destroyed
6 catalyst	**f** constant composition of compound
7 law of conservation of mass	**g** energy transferred during reaction
8 oxidation-reduction (redox) reaction	**h** shorthand representation of reaction
9 heat of reaction (enthalpy)	**i** transfer of electrons

EXERCISES BY TOPIC

Chemical and physical change (9.1)

10 How does a compound differ from **a** a mixture? **b** an element?

11 What takes place in a chemical reaction?

12 Using the atomic theory, explain the basis for the law of definite proportions.

13 Why is the presence of physical change by itself not enough to identify a chemical reaction.

14 Why can't iron be separated from iron(II) sulfide by a magnet?

Chemical equations (9.2)

15 Why are chemical equations useful?

16 What is a catalyst?

17 Translate the following equations into words:

a $S + O_2 \rightarrow SO_2$
b $AgNO_3 + NaCl \rightarrow AgCl\downarrow + NaNO_3$
c $CuSO_4(aq) + Zn(s) \rightarrow ZnSO_4(aq) + Cu(s)$
d $HCl + NaOH \rightarrow NaCl + H_2O$

e $C(s) + H_2O(g) \xrightarrow{\text{catalyst}} H_2(g) + CO(g)$

Conservation of mass (9.3)

18 Using the atomic theory, explain the basis for the conservation of mass in a chemical reaction.

19 What do we mean when we say that an equation is balanced?

20 A chemist starts with 25.0 g of reactants, but ends up with only 15.2 g of products. Did she violate the law of conservation of mass? What might have happened?

Balancing chemical equations (9.4)

21 State whether each of the following equations is balanced:

a $C_3H_8 + O_2 \rightarrow 3CO_2 + H_2O$
b $AgNO_3 + H_2SO_4 \rightarrow Ag_2SO_4 + HNO_3$
c $SO_3 + H_2O \rightarrow H_2SO_4$
d $H_3PO_4 + NaOH \rightarrow Na_3PO_4 + H_2O$

22 Balance any equations that are not balanced in Exercise 21.

23 Balance the following equations:

a $Al + Cl_2 \rightarrow AlCl_3$
b $N_2 + H_2 \rightarrow NH_3$
c $Na + H_2O \rightarrow NaOH + H_2$
d $NO + O_2 \rightarrow NO_2$

24 Balance these equations:

a $C + SO_2 \rightarrow CS_2 + CO$
b $ZnS + O_2 \rightarrow ZnO + SO_2$
c $Na_3PO_4 + CaCl_2 \rightarrow Ca_3(PO_4)_2 + NaCl$
d $Al_2(SO_4)_3 + Ba(NO_3)_2 \rightarrow BaSO_4 + Al(NO_3)_3$

25 Why are the coefficients the only numbers we can change when balancing an equation?

Interpreting balanced equations (9.5)

26 Interpret in terms of molecules the balanced equations of Exercises 21 and 23.

27 Interpret in terms of moles the balanced equations of Exercise 23.

28 Interpret in terms of moles the balanced equations of Exercise 24.

29 Why is it that an equation can be interpreted both in terms of molecules and in terms of moles?

Types of chemical reactions (9.6)

30 Identify the equations in Exercise 21 by type.

31 Identify the equations in Exercise 23 by type.

32 Identify the equations in Exercise 24 by type.

33 What is a redox reaction?

34 Describe how we can use an activity series to predict chemical reactions.

35 What takes place in a double replacement reaction?

Writing chemical equations (9.7)

36 Carbon dioxide can be trapped by reacting it with sodium hydroxide to form sodium carbonate and water. Write a balanced equation for this reaction.

37 Burning the fuel butane, C_4H_{10}, in oxygen produces carbon dioxide and water. Write the balanced equation for this reaction.

38 Impurities in water can be removed by using aluminum hydroxide, which is produced in a double replacement reaction between aluminum sulfate and sodium hydroxide. Write the balanced equation for this reaction.

39 Some antacids contain magnesium hydroxide, which reacts with stomach acid, aqueous hydrogen chloride, in a double replacement reaction. Write the balanced equation for this reaction.

40 Write a balanced equation for the decomposition of hydrogen peroxide, H_2O_2, to water and oxygen.

Energy changes in reactions (9.8)

41 How is energy related to chemical reactions?

42 How does an exothermic reaction differ from an endothermic reaction?

43 Describe the energy changes involved in a combustion reaction.

44 Which types of chemical changes are most likely to take place?

45 What does fuel value measure?

Sources and uses of energy (9.9)

46 Explain how burning a fuel can be used to generate electricity.

47 Describe our major sources and uses of energy.

Important industrial reactions (9.10)

48 Summarize the process by which steel is made.

49 What is the chemical composition of steel?

50 How is sulfuric acid made?

51 How is sodium carbonate produced in the Solvay process?

52 Briefly describe the process by which nitric acid is produced.

ADDITIONAL EXERCISES

53 If someone hands you a substance, how can you tell if it is a compound or a mixture?

54 Translate into words: $2CO_2(g) \leftarrow 2C(s) + 2O_2(g)$

55 Balance the following equations:

a $Al + O_2 \rightarrow Al_2O_3$
b $Ca + H_3PO_4 \rightarrow Ca_3(PO_4)_2 + H_2$
c $Na_2SO_3 + O_2 \rightarrow Na_2SO_4$
d $C_6H_{12}O_6 \text{ (glucose)} \xrightarrow{\text{catalyst}} C_2H_5OH \text{ (ethyl alcohol)} + CO_2$
e $Na_2CO_3 + HCl \rightarrow NaCl + H_2O + CO_2$
f $NH_3 + O_2 \rightarrow NO + H_2O$
g $Al_2(SO_4)_3 + NaOH \rightarrow Al(OH)_3\downarrow + Na_2SO_4$
h $FeS_2 + O_2 \rightarrow Fe_2O_3 + SO_2$
i $Mg(OH)_2 + H_3PO_4 \rightarrow Mg_3(PO_4)_2 + H_2O$
j $C_2H_2 \text{ (acetylene)} + O_2 \rightarrow CO_2 + H_2O$

56 Interpret in terms of moles the balanced equations of Exercise 55.

57 Complete the following equations:

a $CuCl_2 + Zn \rightarrow$?   replacement
b $H_2 + Cl_2 \rightarrow$?   combination
c $HBr + KOH \rightarrow$?   double replacement
d $Ca + O_2 \rightarrow$?   combination
e $LiBr + Cl_2 \rightarrow$?   replacement

58 Write balanced equations for the following:

a zinc sulfide + oxygen → zinc oxide + sulfur dioxide
b aluminum + sulfuric acid → aluminum sulfate + hydrogen
c carbon + sulfur dioxide → carbon disulfide + carbon monoxide
d potassium + water → potassium hydroxide + hydrogen
e potassium nitrate → potassium nitrite + oxygen

59 Explain how the energy for life comes from the sun.

60 Suggest steps we can take to save energy.

61 Balance each of the following equations:

a $Cl_2O_7(g) + H_2O(l) \rightarrow HClO_4(aq)$
b $Ca_3(PO_4)_2(s) + H_2SO_4(aq) \rightarrow CaSO_4(s) + H_3PO_4(aq)$
c $Al_2S_3(s) + H_2O(l) \rightarrow Al(OH)_3(aq) + H_2S(g)$
d $Fe(s) + H_2O(g) \rightarrow Fe_3O_4(s) + H_2(g)$
e $Al_2O_3(s) + HNO_3(aq) \rightarrow Al(NO_3)_3(aq) + H_2O(l)$

62 Interpret in terms of moles the balanced equations of Exercise 61.

63 Classify each of the equations in Exercise 61 according to reaction type.

64 Write and balance the following equations:

a calcium carbonate + sulfuric acid → calcium sulfate + carbon dioxide + water

b potassium iodide + phosphoric acid → hydrogen iodide + potassium phosphate

c sodium oxide + sulfur dioxide + oxygen → sodium sulfate
d barium chloride + lithium phosphate → barium phosphate + lithium chloride

65 Interpret the following equation in terms of **a** molecules and **b** moles:
$2H_2S(g) + 3O_2(g) \rightarrow 2SO_2(g) + 2H_2O(l)$

66 Balance the following equations:

a $Hg(l) + NH_4I(s) \rightarrow HgI_2(s) + H_2(g) + NH_3(g)$
b $Al_2(SO_4)_3(aq) + Ca(NO_3)_2(aq) \rightarrow CaSO_4(s) + Al(NO_3)_3(aq)$
c $Mg_3N_2(s) + H_2O(l) \rightarrow Mg(OH)_2(aq) + NH_3(g)$
d $Al_4C_3(s) + H_2O(l) \rightarrow Al(OH)_3(aq) + CH_4(g)$

67 Interpret the balanced equations in Exercise 66 in terms of moles.

68 Classify the equations in Exercise 66 according to reaction type.

69 Write an equation for the   **a** reaction of oxygen with sulfur dioxide to form sulfur trioxide   **b** decomposition of arsine, AsH_3, into its elements   **c** replacement reaction between calcium and iron(III) sulfide

70 A chocolate drink contains 11 g protein, 36 g carbohydrate, and 10 g fat. If you jog at 5 mi/hr, you burn up 475 kcal per hour. How long must you jog to use up the energy provided by the chocolate drink?

PROGRESS CHART   Check off each item when completed.

_____   Previewed chapter

Studied each section:

_____   **9.1**   Chemical and physical change

_____   **9.2**   Chemical equations

_____   **9.3**   Conservation of mass

_____   **9.4**   Balancing chemical equations

_____   **9.5**   Interpreting balanced equations

_____   **9.6**   Types of chemical reactions

_____   **9.7**   Writing chemical equations

_____   **9.8**   Energy changes in reactions

_____   **9.9**   Sources and uses of energy

_____   **9.10** Important industrial reactions

_____   Reviewed chapter

_____   Answered assigned exercises:

#'s _____

TEN

Calculations based on equations

In the preceding chapter, you learned how a balanced equation shows the relative amounts of reactants and products that take part in a reaction. In Chapter Ten you will learn how to use this information to make predictions about reactions. This chapter presents a method for calculating the amount of product resulting from a certain quantity of reactant. You will also learn the reverse process—finding how much reactant is needed to make a required amount of product. To demonstrate an understanding of Chapter Ten, you should be able to:

1 Explain the basis of stoichiometry.

2 Find the number of moles of a substance in a balanced equation, given the number of moles of another substance.

3 Find the mass (or number of moles) of a substance in a balanced equation, given the number of moles (or mass) of another substance.

4 Find the mass of a substance in a balanced equation, given the mass of another substance.

5 Solve stoichiometry problems using Avogadro's number.

6 Determine which reactant in a reaction is the limiting reagent.

7 Calculate the percent yield of a product in a reaction.

**10.1
STOICHIOMETRY**

The term **stoichiometry** is derived from two Greek words meaning "to measure the elements." It refers to any calculation of the quantities of elements or compounds involved in a chemical reaction. Stoichiometry calculations are extremely important, both for carrying out reactions in the laboratory and for the design of manufacturing processes in industry. Such calculations are based on interpreting chemical equations in terms of numbers of moles of reactants and products.

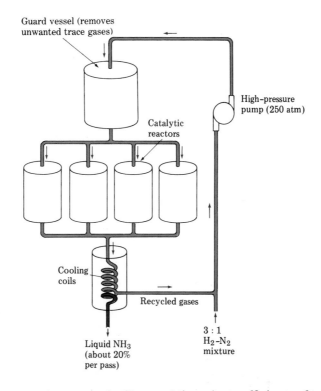

Guard vessel (removes
unwanted trace gases)

High-pressure
pump (250 atm)

Catalytic
reactors

Cooling
coils

Recycled gases

Liquid NH$_3$
(about 20%
per pass)

3 : 1
H$_2$-N$_2$
mixture

FIGURE 10-1

*Ammonia production. In the
Haber process, nitrogen and
hydrogen gas are combined
under high pressure in the
presence of a catalyst.*

As we saw in Chapter Nine, the coefficients of a balanced equation relate the number of moles of substances used up in the reaction to the number of moles of substances formed. This relationship can also be expressed in terms of the masses involved. For example, in the Haber process for the formation of ammonia: $N_2(g) + 3H_2(g) \rightarrow 2NH_3(g)$, one mole of N_2 and three moles of H_2 react to make two moles of NH_3. (See Figure 10-1.) This equation also means that each 28 g of N_2 react with 6 g of H_2 forming 34 g of NH_3, as shown.

$$N_2 + 3H_2 \rightarrow 2NH_3$$

$$1 \text{ mole } N_2 \times \frac{28 \text{ g } N_2}{1 \text{ mole } N_2} = 28 \text{ g } N_2$$

$$3 \text{ moles } H_2 \times \frac{2 \text{ g } H_2}{1 \text{ mole } H_2} = 6 \text{ g } H_2$$

$$2 \text{ moles } NH_3 \times \frac{17 \text{ g } NH_3}{1 \text{ mole } NH_3} = 34 \text{ g } NH_3$$

Note that this method of interpreting the equation demonstrates the law of conservation of mass:

mass of reactants = mass of products
28 g + 6 g   = 34 g

American Institute of Physics/ Niels Bohr Library. Stein Collection.

Before Fritz Haber, ammonia was made from naturally occurring nitrates found mainly in Chile. Along with a fellow German, Carl Bosch, Haber developed a high-pressure synthesis of ammonia from nitrogen and hydrogen. Introduced in 1909, the Haber process was the first commercially important high-pressure chemical process.

The Haber process was vital to Germany during World War I. Ammonia was needed to produce nitric acid, which in turn was used to make explosives. When the British navy cut off all imports, including those of nitrates, the Germans continued to produce ammonia by using nitrogen from the air.

Fritz Haber also helped the German war effort by directing chemical warfare. He introduced the use of poison gas, such as chlorine, as a weapon. Haber won the Nobel Prize for chemistry in 1918, the year the war ended. French, British, and American scientists denounced the award because of Haber's patriotic assistance to the German army.

After the war, Haber became director of the famous Kaiser Wilhelm Institute for Physical Chemistry and Electrochemistry. But when the Nazis rose to power in 1933, they demanded the dismissal of Jewish workers. Haber refused to do this. Instead, he resigned and went into exile in England. In his letter of resignation, he wrote:

For more than 40 years I have selected my collaborators on the basis of their intelligence and their character and not on the basis of their grandmothers, and I am not willing for the rest of my life to change this method which I have found so good.

Haber died of a heart attack a year later.

Reactions of nitrogen and hydrogen always involve exactly these *relative* amounts, although the actual numbers of grams involved may differ from problem to problem. In other words, regardless of how many moles of nitrogen react, three times as many moles of hydrogen are consumed, and two times as many moles of ammonia are produced. All stoichiometric calculations are based on the relationships between the numbers of moles of reactants and products. These relationships follow immediately from the coefficients of the balanced equation.

**10.2
MOLE–MOLE
PROBLEMS**

The simplest type of stoichiometry problem involves finding the number of moles of one substance present in a reaction when we are given the number of moles of another substance. The two substances can both be reactants, or products, or one a reactant and the other a product. To solve such mole–mole problems, we need a conversion factor based on the relative numbers of moles involved. We obtain conversion factors from the coefficients of the balanced equation. For example, in ammonia synthesis, $N_2 + 3H_2 \rightarrow 2NH_3$, possible conversion factors are:

$$\frac{1 \text{ mole } N_2}{3 \text{ moles } H_2} \quad \text{or} \quad \frac{3 \text{ moles } H_2}{1 \text{ mole } N_2}$$

$$\frac{1 \text{ mole } N_2}{2 \text{ moles } NH_3} \quad \text{or} \quad \frac{2 \text{ moles } NH_3}{1 \text{ mole } N_2}$$

$$\frac{3 \text{ moles } H_2}{2 \text{ moles } NH_3} \quad \text{or} \quad \frac{2 \text{ moles } NH_3}{3 \text{ moles } H_2}$$

The first pair of factors relates the amounts of N_2 and H_2. The second pair relates the amounts of N_2 and NH_3, and the third pair relates the amounts of H_2 and NH_3.

Conversion factors for mole–mole problems are called **mole ratios.** They are fractions that relate the numbers of moles of two substances in the reaction. Which mole ratio we use depends on what we are asked to find. For example, suppose that we are asked to find the number of moles of ammonia that can be produced from 250. moles of nitrogen. In this case, we should choose the mole ratio that relates moles of ammonia to moles of nitrogen. We then outline the problem as follows:

UNKNOWN	GIVEN	CONNECTION	
moles NH_3	250. moles N_2	$\dfrac{1 \text{ mole } N_2}{2 \text{ moles } NH_3}$ or	$\dfrac{2 \text{ moles } NH_3}{1 \text{ mole } N_2}$

The connection is derived from the appropriate coefficients of the balanced equation which, in this case, are the coefficients of nitrogen and ammonia.

$$1N_2 + 3H_2 \rightarrow 2NH_3$$

The problem is then set up by the conversion factor method:

$$\text{moles } NH_3 = 250.\ \cancel{\text{moles } N_2} \times \frac{2 \text{ moles } NH_3}{1 \cancel{\text{ mole } N_2}} = 500.\ \text{moles } NH_3$$

The conversion factor, one of the mole ratios, is written so that *moles N_2* cancel, which gives an answer in *moles NH_3*. There is no need to estimate in this simple example; the answer is 500. moles NH_3.

It is always important to check the result. The coefficients in the equation show that one mole of N_2 forms two moles of NH_3. Thus the amount of ammonia produced must be twice as much as the amount of nitrogen consumed. Since we started with 250. moles of N_2, our answer of 500. moles of NH_3 must be correct. Note that in these calculations, the mole ratio is exact and does not limit the number of significant figures in the answer.

Now suppose that we must calculate the number of moles of hydrogen needed to make 0.68 mole of ammonia. This time, we choose the mole ratio that relates moles of hydrogen to moles of ammonia. The connection again follows from the equation

$$N_2 + 3H_2 \rightarrow 2NH_3$$

The problem outline is:

UNKNOWN	GIVEN	CONNECTION	
moles H_2	0.68 mole NH_3	$\dfrac{3 \text{ moles } H_2}{2 \text{ moles } NH_3}$ or	$\dfrac{2 \text{ moles } NH_3}{3 \text{ moles } H_2}$

The problem is set up using a mole ratio as the conversion factor.

$$\text{moles } H_2 = 0.68 \cancel{\text{mole } NH_3} \times \frac{3 \text{ moles } H_2}{2 \cancel{\text{moles } NH_3}}$$

ESTIMATE $7 \times 10^{-1} \cancel{\text{mole } NH_3} \times \dfrac{3 \text{ moles } H_2}{2 \cancel{\text{moles } NH_3}} = 1 \text{ mole } H_2$

CALCULATION   1.0 mole H_2

Again, we must be sure to check the answer. The amount of hydrogen needed is greater than the amount of ammonia formed. This result agrees with the balanced equation.

Example Aluminum forms an oxide coating in water by the following process:

$$2Al(s) \ + \ 3H_2O(l) \rightarrow Al_2O_3(s) \ + \ 6H_2(g)$$

aluminum   water   aluminum   hydrogen
oxide

How many moles of aluminum oxide are formed from 12.3 moles of aluminum?

UNKNOWN	GIVEN	CONNECTION
moles Al$_2$O$_3$	12.3 moles Al	$\dfrac{2 \text{ moles Al}}{1 \text{ mole Al}_2\text{O}_3}$ or $\dfrac{1 \text{ mole Al}_2\text{O}_3}{2 \text{ moles Al}}$

$$\text{moles Al}_2\text{O}_3 \ = \ 12.3 \ \text{moles Al} \times \frac{1 \text{ mole Al}_2\text{O}_3}{2 \text{ moles Al}}$$

ESTIMATE $1 \times 10^1 \ \text{moles Al} \times \dfrac{1 \text{ mole Al}_2\text{O}_3}{2 \text{ moles Al}} = 5 \text{ moles Al}_2\text{O}_3$

CALCULATION   6.15 moles Al$_2$O$_3$

Example Phosphate ions (as in the form of sodium phosphate) can be removed from impure water by adding calcium hydroxide:

$$3Ca(OH)_2(s) \ + \ 2Na_3PO_4(aq) \rightarrow Ca_3(PO_4)_2(s) \ + \ 6NaOH(aq)$$

calcium   sodium   calcium   sodium
hydroxide   phosphate   phosphate   hydroxide

How many moles of calcium hydroxide are needed to remove 235 moles of sodium phosphate?

UNKNOWN	GIVEN	CONNECTION
moles Ca(OH)$_2$	235 moles Na$_3$PO$_4$	$\dfrac{3 \text{ moles Ca(OH)}_2}{2 \text{ moles Na}_3\text{PO}_4}$ or $\dfrac{2 \text{ moles Na}_3\text{PO}_4}{3 \text{ moles Ca(OH)}_2}$

$$\text{moles Ca(OH)}_2 \ = \ 235 \ \text{moles Na}_3\text{PO}_4 \times \frac{3 \text{ moles Ca(OH)}_2}{2 \text{ moles Na}_3\text{PO}_4}$$

$$\text{ESTIMATE} \quad 2 \times 10^2 \text{ moles } \cancel{Na_3PO_4} \times \frac{3 \text{ moles } Ca(OH)_2}{2 \text{ moles } \cancel{Na_3PO_4}}$$

$$= 3 \times 10^2 \text{ moles } Ca(OH)_2 = 300 \text{ moles } Ca(OH)_2$$

CALCULATION $\quad 3.53 \times 10^2 \text{ moles } Ca(OH)_2 = 353 \text{ moles } Ca(OH)_2$

Example Zinc ore (which is mainly zinc sulfide) is "roasted" in oxygen to convert it to the oxide, from which the metal is more easily obtained:

$$2ZnS(s) + 3O_2(g) \rightarrow 2ZnO(s) + 2SO_2(g)$$

zinc sulfide $\quad$ oxygen $\quad$ zinc oxide $\quad$ sulfur dioxide

How many moles of oxygen are needed to form 80.0 moles of zinc oxide?

UNKNOWN	GIVEN	CONNECTION	
moles O_2	80.0 moles ZnO	$\dfrac{3 \text{ moles } O_2}{2 \text{ moles } ZnO}$ or	$\dfrac{2 \text{ moles } ZnO}{3 \text{ moles } O_2}$

$$\text{moles } O_2 = 80.0 \text{ moles } \cancel{ZnO} \times \frac{3 \text{ moles } O_2}{2 \text{ moles } \cancel{ZnO}}$$

$$\text{ESTIMATE} \quad 8 \times 10^1 \text{ moles } \cancel{ZnO} \times \frac{3 \text{ moles } O_2}{2 \text{ moles } \cancel{ZnO}} = 1 \times 10^2 \text{ moles } O_2$$

$$= 100 \text{ moles } O_2$$

CALCULATION $\quad 1.20 \times 10^2 \text{ moles } O_2 = 120. \text{ moles } O_2$

Practice problem The following equation represents the decomposition of sodium hydrogencarbonate (sodium bicarbonate):

$$2NaHCO_3(s) \xrightarrow{\Delta} Na_2CO_3(s) + CO_2(g) + H_2O(l)$$

How many moles of carbon dioxide can be formed from 1.74 moles of sodium hydrogencarbonate?

ANSWER $\quad 0.870 \text{ mole } CO_2$

Practice problem Nitric oxide, NO, can be made in the laboratory by the following reaction:

$$3Cu(s) + 8HNO_3(aq) \rightarrow 3Cu(NO_3)_2(aq) + 2NO(g) + 4H_2O(l)$$

How many moles of copper are needed to react with 0.653 moles of nitric acid, HNO_3?

ANSWER   0.245 moles Cu

10.3
MOLE–MASS
PROBLEMS

A slightly more complicated type of stoichiometry problem involves finding the mass of one substance when the number of moles of another substance is given. We call this type of stoichiometry problem a *mole–mass* problem. In mole–mass problems, it is necessary not only to convert the number of moles of one substance to moles of another, as we do in mole–mole problems, but also to convert moles to grams. Therefore, we need two conversion factors: a mole ratio, and a factor that relates moles to grams.

For example, suppose that we need to know how many grams of ammonia can be synthesized from 100. moles of hydrogen. Again, we use the equation for ammonia synthesis: $N_2 + 3H_2 \rightarrow 2NH_3$. This problem is outlined as follows:

UNKNOWN	GIVEN	CONNECTION	
grams NH_3	100. moles H_2	$\dfrac{3 \text{ moles } H_2}{2 \text{ moles } NH_3}$ or $\dfrac{2 \text{ moles } NH_3}{3 \text{ moles } H_2}$	
		$\dfrac{1 \text{ mole } NH_3}{17.0 \text{ g } NH_3}$ or $\dfrac{17.0 \text{ g } NH_3}{1 \text{ mole } NH_3}$	

As you may recall from Chapter Eight, one mole of a substance has a mass equal to the atomic, molecular, or formula weight of that substance written in grams. Therefore, the conversion factor relating moles to grams is based on the molecular weight of ammonia (17.0 amu):

$$NH_3: \quad 1 \text{ N atom} \times \frac{14.0 \text{ amu}}{1 \text{ N atom}} \quad = 14.0 \text{ amu}$$

$$3 \text{ H atoms} \times \frac{1.01 \text{ amu}}{1 \text{ H atom}} \quad = \underline{3.03 \text{ amu}}$$

molecular weight of NH_3 = 17.0 amu

The molar mass of ammonia is therefore 17.0 g/mole. The mole–mass problem is then set up as follows:

$$\text{grams } NH_3 = 100. \text{ moles } H_2 \times \frac{2 \text{ moles } NH_3}{3 \text{ moles } H_2} \times \frac{17.0 \text{ g } NH_3}{1 \text{ mole } NH_3}$$

The first conversion factor, the mole ratio $\dfrac{2 \text{ moles NH}_3}{3 \text{ moles H}_2}$, converts the given 100. moles of hydrogen into the corresponding number of moles of ammonia. The second conversion factor, the molar mass of NH_3, then converts the number of moles of ammonia into the corresponding mass in grams.

$$\text{moles H}_2 \xrightarrow[\text{factor}]{\text{first}} \text{moles NH}_3 \xrightarrow[\text{factor}]{\text{second}} \text{grams NH}_3$$

Each conversion factor must be arranged so that units cancel properly. In other words, the factor must have the same units in its denominator as are in the numerator of the term to its left. The estimate can now be made:

ESTIMATE $\quad 10^2 \text{ moles H}_2 \times \dfrac{2 \text{ moles NH}_3}{3 \text{ moles H}_2} \times \dfrac{2 \times 10^1 \text{ g NH}_3}{1 \text{ mole NH}_3}$

$$= \dfrac{4 \times 10^3 \text{ g NH}_3}{3} = 1 \times 10^3 \text{ g NH}_3$$

The result of the actual calculation is:

CALCULATION $\quad 1.13 \times 10^3 \text{ g NH}_3 = 1.13 \text{ kg NH}_3$

In this case, checking the answer is not quite as simple as it was in the mole–mole problem. However, we do know that the answer must be smaller than 100 moles times 17 g/mole, or smaller than 1700 g, because the number of moles of ammonia that are formed is smaller than the number of moles of hydrogen that are used up.

The opposite type of problem, which we can call a *mass–mole* problem, is handled in a similar manner. Here we are given the mass in grams of a substance and asked to find the number of moles of a different substance in the reaction.

For example, suppose that we want to know how many moles of nitrogen are needed to make 350. g of ammonia. We outline this problem as follows:

UNKNOWN	GIVEN	CONNECTION		
moles N_2	350. g NH_3	$\dfrac{1 \text{ mole N}_2}{2 \text{ moles NH}_3}$	or	$\dfrac{2 \text{ moles NH}_3}{1 \text{ mole N}_2}$
		$\dfrac{1 \text{ mole NH}_3}{17.0 \text{ g NH}_3}$	or	$\dfrac{17.0 \text{ g NH}_3}{1 \text{ mole NH}_3}$

The problem can then be set up:

$$\text{moles N}_2 = 350. \, \cancel{\text{g NH}_3} \times \frac{1 \, \cancel{\text{mole NH}_3}}{17.0 \, \cancel{\text{g NH}_3}} \times \frac{1 \, \text{mole N}_2}{2 \, \cancel{\text{moles NH}_3}}$$

Here the first conversion factor must convert the given grams of ammonia to moles of ammonia. The second factor then converts moles of ammonia to moles of nitrogen.

$$\text{grams NH}_3 \xrightarrow[\text{factor}]{\text{first}} \text{moles NH}_3 \xrightarrow[\text{factor}]{\text{second}} \text{moles N}_2$$

The order of the two conversion factors is not the same in this problem as it was in the preceding problem because the given information, or the starting point, is not the same.

An estimate can now be made:

ESTIMATE $\quad 4 \times 10^2 \, \cancel{\text{g NH}_3} \times \dfrac{1 \, \cancel{\text{mole NH}_3}}{2 \times 10^1 \, \cancel{\text{g NH}_3}} \times \dfrac{1 \, \text{mole N}_2}{2 \, \cancel{\text{moles NH}_3}}$

$$= \frac{4 \times 10^2}{4 \times 10^1} \, \text{moles N}_2 = 1 \times 10^1 \, \text{moles N}_2 = 10 \, \text{moles N}_2$$

The computed result is

CALCULATION $\quad$ 10.3 moles N_2

To check the answer, we note from the equation that the number of moles of N_2 formed must be less than the moles of NH_3 consumed. The number of moles of N_2 formed was about 10. The amount of ammonia used was 350. g, which is considerably more than 10 moles (10 moles NH_3 = 170. g NH_3).

Example Sodium sulfate, used in the paper industry, is a byproduct of hydrochloric acid production:

$$4\text{NaCl}(s) + 2\text{SO}_2(g) + 2\text{H}_2\text{O}(l) + \text{O}_2(g) \rightarrow 2\text{Na}_2\text{SO}_4(aq) + 4\text{HCl}(aq)$$

sodium chloride $\quad$ sulfur dioxide $\quad$ water $\quad$ oxygen $\quad$ sodium sulfate $\quad$ hydrochloric acid

How many grams of sodium sulfate can be produced from 200. moles of sodium chloride?

UNKNOWN	GIVEN	CONNECTION	
grams Na$_2$SO$_4$	200. moles NaCl	$\dfrac{4 \text{ moles NaCl}}{2 \text{ moles Na}_2\text{SO}_4}$ or	$\dfrac{2 \text{ moles Na}_2\text{SO}_4}{4 \text{ moles NaCl}}$
		$\dfrac{1 \text{ mole Na}_2\text{SO}_4}{142 \text{ g Na}_2\text{SO}_4}$ or	$\dfrac{142 \text{ g Na}_2\text{SO}_4}{1 \text{ mole Na}_2\text{SO}_4}$

$$\text{grams Na}_2\text{SO}_4 = 200. \;\cancel{\text{moles NaCl}} \times \frac{2 \;\cancel{\text{moles Na}_2\text{SO}_4}}{4 \;\cancel{\text{moles NaCl}}} \times \frac{142 \text{ g Na}_2\text{SO}_4}{1 \;\cancel{\text{mole Na}_2\text{SO}_4}}$$

ESTIMATE

$$2 \times 10^2 \;\cancel{\text{moles NaCl}} \times \frac{2 \;\cancel{\text{moles Na}_2\text{SO}_4}}{4 \;\cancel{\text{moles NaCl}}} \times \frac{1 \times 10^2 \text{ g Na}_2\text{SO}_4}{1 \;\cancel{\text{mole Na}_2\text{SO}_4}}$$

$$= 1 \times 10^4 \text{ g Na}_2\text{SO}_4$$

CALCULATION $1.42 \times 10^4 \text{ g Na}_2\text{SO}_4 = 14.2 \text{ kg Na}_2\text{SO}_4$

Example   The following equation represents the reaction of a noble gas compound:

$$2\text{XeF}_2(s) + 2\text{H}_2\text{O}(l) \rightarrow 2\text{Xe}(g) + 4\text{HF}(aq) + \text{O}_2(g)$$
xenon   water   xenon   hydrofluoric   oxygen
difluoride   acid

How many moles of xenon form from 325 g of xenon difluoride?

UNKNOWN	GIVEN	CONNECTION	
moles Xe	325 g XeF$_2$	$\dfrac{2 \text{ moles XeF}_2}{2 \text{ moles Xe}}$ or	$\dfrac{2 \text{ moles Xe}}{2 \text{ moles XeF}_2}$
		$\dfrac{1 \text{ mole XeF}_2}{169 \text{ g XeF}_2}$ or	$\dfrac{169 \text{ g XeF}_2}{1 \text{ mole XeF}_2}$

$$\text{moles Xe} = 325 \;\cancel{\text{g XeF}_2} \times \frac{1 \;\cancel{\text{mole XeF}_2}}{169 \;\cancel{\text{g XeF}_2}} \times \frac{2 \text{ moles Xe}}{2 \;\cancel{\text{moles XeF}_2}}$$

ESTIMATE $3 \times 10^2 \;\cancel{\text{g XeF}_2} \times \dfrac{1 \;\cancel{\text{mole XeF}_2}}{2 \times 10^2 \;\cancel{\text{g XeF}_2}} \times \dfrac{2 \text{ moles Xe}}{2 \;\cancel{\text{moles XeF}_2}}$

$$= 2 \text{ moles Xe}$$

CALCULATION   1.92 moles Xe

Example Steel can be corroded by steam:

$$3Fe(s) + 4H_2O(g) \rightarrow Fe_3O_4(s) + 4H_2(g)$$

iron   water   iron   hydrogen
                       oxide

How many moles of iron have corroded if 28.2 g of iron oxide form?

UNKNOWN	GIVEN	CONNECTION	
moles Fe	28.2 g Fe_3O_4	$\dfrac{3 \text{ moles Fe}}{1 \text{ mole } Fe_3O_4}$ or	$\dfrac{1 \text{ mole } Fe_3O_4}{3 \text{ moles Fe}}$
		$\dfrac{1 \text{ mole } Fe_3O_4}{232 \text{ g } Fe_3O_4}$ or	$\dfrac{232 \text{ g } Fe_3O_4}{1 \text{ mole } Fe_3O_4}$

$$\text{moles Fe} = 28.2 \text{ g } \cancel{Fe_3O_4} \times \frac{1 \text{ mole } \cancel{Fe_3O_4}}{232 \text{ g } \cancel{Fe_3O_4}} \times \frac{3 \text{ moles Fe}}{1 \text{ mole } \cancel{Fe_3O_4}}$$

ESTIMATE $3 \times 10^1 \text{ g } \cancel{Fe_3O_4} \times \dfrac{1 \text{ mole } \cancel{Fe_3O_4}}{2 \times 10^2 \text{ g } \cancel{Fe_3O_4}} \times \dfrac{3 \text{ moles Fe}}{1 \text{ mole } \cancel{Fe_3O_4}}$

$$= 5 \times 10^{-1} \text{ mole Fe}$$

CALCULATION $3.65 \times 10^{-1} \text{ mole Fe} = 0.365 \text{ mole Fe}$

Practice problem Phosphorus pentachloride reacts with water to form phosphoric acid, as shown in the following equation.

$$PCl_5(l) + 4H_2O(l) \rightarrow H_3PO_4(aq) + 5HCl(aq)$$

How many grams of PCl_5 will react with 2.20 moles of water?

ANSWER   115 g PCl_5

Practice problem Hydrogen sulfide can be manufactured by the following process:

$$4S(s) + CH_4(g) + 2H_2O(l) \rightarrow 4H_2S(g) + CO_2(g)$$

How many moles of hydrogen sulfide can be made from 120. g of methane?

ANSWER   30.0 moles of H_2S

Most stoichiometry problems involve mass to mass conversions. In a *mass–mass* problem, we are given the mass of one substance in the reaction, and are asked to find the mass of another substance. Mass–mass problems are common because the amount of a reactant or product is usually determined by measuring its mass. (Gases are an exception; it is usually easier to measure the volume of a gas. This type of problem is discussed in Chapter Eleven.)

Three conversion factors are needed to solve a mass–mass problem. Not only must the number of moles of the two substances be related, but the moles of both substances must also be converted to grams.

Consider the following example: given the equation $N_2 + 3H_2 \rightarrow 2NH_3$, find the number of grams of ammonia produced from 200. g of N_2. This problem is outlined as follows:

UNKNOWN	GIVEN	CONNECTION
grams NH_3	200. g N_2	$\dfrac{1 \text{ mole } N_2}{2 \text{ moles } NH_3}$ or $\dfrac{2 \text{ moles } NH_3}{1 \text{ mole } N_2}$
		$\dfrac{1 \text{ mole } N_2}{28.0 \text{ g } N_2}$ or $\dfrac{28.0 \text{ g } N_2}{1 \text{ mole } N_2}$
		$\dfrac{1 \text{ mole } NH_3}{17.0 \text{ g } NH_3}$ or $\dfrac{17.0 \text{ g } NH_3}{1 \text{ mole } NH_3}$

Notice that an additional conversion factor is required for this problem because the grams of N_2 must first be converted to moles before they can be related to moles of NH_3.

The problem is set up in this way:

$$\text{grams } NH_3 = 200. \text{ g } N_2 \times \frac{1 \text{ mole } N_2}{28.0 \text{ g } N_2} \times \frac{2 \text{ moles } NH_3}{1 \text{ mole } N_2} \times \frac{17.0 \text{ g } NH_3}{1 \text{ mole } NH_3}$$

The first conversion factor converts grams of N_2 to moles of N_2, based on the molecular weight of N_2 in grams. The second factor converts moles of N_2 to an equivalent number of moles of NH_3, using the coefficients of the balanced equation. The third factor then expresses the mass in grams of this number of moles of NH_3.

$$\text{grams } N_2 \xrightarrow[\text{factor}]{\text{first}} \text{moles } N_2 \xrightarrow[\text{factor}]{\text{second}} \text{moles } NH_3 \xrightarrow[\text{factor}]{\text{third}} \text{grams } NH_3$$

(mole ratio)

The key conversion factor in this and in all the other stoichiometry problems is the mole ratio. It is this ratio that contains the fundamental information about relative quantities of substances involved in a reaction. The other conversion factors merely involve changing from moles to grams or from grams to moles.

The result from the example can be estimated as:

ESTIMATE

$$2 \times 10^2 \; \cancel{g \; N_2} \times \frac{1 \; \cancel{mole \; N_2}}{3 \times 10^1 \; \cancel{g \; N_2}} \times \frac{2 \; \cancel{moles \; NH_3}}{1 \; \cancel{mole \; N_2}} \times \frac{2 \times 10^1 \; g \; NH_3}{1 \; \cancel{mole \; NH_3}}$$

$$= \frac{8 \times 10^3 \; g \; NH_3}{3 \times 10^1} = 3 \times 10^2 \; g \; NH_3$$

The calculated result is

CALCULATION $2.43 \times 10^2 \; g \; NH_3 = 243 \; g \; NH_3$

Mass–mass problems are somewhat more difficult to check than mass–mole problems. You should notice in this example, however, that more moles of ammonia are produced than nitrogen consumed. This difference is reflected in their relative masses.

The example just described was based on the relationship between a reactant and a product. However, mass–mass stoichiometry problems can also involve two reactants or two products.

For example, we might be asked to find the number of grams of hydrogen needed to react completely with 50.0 g of nitrogen in the ammonia synthesis: $N_2 + 3H_2 \rightarrow 2NH_3$. This problem is outlined as follows:

UNKNOWN	GIVEN	CONNECTION		
grams H_2	50.0 g N_2	$\dfrac{1 \; mole \; N_2}{28.0 \; g \; N_2}$	or	$\dfrac{28.0 \; g \; N_2}{1 \; mole \; N_2}$
		$\dfrac{1 \; mole \; N_2}{3 \; moles \; H_2}$	or	$\dfrac{3 \; moles \; H_2}{1 \; mole \; N_2}$
		$\dfrac{1 \; mole \; H_2}{2.02 \; g \; H_2}$	or	$\dfrac{2.02 \; g \; H_2}{1 \; mole \; H_2}$

We then set up the problem:

$$\text{grams } H_2 = 50.0 \; \cancel{g \; N_2} \times \frac{1 \; \cancel{mole \; N_2}}{28.0 \; \cancel{g \; N_2}} \times \frac{3 \; \cancel{moles \; H_2}}{1 \; \cancel{mole \; N_2}} \times \frac{2.02 \; g \; H_2}{1 \; \cancel{mole \; H_2}}$$

This problem involves the following three conversions:

$$\text{grams } N_2 \xrightarrow[\text{factor}]{\text{first}} \text{moles } N_2 \xrightarrow[\text{factor}]{\text{second}} \text{moles } H_2 \xrightarrow[\text{factor}]{\text{third}} \text{grams } H_2$$

(mole ratio)

The key step is the conversion of moles of N_2 to moles of H_2 using a mole ratio based on the coefficients of the balanced equation.
We can estimate the answer as follows:

ESTIMATE $\quad 5 \times 10^1 \text{ g } N_2 \times \dfrac{1 \text{ mole } N_2}{3 \times 10^1 \text{ g } N_2} \times \dfrac{3 \text{ moles } H_2}{1 \text{ mole } N_2} \times \dfrac{2 \text{ g } H_2}{1 \text{ mole } H_2}$

$= \dfrac{3 \times 10^2 \text{ g } H_2}{3 \times 10^1} = 1 \times 10^1 \text{ g } H_2 = 10 \text{ g } H_2$

The calculated answer is

CALCULATION $\quad 10.8 \text{ g } H_2$

Although more moles of hydrogen are used, its mass is less than that of the nitrogen consumed because of nitrogen's low molecular weight.

Example Copper can be separated from its sulfide ore by heating in oxygen:

$$Cu_2S(s) + O_2(g) \rightarrow 2Cu(s) + SO_2(g)$$
copper(II)   oxygen   copper   sulfur
sulfide   dioxide

How many grams of copper can be made from 683 g of ore?

UNKNOWN	GIVEN	CONNECTION	
grams Cu	683 g Cu_2S	$\dfrac{1 \text{ mole } Cu_2S}{159 \text{ g } Cu_2S}$ or	$\dfrac{159 \text{ g } Cu_2S}{1 \text{ mole } Cu_2S}$
		$\dfrac{1 \text{ mole } Cu_2S}{2 \text{ moles } Cu}$ or	$\dfrac{2 \text{ moles } Cu}{1 \text{ mole } Cu_2S}$
		$\dfrac{1 \text{ mole } Cu}{63.5 \text{ g } Cu}$ or	$\dfrac{63.5 \text{ g } Cu}{1 \text{ mole } Cu}$

$$\text{grams Cu} = 683 \text{ g } Cu_2S \times \dfrac{1 \text{ mole } Cu_2S}{159 \text{ g } Cu_2S} \times \dfrac{2 \text{ moles } Cu}{1 \text{ mole } Cu_2S} \times \dfrac{63.5 \text{ g } Cu}{1 \text{ mole } Cu}$$

ESTIMATE

$$7 \times 10^2 \text{ g } \cancel{Cu_2S} \times \frac{1 \text{ mole } \cancel{Cu_2S}}{2 \times 10^2 \text{ g } \cancel{Cu_2S}} \times \frac{2 \text{ moles } \cancel{Cu}}{1 \text{ mole } \cancel{Cu_2S}} \times \frac{6 \times 10^1 \text{ g Cu}}{1 \text{ mole } \cancel{Cu}}$$

$$= 4 \times 10^2 \text{ g Cu} = 400 \text{ g Cu}$$

CALCULATION $5.46 \times 10^2 \text{ g Cu} = 546 \text{ g Cu}$

Example   The following reaction takes place during the explosion of black gun-powder:

$$2KNO_3(s) + 4C(s) \rightarrow K_2CO_3(s) + 3CO(g) + N_2(g)$$
potassium   carbon   potassium   carbon   nitrogen
nitrate   carbonate   monoxide

How many grams of potassium nitrate are needed to react with 5.13 g of carbon?

UNKNOWN	GIVEN	CONNECTION	
grams KNO_3	5.13 g C	$\dfrac{1 \text{ mole C}}{12.0 \text{ g C}}$ or $\dfrac{12.0 \text{ g C}}{1 \text{ mole C}}$	
		$\dfrac{2 \text{ moles } KNO_3}{4 \text{ moles C}}$ or $\dfrac{4 \text{ moles C}}{2 \text{ moles } KNO_3}$	
		$\dfrac{1 \text{ mole } KNO_3}{101 \text{ g } KNO_3}$ or $\dfrac{101 \text{ g } KNO_3}{1 \text{ mole } KNO_3}$	

$$\text{grams } KNO_3 = 5.13 \text{ g } \cancel{C} \times \frac{1 \text{ mole } \cancel{C}}{12.0 \text{ g } \cancel{C}} \times \frac{2 \text{ moles } \cancel{KNO_3}}{4 \text{ moles } \cancel{C}} \times \frac{101 \text{ g } KNO_3}{1 \text{ mole } \cancel{KNO_3}}$$

ESTIMATE

$$5 \text{ g } \cancel{C} \times \frac{1 \text{ mole } \cancel{C}}{1 \times 10^1 \text{ g } \cancel{C}} \times \frac{2 \text{ moles } \cancel{KNO_3}}{4 \text{ moles } \cancel{C}} \times \frac{1 \times 10^2 \text{ g } KNO_3}{1 \text{ mole } \cancel{KNO_3}}$$

$$= 3 \times 10^1 \text{ g } KNO_3 = 30 \text{ g } KNO_3$$

CALCULATION $21.6 \text{ g } KNO_3$

Example   In 1774, Joseph Priestley obtained ammonia by heating sal ammoniac (ammonium chloride) with slaked lime (calcium hydroxide):

$$2NH_4Cl(aq) + Ca(OH)_2(aq) \rightarrow 2NH_3(g) + CaCl_2(aq) + 2H_2O(l)$$
ammonium   calcium   ammonia   calcium   water
chloride   hydroxide   chloride

How many grams of calcium hydroxide are needed to make 90.0 g of ammonia?

UNKNOWN	GIVEN	CONNECTION
grams Ca(OH)$_2$	90.0 g NH$_3$	$\dfrac{1 \text{ mole NH}_3}{17.0 \text{ g NH}_3}$ or $\dfrac{17.0 \text{ g NH}_3}{1 \text{ mole NH}_3}$
		$\dfrac{1 \text{ mole Ca(OH)}_2}{2 \text{ moles NH}_3}$ or $\dfrac{2 \text{ moles NH}_3}{1 \text{ mole Ca(OH)}_2}$
		$\dfrac{1 \text{ mole Ca(OH)}_2}{74.1 \text{ g Ca(OH)}_2}$ or $\dfrac{74.1 \text{ g Ca(OH)}_2}{1 \text{ mole Ca(OH)}_2}$

$$\text{grams Ca(OH)}_2 = 90.0 \text{ g NH}_3 \times \frac{1 \text{ mole NH}_3}{17.0 \text{ g NH}_3} \times \frac{1 \text{ mole Ca(OH)}_2}{2 \text{ moles NH}_3} \times \frac{74.1 \text{ g Ca(OH)}_2}{1 \text{ mole Ca(OH)}_2}$$

ESTIMATE $\quad 9 \times 10^1 \text{ g NH}_3 \times \dfrac{1 \text{ mole NH}_3}{2 \times 10^1 \text{ g NH}_3} \times \dfrac{1 \text{ mole Ca(OH)}_2}{2 \text{ moles NH}_3} \times \dfrac{7 \times 10^1 \text{ g Ca(OH)}_2}{1 \text{ mole Ca(OH)}_2}$

$$= 2 \times 10^2 \text{ g Ca(OH)}_2 = 200 \text{ g Ca(OH)}_2$$

CALCULATION $\quad 1.96 \times 10^2 \text{ g Ca(OH)}_2 = 196 \text{ g Ca(OH)}_2$

Practice problem   Magnetite, an iron ore, is formed inside the earth by the following process:

$$3FeO(s) + CO_2(s) \rightarrow Fe_3O_4(s) + CO(g)$$

What mass in grams of carbon dioxide is required to react with 625 g of iron(II) oxide?

ANSWER   127 g CO_2

Practice problem   The following equation represents the decomposition of nitric acid:

$$4HNO_3(aq) \rightarrow 4NO_2(g) + 2H_2O(l) + O_2(g)$$

What mass of oxygen is formed from 153 g of HNO_3?

ANSWER   19.4 g O_2

**10.5
PROBLEMS
INVOLVING
AVOGADRO'S
NUMBER**

Some stoichiometry problems require the use of Avogadro's number to express a quantity in terms of the number of particles, whether they be atoms, molecules, or formula units. This kind of problem is the same as a mass–mass problem except that a conversion is made between moles and numbers of particles instead of between moles and grams.

A typical example is to find the number of ammonia molecules produced from 10.0 g of hydrogen in the reaction $N_2 + 3H_2 \rightarrow 2NH_3$. It is outlined as follows:

UNKNOWN	GIVEN	CONNECTION
molecules NH_3	10.0 g H_2	$\dfrac{1 \text{ mole } H_2}{2.02 \text{ g } H_2}$ or $\dfrac{2.02 \text{ g } H_2}{1 \text{ mole } H_2}$
		$\dfrac{3 \text{ moles } H_2}{2 \text{ moles } NH_3}$ or $\dfrac{2 \text{ moles } NH_3}{3 \text{ moles } H_2}$
		$\dfrac{1 \text{ mole } NH_3}{6.02 \times 10^{23} \text{ molecules } NH_3}$
		or $\dfrac{6.02 \times 10^{23} \text{ molecules } NH_3}{1 \text{ mole } NH_3}$

The additional conversion factor used in this type of problem is the one that shows the number of particles in one mole. In this case, the particles are molecules. This problem is set up as follows:

$$\text{molecules } NH_3 = 10.0 \, \cancel{\text{g } H_2} \times \frac{1 \, \cancel{\text{mole } H_2}}{2.02 \, \cancel{\text{g } H_2}} \times \frac{2 \, \cancel{\text{moles } NH_3}}{3 \, \cancel{\text{moles } H_2}} \times \frac{6.02 \times 10^{23} \text{ molecules } NH_3}{1 \, \cancel{\text{mole } NH_3}}$$

The first conversion factor changes grams of hydrogen to moles of hydrogen. The second factor is the mole ratio that converts moles of hydrogen to moles of ammonia. The third factor uses Avogadro's number to convert moles of ammonia to the number of ammonia molecules.

$$\text{grams } H_2 \xrightarrow[\text{factor}]{\text{first}} \text{moles } H_2 \xrightarrow[\text{factor}]{\text{second}} \text{moles } NH_3 \xrightarrow[\text{factor}]{\text{third}} \text{molecules } NH_3$$
$$\text{(mole ratio)}$$

The answer is estimated to be:

$$1 \times 10^1 \text{ g H}_2 \times \frac{1 \text{ mole H}_2}{2 \text{ g H}_2} \times \frac{2 \text{ moles NH}_3}{3 \text{ moles H}_2} \times \frac{6 \times 10^{23} \text{ molecules NH}_3}{1 \text{ mole NH}_3}$$

$$= \frac{12 \times 10^{24} \text{ molecules NH}_3}{6} = 2 \times 10^{24} \text{ molecules NH}_3$$

The actual result is:

CALCULATION 2.01×10^{24} molecules NH_3

Because 2.01×10^{24} is greater than Avogadro's number, we can see that more than one mole of ammonia molecules is formed in this reaction.

We can also find the number of grams of one substance when we are given the number of particles of another substance.

For example, let us find the mass of nitrogen needed to form 1.00×10^{25} molecules of ammonia, again by the reaction $N_2 + 3H_2 \rightarrow 2NH_3$.

UNKNOWN	GIVEN	CONNECTION
grams N_2	1.00×10^{25} molecules NH_3	$\dfrac{1 \text{ mole N}_2}{2 \text{ moles NH}_3}$ or $\dfrac{2 \text{ moles NH}_3}{1 \text{ mole N}_2}$
		$\dfrac{1 \text{ mole NH}_3}{6.02 \times 10^{23} \text{ molecules NH}_3}$
		or $\dfrac{6.02 \times 10^{23} \text{ molecules NH}_3}{1 \text{ mole NH}_3}$
		$\dfrac{1 \text{ mole N}_2}{28.0 \text{ g N}_2}$ or $\dfrac{28.0 \text{ g N}_2}{1 \text{ mole N}_2}$

This problem is set up in the following way:

$$\text{grams N}_2 = 1.00 \times 10^{25} \text{ molecules NH}_3 \times \frac{1 \text{ mole NH}_3}{6.02 \times 10^{23} \text{ molecules NH}_3} \times$$
$$\frac{1 \text{ mole N}_2}{2 \text{ moles NH}_3} \times \frac{28.0 \text{ g N}_2}{1 \text{ mole N}_2}$$

The first factor converts molecules of ammonia to moles of ammonia. The second factor converts moles of ammonia to moles of nitrogen. Finally, the third factor converts moles of nitrogen to grams of nitrogen.

$$\text{molecules NH}_3 \xrightarrow[\text{factor}]{\text{first}} \text{moles NH}_3 \xrightarrow[\text{factor}]{\text{second}} \text{moles N}_2 \xrightarrow[\text{factor}]{\text{third}} \text{grams N}_2$$
$$\text{(mole ratio)}$$

The estimated result is:

ESTIMATE

$$1 \times 10^{25} \text{ molecules NH}_3 \times \frac{1 \text{ mole NH}_3}{6 \times 10^{23} \text{ molecules NH}_3} \times \frac{1 \text{ mole N}_2}{2 \text{ moles NH}_3}$$
$$\times \frac{3 \times 10^1 \text{ g N}_2}{1 \text{ mole N}_2}$$

$$= \frac{3 \times 10^{26} \text{ g N}_2}{12 \times 10^{23}} = 3 \times 10^2 \text{ g N}_2 = 300 \text{ g N}_2$$

and the computed result is

CALCULATION $2.33 \times 10^2 \text{ g N}_2 = 233 \text{ g N}_2$

Example   In the production of synthetic natural gas (SNG), hydrogen is reacted with carbon monoxide to make methane:

$$3H_2(g) + CO(g) \rightarrow CH_4(g) + H_2O(l)$$
$$\text{hydrogen} \quad \text{carbon} \qquad \text{methane} \quad \text{water}$$
$$\text{monoxide}$$

How many molecules of methane are formed from 240. g of hydrogen?

UNKNOWN	GIVEN	CONNECTION
molecules CH_4	240. g H_2	$\dfrac{1 \text{ mole } H_2}{2.02 \text{ g } H_2}$ or $\dfrac{2.02 \text{ g } H_2}{1 \text{ mole } H_2}$
		$\dfrac{3 \text{ moles } H_2}{1 \text{ mole } CH_4}$ or $\dfrac{1 \text{ mole } CH_4}{3 \text{ moles } H_2}$
		$\dfrac{1 \text{ mole } CH_4}{6.02 \times 10^{23} \text{ molecules } CH_4}$
		or $\dfrac{6.02 \times 10^{23} \text{ molecules } CH_4}{1 \text{ mole } CH_4}$

$$\text{molecules } CH_4 = 240. \text{ g } H_2 \times \frac{1 \text{ mole } H_2}{2.02 \text{ g } H_2} \times \frac{1 \text{ mole } CH_4}{3 \text{ moles } H_2} \times$$
$$\frac{6.02 \times 10^{23} \text{ molecules } CH_4}{1 \text{ mole } CH_4}$$

$$\text{ESTIMATE} \qquad 2 \times 10^2 \ \cancel{g \ H_2} \times \frac{1 \ \cancel{mole \ H_2}}{2 \ \cancel{g \ H_2}} \times \frac{1 \ \cancel{mole \ CH_4}}{3 \ \cancel{moles \ H_2}} \times$$

$$\frac{6 \times 10^{23} \text{ molecules } CH_4}{1 \ \cancel{mole \ CH_4}}$$

$$= 2 \times 10^{25} \text{ molecules } CH_4$$

CALCULATION 2.38×10^{25} molecules CH_4

Example   Sodium thiosulfate is also called photographer's hypo or fixer. It is a fixing agent used to dissolve unchanged silver salts from film negatives after they have been exposed to light. It is made by this process:

$$2Na_2S(aq) \ + \ Na_2CO_3(aq) \ + \ 4SO_2(g) \rightarrow 3Na_2S_2O_3(aq) \ + \ CO_2(g)$$

| sodium | sodium | sulfur | sodium | carbon |
| sulfide | carbonate | dioxide | thiosulfate | dioxide |

How many grams of hypo could be made from 8.33×10^{24} formula units of sodium carbonate?

UNKNOWN	GIVEN	CONNECTION
grams $Na_2S_2O_3$	8.33×10^{24} formula units Na_2CO_3	$\dfrac{1 \text{ mole } Na_2CO_3}{6.02 \times 10^{23} \text{ formula units } Na_2CO_3}$
		or
		$\dfrac{6.02 \times 10^{23} \text{ formula units } Na_2CO_3}{1 \text{ mole } Na_2CO_3}$
		$\dfrac{1 \text{ mole } Na_2CO_3}{3 \text{ moles } Na_2S_2O_3}$
		or $\dfrac{3 \text{ moles } Na_2S_2O_3}{1 \text{ mole } Na_2CO_3}$
		$\dfrac{1 \text{ mole } Na_2S_2O_3}{158 \text{ g } Na_2S_2O_3}$
		or $\dfrac{158 \text{ g } Na_2S_2O_3}{1 \text{ mole } Na_2S_2O_3}$

$$\text{grams} \quad Na_2S_2O_3 \ = \ 8.33 \times 10^{24} \ \cancel{\text{formula units } Na_2CO_3} \times$$

$$\frac{1 \ \cancel{mole \ Na_2CO_3}}{6.02 \times 10^{23} \ \cancel{\text{formula units } Na_2CO_3}} \times \frac{3 \ \cancel{moles \ Na_2S_2O_3}}{1 \ \cancel{mole \ Na_2CO_3}} \times$$

$$\frac{158 \text{ g } Na_2S_2O_3}{1 \ \cancel{mole \ Na_2S_2O_3}}$$

ESTIMATE

$$8 \times 10^{24} \text{ formula units } Na_2CO_3 \times \frac{1 \text{ mole } Na_2CO_3}{6 \times 10^{23} \text{ formula units } Na_2CO_3}$$

$$\frac{3 \text{ moles } Na_2S_2O_3}{1 \text{ mole } Na_2CO_3} \times \frac{2 \times 10^2 \text{ g } Na_2S_2O_3}{1 \text{ mole } Na_2S_2O_3}$$

$$= 8 \times 10^3 \text{ g } Na_2S_2O_3$$

CALCULATION 6.56×10^3 g $Na_2S_2O_3$

Example   Vanadium, used in some types of steel, can be made from its oxide in the following way:

$$V_2O_5(s) + 5Ca(s) \rightarrow 2V(s) + 5CaO(s)$$

vanadium(V) calcium   vanadium calcium
oxide   oxide

How many atoms of calcium are needed to yield 1.24×10^{22} atoms of vanadium?

UNKNOWN	GIVEN	CONNECTION
atoms Ca	1.24×10^{22} atoms V	$\dfrac{5 \text{ moles Ca}}{2 \text{ moles V}}$ or $\dfrac{2 \text{ moles V}}{5 \text{ moles Ca}}$
		$\dfrac{1 \text{ mole V}}{6.02 \times 10^{23} \text{ atoms V}}$
		or $\dfrac{6.02 \times 10^{23} \text{ atoms V}}{1 \text{ mole V}}$
		$\dfrac{1 \text{ mole Ca}}{6.02 \times 10^{23} \text{ atoms Ca}}$
		or $\dfrac{6.02 \times 10^{23} \text{ atoms Ca}}{1 \text{ mole Ca}}$

$$\text{atoms Ca} = 1.24 \times 10^{22} \text{ atoms V} \times \frac{1 \text{ mole V}}{6.02 \times 10^{23} \text{ atoms V}} \times$$

$$\frac{5 \text{ moles Ca}}{2 \text{ moles V}} \times \frac{6.02 \times 10^{23} \text{ atoms Ca}}{1 \text{ mole Ca}}$$

ESTIMATE

$$1 \times 10^{22} \text{ atoms V} \times \frac{1 \text{ mole V}}{6 \times 10^{23} \text{ atoms V}} \times \frac{5 \text{ moles Ca}}{2 \text{ moles V}}$$

$$\times \frac{6 \times 10^{23} \text{ atoms Ca}}{1 \text{ mole Ca}}$$

$$= 3 \times 10^{22} \text{ atoms Ca}$$

CALCULATION 3.10×10^{22} atoms Ca

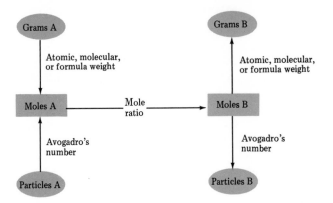

FIGURE 10-2

Stoichiometry. The chart summarizes the conversions in stoichiometry problems. The mole ratio is the key relationship.

Practice problem   Aluminum is recovered from bauxite ore by this process:

$$2Al_2O_3(s) \rightarrow 4Al(s) + 3O_2(g)$$

How many atoms of aluminum are formed from 10.5 g of Al_2O_3?

ANSWER 1.24×10^{23} atoms Al

Practice problem   Iron(III) sulfate can be formed by the following reaction:

$$2Fe(s) + 6H_2SO_4(aq) \rightarrow Fe_2(SO_4)_3(aq) + 3SO_2(g) + 6H_2O(l)$$

What mass of iron(III) sulfate is made from 9.51×10^{22} iron atoms?

ANSWER 31.6 g $Fe_2(SO_4)_3$

Figure 10-2 summarizes the approach used in all of the stoichiometry problems described in this chapter. Substance "A" is given, and is expressed in moles, grams, or particles. Substance "B" is what we are asked to find, and it also can be expressed in moles, grams, or numbers of particles. The key relationship between the two substances is given by their mole ratio.

10.6
LIMITING
REAGENT
PROBLEMS

In most actual chemical reactions that take place in the laboratory, in industry, or in nature, at least one reactant is present in excess and will not be completely used up. For example, if we burn carbon as charcoal, in an open container, there will still be oxygen left in the air after all the carbon is gone. In this case, oxygen is the reactant that is present in excess.

$$C + O_2 \text{ (from air)} \rightarrow CO_2$$
limiting
reagent

Carbon is called the limiting substance or **limiting reagent** because it limits how far the reaction can go. (A reagent is simply a chemical substance that is used for its ability to react with other substances.) When the carbon is completely consumed, the reaction must stop, regardless of how much oxygen is still present. In general, the limiting reagent in a reaction is the reactant used up first. Only if the proportions of reactants mixed correspond exactly to the coefficients in the balanced equation, will all of the reactants be completely consumed. In such cases, the reactants are said to be present in stoichiometric amounts.

To find the limiting reagent, if there is one, in a particular reaction, we must first convert the masses of the reactants into moles. Suppose that we have 60.0 g N_2 mixed with 10.0 g H_2 to form ammonia. Each of these quantities is converted to moles as shown:

$$60.0 \text{ g } N_2 \times \frac{1 \text{ mole } N_2}{28.0 \text{ g } N_2} = 2.14 \text{ moles } N_2$$

$$10.0 \text{ g } H_2 \times \frac{1 \text{ mole } H_2}{2.02 \text{ g } H_2} = 4.95 \text{ moles } H_2$$

Then we compare these amounts to their mole ratio, which is obtained from the balanced equation.

$$N_2 + 3H_2 \rightarrow 2NH_3$$

$$\text{mole ratio: } \frac{1 \text{ mole } N_2}{3 \text{ moles } H_2} \quad \text{or} \quad \frac{3 \text{ moles } H_2}{1 \text{ mole } N_2}$$

The mole ratio shows that there must be three moles of hydrogen molecules present for every mole of nitrogen molecules. Therefore, in order to use up all the nitrogen, there would have to be

$$2.14 \text{ moles } N_2 \times \frac{3 \text{ moles } H_2}{1 \text{ mole } N_2} = 6.42 \text{ moles } H_2$$

But since only 4.95 moles of H_2 are present, all of the nitrogen cannot react. Some nitrogen will be left over. However, all of the H_2 will be consumed, since it requires only:

$$4.95 \text{ moles } H_2 \times \frac{1 \text{ mole } N_2}{3 \text{ moles } H_2} = 1.65 \text{ moles } N_2$$

and there are 2.14 moles N_2 present. *Hydrogen is therefore the limiting reagent*. The number of moles of N_2 that will be left over is $2.14 - 1.65$ moles, or 0.49 moles. The mass of the excess N_2 is:

$$\text{excess } N_2: 0.49 \text{ mole } N_2 \times \frac{28.0 \text{ g } N_2}{1 \text{ mole } N_2} = 14.7 \text{ g } N_2$$

We can summarize this situation as follows:

	Grams initially present	Moles initially present	Moles used up	Moles left over	Grams left over	
N_2	60.0	2.14	1.65	0.49	14.7	excess
H_2	10.0	4.95	4.95	0	0	limiting

Thus, to find the amount of ammonia that is formed, we must base our calculation on hydrogen, the limiting reagent. In other words, the *limiting reagent* must be included in the Connection when we outline the problem. (If we use the reagent that is in excess, the answer will be incorrect.) This problem is outlined as shown.

UNKNOWN	GIVEN	CONNECTION
grams NH_3	4.95 moles H_2	$\dfrac{3 \text{ moles } H_2}{1 \text{ mole } NH_3}$ or $\dfrac{1 \text{ mole } NH_3}{3 \text{ moles } H_2}$
		$\dfrac{1 \text{ mole } NH_3}{17.0 \text{ g } NH_3}$ or $\dfrac{17.0 \text{ g } NH_3}{1 \text{ mole } NH_3}$

Since we already converted grams of H_2 to moles, there is no need to repeat this step. We simply use the result already obtained. We could not solve this problem by using the initial amount of *nitrogen* because not all of the nitrogen is converted to ammonia in the reaction.

Example Metallic plutonium is obtained by a replacement reaction involving calcium:

$$2PuF_3(s) + 3Ca(s) \rightarrow 2Pu(s) + 3CaF_2(s)$$

plutonium(III)   calcium   plutonium   calcium
fluoride   fluoride

If 30.2 g of plutonium fluoride are reacted with 25.0 g of calcium, which is the limiting reagent?

$$30.2 \text{ g } PuF_3 \times \frac{1 \text{ mole } PuF_3}{291 \text{ g } PuF_3} = 0.104 \text{ mole } PuF_3$$

$$25.0 \text{ g } Ca \times \frac{1 \text{ mole Ca}}{40.1 \text{ g } Ca} = 0.623 \text{ mole Ca}$$

$$\text{mole ratio: } \frac{2 \text{ moles PuF}_3}{3 \text{ moles Ca}} \quad \text{or} \quad \frac{3 \text{ moles Ca}}{2 \text{ moles PuF}_3}$$

PuF_3 will be completely gone after it has reacted with

$$0.104 \text{ mole PuF}_3 \times \frac{3 \text{ moles Ca}}{2 \text{ moles PuF}_3} = 0.156 \text{ mole Ca}$$

Since there are 0.623 mole Ca present, the Ca is in excess and PuF_3 is the limiting reagent.

Example   Tarnish, which consists of silver sulfide, can be removed from silverware by heating the tarnished pieces in an aluminum pan filled with water (and a small amount of salt):

$$3Ag_2S(s) + 2Al(s) + 3H_2O(l) \rightarrow 6Ag(s) + Al_2O_3(s) + 3H_2S(g)$$

silver sulfide   aluminum   water   silver   aluminum oxide   hydrogen sulfide

---

Do-it-yourself 10

Fire fighting

A fire is the rapid oxidation of a substance, which produces heat and light. In most fires, the reaction occurs between oxygen in the air and the material that is burning. One way to put out a fire is to cut off the supply of oxygen. In this way, oxygen becomes the *limiting reagent.* The fire goes out when the oxygen is used up.

The oxygen supply can be cut off by smothering the fire with foam, a dry solid, carbon dioxide, or a nonflammable vapor. Baking soda (sodium hydrogencarbonate) is a useful household fire fighter because it is a dry solid and also produces carbon dioxide.

To do: To demonstrate how oxygen can be made the limiting reagent in a combustion reaction, stand a small candle in the middle of a pie plate. (You can secure the candle to the plate with some melted wax.) Fill the plate with water and light the candle. Set a large empty jar over the candle and let it rest on the plate. Can you explain what happens?

To use baking soda as a fire extinguisher, place a lighted match on a metal tray. Sprinkle with baking soda to smother the fire.

To observe the effect of carbon dioxide on a fire, use a wire to hang a lighted candle right-side up in a jar. In a separate pitcher, prepare carbon dioxide by adding a small amount of vinegar to one tablespoon of baking soda. Pour the CO_2 gas from the pitcher into the jar containing the candle.

If the pan contains 40.0 g of exposed aluminum, can 300. g of tarnish be removed?

$$40. \text{ g Al} \times \frac{1 \text{ mole Al}}{27.0 \text{ g Al}} = 1.48 \text{ moles Al}$$

$$300. \text{ g Ag}_2\text{S} \times \frac{1 \text{ mole Ag}_2\text{S}}{248 \text{ g Ag}_2\text{S}} = 1.21 \text{ moles Ag}_2\text{S}$$

$$\text{mole ratio:} \quad \frac{3 \text{ moles Ag}_2\text{S}}{2 \text{ moles Al}} \quad \text{or} \quad \frac{2 \text{ moles Al}}{3 \text{ moles Ag}_2\text{S}}$$

To be completely consumed, the silver sulfide must react with

$$1.21 \text{ moles Ag}_2\text{S} \times \frac{2 \text{ moles Al}}{3 \text{ moles Ag}_2\text{S}} = 0.806 \text{ mole Al}$$

All the tarnish can be removed because there is excess aluminum present.

Example   Chlorine can be produced by the following process:

$$\underset{\substack{\text{manganese(IV)}\\\text{oxide}}}{\text{MnO}_2(s)} + \underset{\substack{\text{hydrochloric}\\\text{acid}}}{4\text{HCl}(aq)} \rightarrow \underset{\substack{\text{manganese(II)}\\\text{chloride}}}{\text{MnCl}_2(s)} + \underset{\text{chlorine}}{\text{Cl}_2(g)} + \underset{\text{water}}{2\text{H}_2\text{O}(l)}$$

If 50.0 g of manganese dioxide reacts with 90.0 g of hydrogen chloride, which reactant is present in excess? How much of this reactant is left over?

$$50.0 \text{ g MnO}_2 \times \frac{1 \text{ mole MnO}_2}{86.9 \text{ g MnO}_2} = 0.575 \text{ mole MnO}_2$$

$$90.0 \text{ g HCl} \times \frac{1 \text{ mole HCl}}{36.5 \text{ g HCl}} = 2.47 \text{ moles HCl}$$

$$\text{mole ratio:} \quad \frac{1 \text{ mole MnO}_2}{4 \text{ moles HCl}} \quad \text{or} \quad \frac{4 \text{ moles HCl}}{1 \text{ mole MnO}_2}$$

Manganese dioxide reacts completely with

$$0.575 \text{ mole MnO}_2 \times \frac{4 \text{ moles HCl}}{1 \text{ mole MnO}_2} = 2.30 \text{ moles HCl}$$

Since there are 2.47 moles HCl, HCl is present in excess and MnO₂ is the limiting reagent. The amount of HCl left over is 2.47 − 2.30 moles, or 0.17 mole.

$$0.17 \text{ mole HCl} \times \frac{36.5 \text{ g HCl}}{1 \text{ mole HCl}} = 6.21 \text{ g HCl}$$

The mass of excess HCl is 6.21 g.

Practice problem   In the conversion of carbon monoxide to carbon dioxide,

$$2CO(g) + O_2(g) \rightarrow 2CO_2(g)$$

75.0 g of CO are mixed with 59.5 g of O_2. Which gas limits the reaction? How much of the other gas in grams is left over?

ANSWER   CO is the limiting reagent; 16.6 g of O_2 remain unreacted.

Practice problem   Acetylene, C_2H_2, can be generated from calcium carbide by the addition of water:

$$CaC_2(s) + 2H_2O(l) \rightarrow Ca(OH)_2(aq) + C_2H_2(g)$$

If 165 g of CaC_2 are mixed with 95.0 g of water, which reagent is limiting? How much of the excess reagent is used up?

ANSWER CaC_2 is limiting; 92.7 g of H_2O are consumed.

10.7 PERCENT YIELD   The amount of product that is calculated in a stoichiometry problem is not usually the same as the amount actually obtained in the reaction. Portions of the product can be lost during the reaction. For example, some of the product may be left on the walls of glassware, or in other places where it cannot be readily recovered. Also, many reactions do not go to completion. Finally, side reactions sometimes take place, which use up reactants without forming the desired product. All of these factors reduce the amount of product that is finally obtained.

After the product of a reaction is isolated, its mass is measured on a balance. This quantity is called the *actual yield* of the reaction. The *theoretical yield* of the reaction is the mass predicted by calculation, based on the assumption that all of the limiting reagent gets converted to product. The theoretical yield would be the same as the actual yield if the reaction and the laboratory technique were perfect. In real life, however, for the reasons just described, the actual yield is always less than the theoretical yield.

The ratio of these two quantities, expressed as percentage, is called the **percent yield**. It is abbreviated as % yield.

$$\% \text{ yield} = \frac{\text{actual yield (g)}}{\text{theoretical yield (g)}} \times 100\%$$

For example, in a particular reaction, if 10.3 g of product are formed, and the calculated amount is 14.2 g, the percent yield is

$$\% \text{ yield} = \frac{10.3 \text{ g}}{14.2 \text{ g}} \times 100\% = 72.5\%$$

Sometimes, the percent yield can be increased by adjusting the reaction conditions. For example, raising the temperature of the reaction, or allowing the reaction to take place for a long time, may increase the amount of product formed.

Example Nitrogen dioxide can be made in the laboratory by heating lead nitrate.

$$2Pb(NO_3)_2(s) \rightarrow 2PbO(s) + 4NO_2(g) + O_2(g)$$

lead(II)   lead(II)   nitrogen   oxygen
nitrate   oxide   dioxide

Box 10

Eutrophication

The basic nutrients needed for plant growth are carbon, nitrogen, and phosphorus. In unpolluted bodies of water, the growth of algae is normally limited because these plants run out of a needed nutrient. Phosphorous in the form of phosphate ions is usually used up first. It is the limiting reagent for plant growth under these conditions.

When dumped into a lake, detergents or other wastes containing phosphorus stimulate an unusual increase in the growth of algae. The algae grow rapidly until one of their nutrients is again exhausted. Then the algae starve and die. The decomposition of algae uses up dissolved oxygen, which is needed by fish and other organisms. Fish begin to die, and the lake gradually fills up with decaying plant and other aquatic life. In addition, the lack of oxygen stimulates the growth of anaerobic bacteria, which produce obnoxious odors. The polluted lake literally chokes to death and becomes a swamp or marsh.

This process is called **eutrophication**, from the Greek word meaning "well nourished." Over thousands of years, lakes normally fill in with decaying plants and animals to become marshes and finally meadows. Artificial enrichment of lakes by pollutants, however, greatly speeds up this normal aging process. Because of their role in eutrophication, phosphates are now banned from detergents in many areas.

If 200. g of lead nitrate are heated, and 120. g of lead oxide are obtained, what is the percent yield for this experiment?

theoretical yield = 135 g PbO (from stoichiometry calculation)

$$\% \text{ yield} = \frac{\text{actual yield}}{\text{theoretical yield}} \times 100\% = \frac{120. \text{ g}}{135 \text{ g}} \times 100\%$$

$$= 88.9\%$$

Example   Molybdenum oxide is used in making certain kinds of steel. It is made from molybdenum sulfide ore:

$$2MoS_2(s) \quad + \; 7O_2(g) \rightarrow 2MoO_3(s) \quad + \; 4SO_2(g)$$
molybdenum(IV)   oxygen   molybdenum(VI)   sulfur
sulfide   oxide   dioxide

If 520. kg of molybdenum oxide are produced from 800. kg of the ore, what is the percent yield of this reaction?

theoretical yield = 720. kg MoO_3 (from stoichiometry calculation)

$$\% \text{ yield} = \frac{\text{actual yield}}{\text{theoretical yield}} \times 100\% = \frac{520. \text{ kg}}{720. \text{ kg}} \times 100\% = 72.2\%$$

Practice problem   The theoretical yield of a reaction is 195 g. The actual yield is 151 g. What is the percent yield?

ANSWER   77.4%

Summary   **10.1** The term stoichiometry refers to any calculation of the quantities of elements or compounds involved in a chemical reaction. These calculations are based on interpreting chemical equations in terms of numbers of moles of reactants and products. The coefficients of the balanced equation indicate the relative numbers of moles of each substance in the reaction.

10.2 The simplest type of stoichiometry problem is a mole–mole conversion. It involves finding the number of moles of one substance in a reaction when the number of moles of another substance is given. To solve a mole–mole problem we use a conversion factor known as the mole ratio. The mole ratio is a fraction that relates the number of moles of two substances involved in the reaction.

10.3 A slightly more complicated stoichiometry problem is the mole–mass problem. It involves finding the mass of one substance when the number of moles of another substance is given. Here a second conversion factor is required: the molar mass. The molar mass is also used to carry out the opposite problem—finding the number of moles of a substance when the mass of another substance in the reaction is given.

10.4 Most stoichiometry problems involve mass to mass conversions. In a mass–mass problem, we are given the mass of one substance in the reaction, and asked to find the mass of a second substance. Three conversion factors are needed to solve a mass–mass problem: the mole ratio and two molar masses. The number of moles of each of the two substances must be related, and the number of moles of each substance must also be related to grams.

10.5 Some stoichiometry problems involve using Avogadro's number to express a quantity in terms of its number of particles. These particles may be either atoms, molecules, or formula units. This type of problem is similar to a mass–mass problem except that a conversion is made between moles and numbers of particles instead of between moles and grams. In this and in all other stoichiometry problems, the key relationship is the mole ratio.

10.6 In most actual chemical reactions, one reactant limits how far the reaction can go. This reactant is known as the limiting reagent because it is the reactant that is used up first. Stoichiometric calculations must be based on the limiting reagent in a reaction.

10.7 The amount of product calculated in a stoichiometry problem is not usually the amount that is actually obtained in the reaction. The percent yield of a reaction is based on the ratio of the actual yield to the predicted theoretical yield.

Exercises

For each term on the left, choose a phrase from the right-hand column that most closely matches its meaning.

1 stoichiometry   **a** gets used up first

2 mole ratio   **b** ratio of actual yield to theoretical yield

3 limiting reagent   **c** conversion factor

4 percent yield   **d** calculation based on equation

EXERCISES BY TOPIC

Stoichiometry (10.1)

5 Describe the basis for stoichiometry calculations.

6 Why is it necessary to start with a balanced equation when doing stoichiometry problems?

7 How is the law of conservation of mass related to stoichiometry?

Mole–mole problems (10.2)

8 What conversion factor is used in mole–mole stoichiometry problems?

9 When magnesium metal burns in air, most is converted to magnesium oxide. A small amount forms magnesium nitride by the following reaction:

$$3Mg(s) + N_2(g) \rightarrow Mg_3N_2(s)$$

a If 0.153 mole of Mg_3N_2 form, how many moles of Mg reacted with N_2?
b How many moles of nitrogen are needed to react with 18.9 moles of magnesium?

10 Aluminum sulfide reacts with water according to the following equation:

$$Al_2S_3(s) + 6H_2O(l) \rightarrow 2Al(OH)_3(s) + 3H_2S(g)$$

a How many moles of H_2S form if 1.75 moles of $Al(OH)_3$ are produced?
b Find the number of moles of Al_2S_3 needed to react with 9.3 moles of H_2O.

11 Hydrogen bromide can be made in the following reaction:

$$PBr_3(l) + 3H_2O(l) \rightarrow 3HBr(aq) + H_3PO_3(aq)$$

a How many moles of HBr can be made from 1.83×10^{-2} mole of PBr_3?
b How many moles of H_3PO_3 form if 0.8234 mole of HBr is produced?

Mole–mass problems (10.3)

12 Excess chlorine can be removed from chlorinated water by passing it through charcoal, which consists of carbon (C):

$$2Cl_2(aq) + C(s) + 2H_2O(l) \rightarrow CO_2(g) + 4HCl(aq)$$

a How many grams of charcoal are needed to remove 10.5 moles of Cl_2?
b How many moles of HCl would be formed from 96.2 g of Cl_2?

13 Potassium chlorate can be used to oxidize nitrite to nitrate, as in the following reaction:

$$KClO_3(s) + 3KNO_2(s) \rightarrow 3KNO_3(s) + KCl(s)$$

a How many moles of KNO_3 can be made from 25.0 g of $KClO_3$?   **b** How many grams of $KClO_3$ are needed to react with 10.1 moles of KNO_2?

14 Phosphorus pentoxide, actually P_4O_{10}, is used as a drying agent because it reacts with water, forming phosphoric acid:

$$P_4O_{10}(s) + 6H_2O(l) \rightarrow 4H_3PO_4(aq)$$

a How many moles of water can be removed by 255 g of P_4O_{10}?   **b** What number of grams of H_3PO_4 are formed from 12.3 moles of H_2O?

Mass–mass problems (10.4)

15 Hydrogen sulfide, released during petroleum refining, can be trapped by reacting it with oxygen:

$$2H_2S(g) + O_2(g) \rightarrow 2S(s) + 2H_2O(l)$$

a How many kilograms of oxygen are needed to react with 350. kg of H_2S?
b How many grams of sulfur are formed from 3.1×10^4 g of H_2S?

16 A stable copper(I) compound can be made by the reaction of Cu(II) ion with iodide ion:

$$2CuSO_4(aq) + 4KI(aq) \rightarrow 2CuI(s) + I_2(s) + 2K_2SO_4(aq)$$

a How many grams of KI are needed to make 120. g of CuI?   **b** What mass of I_2 is formed from 0.153 g of $CuSO_4$?

17 Aluminum metal reduces Sn^{4+} to Sn^{2+}, as in the reaction:

$$3SnCl_4(aq) + 2Al(s) \rightarrow 3SnCl_2(aq) + 2AlCl_3(aq)$$

a How many grams of $SnCl_2$ can be made from 13.6 g of Al?   **b** How many grams of $SnCl_4$ are needed to react with 231 g of Al?

18 The following reaction takes place in the production of lead metal from its ore:

$$PbS(s) + 2PbO(s) \rightarrow 3Pb(s) + SO_2(g)$$

a How many grams of SO_2 form if 625 g of lead are made?   **b** How many grams of Pb are made from 12.2 kg of PbO?

19 Hydrogen bromide can be made by the reaction of sodium bromide with phosphoric acid:

$$3NaBr(s) + H_3PO_4(aq) \rightarrow 3HBr(aq) + Na_3PO_4(aq)$$

a How many grams of H_3PO_4 are needed to make 39.1 g of HBr?   **b** What mass of Na_3PO_4 is formed from 2.2 g of NaBr?

Problems involving Avogadro's number (10.5)

20 Magnesium ion can be precipitated as the hydroxide:

$$MgCl_2(aq) + 2NH_3(aq) + 2H_2O(l) \rightarrow Mg(OH)_2(s) + 2NH_4Cl(aq)$$

a How many formula units of $Mg(OH)_2$ can form from 1.20 moles of NH_3?
b How many grams of H_2O are needed to react with 5.12×10^{25} formula units of $MgCl_2$?

21 Lead sulfide dissolves in hot nitric acid:

$$PbS(s) + 4HNO_3(aq) \rightarrow Pb(NO_3)_2(aq) + S(s) + 2NO_2(g) + 2H_2O(l)$$

a How many grams of NO_2 can be generated from 1.02×10^{22} formula units of PbS?   **b** How many atoms of S are formed if 16.9 g of HNO_3 are reacted?

22 Manganese slowly reacts with water according to the following equation:

$$Mn(s) + 2H_2O(l) \rightarrow Mn(OH)_2(s) + H_2(g)$$

a If 12.5 g of H_2 form, how many water molecules must have reacted? **b** How many atoms of Mn are needed to make 335 g of $Mn(OH)_2$?

Limiting reagent problems (10.6)

23 What is a limiting reagent? Does every reaction have one?

24 Iron(II) compounds change color as a result of the following reaction with oxygen in the air:

$$4FeCl_2(s) + O_2(g) + 4HCl(aq) \rightarrow 4FeCl_3(aq) + 2H_2O(l)$$

a If 50.0 g of $FeCl_2$ is exposed to 21.0 g of oxygen, which is the limiting reagent? **b** What mass of $FeCl_3$ forms when 125 g of $FeCl_2$ contact 85.3 g of O_2?

25 Phosphorus can be made from phosphate rock by the following process:

$$2Ca_3(PO_4)_2(s) + 6SiO_2(s) + 10C(s) \rightarrow P_4(s) + 10CO(g) + 6CaSiO_3(s)$$

a If 235 g of $Ca_3(PO_4)_2$ are reacted with 205 g of C, which reagent is left over, and in how many grams? **b** If 22.1 kg of SiO_2 are reacted with 5.21 kg of C, what mass of P_4 forms?

26 Copper metal can dissolve in nitric acid in the reaction given by this equation:

$$Cu(s) + 4HNO_3(aq) \rightarrow Cu(NO_3)_2(aq) + 2NO_2(g) + 2H_2O(l)$$

a If 1.52 g of metal are placed in contact with 18.5 g of HNO_3, what mass of the excess reagent is used up? **b** How many grams of NO_2 form if 0.513 g of Cu contact 0.981 g of HNO_3?

Percent yield (10.7)

27 What is percent yield? Why is it useful?

28 A reaction whose theoretical yield is 29.3 g produces 8.62 g of the desired compound. What is the percent yield?

29 Iron(III) reacts with sulfide ion according to the following equation:

$$2Fe(NO_3)_3(aq) + 3Na_2S(aq) \rightarrow 2FeS(s) + S(s) + 6NaNO_3(aq)$$

a In a reaction starting with 82.1 g of Na_2S, 32.5 g of FeS form. Find the percent yield. **b** If 112 g of FeS are made from 321 g of $Fe(NO_3)_3$, what is the percent yield?

ADDITIONAL EXERCISES

30 The abrasive silicon carbide, SiC, is made in the following way:

$$SiO_2(s) + 3C(s) \rightarrow SiC(s) + 2CO(g)$$

a How many grams of SiC are made from 1.20 kg of C? **b** How many atoms of C are needed to react with 195 g of SiO_2? **c** If 2.00×10^3 moles of SiC are needed, what mass of SiO_2 should be used? **d** If 10.1 g of SiO_2 are reacted with 8.7 g of C, how much SiC forms?

31 Tetraethyl lead, $Pb(C_2H_5)_4$, an additive in leaded gasoline, is made as follows:

$$4PbNa(s) + 4C_2H_5Cl(g) \rightarrow Pb(C_2H_5)_4(l) + 3Pb(s) + 4NaCl(s)$$

a How many grams of PbNa are needed to make 340. g of $Pb(C_2H_5)_4$? **b** How many millimoles of C_2H_5Cl are needed to react with 0.135 g of PbNa? **c** How many atoms of Pb form from 19.1 kg of C_2H_5Cl? **d** What is the percent yield if 135 g of $Pb(C_2H_5)_4$ are made from 532 g of PbNa?

32 Bromine can be prepared in the lab by this reaction:

$$NaBrO_3(s) + 5NaBr(s) + 3H_2SO_4(aq) \rightarrow 3Br_2(l) + 3Na_2SO_4(aq) + 3H_2O(l)$$

a How many millimoles of bromine can be made from 1.82 g of NaBr? **b** What mass of Na_2SO_4 is formed when 92.1 g of Br_2 are produced? **c** If 7.22×10^{25} Br_2 molecules are needed, what mass of $NaBrO_3$ is required? **d** Find the number of grams left over when 112 g of H_2SO_4 react with 192 g of NaBr (with excess $NaBrO_3$).

33 Aniline, $C_6H_5NH_2$, is used to make many dyes and drugs. It is formed from nitrobenzene, $C_6H_5NO_2$, by the following reaction:

$$C_6H_5NO_2(l) + 3H_2(g) \rightarrow C_6H_5NH_2(l) + 2H_2O(l)$$

a How many kilograms of $C_6H_5NO_2$ are needed to make 2.51×10^4 kg of $C_6H_5NH_2$? **b** How many molecules of $C_6H_5NH_2$ form in part **a**? **c** What is the percent yield if 18.5 kg of $C_6H_5NH_2$ are made from 25.1 kg of $C_6H_5NO_2$? **d** What mass of $C_6H_5NH_2$ will form if 193 kg of $C_6H_5NO_2$ are reacted with 20.1 kg of H_2?

34 Iron can be produced by the reaction:

$$Fe_3O_4(s) + 4H_2(g) \rightarrow 3Fe(s) + 4H_2O(g)$$

a What mass of hydrogen is needed to make 100.0 g of iron? **b** What mass of iron oxide gets used up? **c** How many molecules of water form? **d** If only 85.0 g of iron form, what is the percent yield?

35 The following reaction is used to generate electricity in an automobile battery:

$$Pb(s) + PbO_2(s) + 2H_2SO_4(aq) \rightarrow 2PbSO_4(s) + 2H_2O(l)$$

a Find the mass of lead metal used up when 2.05 g of $PbSO_4$ form. **b** What mass of water is formed? **c** What mass of lead oxide gets consumed? **d** What mass of lead sulfate forms from 15.5 g of PbO_2?

36 Hydrogen chloride can be made by the direct combination of H_2 and Cl_2:

$$H_2(g) + Cl_2(g) \rightarrow 2HCl(g)$$

a What mass of chlorine will combine with 1.38 g of H_2? **b** If 5.00 g of H_2 and 5.00 g of Cl_2 are reacted, which is the limiting reagent? **c** What mass of HCl forms from the reaction described in part **b**?

37 Carbon tetrachloride can be made in the following reaction:

$$C(s) + 2S_2Cl_2(l) \rightarrow CCl_4(l) + 4S(s)$$

a What mass of CCl_4 can be made from 3.65 g of S_2Cl_2?   **b** If 1.51 g of CCl_4 are actually obtained, what is the percent yield?   **c** What mass of carbon is needed to react with 100. g of S_2Cl_2?   **d** What mass of sulfur forms if 1.2 kg of CCl_4 is produced?

38 The Kroll process is used in industry to produce zirconium metal. The reaction involved is:

$$ZrCl_4(s) + 2Mg(s) \rightarrow Zr(s) + 2MgCl_2(s)$$

a What mass of zirconium is obtained from 1.00 kg of $ZrCl_4$?   **b** What mass of $MgCl_2$ forms?   **c** How much magnesium is used up?   **d** What mass of magnesium is needed to react with 250. kg of $ZrCl_4$?

39 The dry cleaning fluid tetrachloroethylene, C_2Cl_4, can be made from acetylene, C_2H_2, as follows:

$$C_2H_2(g) + 3Cl_2(g) + Ca(OH)_2(s) \rightarrow C_2Cl_4(l) + CaCl_2(s) + 2H_2O(l)$$

a What mass of acetylene is needed to make 1.00×10^{27} molecules of tetrachloroethylene?   **b** What mass of chlorine is consumed?   **c** If 1.0 kg of C_2H_2 is reacted with 1.0 kg of Cl_2, how much C_2Cl_4 forms?

PROGRESS CHART   Check off each item when completed.

_____   Previewed chapter

Studied each section:

_____   **10.1** Stoichiometry

_____   **10.2** Mole–mole problems

_____   **10.3** Mole–mass problems

_____   **10.4** Mass–mass problems

_____   **10.5** Problems involving Avogadro's number

_____   **10.6** Limiting reagent problems

_____   **10.7** Percent yield

_____   Reviewed chapter

_____   Answered assigned exercises:

#'s _____

ELEVEN

Gases

In Chapter Eleven, we examine the gaseous state, which we discussed only briefly in Chapter Two. We describe the properties of gases, and explain these properties on the basis of the behavior of molecules. The gas laws are introduced to show relationships between various gas properties. These gas laws are then used in simple calculations. To demonstrate an understanding of Chapter Eleven, you should be able to:

1 Explain the assumptions of the kinetic molecular theory.

2 Define and give units for pressure and atmospheric pressure.

3 Solve problems relating changes in the pressure and volume of a gas.

4 Solve problems relating changes in the pressure and temperature of a gas.

5 Solve problems relating changes in the volume and temperature of a gas.

6 Solve problems relating changes in the pressure, volume, and temperature of a gas.

7 Solve problems relating the number of moles of a gas to its volume.

8 Solve stoichiometry problems involving gaseous reactants or products.

9 Use the ideal gas law to find *P, V, n,* or *T* for a gas.

10 Solve problems involving partial pressures of gases.

11 Describe the major types of air pollutants.

**11.1
KINETIC
MOLECULAR
THEORY OF
GASES**

Even though gas molecules surround us, we are not aware of them because we usually cannot see them or feel them. Nevertheless, we are familiar with many of the properties of gases. They have neither a definite shape nor a definite volume. Gases can expand without limit and can be compressed into very small volumes. The usual densities of gases are extremely low compared to the densities of solids and liquids.

We can explain the gas properties that we observe by recognizing that gases consist of atoms or molecules. For example, consider the ability of gases to be compressed, and consider also their low density. These two properties can be explained if we think of gas molecules as being far apart from each other. In fact, over 99.9% of the volume occupied by the gases in the air consists of empty space. On the average, a gas molecule travels more than 1000 times its own diameter before hitting another molecule.

The fact that gases do not have a definite shape or volume can be explained by the lack of attraction between gas molecules. The molecules are free to move about. The movement of gas molecules also explains **diffusion**, the tendency of a gas to spread out, filling any available space and mixing with other molecules. For example, when gas molecules escape from a perfume bottle, they can be smelled throughout an entire room in a short time. In diffusion, gas molecules spread from regions where there are many molecules to regions where there are fewer.

In the latter half of the 19th century, scientists combined these and other explanations of observed gas properties into the **kinetic molecular theory**. The word kinetic refers to motion. The kinetic molecular theory deals with the motion of the molecules of a gas. It states that the behavior of a gas can be explained on the basis of the following assumptions. (Assumptions are statements taken to be correct without proof.) We accept these assumptions because the theory built upon them is successful in explaining the properties of gases.

1 The molecules of a gas are extremely small and far apart from each other. Most of the volume occupied by a gas consists of empty space.

2 The molecules of a gas are constantly moving in a random, or unpredictable, way.

3 Gas molecules collide with each other and with the walls of the container, producing the pressure exerted by the gas.

4 Collisions of gas molecules are perfectly elastic; no kinetic energy is lost during a collision.

5 Gas molecules do not attract or repel each other.

A gas that obeys these five assumptions perfectly is called an **ideal gas**. (See Figure 11-1.) An ideal gas does not actually exist because no gas obeys the five assumptions perfectly. However, real gases differ only slightly from the ideal gas model. Therefore, the kinetic molecular theory and the concept of an ideal gas are very useful.

The kinetic molecular theory is the basis for the way we think about gases. A gas consists of randomly moving molecules that are far apart. The molecules are constantly moving. *The temperature of a gas is a*

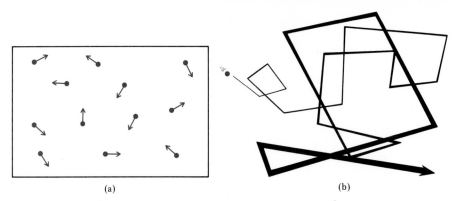

(a)　　　　　　　　　　　　　　　　　　　　(b)

FIGURE 11-1

The kinetic molecular theory of gases. As shown in (a), gas molecules are in constant, random motion. A possible path taken by a single molecule is shown in (b). The changes in direction are caused by collisions with other molecules or with the walls of the container.

measure of the motion of its molecules. A moving object has a kinetic energy that depends on its mass and the square of its velocity (speed):

$$\text{kinetic energy} = \tfrac{1}{2}mv^2$$

where m is mass and v is velocity. Temperature, in kelvins, is directly related to the kinetic energy. The faster the molecules of a gas are moving, the higher the temperature. For example, at room temperature the average speed of oxygen molecules in the air is over 1000 miles/hour. When a gas is heated, its molecules acquire additional speed and energy, and therefore the temperature rises. On the other hand, cooling a gas slows the movement of the gas molecules, causing the temperature to fall.

11.2
PRESSURE

When gas molecules hit the walls of their container or other objects nearby, they push against them, exerting a force. This force, divided by the area over which it is applied is known as pressure, P.

$$P = \text{pressure} = \frac{\text{force}}{\text{area}}$$

The atmosphere surrounding the Earth contains roughly 5×10^{18} tons of air. Due to the Earth's gravity, the weight of the air creates a pressure called the **atmospheric pressure**. The pressure exerted by the atmosphere at sea level (at 0°C) is called 1 atmosphere, abbreviated 1 atm. At higher altitudes, the pressure is smaller because less of the atmosphere is pressing down from above.

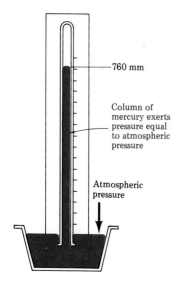

760 mm

Column of
mercury exerts
pressure equal
to atmospheric
pressure

Atmospheric
pressure

FIGURE 11-2

A Torricelli barometer. The barometer is made by filling the long tube with mercury and placing it upside down into a pool of mercury. The height of the mercury column is a direct measure of the atmospheric pressure exerted on the pool of mercury.

Atmospheric pressure can be measured using a barometer, such as the one shown in Figure 11-2. This particular type is called a Torricelli barometer, after the 17th century scientist Evangelista Torricelli.

We can make a Torricelli barometer, by first filling with mercury a long glass tube that is closed at one end. We hold the open end closed to prevent the mercury from spilling out, and then place the open end upside-down in a dish of mercury. Some mercury may flow out. The mercury that remains in the tube is supported by the pressure of the atmosphere on the mercury in the dish.

In this barometer, atmospheric pressure at sea level and 0°C will support a column of mercury that is 760 mm high. Values of pressure can be expressed in millimeters of mercury (mm Hg), which also are called torr, after Torricelli. Thus, 1 atm, 760 mm Hg, and 760 torr, all can be used to describe the pressure of the atmosphere at sea level and 0°C. In weather reports, atmospheric pressure is often expressed in inches of mercury; 1 atm = 29.9 in Hg.

The SI unit for pressure is the pascal, Pa. One torr is equivalent to 133.3 Pa. The atmospheric pressure at sea level and 0°C is 101.3 kPa (101.3 kilopascals). Pressure is sometimes expressed in still another unit, pounds per square inch, lb/in², abbreviated psi (pronounced P-S-I). The pressure of the atmosphere at sea level and 0°C is 14.7 psi. Tire pressure is usually expressed in psi. For example, a typical automobile tire contains air at a pressure of about 30 psi greater than the atmospheric pressure. Table 11-1 summarizes the various units used to express values of pressure.

TABLE 11-1 *Units for pressure*

Unit	Symbol	Value of atmospheric pressure in that unit[a]
atmosphere	atm	1 atm
centimeters of mercury	cm Hg	76 cm Hg
inches of mercury	in Hg	29.9 in Hg
millimeters of mercury	mm Hg (torr)	760 mm Hg (760 torr)
pascals	Pa	101.3 kPa
pounds per square inch	psi (lb/in²)	14.7 psi

[a]At sea level and O°C.

Gas pressures other than that of the atmosphere are measured in the laboratory using a manometer. The manometer shown in Figure 11-3 measures an unknown pressure by comparing it with atmospheric pressure. The difference in height of the mercury columns in the two arms of the manometer is used to find the pressure of the gas, as shown in the diagram.

When a gas is purchased for laboratory use, it is usually contained in a heavy metal cylinder or tank under extremely high pressure (over 100 atm). A regulator, such as the one shown in Figure 11-4, is attached to the tank to reduce this large pressure to a lower value at which the gas can be used. This regulator has two gauges. One gauge shows the tank pressure, and the other gauge gives the reduced pressure of the gas leaving the regulator. Table 11-2 lists various commercially available gases and their properties.

Box 11

Gravity and the atmosphere

Our atmosphere, the mixture of gases that surrounds the Earth, is held in place by gravity. In order for gas molecules to overcome the Earth's gravitational attraction, they need to move at tremendous speeds. Consider what happens when we throw a stone upwards. It reaches a certain height and then is pulled back to the ground by gravity. We would have to give the stone a velocity of 11 kilometers per second, or 25,000 miles per hour, to make it overcome the gravitational attraction between itself and the Earth. This velocity is called the Earth's escape velocity.

The molecules in our atmosphere, mostly N_2 and O_2, move at speeds much lower than the escape velocity and therefore cannot break away. When the Earth was formed, lighter gases such as hydrogen and helium were probably present in the atmosphere in large amounts. But these gases soon escaped, because light molecules move much faster than heavy molecules at the same temperature and are able to exceed the Earth's escape velocity.

The moon is much less massive than the Earth and as a result has a weaker gravitational attraction. The escape velocity from the surface of the moon is only 2.4 kilometers per second, or 5400 miles per hour. N_2 and O_2 move at higher speeds than this, and are able to escape. The moon therefore has lost even the relatively heavier gases, and has no atmosphere.

The planet Jupiter, on the other hand, is much more massive than the Earth, and has a much stronger gravitational attraction. The escape velocity from Jupiter is 59 kilometers per second or 134,000 miles per hour. Because of Jupiter's strong gravitational attraction, and high escape velocity, its atmosphere still contains the hydrogen and helium molecules, as well as heavier molecules, that were present when the planet formed.

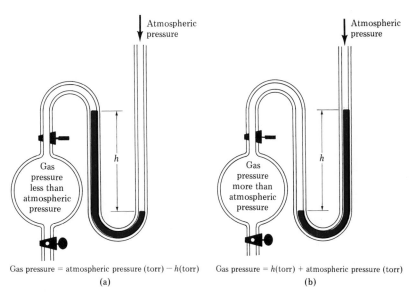

Gas pressure = atmospheric pressure (torr) − h(torr)

(a)

Gas pressure = h(torr) + atmospheric pressure (torr)

(b)

FIGURE 11-3

Manometer. A manometer measures the pressure exerted by a gas. The pressure is read from the difference in the height of the mercury in the two arms of the tube.

FIGURE 11-4

A gas pressure regulator. The right-hand gauge reads the pressure in the gas cylinder to which the regulator is attached. The left-hand gauge reads the reduced pressure, which is controlled by the flow control valve. (As gas is let out of the cylinder, pressure is reduced.) (Photo courtesy of the Matheson Division of Searle Medical Products, USA, Inc.)

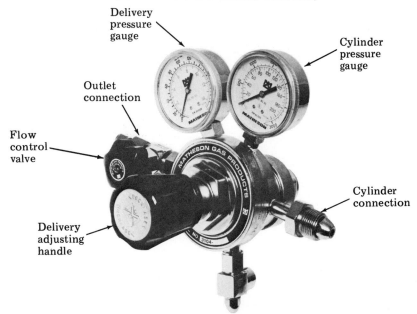

Delivery pressure gauge

Cylinder pressure gauge

Outlet connection

Flow control valve

Delivery adjusting handle

Cylinder connection

Robert Boyle was the leading scientist of his time. He was one of the first to use experimental techniques to study matter. For example, he used the method of collecting a gas by displacing water in an inverted flask.

Boyle's most important experiment involved measuring the volume of a gas at various pressures. He trapped air in the closed end of a tube shaped like the letter J by pouring mercury into the open end (see figure below). Boyle then added more mercury to increase the pressure on the trapped air, and measured the new volume. He made a total of 25 measurements of pressure and volume. Boyle found that the pressure and volume were inversely related, and that they obeyed the following equation: $P \times V$ = constant. Now known as Boyle's law, this equation represents one of the first expressions of a scientific principle in mathematical form.

Boyle separated chemistry from alchemy and established chemistry as a separate science. Since the time of Aristotle, people had believed that there were four elements: earth, water, fire, and air. Boyle introduced a new definition based on experiment: an element is a substance that cannot be broken down into still simpler substances.

No one before Boyle had devoted his life to performing experiments and publishing the results. In fact, Boyle began the practice of recording his experiments thoroughly so that others could repeat them. This practice is essential to modern science. It permits scientists to test and build upon the results of others. Because of his reliance on experimentation, Boyle has been called the father of the experimental method.

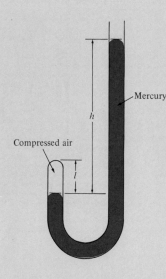

Boyle's apparatus. Boyle determined the volume of trapped air by measuring length (l) at different pressures. He increased the pressure by adding more mercury to the tube, increasing its height (h).

11.3
PRESSURE
AND VOLUME

The relationship between the pressure and the volume of a gas was discovered by Robert Boyle in 1660. Boyle found that when the pressure applied to a gas increases, the volume of the gas decreases if the temperature and the amount of gas are kept the same. In this experi-

TABLE 11-2 *Properties of common gases*

Gas	Formula	Color	Odor	Other properties	Examples of use
acetylene	C_2H_2	colorless	garlic-like	flammable	welding, organic synthesis
ammonia	NH_3	colorless	pungent, irritating	toxic, flammable	fertilizer production
argon	Ar	colorless	odorless	nonreactive	nonreactive atmosphere
carbon dioxide	CO_2	colorless	odorless	mildly toxic	refrigeration, carbonated drinks, respiratory stimulant
carbon monoxide	CO	colorless	odorless	toxic, flammable	synthesis, fuel
chlorine	Cl_2	yellow-green	pungent, irritating	highly toxic, corrosive	synthesis
cyclopropane	C_3H_6	colorless	pungent	flammable, narcotic	anesthetic
ethane	C_2H_6	colorless	odorless	flammable	organic synthesis
ethylene	C_2H_4	colorless	sweet	flammable	organic synthesis, anesthetic
fluorine	F_2	pale yellow	sharp	highly toxic, corrosive	oxidizer, synthesis
freon-12	CCl_2F_2	colorless	ether-like	nonflammable	refrigerant
helium	He	colorless	odorless	nonreactive	balloons, breathing mixtures
hydrogen	H_2	colorless	odorless	flammable	fuel, reducing agent, synthesis
hydrogen chloride	HCl	colorless	suffocating	highly toxic, corrosive	synthesis
hydrogen sulfide	H_2S	colorless	rotten egg	highly toxic, flammable	precipitating sulfides of metals
methane	CH_4	colorless	odorless	flammable	raw material, fuel
nitrogen	N_2	colorless	odorless	relatively nonreactive	synthesis of nitrogen compounds
nitrogen dioxide	N_2O_4	brown	slightly irritating	highly toxic, corrosive	nitric acid synthesis, catalyst
nitrous oxide	N_2O	colorless	sweet	explosive (with air)	anesthetic
oxygen	O_2	colorless	odorless	highly oxidizing	breathing mixtures, combustion, oxidizing agent
propane	C_3H_8	colorless	faintly disagreeable	flammable	organic synthesis
sulfur dioxide	SO_2	colorless	pungent, irritating	highly toxic	synthesis of sulfuric acid, bleaching

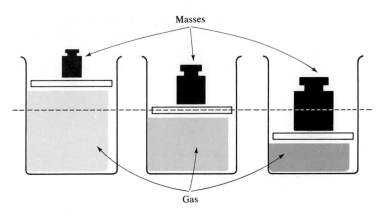

Masses

Gas

FIGURE 11-5

Pressure and volume. As the pressure produced by the mass increases, the volume of the gas decreases (at constant temperature).

ment, which is shown in Figure 11-5, the pressure and volume are *variables* because they can take on various values. The amount of the gas and its temperature are *constants* because they do not change.

Because much of the volume occupied by a gas consists of empty space, we can squeeze the gas into a smaller volume by exerting pressure on it. For example, if we place a balloon in a high-pressure chamber, the balloon will become smaller. On the other hand, if we decrease the pressure on a gas, the gas will expand as a result of molecular motion.

If we reduce the volume of a gas, the pressure of the gas increases. For example, the piston in an automobile engine compresses gas and decreases its volume. The molecules of the gas now strike a smaller area with the same force, increasing the pressure. If the volume of the gas is increased, on the other hand, the pressure will decrease because the impact of the gas molecules becomes spread out.

Thus, *pressure and volume are inversely related.* If one quantity increases, the other quantity decreases when the amount of gas and the temperature are held constant. This relationship is often referred to as **Boyle's law**. Appendix C.4 shows this relationship in the form of a graph.

Mathematically, Boyle's law can be expressed in several different ways. For example:

$$P \propto \frac{1}{V} \quad \text{or} \quad V \propto \frac{1}{P}$$

These two equivalent relationships show in mathematical terms that pressure and volume are inversely related. The symbol $\propto$ means "is proportional to." It shows how one variable depends on another. Another way to represent Boyle's law is to say that when either the pressure or the volume is changed, the product $P \times V$ stays the same.

$$P \times V = \text{constant}$$

For example, if the volume is doubled, the pressure becomes one-half its original value because the product $P \times V$ must remain constant. Thus when the amount of gas and temperature are fixed, the following relationship holds for any change in either pressure or volume.

$$P(\text{initial}) \times V(\text{initial}) = P(\text{final}) \times V(\text{final})$$

We can write this equation in simpler form by letting the subscript 1 represent the initial values of pressure and volume, and the subscript 2 represent the final values:

$$P_1V_1 = P_2V_2$$

(When quantities are written immediately next to each other, it is understood that they are to be multiplied.)

Now let's use Boyle's law to solve a problem. Consider a gas that has an initial pressure of 2.0 atm. Its volume is changed (at constant temperature) from 5.0 liters (dm^3) to 10.0 liters (dm^3). Using the relationship $P_1V_1 = P_2V_2$, the final pressure of the gas can be found:

UNKNOWN	GIVEN	CONNECTION
P_2	$P_1 = 2.0$ atm $V_1 = 5.0$ L $V_2 = 10.0$ L	$P_1V_1 = P_2V_2$

Do-it-yourself 11

Blow it up

A balloon is a bag filled with gas. Balloons that could be used for traveling high in the sky were invented in the late 18th century. They rise because the gas inside is less dense than the surrounding air.

A small gas heater can be used to increase the temperature of the air in a balloon that is open at the bottom. We call such a balloon a hot-air balloon. The trapped air expands as its temperature rises. As the gas volume increases, the density decreases, and the balloon rises. A gas heater is unnecessary in hydrogen or helium balloons, because the hydrogen or helium gas is already less dense than the nitrogen and oxygen gas in the surrounding air.

You can use a toy balloon to demonstrate some of the properties of a gas.

To do: Blow up a balloon. Can you explain why its size has increased?

Place the inflated balloon in the refrigerator for a few minutes. Why does it shrink in size?

The balloon will gradually become smaller at room temperature, even if the neck is kept tightly closed. Can you explain this observation, using the kinetic molecular theory?

The key step in solving such a problem is the proper identification of the variables. P_1 is the initial value of the pressure, 2.0 atm. P_2, the final pressure, is unknown. V_1, the initial volume, is 5.0 L, and V_2, the final volume of the gas, is 10.0 L. The set-up then becomes:

$$P_1V_1 = P_2V_2$$

$$(2.0 \text{ atm})(5.0 \text{ L}) = P_2(10.0 \text{ L})$$

$$\frac{(2.0 \text{ atm})(5.0\, \cancel{L})}{(10.0\, \cancel{L})} = P_2$$

$$1.0 \text{ atm} = P_2$$

The final pressure of the gas, 1.0 atm, is one-half the original pressure because the volume was doubled.

This problem can be solved also by using the conversion factor method. We rearrange the equation $P_1V_1 = P_2V_2$ as follows:

$$P_2 = P_1 \times \frac{V_1}{V_2}$$

This equation shows that the final pressure (P_2) is found by multiplying the initial pressure (P_1) with a conversion factor that is based on the ratio of the volumes (V_1/V_2). However, unlike the other conversion factors we have used, the factor V_1/V_2 has the same units in the numerator and denominator. Thus, the cancellation of units cannot be used to insure that the conversion factor is set up properly. The only way to check the set-up is to make sure that if the volume increases, as it does in this case, the ratio must be less than one. When this ratio is less than one (5.0 L/10.0 L), the new pressure will be smaller than the original pressure. Use this method only when you are sure that you understand the inverse mathematical relationship between pressure changes and volume changes.

Example A sample of methane initially has a pressure of 500. torr and a volume of 2.35 L. If its pressure is changed (at constant temperature) to 400. torr, what is the new volume of the methane?

UNKNOWN	GIVEN	CONNECTION
V_2	$P_1 = 500.$ torr $P_2 = 400.$ torr $V_1 = 2.35$ L	$P_1V_1 = P_2V_2$

$$P_1V_1 = P_2V_2$$

$$(500.\ \text{torr})(2.35\ \text{L}) = (400.\ \text{torr})(V_2)$$

$$\frac{(500.\ \text{torr})(2.35\ \text{L})}{(400.\ \text{torr})} = V_2$$

Conversion factor method: $V_2 = 2.35\ \text{L} \times \dfrac{500.\ \text{torr}}{400.\ \text{torr}}$

ESTIMATE $\qquad \dfrac{(5 \times 10^2\ \text{torr})(2\ \text{L})}{4 \times 10^2\ \text{torr}} = 3\ \text{L}$

CALCULATION $\quad$ 2.94 L

Example A certain amount of neon now has a pressure of 80.2 kPa and a volume of 890. mL. If the original volume of the sample was 520. mL, what pressure did it originally have?

UNKNOWN	GIVEN	CONNECTION
P_1	$P_2 = 80.2$ kPa $V_1 = 520.$ mL $V_2 = 890.$ mL	$P_1V_1 = P_2V_2$

$$P_1V_1 = P_2V_2$$

$$P_1\ (520.\ \text{mL}) = (80.2\ \text{kPa})(890.\ \text{mL})$$

$$P_1 = \frac{(80.2\ \text{kPa})(890.\ \text{mL})}{520\ \text{mL}}$$

Conversion factor method: $P_1 = 80.2\ \text{kPa} \times \dfrac{890.\ \text{mL}}{520.\ \text{mL}}$

ESTIMATE $\qquad \dfrac{(8 \times 10^1\ \text{kPa})(9 \times 10^2\ \text{L})}{5 \times 10^2\ \text{L}} = 1 \times 10^2\ \text{kPa}$

CALCULATION $\quad$ 1.37×10^2 kPa

Example Nitrogen gas, originally at 2.54 atm, is now at a pressure of 1.53 atm and occupies 30.5 dm³. Find its original volume.

UNKNOWN	GIVEN	CONNECTION
V_1	$P_1 = 2.54$ atm $P_2 = 1.53$ atm $V_2 = 30.5$ dm³	$P_1V_1 = P_2V_2$

$$P_1V_1 = P_2V_2$$

$$(2.54 \text{ atm})V_1 = (1.53 \text{ atm})(30.5 \text{ dm}^3)$$

$$V_1 = \frac{(1.53 \text{ atm})(30.5 \text{ dm}^3)}{2.54 \text{ atm}}$$

Conversion factor method: $V_1 = 30.5 \text{ dm}^3 \times \dfrac{1.53 \text{ atm}}{2.54 \text{ atm}}$

ESTIMATE $\quad \dfrac{(2 \text{ atm})(3 \times 10^1 \text{ dm}^3)}{3 \text{ atm}} = 20 \text{ dm}^3$

CALCULATION $\quad 18.4 \text{ dm}^3$

As shown in Figure 11-6, the process of breathing illustrates the pressure-volume relationship given by Boyle's law. When you breathe in, a downward movement of your diaphragm increases the volume of your chest cavity. This change in volume causes the pressure inside to decrease to about 3 torr below the normal atmospheric pressure. Because the air has a greater pressure outside the body, it is pushed into the lungs. Notice that a gas always moves from a region of higher pressure to one of lower pressure. When you breathe out, the diaphragm returns to its resting position, and your chest cavity returns to its normal size. This decrease in volume increases the pressure of the air inside the lungs, also by about 3 torr, forcing it out into the atmosphere again. You normally inhale and exhale about 500 mL of air by this process 12 times each minute.

FIGURE 11-6

Breathing. Movement of the rib cage and diaphragm changes the volume of the chest cavity, and therefore changes the pressure inside. During inhalation, the volume increases and the pressure decreases. The atmospheric pressure is then greater than the pressure inside the chest, and air is able to flow into the lungs.

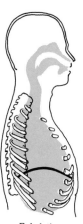

Exhalation
(Resting position)

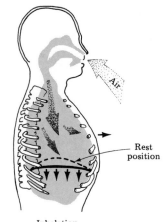

Rest position

Inhalation
(Rib cage moves up and out;
diaphragm moves down)

**11.4
PRESSURE AND
TEMPERATURE**

The pressure of a gas depends on its temperature as well as on its volume. When the volume of a fixed amount of gas is held constant as shown in Figure 11-7, *the pressure is directly proportional to the temperature in kelvins:*

$$P \propto T \text{ (kelvin)}$$

An increase in temperature is nothing more than an increase in the average kinetic energy of the molecules. As the molecules move faster, they hit the walls of the container with greater force, and this force increases the gas pressure. As the temperature decreases, the molecules have less energy and the pressure that they exert decreases also. At constant volume, then, the pressure and temperature of a fixed quantity of gas change in the same manner. They both go up or both go down.

The relationship between pressure and temperature is something to keep in mind if you drive a car. We mentioned that the pressure within an automobile tire is about 30 psi greater than the atmospheric pressure. A tire may have this pressure when you start to drive, but after several hours of driving the tire can become quite hot. The pressure inside the tire will increase and may reach 35 psi. You need not be alarmed, because tires are designed to withstand this rise in pressure. However, you should not reduce the pressure in a hot tire back to 30 psi. If you did so, the pressure could drop *below* 30 psi when the tire cools down. Driving when a tire is underinflated may damage the tire.

The direct relation between pressure and temperature differs from the inverse relation between pressure and volume. Because pressure is directly proportional to temperature, a change in one quantity will cause an equivalent change in the other quantity. In other words, the ratio P/T stays the same for a gas with constant volume.

$$\frac{P}{T} = \text{constant}$$

FIGURE 11-7

Pressure and temperature. The temperature of a gas in kelvins and its pressure are directly related (at constant volume). Thus, as the temperature increases, the pressure of the gas also increases.

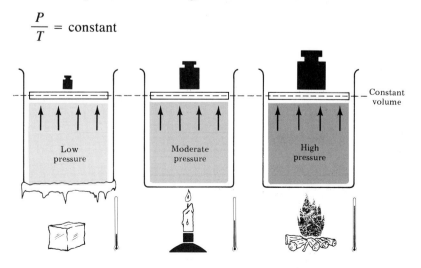

We can also represent this relationship as a proportion:

$$\frac{P(\text{initial})}{T(\text{initial})} = \frac{P(\text{final})}{T(\text{final})}$$

Or in shorthand notation:

$$\frac{P_1}{T_1} = \frac{P_2}{T_2}$$

For this relationship to hold between pressure and temperature, the volume and amount of gas must remain constant. Furthermore, the temperature must be expressed in kelvins. Only the temperature in kelvins is proportional to the energy and therefore proportional to the pressure exerted by the molecules.

Suppose that a gas at a pressure of 2.10 atm and a temperature of 300. K is heated to 400. K at constant volume. The new pressure can be found using the equation relating pressure and temperature.

UNKNOWN	GIVEN	CONNECTION
P_2	$P_1 = 2.10$ atm $T_1 = 300.$ K $T_2 = 400.$ K	$\dfrac{P_1}{T_1} = \dfrac{P_2}{T_2}$

$$\frac{P_1}{T_1} = \frac{P_2}{T_2}$$

$$\frac{2.10 \text{ atm}}{300. \text{ K}} = \frac{P_2}{400. \text{ K}}$$

To find P_2, cross multiply the terms of the proportion and solve for P_2.

$$(300. \text{ K})P_2 = (2.10 \text{ atm})(400. \text{ K})$$

$$P_2 = \frac{(2.10 \text{ atm})(400. \text{ K})}{300. \text{ K}}$$

This problem also can be set up by the conversion factor method. However, as we saw in the previous section, that method is somewhat more difficult since the units in the numerator and denominator of the factor are identical.

$$P_2 = P_1 \times \frac{T_2}{T_1} = 2.10 \text{ atm} \times \frac{400. \text{ K}}{300. \text{ K}}$$

Because the temperature increases in this problem, we know that the pressure must also increase. Therefore, the ratio of temperatures in the conversion factor must be greater than one. The factor

$$\frac{T_2}{T_1} = \frac{400.\ K}{300.\ K}$$

is therefore correct. The problem can now be solved.

ESTIMATE $\quad \dfrac{(2\ atm)(4 \times 10^2\ \cancel{K})}{3 \times 10^2\ \cancel{K}} = 3\ atm$

CALCULATION $\quad$ 2.80 atm

The new pressure is larger than the original pressure since the temperature was increased.

Example A sample of carbon dioxide gas originally at 620. torr and 35.0°C is heated to 96.5°C, while keeping the volume constant. Find the final pressure.

UNKNOWN	GIVEN	CONNECTION
P_2	$P_1 = 620.\ torr$ $t_1 = 35.0°C$ $t_2 = 96.5°C$	$\dfrac{P_1}{T_1} = \dfrac{P_2}{T_2}$

The temperatures in this problem are given in degrees Celsius, and must first be converted to kelvins. We let t_1 and t_2 stand for the two temperatures in degrees Celsius, and T_1 and T_2 represent temperature in kelvins. We then find T_1 and T_2 as follows:

$$K = °C + 273$$

$$T_1 = t_1 + 273 = 35.0 + 273 = 308\ K$$

$$T_2 = t_2 + 273 = 96.5 + 273 = 370.\ K$$

Now the problem can be solved:

$$\frac{P_1}{T_1} = \frac{P_2}{T_2}$$

$$\frac{620.\ torr}{308\ K} = \frac{P_2}{370.\ K}$$

$$(308 \text{ K})P_2 = (620. \text{ torr})(370. \text{ K})$$

$$P_2 = \frac{(620. \text{ torr})(370. \cancel{K})}{308 \cancel{K}}$$

Conversion factor method: $P_2 = 620. \text{ torr} \times \dfrac{370. \cancel{K}}{308 \cancel{K}}$

ESTIMATE $\dfrac{(6 \times 10^2 \text{ torr})(4 \times 10^2 \cancel{K})}{3 \times 10^2 \cancel{K}} = 8 \times 10^2 \text{ torr}$

CALCULATION 744 torr

Example Cyclopropane gas, originally at 453 K, is cooled to 189 K at constant volume. Its pressure is now 1.09 atm. What was the initial pressure?

UNKNOWN	GIVEN	CONNECTION
P_1	$P_2 = 1.09 \text{ atm}$ $T_1 = 453 \text{ K}$ $T_2 = 189 \text{ K}$	$\dfrac{P_1}{T_1} = \dfrac{P_2}{T_2}$

$$\frac{P_1}{T_1} = \frac{P_2}{T_2}$$

$$\frac{P_1}{453 \text{ K}} = \frac{1.09 \text{ atm}}{189 \text{ K}}$$

$$P_1(189 \text{ K}) = (453 \text{ K})(1.09 \text{ atm})$$

$$P_1 = \frac{(453 \cancel{K})(1.09 \text{ atm})}{189 \cancel{K}}$$

Conversion factor method: $P_1 = 1.09 \text{ atm} \times \dfrac{453 \cancel{K}}{189 \cancel{K}}$

ESTIMATE $\dfrac{(5 \times 10^2 \cancel{K})(1 \text{ atm})}{2 \times 10^2 \cancel{K}} = 3 \text{ atm}$

CALCULATION 2.61 atm

Example A sample of chlorine gas increases in pressure from 75.0 kPa to 146.0 kPa when it is heated to 65.0°C at constant volume. What was the original temperature of this gas?

UNKNOWN	GIVEN	CONNECTION
T_1	$P_1 = 75.0$ kPa $P_2 = 146.0$ kPa $t_2 = 65.0°C$	$\dfrac{P_1}{T_1} = \dfrac{P_2}{T_2}$

$$T_2 = t_2 + 273 = 65.0 + 273 = 338 \text{ K}$$

$$\frac{P_1}{T_1} = \frac{P_2}{T_2}$$

$$\frac{75.0 \text{ kPa}}{T_1} = \frac{146.0 \text{ kPa}}{338 \text{ K}}$$

$$T_1(146.0 \text{ kPa}) = (75.0 \text{ kPa})(338 \text{ K})$$

$$T_1 = \frac{(75.0 \text{ kPa})(338 \text{ K})}{146.0 \text{ kPa}}$$

Conversion factor method: $T_1 = 338 \text{ K} \times \dfrac{75.0 \text{ kPa}}{146.0 \text{ kPa}}$

ESTIMATE $\quad \dfrac{(8 \times 10^1 \text{ kPa})(3 \times 10^2 \text{ K})}{1 \times 10^2 \text{ kPa}} = 2 \times 10^2 \text{ K}$

CALCULATION $\quad$ 174 K or $-99°C$

The direct relationship between pressure and temperature is the reason that spray cans contain the warning "Do not incinerate." Heating a spray can will increase the pressure of the gas inside, which may cause the can to explode. For the same reason, cylinders of compressed gas generally should not be stored in locations where the temperature can reach 120°F (49°C) or more.

The pressure-temperature relationship has important applications. For example, the autoclave, a device that sterilizes medical and biological supplies is made to take advantage of this relationship. A chamber with a fixed volume is filled with steam at high pressure (up to 15.5 psi). At such pressures, the temperature of steam increases from 100°C to more than 120°C. Microorganisms are more readily destroyed at these high temperatures.

11.5
VOLUME AND
TEMPERATURE

When the pressure and amount of a gas are held constant, *its volume depends directly on the temperature in kelvins,* as shown in Figure 11-8. When a fixed amount of gas held at a constant pressure is heated, the molecules become more energetic and the gas expands. If the gas is cooled, the molecules lose kinetic energy and the volume decreases.

FIGURE 11-8

Volume and temperature. The temperature of a gas in kelvins and its volume are directly related (at constant pressure). Thus, increasing the temperature increases the volume of the gas.

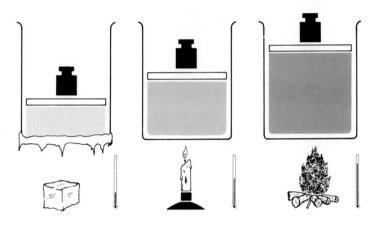

For example, a balloon expands when heated and contracts when cooled. Thus, volume and temperature, like pressure and temperature, change in the same direction. Because this relationship between pressure and temperature was observed in 1787 by Jacques Alexandre Charles, it is often called Charles' law.

The direct relationship between volume and temperature in kelvins for a gas at constant pressure is the basis for gas thermometers, such as the one shown in Figure 11-9. The position of the mercury plug indicates the volume of gas inside the thermometer. As the temperature of the trapped gas rises, the volume of the gas expands, pushing the mercury towards the open end of the tube. If the temperature falls, then the gas contracts, and the mercury moves toward the closed end. The thermometer can be calibrated by marking the position of the mercury at several known temperatures. Gas thermometers are among the most accurate thermometers in existence today.

At constant pressure, the volume of a gas is directly proportional to the kelvin temperature.

$$V \propto T \text{ (kelvin)}$$

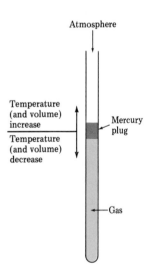

FIGURE 11-9

A gas thermometer. The operation of this thermometer is based on the direct relation between volume and temperature. The position of the mercury plug depends on the volume, and therefore on the temperature, of the trapped gas.

Since the volume changes with the temperature, the ratio V/T stays constant.

$$\frac{V}{T} = \text{constant}$$

For a change in either volume or temperature at constant pressure and fixed amount of gas, the following equation must hold.

$$\frac{V(\text{initial})}{T(\text{initial})} = \frac{V(\text{final})}{T(\text{final})}$$

Using abbreviated notation, this equation is:

$$\frac{V_1}{T_1} = \frac{V_2}{T_2}$$

The following problem illustrates the use of this equation. A gas with a volume of 1.50 L at 298 K is heated at constant pressure to 398 K. Using these three pieces of information and the equation relating volume and temperature, we can determine the new volume.

UNKNOWN	GIVEN	CONNECTION
V_2	$V_1 = 1.50$ L $T_1 = 298$ K $T_2 = 398$ K	$\dfrac{V_1}{T_1} = \dfrac{V_2}{T_2}$

$$\frac{V_1}{T_1} = \frac{V_2}{T_2}$$

$$\frac{1.50 \text{ L}}{298 \text{ K}} = \frac{V_2}{398 \text{ K}}$$

$$(298 \text{ K})V_2 = (1.50 \text{ L})(398 \text{ K})$$

$$V_2 = \frac{(1.50 \text{ L})(398 \text{ K})}{298 \text{ K}}$$

Again, if we had used the conversion factor method, we would have reached the same point since

$$V_2 = V_1 \times \frac{T_2}{T_1} = 1.50 \text{ L} \times \frac{398 \text{ K}}{298 \text{ K}}$$

The temperature factor must be greater than one because the volume increases in this example. The problem can now be solved.

ESTIMATE $\qquad \dfrac{(2 \text{ L})(4 \times 10^2 \text{ K})}{3 \times 10^2 \text{ K}} = 3 \text{ L}$

CALCULATION $\quad$ 2.00 L

As we should have expected, the new volume is greater than the original volume because the gas temperature was increased.

Example Nitrous oxide has a volume of 853 dm^3 at 20.1°C. If the gas is held at constant pressure, at what temperature will the volume become 260. dm^3?

UNKNOWN	GIVEN	CONNECTION
T_2	$V_1 = 853\ dm^3$ $t_1 = 20.1°C$ $V_2 = 260.\ dm^3$	$\dfrac{V_1}{T_1} = \dfrac{V_2}{T_2}$

(Remember to convert the temperature to kelvins.)

$$T_1 = t_1 + 273 = 20.1 + 273 = 293\ K$$

$$\frac{V_1}{T_1} = \frac{V_2}{T_2}$$

$$\frac{853\ dm^3}{293\ K} = \frac{260.\ dm^3}{T_2}$$

$$(853\ dm^3)T_2 = (293\ K)(260.\ dm^3)$$

$$T_2 = \frac{(293\ K)(260.\ \cancel{dm^3})}{853\ \cancel{dm^3}}$$

Conversion factor method: $T_2 = 293\ K \times \dfrac{260.\ \cancel{dm^3}}{853\ \cancel{dm^3}}$

ESTIMATE $\quad \dfrac{(3 \times 10^2\ K)(3 \times 10^2\ \cancel{dm^3})}{9 \times 10^2\ \cancel{dm^3}} = 1 \times 10^2\ K$

CALCULATION $\quad$ 89.4 K or $-184°C$

Example When the volume of gas in a particular thermometer is 5.30 mL, the temperature reads 295 K. What was the volume of the gas when the temperature read 220. K?

UNKNOWN	GIVEN	CONNECTION
V_1	$T_1 = 220.\ K$ $V_2 = 5.30\ mL$ $T_2 = 295\ K$	$\dfrac{V_1}{T_1} = \dfrac{V_2}{T_2}$

$$\frac{V_1}{T_1} = \frac{V_2}{T_2}$$

$$\frac{V_1}{220.\ K} = \frac{5.30\ mL}{295\ K}$$

$$V_1(295\ K) = (220.\ K)(5.30\ mL)$$

$$V_1 = \frac{(220. \cancel{K})(5.30 \text{ mL})}{295 \cancel{K}}$$

Conversion factor method: $V_1 = 5.30 \text{ mL} \times \dfrac{220. \cancel{K}}{295 \cancel{K}}$

ESTIMATE $\quad \dfrac{(2 \times 10^2 \cancel{K})(5 \text{ mL})}{3 \times 10^2 \cancel{K}} = 3 \text{ mL}$

CALCULATION $\quad$ 3.95 mL

Example $\quad$ A 10.2 L sample of ethane increases in volume to 35.2 L when the temperature is raised to 89.4°C at constant pressure. What was the original temperature of the gas?

UNKNOWN	GIVEN	CONNECTION
T_1	$V_1 = 10.2 \text{ L}$ $V_2 = 35.2 \text{ L}$ $t_2 = 89.4°C$	$\dfrac{V_1}{T_1} = \dfrac{V_2}{T_2}$

$$T_2 = t_2 + 273 = 89.4 + 273 = 362 \text{ K}$$

$$\frac{V_1}{T_1} = \frac{V_2}{T_2}$$

$$\frac{10.2 \text{ L}}{T_1} = \frac{35.2 \text{ L}}{362 \text{ K}}$$

$$T_1 (35.2 \text{ L}) = (10.2 \text{ L})(362 \text{ K})$$

$$T_1 = \frac{(10.2 \cancel{L})(362 \text{ K})}{35.2 \cancel{L}}$$

Conversion factor method: $T_1 = 362 \text{ K} \times \dfrac{10.2 \cancel{L}}{35.2 \cancel{L}}$

ESTIMATE $\quad \dfrac{(1 \times 10^1 \cancel{L})(4 \times 10^2 \text{ K})}{4 \times 10^1 \cancel{L}} = 1 \times 10^2 \text{ K}$

CALCULATION $\quad$ 105 K or −168°C

11.6
 COMBINED GAS LAW $\quad$ The relationship between pressure, volume, and temperature can be combined into a single equation, which we call the **combined gas law:**

$$\frac{P(\text{initial}) \times V(\text{initial})}{T(\text{initial})} = \frac{P(\text{final}) \times V(\text{final})}{T(\text{final})}$$

Using the subscript 1 for initial conditions and 2 for final conditions, the combined gas law can be written as:

$$\frac{P_1V_1}{T_1} = \frac{P_2V_2}{T_2}$$

This equation brings together the three relationships described in the preceding sections. The combined gas law can be used as long as the amount of gas stays constant when the change takes place.

Notice that each of the gas equations involving two variables can be derived from this one combined equation. For example, if the temperature remains constant, then $T_1 = T_2$ and the combined equation reduces to $P_1V_1 = P_2V_2$. This is the relationship between pressure and volume that we discussed in Section 11.3. Similarly, if we hold the volume constant, then $V_1 = V_2$, and the combined equation reduces to $P_1/T_1 = P_2/T_2$. This result is the equation we discussed in Section 11.4. Finally, we obtain the equation relating volume and temperature (Section 11.5) by holding the pressure constant. This time, the combined equation reduces to $V_1/T_1 = V_2/T_2$.

The combined gas law is used in problems where more than one change occurs. For example, both temperature and pressure may vary. Suppose that 2.50 L of a gas at 200. K and 1.50 atm is heated to 300. K and allowed to reach a new pressure of 2.00 atm. The new volume of the gas can be calculated by using the combined gas law.

UNKNOWN	GIVEN	CONNECTION
V_2	$P_1 = 1.50$ atm $V_1 = 2.50$ L $T_1 = 200.$ K $P_2 = 2.00$ atm $T_2 = 300.$ K	$\dfrac{P_1V_1}{T_1} = \dfrac{P_2V_2}{T_2}$

$$\frac{P_1V_1}{T_1} = \frac{P_2V_2}{T_2}$$

$$\frac{(1.50 \text{ atm})(2.50 \text{ L})}{(200. \text{ K})} = \frac{(2.00 \text{ atm})V_2}{300. \text{ K}}$$

$$(200. \text{ K})(2.00 \text{ atm})V_2 = (1.50 \text{ atm})(2.50 \text{ L})(300. \text{ K})$$

$$V_2 = \frac{(1.50 \text{ a\tm})(2.50 \text{ L})(300. \text{ K})}{(200. \text{ K})(2.00 \text{ a\tm})}$$

Notice that this problem could also be set up by using two conversion factors, one for the temperature change, and one for the pressure change. The temperature factor must be larger than one (300. K/200. K) because increasing the temperature makes the volume increase. But the pressure factor must be less than one (1.50 atm/2.00 atm) because increasing the pressure makes the volume decrease.

$$V_2 = V_1 \times \frac{T_2}{T_1} \times \frac{P_1}{P_2} = 2.50 \text{ L} \times \frac{300. \cancel{K}}{200. \cancel{K}} \times \frac{1.50 \cancel{\text{ atm}}}{2.00 \cancel{\text{ atm}}}$$

The answer to this problem is the following:

ESTIMATE $\dfrac{(2 \cancel{\text{ atm}}) (3 \text{ L}) (3 \times 10^2 \cancel{K})}{(2 \times 10^2 \cancel{K}) (2 \cancel{\text{ atm}})} = 5 \text{ L}$

CALCULATION   2.81 L

Notice that this problem can also be solved in two separate steps, one involving pressure and volume, and the other involving temperature and volume.

(1) $P_1 V_1 = P_2 V_2$ (at 200 K)

$(1.50 \text{ atm})(2.50\text{L}) = (2.00 \text{ atm})V_2$

$$V_2 = \frac{(1.50 \cancel{\text{ atm}})(2.50 \text{ L})}{2.00 \cancel{\text{ atm}}} = 1.88 \text{ L}$$

(2) $\dfrac{V_1}{T_1} = \dfrac{V_2}{T_2}$

$$\frac{1.88 \text{ L}}{200. \text{ K}} = \frac{V_2}{300. \text{ K}}$$

$$V_2 = \frac{(1.88 \text{ L})(300. \cancel{K})}{200. \cancel{K}} = 2.82 \text{ L}$$

In this two-step method, we first calculate the effect of changing the pressure. Then the resulting volume is used in the next step to calculate the effect of changing the temperature. The two-step method shows the effects of changing each variable more clearly than does the combined gas law. However, the combined gas law calculation is simpler since it involves only one step.

Example   A certain amount of oxygen has a volume of 540. mL at 700. torr and 250. K. Find its pressure if the volume is changed to 800. mL and the temperature is changed to 300. K.

UNKNOWN	GIVEN	CONNECTION
P_2	$P_1 = 700.$ torr $V_1 = 540.$ mL $T_1 = 250.$ K $V_2 = 800.$ mL $T_2 = 300.$ K	$\dfrac{P_1 V_1}{T_1} = \dfrac{P_2 V_2}{T_2}$

$$\frac{P_1 V_1}{T_1} = \frac{P_2 V_2}{T_2}$$

$$\frac{(700.\ \text{torr})(540.\ \text{mL})}{250.\ \text{K}} = \frac{P_2(800.\ \text{mL})}{300.\ \text{K}}$$

$$(250.\ \text{K})(P_2)(800.\ \text{mL}) = (700.\ \text{torr})(540.\ \text{mL})(300.\ \text{K})$$

$$P_2 = \frac{(700.\ \text{torr})(540.\ \cancel{\text{mL}})(300.\ \cancel{\text{K}})}{(250.\ \cancel{\text{K}})(800.\ \cancel{\text{mL}})}$$

Conversion factor method: $P_2 = 700.\ \text{torr} \times \dfrac{540.\ \cancel{\text{mL}}}{800.\ \cancel{\text{mL}}} \times \dfrac{300.\ \cancel{\text{K}}}{250.\ \cancel{\text{K}}}$

ESTIMATE $\quad \dfrac{(7 \times 10^2\ \text{torr})(5 \times 10^2\ \cancel{\text{mL}})(3 \times 10^2\ \cancel{\text{K}})}{(3 \times 10^2\ \cancel{\text{K}})(8 \times 10^2\ \cancel{\text{mL}})} = 4 \times 10^2\ \text{torr}$

CALCULATION $\quad$ 567 torr

Example A hydrogen gas sample has a volume of 350. dm³ at 25.0°C and 1.25 atm. Find its temperature if the volume is changed to 480. dm³ at 1.10 atm.

UNKNOWN	GIVEN	CONNECTION
T_2	$P_1 = 1.25$ atm $V_1 = 350.$ dm³ $t_1 = 25.0°C$ $P_2 = 1.10$ atm $V_2 = 480.$ dm³	$\dfrac{P_1 V_1}{T_1} = \dfrac{P_2 V_2}{T_2}$

$$T_1 = t_1 + 273 = 25.0 + 273 = 298\ \text{K}$$

$$\frac{P_1 V_1}{T_1} = \frac{P_2 V_2}{T_2}$$

$$\frac{(1.25 \text{ atm})(350. \text{ dm}^3)}{298 \text{ K}} = \frac{(1.10 \text{ atm})(480. \text{ dm}^3)}{T_2}$$

$$(1.25 \text{ atm})(350. \text{ dm}^3)T_2 = (298 \text{ K})(1.10 \text{ atm})(480. \text{ dm}^3)$$

$$T_2 = \frac{(298 \text{ K})(1.10 \text{ atm})(480. \text{ dm}^3)}{(1.25 \text{ atm})(350. \text{ dm}^3)}$$

Conversion factor method: $T_2 = 298 \text{ K} \times \dfrac{1.10 \text{ atm}}{1.25 \text{ atm}} \times \dfrac{480. \text{ dm}^3}{350. \text{ dm}^3}$

ESTIMATE $\qquad \dfrac{(3 \times 10^2 \text{ K})(1 \text{ atm})(5 \times 10^2 \text{ dm}^3)}{(1 \text{ atm})(4 \times 10^2 \text{ dm}^3)} = 4 \times 10^2 \text{ K}$

CALCULATION   360. K or 87°C

Example   A gas has a volume of 2.31 L at 430. K and 60.2 kPa. What was the pressure of this gas when it occupied 1.65 L at 328 K?

UNKNOWN	GIVEN	CONNECTION
P_1	$V_1 = 1.65 \text{ L}$ $T_1 = 328 \text{ K}$ $P_2 = 60.2 \text{ kPa}$ $V_2 = 2.31 \text{ L}$ $T_2 = 430. \text{ K}$	$\dfrac{P_1V_1}{T_1} = \dfrac{P_2V_2}{T_2}$

$$\frac{P_1V_1}{T_1} = \frac{P_2V_2}{T_2}$$

$$\frac{P_1(1.65 \text{ L})}{328 \text{ K}} = \frac{(60.2 \text{ kPa})(2.31 \text{ L})}{430. \text{ K}}$$

$$P_1 (1.65 \text{ L})(430. \text{ K}) = (328 \text{ K})(6.02 \text{ kPa})(2.31 \text{ L})$$

$$P_1 = \frac{(328 \text{ K})(6.02 \text{ kPa})(2.31 \text{ L})}{(1.65 \text{ L})(430. \text{ K})}$$

Conversion factor method: $P_1 = 6.02 \text{ kPa} \times \dfrac{2.31 \text{ L}}{1.65 \text{ L}} \times \dfrac{328 \text{ K}}{430. \text{ K}}$

ESTIMATE $\qquad \dfrac{(3 \times 10^2 \text{ K})(6 \text{ kPa})(2 \text{ L})}{(2 \text{ L})(4 \times 10^2 \text{ K})} = 5 \text{ kPa}$

CALCULATION   6.43 kPa

11.7
RELATING
MOLES TO
VOLUME

All of the relationships that we have discussed so far deal with a fixed amount of gas. In other words, we have purposely ignored relationships between the *amount* of gas and the pressure, volume and temperature. But pressure, volume, and temperature depend not only on each other, but also on the number of gas molecules present. For example, as we blow up a balloon, the number of air molecules inside the balloon increases. Because the balloon is flexible, its volume can increase. Therefore, as the number of gas molecules increases (at constant temperature and pressure), the volume of gas inside the balloon increases and the balloon expands.

The situation is somewhat different when we pump up a tire. Pumping increases the number of air molecules inside the tire, but unlike the volume of a balloon, the volume of the tire is fixed. Therefore, as the number of molecules increases, the pressure of the gas in the tire increases (at constant volume and temperature).

A vacuum pump (see Figure 11-10) removes gas molecules from an attached container that has a constant volume. This removal of molecules causes the pressure inside the system to decrease, and creates a vacuum. A "high vacuum" is actually a very low pressure, such as 10^{-6} torr, and means that very few gas molecules are present.

A medicine dropper is filled by creating a vacuum. Squeezing the flexible bulb forces air molecules out of the dropper, lowering the pressure inside. When the bulb is released, the higher air pressure outside forces liquid into the dropper.

The volume of a gas can be measured more easily than the mass. As a result, the relationship between the volume and the amount of gas is especially important. At constant temperature and pressure, *the volume of a gas is directly proportional to the number of moles of gas present.*

FIGURE 11-10

A vacuum pump. The moving vanes trap air molecules coming through the inlet and force them out through the outlet.

$$V \propto n$$

This relationship, known as Avogadro's law, means that equal volumes of all gases (at the same T and P) contain the same number of atoms or molecules. Thus, at a particular temperature and pressure, one mole of any gas has a definite, fixed volume.

The values 273 K (0°C) and 1 atm (760 torr, 101.3 kPa) have been chosen as the **standard temperature and pressure**, abbreviated **STP**. Under these standard conditions, the 6.02×10^{23} atoms or molecules in one mole of any gas occupy a volume of 22.4 liters (dm³), which is approximately the volume of a basketball. This value is called the **molar volume** at STP. Thus, *the volume of 1 mole (molar volume) of any gas at STP = 22.4 L.*

The molar volume at STP can be used to convert between the volume of a gas in liters (or cubic decimeters) and the amount of gas in moles (or in grams if the molecular weight is known).

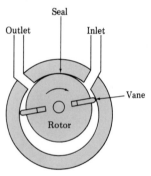

Seal

Outlet   Inlet

Vane

Rotor

Example   What is the volume occupied by 3.65 moles of argon at STP?

UNKNOWN	GIVEN	CONNECTION
liters Ar	3.65 moles Ar	$\dfrac{22.4 \text{ L Ar}}{1 \text{ mole Ar}}$ or $\dfrac{1 \text{ mole Ar}}{22.4 \text{ L Ar}}$

$$\text{liters Ar} = 3.65 \text{ moles Ar} \times \frac{22.4 \text{ L Ar}}{1 \text{ mole Ar}}$$

ESTIMATE $4 \text{ moles Ar} \times \dfrac{2 \times 10^1 \text{ L Ar}}{1 \text{ mole Ar}} = 8 \times 10^1 \text{ L Ar}$

CALCULATION   81.8 L Ar

Example   A sample of helium has a volume of 960. mL. How many moles of gas are present at STP?

UNKNOWN	GIVEN	CONNECTION
moles He	960. mL He	$\dfrac{22.4 \text{ L He}}{1 \text{ mole He}}$ or $\dfrac{1 \text{ mole He}}{22.4 \text{ L He}}$
		$\dfrac{1 \text{ mL}}{10^{-3} \text{ L}}$ or $\dfrac{10^{-3} \text{ L}}{1 \text{ mL}}$

$$\text{moles He} = 960. \text{ mL He} \times \frac{10^{-3} \text{ L}}{1 \text{ mL}} \times \frac{1 \text{ mole He}}{22.4 \text{ L He}}$$

ESTIMATE $1 \times 10^3 \text{ mL He} \times \dfrac{10^{-3} \text{ L}}{1 \text{ mL}} \times \dfrac{1 \text{ mole He}}{2 \times 10^1 \text{ L He}}$
$= 5 \times 10^{-2} \text{ mole He}$

CALCULATION   0.0429 mole He

Example   What is the volume at STP of 10.5 g of N_2?

UNKNOWN	GIVEN	CONNECTION
liters N_2	10.5 g N_2	$\dfrac{1 \text{ mole } N_2}{28.0 \text{ g } N_2}$ or $\dfrac{28.0 \text{ g } N_2}{1 \text{ mole } N_2}$
		$\dfrac{1 \text{ mole } N_2}{22.4 \text{ L } N_2}$ or $\dfrac{22.4 \text{ L } N_2}{1 \text{ mole } N_2}$

$$\text{liters N}_2 \;=\; 10.5 \text{ g } \cancel{\text{N}_2} \times \frac{1 \text{ mole } \cancel{\text{N}_2}}{28.0 \text{ g } \cancel{\text{N}_2}} \times \frac{22.4 \text{ L N}_2}{1 \text{ mole } \cancel{\text{N}_2}}$$

ESTIMATE $\quad 1 \times 10^1 \text{ g } \cancel{\text{N}_2} \times \dfrac{1 \text{ mole } \cancel{\text{N}_2}}{3 \times 10^1 \text{ g } \cancel{\text{N}_2}} \times \dfrac{2 \times 10^1 \text{ L N}_2}{1 \text{ mole } \cancel{\text{N}_2}} = 7 \text{ L N}_2$

CALCULATION $\quad 8.40 \text{ L N}_2$

The molar volume can also be used to find the density of a gas at STP from its molecular weight. Molar volume serves as a conversion factor that changes molar mass (grams per mole) into grams per liter. The following example illustrates this type of conversion.

Example Find the density in grams per liter at STP of the gaseous anesthetic cyclopropane, C_3H_6.

$$\text{molecular weight C}_3\text{H}_6 = 42.0 \text{ amu}$$

$$1 \text{ mole C}_3\text{H}_6 = 42.0 \text{ g}$$

$$\text{molar mass C}_3\text{H}_6 = \frac{42.0 \text{ g C}_3\text{H}_6}{1 \text{ mole C}_3\text{H}_6}$$

UNKNOWN	GIVEN	CONNECTION	
density C_3H_6 (grams/liter)	$\dfrac{42.0 \text{ g C}_3\text{H}_6}{1 \text{ mole C}_3\text{H}_6}$	$\dfrac{1 \text{ mole C}_3\text{H}_6}{22.4 \text{ L C}_3\text{H}_6}$ or	$\dfrac{22.4 \text{ L C}_3\text{H}_6}{1 \text{ mole C}_3\text{H}_6}$

$$\frac{\text{grams C}_3\text{H}_6}{\text{liter C}_3\text{H}_6} = \frac{42.0 \text{ g C}_3\text{H}_6}{1 \text{ mole } \cancel{\text{C}_3\text{H}_6}} \times \frac{1 \text{ mole } \cancel{\text{C}_3\text{H}_6}}{22.4 \text{ L C}_3\text{H}_6}$$

ESTIMATE $\quad \dfrac{4 \times 10^1 \text{ g C}_3\text{H}_6}{1 \text{ mole } \cancel{\text{C}_3\text{H}_6}} \times \dfrac{1 \text{ mole } \cancel{\text{C}_3\text{H}_6}}{2 \times 10^1 \text{ L C}_3\text{H}_6} = 2 \dfrac{\text{g C}_3\text{H}_6}{\text{L C}_3\text{H}_6} = 2 \text{ g/L}$

CALCULATION $\quad 1.88 \dfrac{\text{g C}_3\text{H}_6}{\text{L C}_3\text{H}_6} = 1.88 \text{ g/L}$

The density of cyclopropane at STP is 1.88 g/L.

The density of a gas at STP can often be determined in the laboratory. The molecular weight of the gas can be calculated once its density is known. In this case, the molar volume serves as a conversion factor to change grams per liter into grams per mole.

Example In a laboratory experiment, you find that the density of an unknown gas is 0.759 g/L at STP. What is its molecular weight?

UNKNOWN	GIVEN	CONNECTION
g/mole gas	0.759 g/L	$\dfrac{1 \text{ mole gas}}{22.4 \text{ L gas}}$ or $\dfrac{22.4 \text{ L gas}}{1 \text{ mole gas}}$

$$\text{g/mole gas} = \frac{0.759 \text{ g gas}}{1 \text{ L gas}} \times \frac{22.4 \text{ L gas}}{1 \text{ mole gas}}$$

ESTIMATE $\dfrac{8 \times 10^{-1} \text{ g gas}}{1 \text{ L gas}} \times \dfrac{2 \times 10^{1} \text{ L gas}}{1 \text{ mole gas}} = 2 \times 10^{1} \dfrac{\text{g gas}}{\text{mole gas}}$

$= 2 \times 10^{1} \text{ g/mole}$

CALCULATION $17.0 \dfrac{\text{g gas}}{\text{mole gas}} = 17.0 \text{ g/mole}$

molecular weight $= 17.0$ amu

11.8 STOICHIOMETRY INVOLVING GASES

In the early 19th century, Joseph Louis Gay-Lussac made an important discovery. He found that when gases react at constant temperature and pressure, the volumes of the gases involved are always related by simple whole numbers. For example, Gay-Lussac found that when one volume of hydrogen reacted with one volume of chlorine, two volumes of hydrogen chloride were formed. He also noticed that when two volumes of hydrogen reacted with one volume of oxygen, two volumes of water were formed. These observations are explained by the law of combining volumes. This law is based on the fact that the numbers of particles present in equal volumes of gases are the same (at constant temperature and pressure), and that atoms or molecules combine in simple whole number ratios.

The coefficients of a balanced equation for the reaction of gases at constant temperature and pressure thus can be interpreted in terms of volumes. For example

$$N_2(g) + 3H_2(g) \rightarrow 2NH_3(g)$$
nitrogen hydrogen ammonia

One volume of nitrogen molecules reacts with *three volumes* of hydrogen molecules to form *two volumes* of ammonia molecules.

If this reaction were carried out at STP, the volumes involved would be:

$$N_2 \qquad\qquad + 3H_2 \qquad\qquad \rightarrow 2NH_3$$

1 mole N_2	3 moles H_2	2 moles NH_3
$\times$ 22.4 L/mole	$\times$ 22.4 L/mole	$\times$ 22.4 L/mole
= 22.4 L N_2	= 67.2 L H_2	= 44.8 L NH_3

Stoichiometry problems involving volumes of gases can now be solved, as illustrated in the following example.

Example Hydrogen gas is produced industrially from methane by the following process:

$$CH_4(g) + 2H_2O(g) \rightarrow CO_2(g) + 4H_2(g)$$
methane   water   carbon   hydrogen
                         dioxide

What volume of hydrogen at STP can be made from 5.00 moles of methane?

UNKNOWN	GIVEN	CONNECTION
liters H_2	5.00 moles CH_4	$\dfrac{1 \text{ mole } CH_4}{4 \text{ moles } H_2}$ or $\dfrac{4 \text{ moles } H_2}{1 \text{ mole } CH_4}$
		$\dfrac{1 \text{ mole } H_2}{22.4 \text{ L } H_2}$ or $\dfrac{22.4 \text{ L } H_2}{1 \text{ mole } H_2}$

$$\text{liters } H_2 = 5.00 \text{ moles } CH_4 \times \frac{4 \text{ moles } H_2}{1 \text{ mole } CH_4} \times \frac{22.4 \text{ L } H_2}{1 \text{ mole } H_2}$$

ESTIMATE

$$5 \text{ moles } CH_4 \times \frac{4 \text{ moles } H_2}{1 \text{ mole } CH_4} \times \frac{2 \times 10^1 \text{ L } H_2}{1 \text{ mole } H_2} = 4 \times 10^2 \text{ L } H_2$$

CALCULATION   448 L H_2

Example Hydrogen sulfide can be converted to sulfur dioxide by oxidation:

$$2H_2S(g) + 3O_2(g) \rightarrow 2SO_2(g) + 2H_2O(g)$$
hydrogen   oxygen   sulfur   water
sulfide   dioxide

How many liters of oxygen are required to react completely with 12.5 L of hydrogen sulfide at STP?

UNKNOWN	GIVEN	CONNECTION
liters O_2	12.5 L H_2S	$\dfrac{1 \text{ mole } H_2S}{22.4 \text{ L } H_2S}$ or $\dfrac{22.4 \text{ L } H_2S}{1 \text{ mole } H_2S}$
		$\dfrac{2 \text{ moles } H_2S}{3 \text{ moles } O_2}$ or $\dfrac{3 \text{ moles } O_2}{2 \text{ moles } H_2S}$
		$\dfrac{1 \text{ mole } O_2}{22.4 \text{ L } O_2}$ or $\dfrac{22.4 \text{ L } O_2}{1 \text{ mole } O_2}$

$$\text{liters } O_2 = 12.5 \text{ L } H_2S \times \frac{1 \text{ mole } H_2S}{22.4 \text{ L } H_2S} \times \frac{3 \text{ moles } O_2}{2 \text{ moles } H_2S} \times \frac{22.4 \text{ L } O_2}{1 \text{ mole } O_2}$$

ESTIMATE

$$1 \times 10^1 \text{ L } H_2S \times \frac{1 \text{ mole } H_2S}{2 \times 10^1 \text{ L } H_2S} \times \frac{3 \text{ moles } O_2}{2 \text{ moles } H_2S} \times \frac{2 \times 10^1 \text{ L } O_2}{1 \text{ mole } O_2}$$
$$= 2 \times 10^1 \text{ L } O_2$$

CALCULATION   18.8 L O_2

(Note: Using the mole ratio alone, the volume of O_2 could be calculated directly as $3/2 \times 12.5 \text{ L} = 18.8 \text{ L}$.)

Example   Nitric oxide and ammonia react in the catalytic converter of an automobile to form harmless nitrogen gas.

$$6NO(g) + 4NH_3(g) \rightarrow 5N_2(g) + 6H_2O(g)$$

nitric   ammonia   nitrogen   water
oxide

What volume of nitrogen at STP can be made from 80.0 g of nitric oxide?

UNKNOWN	GIVEN	CONNECTION
liters N_2	80.0 g NO	$\dfrac{1 \text{ mole NO}}{30.0 \text{ g NO}}$ or $\dfrac{30.0 \text{ g NO}}{1 \text{ mole NO}}$
		$\dfrac{6 \text{ moles NO}}{5 \text{ moles } N_2}$ or $\dfrac{5 \text{ moles } N_2}{6 \text{ moles NO}}$
		$\dfrac{1 \text{ mole } N_2}{22.4 \text{ L } N_2}$ or $\dfrac{22.4 \text{ L } N_2}{1 \text{ mole } N_2}$

$$\text{liters N}_2 = 80.0 \, \cancel{\text{g NO}} \times \frac{1 \, \cancel{\text{mole NO}}}{30.0 \, \cancel{\text{g NO}}} \times \frac{5 \, \cancel{\text{moles N}_2}}{6 \, \cancel{\text{moles NO}}} \times \frac{22.4 \, \text{L N}_2}{1 \, \cancel{\text{mole N}_2}}$$

ESTIMATE

$$8 \times 10^1 \, \cancel{\text{g NO}} \times \frac{1 \, \cancel{\text{mole NO}}}{3 \times 10^1 \, \cancel{\text{g NO}}} \times \frac{5 \, \cancel{\text{moles N}_2}}{6 \, \cancel{\text{moles NO}}} \times \frac{2 \times 10^1 \, \text{L N}_2}{1 \, \cancel{\text{mole N}_2}}$$
$$= 4 \times 10^1 \, \text{L N}_2$$

CALCULATION   49.8 L N$_2$

11.9
IDEAL
GAS LAW

The four variables discussed in this chapter—pressure (*P*), volume (*V*), temperature (*T*), and moles (*n*) of gas—are all related to one another. As we saw earlier, the relationships between certain pairs of these variables, written either as a product (like *P* × *V*) or as a ratio (like *V/T*), are constant when the other two variables are fixed. These relationships can be combined into a single expression, which holds true under all conditions for an ideal gas.

$$\frac{PV}{nT} = \text{constant} = R$$

The constant in this equation is called the gas constant, and is given the symbol *R*. The equation can be rearranged in the following way:

$$PV = nRT$$

This equation is called the **ideal gas law**. It states that *the pressure times the volume of a gas is equal to the number of moles of gas times the gas constant times the temperature in kelvins.*

The value of the gas constant *R* can be found by inserting into the ideal gas law equation the values of *P*, *V*, *n*, and *T* for one mole of any gas at STP, and solving for *R*. Thus for one mole of gas at STP: $n = 1.00$, $P = 1.00$ atm, $T = 273$ K, and $V = 22.4$ L.

$$R = \frac{PV}{nT} = \frac{(1.00 \, \text{atm})(22.4 \, \text{L})}{(1.00 \, \text{mole})(273 \, \text{K})} = 0.0821 \, \frac{\text{(atm) (L)}}{\text{(mole) (K)}}$$

The value of *R* depends on the units chosen for pressure and volume. For example, in SI units, where pressure is expressed in kilopascals and volume is expressed in cubic decimeters, we get:

$$R = \frac{PV}{nT} = \frac{(101.3 \, \text{kPa}) (22.4 \, \text{dm}^3)}{(1.00 \, \text{mole})(273 \, \text{K})} = 8.31 \, \frac{\text{(kPa)(dm}^3)}{\text{(mole)(K)}}$$

Other values of R are:

62.36 (liter)(torr)/(mole)(K)

1.987 cal/(mole)(K)

8.314 J/(mole)(K)

Keep in mind that all these expressions for R are equivalent. The numerical values are different only because different units are used.

The ideal gas law, $PV = nRT$, is very important because it allows us to determine any one of the four variables (P, V, T, or n) when the other three variables are known.

Known values	Value that can be found
P, V, T	$n = \dfrac{PV}{RT}$
V, T, n	$P = \dfrac{nRT}{V}$
P, T, n	$V = \dfrac{nRT}{P}$
P, V, n	$T = \dfrac{PV}{nR}$

When we use the ideal gas law, we must be careful to write the quantities with the proper units and to select the appropriate value for R. For example, suppose that a sample of carbon monoxide contains 2.50 moles and has a pressure of 0.890 atm at 298 K. The missing variable, volume, can be found as follows.

UNKNOWN	GIVEN	CONNECTION
V, liters CO	$P = 0.890$ atm $T = 298$ K $n = 2.50$ moles	$PV = nRT$

Since volume is the unknown, the ideal gas law is solved for V (by dividing both sides of the equation by P).

$$\frac{\cancel{P}V}{\cancel{P}} = \frac{nRT}{P}$$

$$V = \frac{nRT}{P}$$

The given values and the correct value for R are substituted into the equation. Note that the units for R must correspond exactly to the units for P, V, n and T in the example.

$$V = \frac{(2.50 \text{ moles})\left(0.0821 \frac{\text{atm L}}{\text{mole K}}\right)(298 \text{ K})}{0.890 \text{ atm}}$$

We must write out the units for all quantities, especially R. Only then can we check to make sure that the units cancel properly to give the units of the desired answer. In this case, the answer should be in liters. The volume of carbon monoxide is

ESTIMATE $\qquad \frac{(3 \text{ moles})\left(8 \times 10^{-2} \frac{\text{atm L}}{\text{mole K}}\right)(3 \times 10^2 \text{ K})}{9 \times 10^{-1} \text{ atm}} = 8 \times 10^1 \text{ L CO}$

CALCULATION   68.6 L CO

Example   How many moles of krypton are present in a sample with a volume of 8.95 dm³ at 30.5°C and 125 kPa?

UNKNOWN	GIVEN	CONNECTION
n, moles Kr	$P = 125$ kPa $V = 8.95$ dm³ $t = 30.5°$C	$PV = nRT$

$T = t + 273 = 30.5 + 273 = 304$ K

$$n = \frac{PV}{RT}$$

$$n = \frac{(125 \text{ kPa})(8.95 \text{ dm}^3)}{\left(8.31 \frac{\text{kPa dm}^3}{\text{mole K}}\right)(304 \text{ K})}$$

Note: After cancelling units, only the unit 1/mole remains in the denominator. Therefore, the units of the answer must be

$$\frac{1}{\frac{1}{\text{mole}}} = 1 \div \frac{1}{\text{mole}} = 1 \times \frac{\text{mole}}{1} = \text{mole}$$

ESTIMATE $\qquad \frac{(1 \times 10^2 \text{ kPa})(9 \text{ dm}^3)}{\left(8 \frac{\text{kPa dm}^3}{\text{mole K}}\right)(3 \times 10^2 \text{ K})} = 4 \times 10^{-1} \text{ mole Kr}$

CALCULATION   0.443 mole Kr

Example What is the pressure in atmospheres of 0.250 mole of ammonia if the volume is 950. mL at 310. K?

UNKNOWN	GIVEN	CONNECTION
P, atm NH$_3$	$V = 950.$ mL $T = 310.$ K $n = 0.250$ mole	$PV = nRT$

$$950. \text{ mL} \times \frac{10^{-3} \text{ L}}{1 \text{ mL}} = 0.950 \text{ L}$$

$$P = \frac{nRT}{V}$$

$$P = \frac{(0.250 \text{ mole})\left(0.0821 \dfrac{\text{atm L}}{\text{mole K}}\right)(310. \text{ K})}{0.950 \text{ L}}$$

(Note: If the volume were left in milliliters, the units would not cancel in this equation.)

ESTIMATE

$$\frac{(3 \times 10^{-1} \text{ mole})\left(8 \times 10^{-2} \dfrac{\text{atm L}}{\text{mole K}}\right)(3 \times 10^2 \text{ K})}{1 \text{ L}} = 7 \text{ atm NH}_3$$

CALCULATION   6.70 atm NH$_3$

11.10 AIR AND PARTIAL PRESSURE Air is a mixture of gases, as shown in Table 11-3. Over 78% of this mixture is nitrogen, N$_2$, which is harmless and relatively unreactive. Oxygen, O$_2$, makes up about 21% of the air, and is essential to life. The noble gas argon, Ar, is roughly 1% of the total volume of air. Carbon dioxide, CO$_2$, which we exhale, has a concentration of only 0.03%. A large number of other gases are present in even smaller amounts in clean dry air.

In air, and in any other mixture of gases, the pressure of each gas contributes to the total pressure (P) of the mixture. The pressure of each gas is called its **partial pressure** (p). *The total pressure is simply the sum of the partial pressures of all the gases present.* This rela-

TABLE 11-3

Composition of dry air

Gas	Percentage (by volume)
nitrogen (N_2)	78.084
oxygen (O_2)	20.946
argon (Ar)	0.934
carbon dioxide (CO_2)	0.033
trace gases	0.003

[a]Neon, helium, methane, krypton, hydrogen, nitrous oxide, carbon monoxide, xenon, ozone, ammonia, nitrogen dioxide, nitric oxide, sulfur dioxide, hydrogen sulfide.

tionship is known as **Dalton's law of partial pressures** (named after John Dalton). Partial pressures of the gases present in air are given in Table 11-4. As shown, the sum of these partial pressures equals the total pressure of the atmosphere (1 atm, 760 torr, or 101.3 kPa) at sea level.

$$P_{atm} = p_{N_2} + p_{O_2} + p_{Ar} + p_{CO_2} + p_{trace\ gases}$$

The greater the amount of a particular gas in the mixture, the greater is the partial pressure of that gas. (This result follows from the ideal gas law written in the form: $P = (RT/V)n$. At constant T and V, P is directly proportional to n.) In other words, the more molecules of the gas that are present, the more pressure is exerted by that gas. Thus, because nitrogen makes up most of the air, the largest contribution to the atmospheric pressure comes from nitrogen. (See Table 11-4.)

The exchange of gases between our bodies and the air, called respiration, depends on differences in partial pressures. Oxygen moves from the air, where its partial pressure is relatively high (159 torr) to the blood in the lungs where its partial pressure is lower (116 torr). As the blood circulates, oxygen is transferred to the tissues, where the partial pressure of oxygen is still lower (35 torr). Carbon dioxide, which is the waste product of respiration, moves in the opposite direction. It is transferred from the tissues into the blood and is finally exhaled into the atmosphere, again as a result of differences in partial pressures.

The partial pressure of oxygen in the air at normal atmospheric pressure is 159 torr. Therefore, a person can be quite comfortable in a pure oxygen atmosphere if the pressure is kept at about 159 torr. High-flying pilots in a nonpressurized cabin are comfortable breathing pure oxygen at altitudes of 40,000 feet because the atmospheric pressure there is about 150 torr.

Deep sea divers breathe air under increased pressure. For example, at depths of 120 feet they breath air at five times normal pressure. If

TABLE 11-4 *Partial pressures of the gases in the atmosphere*

Gas	Partial pressure (atm)	Partial pressure (torr)	Partial pressure (kPa)
nitrogen	0.78084	593.44	79.119
oxygen	0.20946	159.19	21.224
argon	0.00934	7.09	0.945
carbon dioxide	0.00033	0.25	0.033
trace gases	0.00003	0.03	0.004
total pressure	1.00000	760.00	101.325

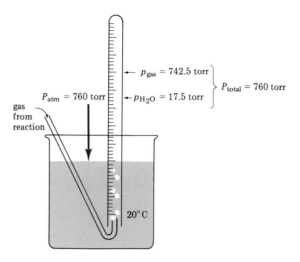

FIGURE 11-11

Dalton's law of partial pressures. The total pressure exerted by a wet gas (collected here inside the inverted tube) equals the partial pressure of the dry gas (p_{gas}) plus the vapor pressure of water (p_{H_2O}) at that temperature.

regular air were used in divers' tanks, the partial pressure of the oxygen would be 750 torr, or about 1 atm, at these depths. This high oxygen pressure can damage a diver's health. Therefore deep sea divers breathe a mixture that is 96% nitrogen and only 4% oxygen. The partial pressure of oxygen is much less in such a mixture. (Actually, nitrogen under high pressure can prove dangerous and often helium is used as a replacement.)

Gases produced in the laboratory are often collected by being bubbled into a tube or bottle filled with water, as illustrated in Figure 11-11. When the gas enters the tube, it displaces the water (assuming that the gas does not dissolve in water). Due to vaporization (explained in the following chapter), part of the water mixes with the gas as water vapor. The total pressure of the mixture is the sum of the partial pressure of the gas and the partial pressure of the water vapor.

$$P_{total} = p_{gas} + p_{H_2O}$$

Table 11-5 gives values for the partial pressure of water vapor at various temperatures. The appropriate value from the table is subtracted from the total pressure (called the wet pressure) to find the partial pressure of the dry gas.

$$p_{gas} = P_{total} - p_{H_2O}$$

Since most gases that are collected in the laboratory are at atmospheric pressure, the total pressure can be read from a barometer. The gas pressure can then be corrected for the presence of water vapor, as we just described.

TABLE 11-5 *Vapor pressure of water as a function of temperature*

Temperature (°C)	Vapor pressure (torr)	Vapor pressure (kPa)
0	4.6	0.61
5	6.5	0.87
10	9.2	1.23
15	12.8	1.71
20	17.5	2.33
25	23.8	3.17
30	31.8	4.24
50	92.5	12.33
70	233.7	31.16
75	289.1	38.63
80	355.1	47.34
85	433.6	57.81
90	525.8	70.10
95	633.9	84.51
100	760.0	101.32

Example Oxygen, generated by the decomposition of potassium chlorate, was collected by the displacement of water. The atmospheric pressure during the experiment was 750. torr and the temperature was 25°C. Find the pressure of dry oxygen that was collected in a 0.250 L flask.

UNKNOWN	GIVEN	CONNECTION
p_{O_2}, torr	$P_{total} = 750.$ torr $p_{H_2O} = 23.8$ torr (from Table 11-5)	$P_{total} = p_{O_2} + p_{H_2O}$

$$p_{O_2} = P_{total} - p_{H_2O} = 750.\ torr - 23.8\ torr = 726\ torr$$

Once the gas pressure is corrected for the presence of water vapor, it can then be used in further calculations such as those worked out in previous sections.

FIGURE 11-12

Air pollution. The manufacturing plant in this photograph is located in Lehigh, Pennsylvania. (Photo courtesy of the Sierra Club.)

**11.11
AIR POLLUTION**

An air pollutant is a substance that changes the properties of clean air, causing harm to people, animals, plants, and even buildings and materials. Over 300 million tons of air pollutants are added to the atmosphere above the United States each year (see Figure 11-12). Table 11-6 lists the amount and source of each type of air pollutant. Although some pollutants enter the atmosphere by natural processes (such as forest fires), these are usually spread out over a large area and have much less effect than do pollutants generated by people living or working in heavily populated areas. Note from Table 11-6 that transportation is the largest single source of air pollution.

Carbon monoxide, CO, is the most abundant air pollutant. A principal source of CO is the incomplete combustion by motor vehicles of the carbon compounds present in gasoline.

TABLE 11-6 *Air pollution in the U.S. (millions of tons per year)*[a]

Source	Carbon monoxide	Nitrogen oxides	Sulfur oxides	Hydrocarbons	Particulates
nature	?	?	4.2	30.7	?
stationary combustion	1.1	11.0	22.1	0.4	7.1
transportation	111.5	11.2	1.1	19.8	0.8
industrial processes	12.0	0.2	7.5	5.5	14.4
miscellaneous	26.1	2.4	0.4	11.2	12.8

[a]From *Energy/Environment Fact Book*, U.S. Environmental Protection Agency, 1978.

$$2C + O_2(g) \rightarrow 2CO(g)$$
carbon oxygen carbon
               monoxide

High concentrations of CO build up in the air wherever there is heavy automobile traffic. It's not surprising then that taxi drivers are exposed to more carbon monoxide than people in any other occupation. Carbon monoxide binds to hemoglobin, the oxygen carrier in the blood, over 200 times more strongly than oxygen. The CO replaces oxygen in some hemoglobin molecules, forming carboxyhemoglobin instead of oxy-hemoglobin. The effects of carbon monoxide on a person depend on the concentration of carboxyhemoglobin in the blood, which in turn depends on the amount of CO in the inhaled air. When the level of carboxyhemoglobin in the blood is between 1 and 2% of the total hemoglobin, it can cause behavioral changes. Concentrations of 2 to 5% affect the central nervous system. At concentrations of 5 to 10%, the heart and lungs are affected. When the level of carboxyhemoglobin reaches 10%, it can cause headache and drowsiness. As the percentage in the blood increases above 30%, it can cause coma, respiratory failure, and eventual death due to hypoxia (lack of oxygen), because needed oxygen is replaced by CO and cannot reach the tissues. Each year people die from carbon monoxide poisoning by foolishly remaining in unventilated areas (like garages) with their automobile engines running. People who smoke a pack of cigarettes a day have a carboxy-hemoglobin blood level of about 5%.

Nitrogen oxides are formed by the combustion of gasoline in motor vehicles as well as by combustion of natural gas and coal. Nitrogen, N_2, is oxidized at the high temperatures (1300-2500°C) produced in these processes.

$$N_2(g) + O_2(g) \rightarrow 2NO(g)$$
nitrogen   oxygen   nitric
                  oxide

$$2NO(g) + O_2(g) \rightarrow 2NO_2(g)$$
nitric   oxygen   nitrogen
oxide   dioxide

Over 90% of the nitrogen oxides (NO and NO_2) in the atmosphere consist of nitric oxide, NO. At high concentrations, nitrogen oxides can affect the human respiratory tract, cause dyes to fade, and cause metals to corrode.

Sulfur oxides are formed in large amounts from coal combustion in power plants that generate electricity. Sulfur, present in coal at concentrations of 0.2 to 7%, is oxidized during combustion.

$$S(s) + O_2(g) \rightarrow SO_2(g)$$
sulfur   oxygen   sulfur
                 dioxide

Smaller amounts of sulfur oxides are formed during the refining or smelting of metals. When metals are refined, sulfide ores, such as copper sulfide, are oxidized:

$$2CuS(s) + 3O_2(g) \rightarrow 2CuO(s) + 2SO_2(g)$$

| copper(II) sulfide | oxygen | copper(II) oxide | sulfur dioxide |

About 1 to 10% of the sulfur dioxide produced by either process can react further with oxygen to generate sulfur trioxide.

$$2SO_2(g) + O_2(g) \rightarrow 2SO_3(s)$$

| sulfur dioxide | oxygen | sulfur trioxide |

Both SO_2 and SO_3 can react with water to make corrosive sulfuric acid.

$$2SO_2(g) + 2H_2O(l) + O_2(g) \rightarrow 2H_2SO_4(aq)$$

| sulfur dioxide | water | oxygen | sulfuric acid |

$$SO_3(s) + H_2O(l) \rightarrow H_2SO_4(aq)$$

| sulfur trioxide | water | sulfuric acid |

Sulfur oxides irritate the human respiratory system and cause constriction of the air pathways. These effects are most dangerous for elderly people, and for those persons with chronic respiratory disease or asthma. Like nitrogen oxides, sulfur oxides also cause corrosion of metals.

Hydrocarbons (discussed in Chapter Eighteen) are compounds containing only carbon and hydrogen. A large number of compounds are included in this category. For example, gasoline is a mixture of hydrocarbons. The gasoline used in automobiles is a major source of hydrocarbons in the air. Some of this gasoline simply evaporates, and some is released unburned from the engines. Harmful effects result not from the hydrocarbons directly but from secondary pollutants, formed in the atmosphere by photochemical (light-requiring) reactions involving hydrocarbons. These pollutants, called photochemical oxidants, are even stronger oxidizing agents than oxygen, O_2. An example is PAN, peroxyacetylnitrate ($CH_3CO_3NO_2$). (Ozone, O_3, is also classified with the photochemical oxidants, although it is not formed from hydrocarbons.) These oxidants mix with other pollutants to form photochemical smog ("smoke-fog"), which irritates eyes and lungs. Photochemical oxidants also damage materials such as rubber and textiles.

The final category of air pollutants is the particulates. Particulates are not gases, but are tiny, solid particles and droplets of liquid. Examples of particulates in the atmosphere are sea salt, and the nitrates, sulfates, and oxides of many elements. Some particulates are generated during industrial processes, and by coal combustion, which produces "ash." Their effect on the human respiratory system depends on the size of the particles. If the diameter of a particle is less than 5 μm, it can enter the lungs and interfere with normal respiration. Particulates can carry pollutants into the body or can be toxic themselves. For example, particles of the metals antimony, beryllium, bismuth, cadmium, lead, mercury, nickel, and tin are toxic. Certain particulates cause specific diseases. For example, asbestosis is caused by the inhalation of asbestos particles, silicosis by inhalation of silicon dioxide particles, and "black lung" disease by breathing coal dust.

Summary   **11.1** The kinetic molecular theory deals with the motion of the molecules or atoms of a gas. Gas molecules are constantly moving in a random manner, hitting each other and the walls of their container. The temperature of a gas is a measure of the kinetic energy of its molecules.

11.2 The collisions of gas molecules with a surface create pressure. The Earth's atmosphere exerts a pressure called atmospheric pressure. Atmospheric pressure is equal to 1 atm or 760 torr at sea level and 0°C.

11.3 The pressure and volume of a gas are inversely related (for a constant amount of gas at a fixed temperature). If one quantity increases, the other quantity decreases. Thus, the product of the pressure and volume remains constant.

11.4 Pressure and temperature (for a fixed amount of gas at constant volume) are directly related. A change in one quantity causes an equivalent change in the other. Thus the ratio of the pressure and temperature remains constant. Temperature must be expressed in kelvins for this relationship to hold.

11.5 Volume and temperature (for a fixed amount of gas at constant pressure) are also directly related. The ratio of volume and temperature remains the same when one of these two variables is changed.

11.6 The relationships between pressure, volume, and temperature can be combined into a single equation, called the combined gas law $\left(\dfrac{P_1 V_1}{T_1} = \dfrac{P_2 V_2}{T_2} \right)$. This equation is used in problems that involve changes in any two of these variables.

11.7 Volume and pressure are each directly related to the number of moles of gas present. The molar volume of any gas at STP (0°C and 1 atm) is 22.4 L.

11.8 Stoichiometry problems can be solved using the volumes of reactants or products. The coefficients of a balanced equation involving gases can be interpreted in terms of volumes.

11.9 The ideal gas law, $PV = nRT$, relates the four variables: pressure, volume, number of moles, and temperature. R is the gas constant. Any one variable can be calculated if the other three variables are known.

11.10 Air is a mixture of gases consisting of about 78% N_2 and 21% O_2. In a mixture of gases, the total pressure is equal to the sum of the partial pressures of all the gases present.

11.11 Air pollutants change the properties of clean air and can have harmful effects on people, animals, plants, and materials. The most common pollutants are carbon monoxide, nitrogen oxides, sulfur oxides, hydrocarbons, and particulates.

Exercises

KEY-WORD MATCHING EXERCISE For each term on the left, choose a phrase from the right-hand column that most closely matches its meaning.

1 kinetic molecular theory   **a** $PV = nRT$

2 pressure   **b** pressure in a mixture

3 STP   **c** 273 K and 1 atm

4 ideal gas law   **d** based on random movement of molecules

5 partial pressure   **e** defined as force/area

6 molar volume   **f** 22.4 L

EXERCISES BY TOPIC **Kinetic molecular theory (11.1)**

7 Summarize the kinetic molecular theory.

8 What is an ideal gas?

9 Explain diffusion in molecular terms.

10 What is the meaning of the temperature of a gas?

Pressure (11.2)

11 What is the molecular explanation for the pressure of a gas?

12 What causes atmospheric pressure?

13 Describe the operation of **a** a Torricelli barometer **b** a manometer

14 A gas in a cylinder has a pressure of 120. atm. Convert this value into **a** kPa **b** torr **c** psi

15 A barometer reads 750. torr. Convert this value into **a** atm **b** kPa

16 According to a weather report, the atmospheric pressure on a certain day was 28.5 in Hg. Express this value in **a** atm **b** kPa

Pressure and volume (11.3)

17 A nitrous oxide sample originally had a pressure of 620. torr and a volume of 10.5 L. If the pressure is now 875 torr, what is the volume?

18 A fixed amount of helium had an initial volume of 400. mL and a final volume of 920. mL. If the initial pressure was 1.35 atm, find the final pressure.

19 A gas expanded to 6.20 L when the pressure dropped from 160. kPa to 135 kPa. What was its original volume?

20 Fifty grams of cyclopropane decreased in volume from 821 cm^3 to 358 cm^3. If the final pressure was 0.79 atm, what was the initial pressure?

21 Explain how breathing depends on Boyle's law.

Pressure and temperature (11.4)

22 What happens to the pressure in an automobile tire in cold weather? Explain.

23 A gas has a pressure of 720. torr at 250. K. Find its pressure at 322. K.

24 Ethylene has a pressure of 1.32 atm at 25.5°C. At what temperature is its pressure equal to 1.00 atm?

25 A gas at 87°C and 150. kPa is cooled to −12°C. Find its pressure at this temperature.

26 Explain in molecular terms why increasing the temperature causes the pressure of a gas to increase (at constant volume).

Volume and temperature (11.5)

27 How does a gas thermometer work?

28 An ethane sample at 280. K has a volume of 1.56 L. Find its volume at 380. K.

29 A sample of oxygen with a volume of 730. cm^3 is cooled from 90.°C to −10.°C. Find its new volume.

30 If a gas expands from 25.2 mL to 89.7 mL when the temperature is increased to 125°C, find its original temperature.

31 Gaseous helium occupies 15.3 L after its temperature decreases from 456 K to 251 K. What was its initial volume?

Combined gas law (11.6)

32 A carbon dioxide sample has a volume of 820. mL at 20°C and 1.05 atm. Find its volume at 65°C and 1.11 atm.

33 In a gas cylinder, hydrogen chloride has a pressure of 135 atm at 20.°C. If all of the gas is released, it occupies 327 L at 15°C and 1.00 atm. Find the volume of the cylinder.

34 At −5.00°C, 125 mL of argon has a pressure of 110.2 kPa. If the gas expands to 393 mL at 38.2°C, what will its pressure then be?

35 At 58.1°C, 530. cm^3 of chlorine has a pressure of 645 torr. At what temperature will it have a volume of 892 cm^3 and a pressure of 720 torr?

Relating moles to volume (11.7)

36 How does a mechanical vacuum pump work?

37 In terms of gas properties, what happens when you pump up a tire?

38 What volume at STP is occupied by 0.875 moles of fluorine?

39 How many moles are present in 18.3 L of ammonia at STP?

40 What volume at STP is occupied by 12.2 g of xenon?

Stoichiometry involving gases (11.8)

41 Coal contains impurities such as iron(II) sulfide that release sulfur dioxide when burned:

$$4FeS_2(s) + 11O_2(g) \rightarrow 2Fe_2O_3(s) + 8SO_2(g)$$

a How many liters of SO_2 at STP are formed from 100. g of FeS_2? **b** How many liters of O_2 are needed to react with 2.81 kg of FeS_2?

42 Sulfur dioxide can be trapped in smokestacks by reacting it with calcium carbonate:

$$2SO_2(g) + 2CaCO_3(s) + O_2(g) \rightarrow 2CaSO_4(s) + 2CO_2(g)$$

a How many grams of $CaCO_3$ are needed to react with 155 L of SO_2 at STP? **b** How many liters of CO_2 are produced from 1.38×10^3 g of SO_2?

43 Ammonia can be converted to urea, $CO(NH_2)_2$, for use as a fertilizer:

$$2NH_3(g) + CO_2(g) \rightarrow CO(NH_2)_2(s) + H_2O(l)$$

a How many liters of CO_2 at STP are needed to react with 751 L of NH_3? **b** How many grams of NH_3 would be needed to react with 2.73 L of CO_2?

44 State the law of combining volumes and explain it in molecular terms.

Ideal gas law (11.9)

45 Why is the ideal gas law useful?

46 A sample of boron trifluoride, BF_3, has a volume of 18.1 L at 98.1°C and 750. torr. What quantity of gas is present?

47 Find the volume of 6.25 moles of ether if its pressure is 1.21 atm at 20.°C.

48 At −5.12°C, 52 g of H_2 occupy 933 cm^3. What is its pressure?

49 At what temperature will 88 g of N_2 have a volume of 12.5 L and a pressure of 1.10 atm?

Air and partial pressure (11.10)

50 Describe the composition of air.

51 Why must a correction be made when you collect a gas over water?

52 The partial pressures of the components of a gaseous mixture are: Ar, 120. torr; CO_2, 290. torr; Xe, 50. torr; and N_2, 430. torr. What is the total pressure of this mixture?

53 Hydrogen is collected over water at 30.°C. If the atmospheric pressure is 755 torr, what is the pressure of dry H_2 collected?

54 At 25.0°C and 750. torr, 25.0 g of O_2 were collected. What volume of dry oxygen was collected?

55 In a helium-oxygen mixture with a pressure of 828 torr, helium has a partial pressure of 655 torr. Find the partial pressure of oxygen.

Air pollution (11.11)

56 What are the major types of air pollutants?

57 Describe the sources and effects of: **a** particulates **b** CO **c** nitrogen oxides **d** sulfur oxides **e** hydrocarbons

58 Why does heavy traffic result in high CO levels in drivers?

59 What is photochemical smog?

ADDITIONAL EXERCISES

60 Explain why each assumption of the kinetic molecular theory is reasonable.

61 Convert 0.899 atm to **a** kPa **b** torr **c** psi **d** cm Hg **e** in Hg

62 Convert 782 torr to **a** atm **b** mm Hg **c** psi **d** in Hg **e** kPa

63 Carbon dioxide reacts with magnesium by the following reaction (which prevents the use of CO_2 fire extinguishers on magnesium fires): $CO_2(g)$ + $2Mg(s) \rightarrow 2MgO(s) + C(s)$. How many grams of MgO will be formed from 625 mL of CO_2 at STP?

64 Cyclopropane at 719 torr and 25.1°C is heated to 100.0°C. What is the value of the pressure?

65 A sample of nitrous oxide occupies 1.98 L at 5.2°C and 0.88 atm. What is its volume at 45°C and 0.75 atm?

66 Find the temperature of an oxygen sample if 0.21 kg occupied 18.2 L at 0.621 atm.

67 Find the volume of 6.92 g of helium at STP.

68 A gas at 21.0 psi occupies 18.1 dm^3. Find the volume at 15.1 psi.

69 Nitrogen is collected over water at 20.°C and 775 torr. If the volume is 50.1 mL, what mass of N_2 is collected?

70 A balloon has a volume of 10.32 dm^3 at 31.0°C. Find its volume at -5.21°C.

71 Starting with the combined gas law, derive the relation:

a $\dfrac{P_1}{T_1} = \dfrac{P_2}{T_2}$ **b** $\dfrac{V_1}{T_1} = \dfrac{V_2}{T_2}$

72 Ammonia and chlorine react in the following way:

$$3Cl_2(s) + 8NH_3(g) \rightarrow 6NH_4Cl(s) + N_2(s)$$

How many liters of ammonia at STP are needed to react with 12.1 kg of Cl_2?

73 Find the molecular weight of a gas if its density is 0.88 g/L at STP.

74 A laboratory barometer reads 745 torr. Express this pressure in **a** psi **b** mm Hg **c** in Hg **d** Pa **e** atm

75 Samples of oxygen and of radon are both at 35°C. In which gas are the molecules moving faster? Explain your answer.

76 What is the density of gaseous XeF_2 at STP?

77 Carbon dioxide can be generated from the reaction of calcium carbonate and hydrochloric acid:

$$CaCO_3(s) + 2HCl(aq) \rightarrow CaCl_2(aq) + CO_2(g) + H_2O(l)$$

What volume of CO_2 at STP can be obtained from 50.0 g of $CaCO_3$?

78 The pressure gauge on a steel tank reads 298 torr at 25°C. At what temperature would it read 1000. torr?

79 The volume of a 100-watt bulb is 130. cm^3. If the bulb contains 3.0×10^{-3} mole N_2, what is the pressure inside the bulb at 20.°C?

80 A train containing 1.28×10^6 lb of liquid chlorine was involved in an accident. If all the chlorine were converted to a gas at 740. torr and 15°C, what volume would it occupy?

81 A radio announcer says: "The average temperature yesterday was 15°C, but today's temperature will reach 30°C; twice as hot." Is this statement correct? Explain.

82 A sample of gas that occupies 270. mL at 740. torr and 98°C has a mass of 0.276 g. Is it methyl alcohol, CH_3OH, or ethyl alcohol, C_2H_5OH?

83 Oxygen was produced in the laboratory by the decomposition of potassium chlorate:

$$2KClO_3(s) \rightarrow 2KCl(s) + 3O_2(g)$$

A volume of 550. mL of O_2 was collected over water at 743 torr and 21°C. If the vapor pressure of water is 19 torr at 21°C, how many moles of O_2 were collected?

84 The liquid fuel in a pocket lighter has a mass of 14.70 mg. When it evaporates, its vapor occupies 6.76 mL at 120.°C and 740. torr. What is the molecular weight of the fuel?

PROGRESS CHART Check off each item when completed.

_____ Previewed chapter

Studied each section:

_____ **11.1** Kinetic molecular theory of gases

_____ **11.2** Pressure

_____ **11.3** Pressure and volume

_____ **11.4** Pressure and temperature

_____ **11.5** Volume and temperature

_____ **11.6** Combined gas law

_____ **11.7** Relating moles to volume

_____ **11.8** Stoichiometry involving gases

_____ **11.9** Ideal gas law

_____ **11.10** Air and partial pressure

_____ **11.11** Air pollution

_____ Reviewed chapter

_____ Answered assigned exercises:

    #'s _____

Review Exercises for Chapters Nine–Eleven

1 How does a mixture of iron and oxygen differ from a compound of iron and oxygen?

2 Balance the following equations:

a $HNO_3(aq) \rightarrow NO_2(g) + H_2O(l) + O_2(g)$
b $Fe(s) + H_2SO_4(aq) \rightarrow Fe_2(SO_4)_3(s) + SO_2(g) + H_2O(l)$

3 Interpret in terms of moles the balanced equations of Exercise 2.

4 Label each of the following reactions as either combination, decomposition, replacement, or double replacement:

a $2SO_2(g) + 4NH_3(aq) + 2H_2O(l) + O_2(g) \rightarrow 2(NH_4)_2SO_4(s)$
b $C_6H_{12}(l) \rightarrow C_6H_6(l) + 3H_2(g)$
c $FeO(s) + C(s) \rightarrow Fe(s) + CO(g)$
d $3Ca(OH)_2(aq) + 2Na_3PO_4(aq) \rightarrow Ca_3(PO_4)_2(s) + 6NaOH(aq)$

5 Explain how the formation of an ionic compound from elements must involve a redox reaction.

6 Write a balanced chemical equation for: **a** a double replacement reaction between iron(II) sulfide and copper(I) oxide   **b** a replacement reaction between vanadium(V) oxide and calcium   **c** a synthesis reaction between carbon monoxide and oxygen.

7 What is an exothermic reaction?

8 Hydrogen cyanide can be prepared from methane, ammonia, and oxygen:

$$2CH_4(g) + 2NH_3(g) + 3O_2(g) \rightarrow 2HCN(g) + 6H_2O(l)$$

How many grams of HCN can be made from 1.25 kg of CH_4?

9 How many molecules of O_2 are needed to react with 117 g of NH_3 in the HCN synthesis given in Exercise 8?

10 What mass of O_2 is needed to react with 541 g of CH_4 in the HCN synthesis?

11 If 21.5 g of NH_3 react with 53.7 g of O_2 in the HCN synthesis, **a** what mass of HCN is formed?   **b** what mass of excess reagent is left over?

12 If 199 g of HCN are produced from 351 g of CH_4, what is the percent yield?

13 Give evidence for the idea that gas molecules are constantly, randomly moving.

14 Convert 769 torr to: **a** atm   **b** kPa

15 Acetylene gas, C_2H_2, has a pressure of 0.799 atm in a volume of 1.2 L at 28°C. How many grams are present?

16 A gas at 500. torr and $-12°C$ is heated to 100.°C. Find its new pressure.

17 If 120. cm^3 of a gas at 5.12 kPa are expanded to 641 cm^3 at constant temperature, what is the final pressure?

18 How many liters of methane are needed to make 1.00 kg of HCN (at STP) in the reaction given in Exercise 8?

19 If a gas occupies 18.2 dm^3 at 91°F and 251. torr, what is its volume at STP?

20 Oxygen is collected by water displacement at 15°C when the atmospheric pressure is 0.951 atm. If 0.21 mole is produced, what is the volume of the oxygen?

TWELVE

Water:
the properties
of a liquid

Water is the most abundant chemical compound, both on the surface of the Earth and inside our bodies. The Earth holds over 10^9 km^3 of water, and our bodies are composed of about 60% water. This chapter is devoted to a discussion of this important compound. Because water commonly exists in the liquid state, much of what we say about water is also true of liquids in general. Chapter Twelve describes the properties of water, the energies involved in changes of state, and related topics such as water pollution and treatment. To demonstrate an understanding of Chapter Twelve, you should be able to:

1 Explain the kinetic molecular theory of liquids and solids.

2 Describe the structure of water and the formation of hydrogen bonds.

3 Explain how the density of water varies with temperature, and define specific gravity.

4 Describe the process of vaporization and explain the low vapor pressure of water.

5 Define heat capacity and specific heat.

6 Explain the energy changes involved in conversions between ice, liquid water, and water vapor.

7 Explain surface tension and the formation of a meniscus in terms of cohesion or adhesion.

8 Write the names and formulas for compounds containing water of hydration.

9 Identify the major sources of water pollution.

10 Describe the major methods used in water treatment.

11 Define hard water and ways of converting it to soft water.

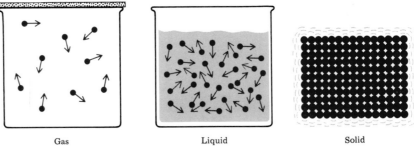

Gas   Liquid   Solid

FIGURE 12-1

Comparing a gas, a liquid, and a solid. In a gas, the atoms or molecules are far apart and move freely. In a liquid, the atoms or molecules are much closer together, but still move freely. In a solid, the atoms or molecules are still closer together and vibrate in fixed positions.

**12.1
KINETIC THEORY
OF LIQUIDS
AND SOLIDS**

Water can exist as a liquid, a solid (ice), and a gas (steam or water vapor). We already learned about gases in Chapter Eleven. To understand the properties of water, we must now also examine the liquid and solid states. The molecules of a liquid or solid are much closer together than the molecules of a gas, as shown in Figure 12-1. In other words, liquids and solids are denser than gases. This greater density means that the same amount of matter occupies a considerably smaller volume. For example, one gram of water occupies more than one liter as a gas but occupies only one cubic centimeter as a liquid or solid. Because there is less empty space in a liquid or a solid than there is in a gas, the properties of liquids and solids depend more on the nature of their individual molecules and how they interact than do the properties of gases.

In a solid, the atoms, ions, or molecules cannot move about freely. Instead, they vibrate around relatively fixed positions that are very close together. Strong attractive forces give solids their rigid shape. Most solids consist of regular repeating arrangements of atoms or ions, like crystals of sodium chloride (see Figure 6-6).

Liquids have properties that range between those of gases and solids. The molecules of a liquid are thousands of times closer together than the molecules of a gas. However, they are not as close together as the atoms, ions, or molecules of a solid. Because of attractive forces between the molecules, a liquid has definite boundaries and a definite volume. Like solids, liquids cannot be compressed very much because the molecules are already so close together. But, as in a gas, the molecules of a liquid are still in rapid motion, colliding or sliding past each other. A liquid therefore flows rather than holds a definite shape. Diffusion takes place within liquids, although not as quickly as it does within gases. The liquid state is the ideal medium for carrying out chemical reactions because the reactant molecules or ions move about, but are close enough to collide frequently.

Smithsonian Institution: Photo # 72-33

Henry Cavendish fit the stereotype of the mad scientist rather well. He was absentminded, shy, and as one fellow scientist commented, he "probably uttered fewer words than any other man."

At the age of 40 he inherited a vast fortune. But he ignored his wealth and lived a relatively simple life. He used his money only to equip his home with a laboratory. He was so uncomfortable with women that he built a private entrance to his house, so that he could come and go without seeing any of the maids. The only pictures of Cavendish were drawn secretly because he refused to sit for a formal portrait.

Cavendish was interested only in his research. He literally spared himself no pains in his work. During electrical experiments, for example, he would measure the strength of a current by shocking himself and then recording the degree of pain.

This strange English scientist is generally credited with the discovery of "inflammable air," later named hydrogen by Lavoisier. By burning "inflammable air" and producing water, Cavendish demonstrated that water is a combination of the elements hydrogen and oxygen.

Cavendish also performed experiments in which he measured the specific heats of substances, and the mass of the Earth (the "Cavendish experiment").

Despite his great achievements, Cavendish was never interested in publishing his results or receiving credit for them. But today, the Cavendish Physical Laboratory in Cambridge, England, one of the most famous laboratories in the world, is named in his honor.

**12.2
STRUCTURE
OF WATER**

(a)

(b)

FIGURE 12-2

Water, H₂O. The ball-and-stick model (a) shows the way in which the atoms are connected. The space-filling model (b) shows the shape of the molecule more realistically. (Photo by Al Green.)

The water molecule consists of two hydrogen atoms covalently bonded to one oxygen atom. The angle formed by the two hydrogen atoms is 105°. Two different models of a water molecule are shown in Figure 12-2. The ball and stick model indicates the way that the atoms are connected. However, the space-filling model presents the shape of the molecule more accurately. The Lewis structure of a water molecule is the following:

> xx
> H⦂O⦂x   Lewis structure
> ⦂x   of a water molecule
> H

Although the Lewis structure does not show the geometry of the molecule very well, it clearly shows the electron pair that is shared between each hydrogen atom and the central oxygen atom. The bonds are formed from the overlap of hydrogen 1s orbitals with oxygen 2p orbitals.

Because oxygen is more electronegative than hydrogen, the electrons from the hydrogen atoms are pulled toward the oxygen atom. These polar covalent bonds create an unequal distribution of electrical charge within the molecule. The hydrogen end of the water molecule has a partial positive charge ($\delta+$, read "delta plus") because there is a relative shortage of electrons in this part of the molecule. The oxygen end of the molecule has an excess of electrons and therefore has a partial negative charge ($\delta-$, read "delta minus"). This separation of charge within the molecule makes the water molecule polar. The molecule as a whole, however, remains neutral.

$\delta+$ H
  \
    O $\delta-$ | H "end" (+ −) O "end"
  /
$\delta+$ H

the polar nature of a water molecule

Since opposite charges attract, the partially negative oxygen end of one water molecule will attract a partially positive hydrogen atom of another water molecule. This rather weak but important force of attraction is called a **hydrogen bond**. As shown in Figure 12-3, the hydrogen atom forms a bridge between two water molecules. The hydrogen atom is attached to one oxygen atom by a covalent bond and to a second oxygen atom by a hydrogen bond (shown by the dotted line).

Hydrogen bonds are much weaker than covalent bonds. The O—H bond energy of a hydrogen bond is only about 7 kcal/mole (29 kJ/mole); the bond energy of a covalent O—H bond is 111 kcal/mole (464 kJ/mole). Nevertheless, each water molecule forms hydrogen bonds to

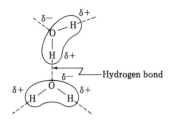

FIGURE 12-3

Hydrogen bonding between water molecules. The hydrogen atom of one water molecule is weakly attracted by the oxygen atom of another water molecule.

three or four other molecules, and these weak attractive forces greatly influence the properties of liquid water. Hydrogen bonding does not occur in most other liquids. The ability of water molecules to stick together through hydrogen bonding gives water special properties, as we will see.

**12.3
DENSITY OF
WATER AND
SPECIFIC
GRAVITY**

The density of water is about 1 g/mL (or 1 g/cm³). The density varies with temperature, however. For example, at 20°C the density of water is 0.998 g/mL. When the temperature is lowered, the density increases slightly as the volume contracts, until a maximum density of 1.000 g/mL is reached at 4°C. At this temperature, the water molecules are as tightly packed together as they can get.

If the temperature is lowered still more, something interesting happens: the density *decreases*. The water freezes and becomes a solid, called ice. The density of ice is only 0.917 g/mL. It is lower than the density of water at 4°C or even at 20°C. Water molecules in ice are arranged in a rigid, open structure shown in Figure 12-4 that takes up more space than the disorganized liquid form. Thus, when water freezes it expands and becomes less dense than liquid water. Filled water pipes can burst when the temperature falls below 0°C because the water in the pipes expands upon freezing. Similarly, cracks in rocks on the Earth's surface may widen when water freezes and expands within them.

Ice floats because it is less dense than liquid water. If water behaved like most liquids do, becoming denser as it froze, ice would sink in liquid water. The bottoms of rivers, lakes, and oceans would fill with ice that sank after it had formed at the surface during the winter. Spring and summer warmth would stop further ice formation, but probably would not melt the ice on the bottom. After many years, our natural bodies of water would be locked in ice and could contain no living things. But fortunately, water expands on freezing and ice floats. Therefore, fish and other water life can continue to exist undisturbed beneath the surface.

Water density is used as the basis for a relative scale of densities. The ratio of the density of a substance to that of water is known as the **specific gravity** of that substance.

$$\text{specific gravity} = \frac{\text{density of substance}}{\text{density of water}}$$

Because it gives the density of a substance relative to the density of water, specific gravity is also known as relative density. For precise work, the temperatures corresponding to the two density values are given, since density varies with temperature.

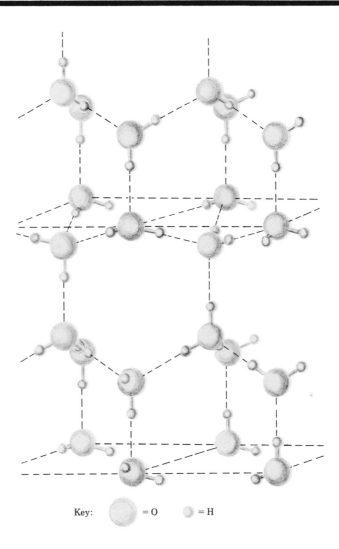

FIGURE 12-4

Ice. The water molecules in ice form a honeycomb-like structure through hydrogen bonding.

Key: ⬤ = O ○ = H

In general, the value used for water is 1.000 g/mL, its density at 4°C. Since dividing by one leaves a number unchanged, the specific gravity of a substance has about the same numerical value as the density of that substance (in g/mL or g/cm³). But specific gravity has no units because the units in the numerator and denominator of the ratio are the same and therefore cancel.

Specific gravity compares the mass of a substance to the mass of the same volume of water. Thus specific gravity can also be defined as:

$$\text{specific gravity} = \frac{\text{mass of substance}}{\text{mass of equal volume of water}}$$

Consider the mass of two cubic meters (2 m³) of gold, which is 38,600 kg. The mass of an equal volume of water is 2000 kg. The specific gravity of gold can be calculated from this information:

$$\text{specific gravity of gold} = \frac{\text{mass of gold}}{\text{mass of equal volume of water}} = \frac{38{,}600 \text{ kg}}{2000 \text{ kg}} = 19.3$$

The specific gravity can also be calculated if the density is known. For example, the density of mercury is 13.6 g/mL. Using 1.00 g/mL for the density of water, the specific gravity of mercury is:

$$\text{specific gravity of mercury} = \frac{\text{density of mercury}}{\text{density of water}} = \frac{13.6 \text{ g/mL}}{1.00 \text{ g/mL}} = 13.6$$

Notice again that the specific gravity of mercury has no units. The value is numerically equal to the density measured in g/mL.

We can measure specific gravity with a hydrometer. A hydrometer is a glass bulb with markings on its neck. The bulb is allowed to float in the liquid that is being measured. The operation of the hydrometer is based on Archimede's principle, which states that a floating object displaces a volume of liquid having a mass equal to the mass of the object. Since the hydrometer has a fixed mass, the volume of liquid

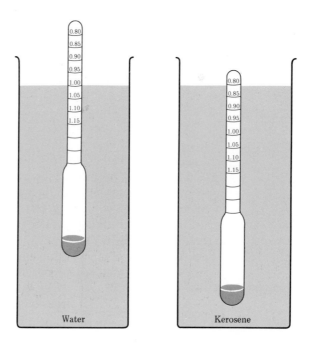

FIGURE 12-5

Hydrometers. The depths to which the weighted glass bulbs sink depend on the specific gravity of the liquids. Water has a specific gravity of 1.0, while kerosene has a specific gravity of about 0.80.

FIGURE 12-6

A urinometer. This hydrometer is used to measure the specific gravity of urine samples. (Photo by Al Green.)

displaced, and therefore how far the bulb sinks, depends on the specific gravity of the liquid.

Consider a hydrometer that has a mass of 20 g. It will have to displace 20 mL of water to float because water has a density of about 1 g/mL. To float in kerosene, which has a density of 0.8 g/mL, it would have to displace 25 mL of the liquid because 25 mL of kerosene has a mass of 20 g. Kerosene has a lower density, and lower specific gravity, than water. (See Figure 12-5.)

A urinometer is a special type of hydrometer used to measure the specific gravity of urine (see Figure 12-6). The normal range for the specific gravity of urine is 1.008 to 1.030.

**12.4
VAPORIZATION**

Because the molecules in a liquid are in constant motion, some may have enough kinetic energy to escape from the surface and become gas molecules. This process is called **vaporization**, or, when it takes place from an open container, **evaporation**. The gaseous form of a substance that normally exists as a liquid or solid is known as that substance's **vapor**. For example, when liquid water vaporizes, water vapor forms.

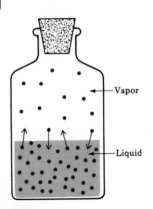

FIGURE 12-7

Vaporization. In a closed container, the rate at which molecules return to the liquid state becomes equal to the rate at which molecules leave the liquid state. Under these conditions, the pressure exerted by the gaseous form of the liquid is called the vapor pressure.

Vaporization in a closed container is shown in Figure 12-7. As the amount of vapor increases, some of the molecules that escaped from the liquid are recaptured when they hit the liquid surface. The chance of gas molecules rejoining the liquid increases as more and more molecules enter the vapor. After a while, the rate at which molecules escape from the liquid and enter the vapor becomes equal to the rate at which molecules are recaptured by the liquid. For example, if a million molecules escape from the liquid per second, another million molecules per second return. This *balance between opposing processes* is called **equilibrium**. When equilibrium is reached, there is no longer any observable change taking place in the system. At equilibrium, the amount of vapor and the amount of liquid are fixed, although they are not necessarily equal. Molecules continue to move between the liquid and the vapor, but in opposite directions at equal rates.

The pressure exerted by a vapor at equilibrium is called the **vapor pressure**. Some representative values of vapor pressure are listed in Table 12-1. The vapor pressure of a substance reflects the attraction between molecules in the liquid state. For example, a strong attraction makes it difficult for molecules to escape from the liquid, resulting in a low vapor pressure. Water has a low vapor pressure because hydrogen bonding between water molecules makes it difficult for them to leave the surface.

In a liquid, as in a gas, temperature is a measure of the kinetic energy of the molecules. The rate of vaporization therefore depends on the temperature. At higher temperatures, molecules have more energy and can escape more easily from the liquid to form vapor. Table 11-5 (in the preceding chapter) shows how the vapor pressure of water increases with temperature.

When you exercise heavily, your body temperature increases, causing you to perspire. Those water molecules in perspiration that have the greatest kinetic energy escape by evaporating from your skin. (As explained in the following sections, about 540 cal of heat are lost from the body for each gram of water that evaporates.) The molecules left

TABLE 12-1 *Vapor pressures of liquids*

| Liquid | Vapor pressure, 20°C | |
	(torr)	*(kPa)*
ether	436	58.1
acetone	177	23.6
chloroform	145	19.3
carbon tetrachloride	87	11.6
alcohol (ethyl)	43	5.7
water	18	2.4

behind on your skin as liquid water have less kinetic energy on the average than the molecules that escape. Since they have less kinetic energy, the temperature of this remaining water is lower. Therefore heat has been removed and your body feels cooler.

More rapid cooling of the body can be achieved through an alcohol sponge bath. Alcohol evaporates more readily than water because the attractive forces between its molecules are smaller. (Compare the vapor pressure of alcohol and water in Table 12-1.) Thus, alcohol evaporates faster and removes heat from the body more quickly than does perspiration. Even more rapid cooling is achieved by using a spray of ethyl chloride under pressure. The liquid droplets hit the skin and evaporate immediately. The cooling effect of ethyl chloride spray is used by dermatologists to freeze the skin. Dermatologists freeze warts and moles before removing them in order to prevent bleeding and to numb the pain.

Certain medications contain alcohol or other liquids that have relatively high vapor pressures. These preparations therefore should be stored in tightly closed containers. Otherwise, evaporation of the liquid will change the medicine's strength.

Box 12

Humidity

On days when the air feels damp, we complain that the humidity is high, or that the day is humid. Humidity refers to water vapor in the air. Under most conditions, the air in an average-sized room (10 ft by 10 ft with a 9 ft ceiling) contains about one cup of water in the form of vapor.

There is a limit to the amount of water vapor that air can hold. The maximum amount of water vapor that the air can hold decreases as the temperature falls. For this reason, water vapor from humid air condenses onto a glass containing an iced drink. When the maximum amount of vapor is reached in the atmosphere, clouds form, and then rain or snow may fall.

You probably feel uncomfortable when the humidity is high. Because the air on a humid day already contains many water molecules, perspiration collects on your skin instead of evaporating. When humidity is low, on the other hand, your skin becomes dry because perspiration evaporates readily.

Humidity can be measured in several different ways. One method makes use of the color change between an anhydrous compound and its hydrated form. Cobalt(II) chloride is blue when anhydrous and red when hydrated as $CoCl_2 \cdot 6H_2O$. A humidity indicator can be made from a cloth or paper that is soaked in water containing cobalt(II) chloride and then allowed to dry. The cloth or paper will appear bluish when the humidity is low and reddish when it is high.

The **heat capacity** of a substance relates the change in temperature of that substance to the amount of heat it absorbs.

$$\text{heat capacity} = \frac{\text{heat absorbed}}{\text{temperature change}}$$

A liquid with a high heat capacity requires a relatively large amount of energy to cause an increase in its temperature. Of all the common liquids, water has the highest heat capacity. To raise the temperature of one mole of water by one degree Celsius (or one kelvin), the amount of energy required is 18.0 cal (or 75.3 J). The molar heat capacity of water is therefore 18.0 cal/(mole)(°C) or 75.3 J/(mole)(°C). These units are read as calories (or joules) per mole per degree Celsius.

Heat capacity is also expressed in terms of the amount of heat needed to raise the temperature of *one gram* of the substance by one degree Celsius. The heat capacity per gram of substance is called the **specific heat** (or specific heat capacity). Because the calorie is defined as the amount of energy needed to raise the temperature of one gram of water by one degree Celsius (from 14.5°C to 15.5°C), the specific heat capacity of water is simply 1.00 cal/(g)(°C) or 4.18 J/(g)(°C). These units are read at calories (or joules) per gram per degree Celsius.

The following equation shows the relationship between heat and temperature change in mathematical form:

$$q = cm\Delta T$$

Here, q is the amount of energy (in calories or joules), c is the specific heat, m is the mass (in grams), and ΔT is the temperature change (in degrees Celsius or kelvins). Thus, the amount of energy needed to raise the temperature of 1.00 gram of water by 1.00°C is:

$$q = 1.00 \; \frac{\text{cal}}{(g)(°C)} \times 1.00 \, g \times 1.00 °C$$

$$= 1.00 \text{ cal}$$

Example A beaker containing 200. g of water is heated. The water temperature rises from 20.°C to 80.°C. How many calories were required?

UNKNOWN	GIVEN	CONNECTION
q	$m = 200.$ g $\Delta T = 80.°C - 20.°C = 60.°C$ $c = 1.00$ cal/(g)(°C)	$q = cm\Delta T$

$$q = cm\Delta T$$

$$q = 1.00\ \frac{cal}{(g)(°C)} \times 200.\ g \times 60°C$$

ESTIMATE $\quad 1\ \dfrac{cal}{(g)(°C)} \times 2 \times 10^2\ g \times 6 \times 10^1\ °C = 1 \times 10^4\ cal$

$\qquad\qquad = 10$ kcal

CALCULATION $\quad 1.2 \times 10^4$ cal $= 12$ kcal

Water has a higher specific heat than most other substances. In fact, the specific heat of water is twice as great as that of many liquids and over ten times as great as that of many solids. For example, some typical values of specific heat are:

alcohol	0.59 cal/(g)(°C)	2.5 J/(g)(°C)
glass	0.09 cal/(g)(°C)	0.4 J/(g)(°C)
gold	0.03 cal/(g)(°C)	0.1 J/(g)(°C)

The high specific heat of water helps to maintain your body temperature. Because relatively large amounts of heat energy are needed to change the temperature of water, your body (which is about 60% water) maintains a fairly constant temperature. For the same reason, the huge quantity of water on Earth prevents large temperature variations between night and day. Water is used in a hot water bottle as a heat pack for a similar reason—it changes temperature relatively slowly.

**12.6
CHANGES
OF STATE**

Water exists in all three physical states: solid, liquid, and gas. Water is a liquid at room temperature and normal atmospheric pressure. At one atmosphere and temperatures less than 0°C (273 K), water is a solid (ice). At one atmosphere and temperatures above 100°C (373 K), it is a gas (water vapor or steam). These three forms of water are chemically the same; only the arrangement of the water molecules is different. Adding or removing energy as heat converts water from one state to another.

As liquid water is heated, its temperature rises and its vapor pressure increases. If the atmospheric pressure is 1 atm (760 torr or 101.3 kPa), the vapor pressure becomes equal to the pressure of the atmosphere at 100°C. Boiling, which is the formation of vapor bubbles throughout the liquid, takes place at this temperature, and 100°C is called the **boiling point.** As water boils, some of the resulting vapor may be cooled enough by the surrounding air to condense into tiny droplets of liquid water. The presence of these droplets makes steam visible to us.

If the atmospheric pressure is less than one atmosphere, the boiling point decreases. It decreases because the vapor pressure becomes equal to the lower atmospheric pressure at a lower temperature. The molecules require less energy, and less heat is needed. Thus, in Denver, the "mile-high city," where the atmospheric pressure is only 630 torr (84 kPa), water boils at about 95°C (see Table 11-5). Cooking times are therefore somewhat longer in Denver than in most other parts of the United States.

By increasing the external pressure, the boiling point of water can be raised. This process takes place in a pressure cooker, which is a sealed container of fixed volume. If a pressure cooker containing water is heated, water molecules that escape from the liquid are trapped in the container. Thus, pressure above the liquid builds up. When the vapor pressure increases to 1000 torr (about 5 psi greater than normal atmospheric pressure), water boils at 108°C.

When we supply energy continuously to a quantity of water, its temperature rises until the boiling point is reached. At the boiling point, the temperature stops rising even though the energy that is supplied is still being absorbed. Instead of raising the temperature of the liquid, the absorbed energy is now used to separate the water molecules, thus converting the liquid to a gas. The energy needed to convert a liquid *at its boiling point* to a gas is called the **heat of vaporization.** For water, the heat of vaporization is 9.72 kcal/mole or 40.7 kJ/mole. The heat of vaporization may also be expressed as 540. cal/g or 2.259 kJ/g. Thus, even after water has reached 100°C, an additional 540 calories are required to convert each gram of water to steam. This value is high compared to that of other liquids because the hydrogen bonding between water molecules makes them difficult to separate. In contrast, alcohol (ethanol) has a heat of vaporization of 204 cal/g, and chloroform has a heat of vaporization of 59 cal/g.

Consider the amount of energy needed to boil away a cup of water. A cup of water at 20°C has a mass of 240. g. To raise its temperature by 80°C to 100°C, the number of calories needed can be calculated using the specific heat of water.

$$q_1 = cm\Delta T$$

$$q_1 = 1.00 \frac{cal}{(g)(°C)} \times 240. g \times 80. °C$$

$$q_1 = 1.9 \times 10^4 \, cal = 19 \, kcal$$

Now to convert the water completely from the liquid state at 100°C to vapor at 100°C, the additional number of calories needed is found from the heat of vaporization. If 540 calories are needed to convert one gram, then

$$q_2 = 540. \frac{cal}{g} \times 240. g$$

$$q_2 = 1.30 \times 10^5 \text{ cal} = 130. \text{ kcal}$$

The total energy required is the sum of the number of calories needed for each process: $q_1 + q_2 = 19$ kcal $+ 130$ kcal $= 149$ kcal. As you can see, most of the needed energy is used for the vaporization of the water.

Because of the law of conservation of energy, the same amount of heat, 540. cal/g or 2.26 kJ/g must be released when vapor condenses back to water at 100°C. The **heat of condensation** *is equal and opposite to the heat of vaporization.*

Vaporization is endothermic, absorbing heat, while condensation is exothermic, releasing heat. The large value for the heat of condensation explains why you can be badly burned by hot water vapor or steam. Of course, steam is less dense than boiling water so that the amount you might contact in an accident would be less. But each gram of steam causes a more serious burn because of the extra thermal energy it contains.

If heat is supplied to a solid like ice, the temperature of the solid will increase until it reaches the **melting point**. At this point, the temperature stops rising and the solid gradually becomes a liquid. The melting point for ice at 1 atm is 0°C (273 K). The amount of energy needed to melt the solid is called the **heat of fusion** (fusion means melting). For ice, the heat of fusion is 1.44 kcal/mole (6.01 kJ/mole), or 80. cal/g (333 J/g). Again, the value for water is unusually high because of the strong attractive forces between water molecules. An ice pack is therefore effective in removing heat from its surroundings. Each gram of ice absorbs 80. cal as it melts, cooling the part of the body it touches.

Consider what happens when you take 100 g (about 4 oz) of ice at 0°C in a glass and add 240 g of water (about 10 oz) at 30°C. The ice absorbs heat and melts, cooling the water. After 90. g of ice has melted, $q = 80.$ cal/g $\times 90.$ g $= 7.2 \times 10^3$ cal, or 7200 cal have been absorbed. This absorption of heat reduces the temperature of the warm water.

$$7.2 \times 10^3 \text{ cal} = 1.00 \frac{cal}{(g)(°C)} \times 240. g \times \Delta T$$

$$\frac{7.2 \times 10^3 \text{ cal}}{1.00 \frac{cal}{(g)(°C)} \times 240. g} = \Delta T$$

$$30°C = \Delta T$$

ASIDE . . .

The human body is composed of 60% to 70% water by weight.

The temperature is around 0°C, and the drink is now cold. Moreover, 10 g of ice still remain. This remaining ice will slowly melt as the glass of water absorbs heat from its surroundings, and in doing so it will continue to keep the water cold.

FIGURE 12-8

Conversions between the three states of water. Heat is required for fusion and vaporization. Heat is released by condensation and crystallization.

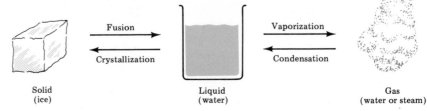

Solid
(ice)

Fusion

Crystallization

Liquid
(water)

Vaporization

Condensation

Gas
(water or steam)

Freezing, the conversion of a liquid to a solid, must release the same amount of energy that is absorbed in the opposite process, melting. This energy is called the **heat of crystallization**. For water, the energy released when liquid water forms ice must be 1.44 kcal/mole or 6.01 kJ/mole, or 80. cal/g or 333 J/g. The term **freezing point** is often used instead of melting point when we talk about this reverse process, even though the values of the freezing point and the melting point are identical.

Figure 12-8 summarizes the conversions among the three states of water. These conversions also apply to most other substances. A few substances, like iodine, can be converted directly from the solid to a vapor without passing through the liquid state. This process is called **sublimation**. Energy is needed for the endothermic processes of fusion and vaporization. The reverse processes, crystallization and condensation, are exothermic and release energy.

Figure 12-9 is a phase diagram for water. It shows how the three states, or phases, of water depend on temperature and pressure. We

FIGURE 12-9

Phase diagram for water. This diagram indicates the conditions of pressure and temperature under which each state of water exists. For example, follow the horizontal line at 760 torr (1 atm) from left to right. At 0°C and 1 atm, solid ice changes to liquid water. At 100°C and 1 atm, liquid water changes to water vapor or steam.

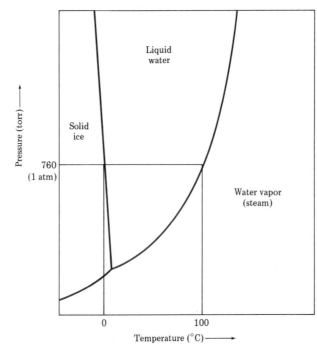

have been concerned primarily with the changes of state at 1 atm. Note on this diagram that at 1 atm (760 torr), water exists as solid ice below 0°C, as liquid water between 0°C and 100°C, and as water vapor (steam) above 100°C.

12.7
COHESION
AND ADHESION

Attraction between water molecules through hydrogen bonding results in a property called **cohesion**. Cohesion is the tendency of parts of a substance to cling together. In a water droplet, for example, the molecules pull toward each other. In doing so, they decrease the amount of exposed surface of the liquid.

The molecules at the surface of a liquid are attracted by the many molecules below the surface. However, there are few, if any, molecules attracting them from above. Therefore, surface molecules are pulled inward. This inward attraction creates **surface tension**. The water surface in a flask or jar, for example, is like a stretched piece of rubber. Although a razor blade or needle will normally sink in water, we can actually place one on top of the water if we are careful not to disturb

Do-it-yourself 12

On the surface

Surface tension is the tendency of a liquid to reduce its exposed surface to the smallest possible area. This tendency results from cohesion, the attractive forces between the molecules of the liquid. Water has a high surface tension because of the hydrogen bonds that form between adjacent water molecules.

Detergents reduce the surface tension of water by breaking up the hydrogen bonds. Water molecules are attracted to detergent molecules instead of to each other. As a result, water becomes "wetter," or better able to attach to other substances and make them wet. This property improves the cleaning action of detergents.

You can use a razor blade to demonstrate water's high surface tension. A razor blade is made of steel, which is much denser than water. You therefore would expect a razor blade to sink when placed in water. However, due to surface tension, the razor blade can be made to float.

To do: Carefully place a razor blade flat onto the surface of water in a bowl. An easy way to do this is to first put the blade on a small piece of paper towel. Then place the towel and blade onto the water. The absorbent towel sinks, but the razor blade floats! It is supported by the surface tension of water.

Now add some liquid detergent to the water. What happens to the razor blade?

Take a sheet of notebook paper and write a word on it using a piece of hard soap. Dip the paper completely under water and lift it out. Can you explain what happens?

the surface. Insects called water striders, or water skaters, are so light that they can move over the surface of calm water without sinking. The surface tension of water is enough to support them.

The attraction of one substance for *other* substances is called **adhesion**. Capillary action, the rising of water in a small-diameter (capillary) tube, is based on adhesion, because water molecules are attracted to the inside surface of the tube. Cotton, wicks, and paper towels absorb water by capillary action. Water rises in narrow channels in these materials. Similarly, capillary action causes water to be distributed in soil, providing nutrients to plants.

If we examine a glass tube that contains water, we notice that the outer edges of the water surface are higher than the middle. Water adheres to the inner surface of the glass tube. The water surface forms a curve or **meniscus** (pronounced meh-niss' kus), shown in Figure 12-10(a). The meniscus becomes more noticeable in tubes with smaller diameters. When measuring the level of water in a tube, we must always read the value at the lowest point, the bottom of the meniscus, to avoid the distortion caused by adhesion.

A water meniscus is said to be concave because it is lowest in the center. However, as shown in Figure 12-10(b), mercury forms a meniscus that is highest at the center. Its meniscus is said to be convex. Mercury forms a convex meniscus because mercury atoms have a greater attraction for each other than for the glass surface of the tube. The surface tension of mercury is even greater than that of water. Thus in mercury, the force of cohesion is stronger than that of adhesion. When measuring the height of a mercury column, as in a Torricelli barometer, we must take our reading at the highest point of the meniscus.

FIGURE 12-10

A water meniscus (a) and a mercury meniscus (b). The water meniscus is concave, while the mercury meniscus is convex. Thus, the amount of water is read at the bottom of the meniscus, and the amount of mercury is read at the top of the meniscus. (Photos by Al Green.)

(a)   (b)

A liquid is said to "wet" a solid when the attraction between the molecules of the liquid and the solid is greater than the internal cohesive forces in the liquid. Certain substances known as "wetting agents" or surfactants (surface-active-agents) reduce the cohesive forces between molecules. This reduction of cohesive forces lowers the surface tension and increases the ability of the liquid to wet a solid surface by adhesion. Surfactants are commonly used in laundry products because they make it easier for water to penetrate fabrics.

**12.8
WATER OF
HYDRATION**

Hydration is the process by which water molecules react as single units with certain substances. In other words, the oxygen-hydrogen bonds of H_2O molecules are not broken. These reactions produce compounds called **hydrates,** which contain water in definite proportions. The water in a hydrate is called the **water of hydration**.

Sometimes several different hydrates exist for the same substance. Sodium sulfate, for example, forms hydrates containing one molecule of water, $Na_2SO_4 \cdot H_2O$; hydrates containing seven molecules of water, $Na_2SO_4 \cdot 7H_2O$; and hydrates containing ten molecules of water, $Na_2SO_4 \cdot 10H_2O$. Note that the formula of a hydrate is written with a raised period followed by the number of attached water molecules. The name of a hydrate is written in either of the following two ways:

$Na_2SO_4 \cdot H_2O$

sodium sulfate monohydrate

sodium sulfate 1-water

The term **anhydrous** refers to the original substance with no water of hydration. For example, anhydrous sodium sulfate is Na_2SO_4. Other examples of the names and formulas of hydrates are given in Table 12-2.

The water molecules in hydrates are weakly held. The removal of water, called **dehydration,** can therefore be accomplished simply by heating. Certain compounds lose their water of hydration merely on

TABLE 12·2 *Examples of hydrates*

Formula	*Name*
$CaSO_4 \cdot 2H_2O$	calcium sulfate dihydrate (calcium sulfate 2-water)
$CuSO_4 \cdot 5H_2O$	copper sulfate pentahydrate (copper sulfate 5-water)
$FeSO_4 \cdot 7H_2O$	iron(II) sulfate heptahydrate (iron(II) sulfate 7-water)
$Na_2CO_3 \cdot 10H_2O$	sodium carbonate decahydrate (sodium carbonate 10-water)

exposure to air. This process is called **efflorescence**. For example, sodium sulfate decahydrate ($Na_2SO_4 \cdot 10H_2O$) is an efflorescent compound. It has a significant vapor pressure (about 10 torr) at average room temperature and humidity.

Hygroscopic compounds are compounds that pick up water vapor from the air. If enough water is absorbed so that the compound actually dissolves, the process is called **deliquescence**. Certain anhydrous compounds that are hygroscopic, such as $CaCl_2$, $MgSO_4$, and Na_2SO_4, are used as drying agents. As these compounds form their respective hydrates, water is removed from the surroundings.

Plaster of paris, which is used in making plaster casts, is a hydrate with well-known properties. It is a form of calcium sulfate that can be represented by the formula $CaSO_4 \cdot \frac{1}{2}H_2O$ or $(CaSO_4)_2 \cdot H_2O$. This compound contains one water molecule associated with every two formula units of calcium sulfate. When mixed with water, additional hydration occurs, forming the very hard, white substance called gypsum, $CaSO_4 \cdot 2H_2O$.

$$(CaSO_4)_2 \cdot H_2O(s) + 3H_2O(l) \rightarrow 2CaSO_4 \cdot 2H_2O(s)$$
plaster of paris   water   gypsum

The plaster of paris itself is produced by the reverse process, the dehydration of natural gypsum.

12.9 WATER POLLUTION

Only about 0.003% of the total world water supply is available for drinking. Over 97% of the world's water is saltwater in the oceans, and most of the fresh water is trapped as polar ice and glaciers. The water supply is constantly redistributing itself, evaporating into the atmosphere and then returning as rain and snow. This movement of water is called the **water cycle**.

Natural water is not pure, but contains dissolved gases and minerals, which provide its taste. These impurities are generally harmless. However, additional impurities that are harmful sometimes enter the water. These impurities, which prevent the normal uses of water, are called **water pollutants**.

Certain trace elements in water, such as iron, copper, iodine, and zinc, are essential nutrients. Others, particularly the ''heavy'' (high density) metals, are harmful. Mercury, used in making electrical equipment and in the chemical industry, may be dumped or may leak into a body of water. The element mercury and its compounds can affect the human nervous system, causing irritability, paralysis, blindness, or insanity. (At one time, hat makers used mercury compounds in their work and suffered from mercury poisoning. The character of the Mad Hatter in ''Alice in Wonderland'' may be based on that unfortunate practice.) Mercury that is discharged by an industrial plant into a stream may eventually become concentrated in fish. Birds or people who then eat these contaminated fish will suffer the ill effects (see Figure 12-11).

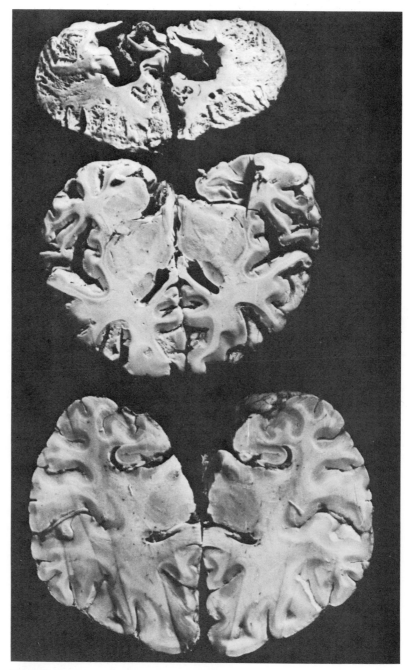

FIGURE 12-11

Mercury poisoning. Mercury damages the central nervous system and is therefore a dangerous water pollutant. Shown here (from top to bottom) are brain sections from a 7-year-old boy who died after 4 years of mercury poisoning; an 8-year-old girl who died after 2 years and 9 months of mercury poisoning; a 30-year-old man who did not suffer mercury poisoning, and had a normal brain and nervous system. (Copyright © 1975 by Aileen Mioko Smith from Minamatta. Photo by W. Eugene Smith and Aileen Mioko Smith.)

Lead, used for making automobile batteries, is another toxic metal which sometimes leaks or is dumped into water. It can cause paralysis and brain damage if ingested. Lead chromate was once used in indoor paints, but this use is now prohibited. (Old buildings with lead chromate paint still pose health hazards to young children, who may eat chips of peeling paint.) Cadmium, used in metal plating, can cause high blood pressure, and kidney and red blood cell damage. In Japan, the contamination of water with cadmium was responsible for "itai itai" ("ouch ouch") disease, which involves painful bone fractures.

A large number of organic compounds are potential water pollutants. Pesticides, which are used to destroy insects, weeds, molds, or worms, may find their way into the water supply. The most abundant group of pesticides in the environment are the chlorinated hydrocarbons. (These compounds consist of molecules that contain carbon, hydrogen, and chlorine atoms.) They can be harmful to fish, birds, and possibly humans. An example of a chlorinated hydrocarbon is DDT (*dichlorodiphenyltrichloroethane*). The use of DDT is now restricted. A related group of compounds are known as PCBs, or polychlorinated biphenyls. These were once used in electrical equipment as insulating fluid. PCBs seem to have harmful effects similar to those of DDT.

Synthetic detergents consist of organic molecules that contain a polar and a nonpolar end. As illustrated in Figure 12-12, they are good cleaning agents because the nonpolar end attracts oily or greasy dirt while the polar end dissolves in water. To prevent the buildup of detergent foam in water supplies, detergent molecules must be biodegradable. A biodegradable substance is one that can be decomposed by naturally occuring microorganisms. Detergents usually contain additional substances called "builders." Builders remove minerals that prevent sudsing. The use of phosphates as builders may promote eutrophication, as we discussed in Box 10.

Oil, which contains a mixture of hydrocarbons, is discharged into

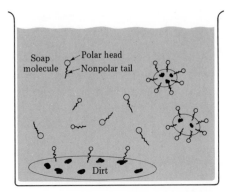

FIGURE 12-12

How soap works. The nonpolar tail of the soap molecule dissolves in greasy dirt, which is also nonpolar. The polar head of the molecule dissolves in the water and pulls the dirt off the surface being cleaned.

water during normal ship and tanker operations, as well as through accidental oil spills. It directly damages water birds and plant life by sticking to them. Oil also harms aquatic life indirectly by reducing the amount of light and oxygen that enters the water. Long-term effects of oil spills may result from the absorption by organisms of the organic components of oil.

Plants and animals that live in water need dissolved oxygen. Certain pollutants, called oxygen-demanding wastes, reduce the amount of dissolved oxygen to levels that can no longer sustain life. These pollutants support microorganisms in the water that use up oxygen while they decompose waste. Some examples of oxygen-demanding waste are sewage, farm runoffs, and wastes from food processing and the paper industry.

Heat, or thermal pollution, speeds up the disappearance of dissolved oxygen because at higher temperatures, less oxygen can dissolve in water. Also, because heat increases the metabolism of organisms in the water, these organisms use up the remaining oxygen at a faster rate. Aquatic life cycles can also be upset by thermal pollution. Eggs may hatch prematurely, for example. Thermal pollution can be caused by the discharge of hot water by industrial and power plants.

Radioactive pollutants are substances that spontaneously emit radiation (by processes described in Chapter Seventeen). A common source of radioactive pollution is uranium "tailings," the leftovers from uranium mining operations.

Pathogens, which are disease-causing organisms, are another type of water pollutant. They enter the water supply through the discharge of contaminated feces or urine. This contamination occurs most commonly in places with poor sanitary facilities. Intestinal tract infections such as typhoid, dysentery, or cholera may result from water-carried pathogens.

**12.10
WATER
TREATMENT**

To prevent pollutants from being released into the water supply, sewage and industrial wastewater must undergo treatment, or purification. **Primary treatment**, illustrated in Figure 12-13(a), removes about one-half of the suspended solids in the water and reduces the oxygen demand by approximately one-third. In the first step of primary treatment large objects, such as wood and other floating substances, are screened out. Then, in a settling process, solids known as grit (sand, gravel, etc.) are removed by gravity. Finally, smaller suspended solids settle out in a sedimentation tank, forming a solid mass called sludge at the bottom of the tank.

Sedimentation can be speeded up by coagulation, the formation of larger particles that bring down the suspended impurities. Aluminum

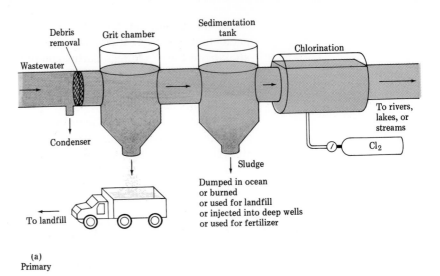

(a)
Primary

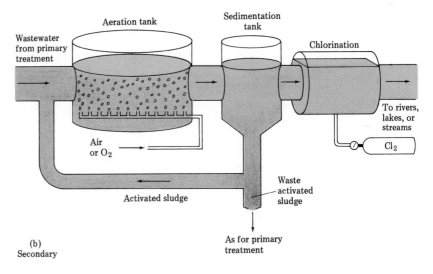

FIGURE 12-13

Water purification. In primary treatment (a), solids are removed and chemicals like chlorine, Cl$_2$, may be added. In secondary treatment (b), most oxygen-demanding wastes are removed.

(b)
Secondary

hydroxide, a gelatinous compound, is often used to trap the smaller particles. It is formed from aluminum sulfate:

$$\text{Al}_2(\text{SO}_4)_3(s) + 3\text{Na}_2\text{CO}_3(s) + 3\text{H}_2\text{O}(l)$$

aluminum   sodium   water
sulfate   carbonate

$$\rightarrow 2\text{Al}(\text{OH})_3(s) + 3\text{Na}_2\text{SO}_4(aq) + 3\text{CO}_2(g)$$

aluminum   sodium   carbon
hydroxide   sulfate   dioxide

Filtration may follow coagulation, thus removing the precipated solids.

If the water will not be purified further, it is chlorinated with Cl_2 to destroy disease-causing bacteria, and then discharged.

Secondary treatment further removes suspended solids and decreases the oxygen demand by up to 90%. The waste is exposed to bacteria that decompose it in the presence of oxygen. One method of secondary treatment, shown in Figure 12-13(b), involves mixing the waste in a tank with activated sludge, the sediment from previous batches of waste that has developed a large population of microorganisms. In a process called aeration, air or oxygen is pumped through this mixture. Another method of secondary treatment uses a "trickling filter," which contains several feet of small stones. Microorganisms collect on the stones and decompose the waste as its passes through.

Tertiary treatment, or advanced wastewater treatment, is sometimes used to remove specific pollutants that remain after secondary treatment. Usually, these pollutants are dissolved inorganic or organic compounds. Organic molecules can be removed by adsorption onto the surface of activated charcoal. Organic pollutants can also be eliminated by oxidation with ozone, O_3, or hydrogen peroxide, H_2O_2. The removal of inorganic compounds depends on their composition. Phosphates, for example, can be precipitated as $AlPO_4$ by adding aluminum sulfate (as in the coagulation reaction).

12.11
HARD WATER

Hard water is water that contains calcium ions (Ca^{2+}) and magnesium ions (Mg^{2+}) in relatively large amounts. (Soap cannot form suds in hard water because the soap molecules precipitate in the presence of Ca^{2+} or Mg^{2+} ions.) When the accompanying anion is hydrogencarbonate, HCO_3^-, we say that the hardness is temporary. Temporary hardness can be removed by boiling, which causes calcium and magnesium carbonate to precipitate.

$$Ca^{2+}(aq) + 2HCO_3^-(aq) \rightarrow CaCO_3(s) + CO_2(g) + H_2O(l)$$

calcium ion   hydrogen-carbonate ion   calcium carbonate   carbon dioxide   water

These insoluble carbonate deposits, called boiler scale, are often found in pipes (see Figure 12-14), boilers, and tea kettles.

When hard water contains sulfate (SO_4^{2-}), or other anions that do not precipitate the Ca^{2+} or Mg^{2+} ions upon boiling, we say that the hardness is "permanent."

As part of tertiary wastewater treatment, "permanent" hardness can be removed by the lime-soda process. Calcium hydroxide (hydrated lime) and sodium carbonate (soda) cause the formation of calcium carbonate and magnesium hydroxide precipitates. The water left behind is soft water, which no longer contains Ca^{2+} and Mg^{2+} ions.

FIGURE 12-14

Hard water scale. This pipe is clogged by deposits of insoluble calcium and magnesium compounds. (Photo courtesy of Culligan Water Institute.)

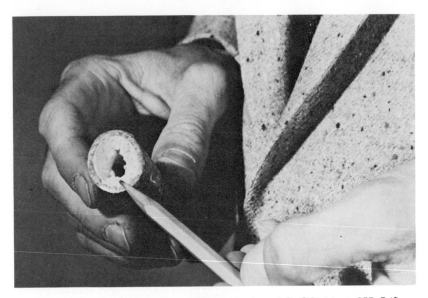

$$Ca^{2+}(aq) + Ca(OH)_2(s) + 2HCO_3^-(aq) \rightarrow 2CaCO_3(s) + 2H_2O(l)$$

calcium   calcium   hydrogen-   calcium   water
ion   hydroxide   carbonate   carbonate
                           ion

$$Mg^{2+}(aq) + Ca(OH)_2(s) + Na_2CO_3(s)$$

magnesium   calcium   sodium
ion   hydroxide   carbonate

$$\rightarrow Mg(OH)_2(s) + CaCO_3(s) + 2Na^+(aq)$$

magnesium   calcium   sodium
hydroxide   carbonate   ion

Another method for softening water is to pass it through an ion exchanger. An ion exchanger is a substance that can replace ions in water

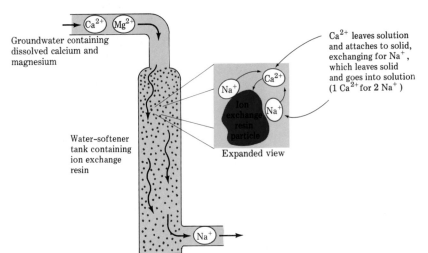

Groundwater containing dissolved calcium and magnesium

Ca^{2+} leaves solution and attaches to solid, exchanging for Na^+, which leaves solid and goes into solution (1 Ca^{2+} for 2 Na^+)

Water–softener tank containing ion exchange resin

Ion exchange resin particle

Expanded view

FIGURE 12-15

How a water softener works. Water softeners are ion exchangers. They replace the Ca^{2+} and the Mg^{2+} ions in the water with Na^+ ions.

with different ions initially present on the exchanger, without itself dissolving. (See Figure 12-15.) When hard water passes through a cation exchanger, for example, the exchanger picks up the Mg^{2+} and Ca^{2+} ions, and releases Na^+ ions into the water in their place. Home water softeners are ion exchangers. Therefore, people who must restrict their intake of Na^+ ions, (heart patients for example), should not drink softened water since it may contain significant amounts of Na^+ ions.

In the laboratory, water can be purified by passing it through both a cation and an anion exchanger to produce de-ionized water. Another method of removing ions and dissolved substances is distillation, shown in Figure 12-16. Water is boiled and the vapor is condensed by running cold water around a tube through which the vapor passes. Nonvolatile impurities (those that do not vaporize readily) are left behind in the original flask.

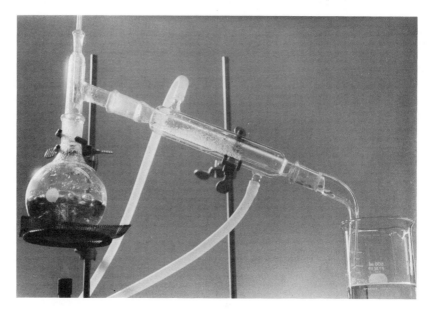

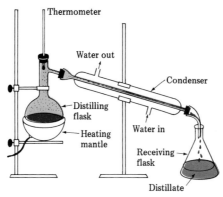

FIGURE 12-16

Distillation. A liquid is vaporized by heating and is then condensed by cooling. Impurities are left behind in the distilling flask. (Photo by Al Green.)

Summary

12.1 In a liquid, the molecules are much closer together then they are in a gas. The molecules are in motion but are restricted by forces of attraction. In a solid, the atoms, ions, or molecules are even closer together and cannot move around. Instead, they vibrate about fixed positions.

12.2 The water molecule is polar. The oxygen end has a partial negative charge, and the hydrogen end has a partial positive charge. Water molecules are attracted to each other through hydrogen bonds.

12.3 The density of water is about 1 g/mL. Specific gravity is a relative scale of densities for solids and liquids that is based on the ratio of the density of a substance to the density of water.

12.4 Vaporization is the escape of molecules from the surface of a liquid. The pressure exerted by a vapor at equilibrium is called the vapor pressure. Water has a low vapor pressure as a result of hydrogen bonding between its molecules.

12.5 The heat capacity of a substance relates the change in temperature of the substance to the amount of heat that it absorbs. Water has a high heat capacity; a large amount of energy is needed to raise its temperature.

12.6 Conversions among the three states of water involve energy changes. Energy is needed for the endothermic processes of fusion and vaporization. Energy is released by the exothermic reverse processes, crystallization and condensation.

12.7 Water tends to form drops because of cohesion, the attraction between its molecules. Water also exhibits adhesion, the attraction of molecules for other substances, such as the inner wall of a glass tube.

12.8 Hydration is a process in which water molecules react as single units. Compounds called hydrates form. Hydrates contain water in definite proportions. The water in a hydrate is called the water of hydration.

12.9 Water pollutants are impurities that prevent the normal use of water. They include: heavy metals, pesticides, detergents, oil, oxygen-demanding wastes, heat, and pathogens.

12.10 Water can be treated to remove pollutants. Primary treatment removes suspended solids, and secondary treatment removes oxygen-demanding wastes. Tertiary treatment may be used to remove specific pollutants that remain after secondary treatment.

12.11 Water is said to be hard when it contains Ca^{2+} and Mg^{2+} ions in large amounts. Soap cannot form suds in hard water because the calcium and magnesium ions cause soap molecules to precipitate. Hard water can be softened by passing it through an ion exchanger.

Exercises

KEY-WORD MATCHING EXERCISE

For each term on the left, choose a phrase from the right-hand column that most closely matches its meaning.

1 hydrogen bonding		**a**	picks up water vapor
2 specific gravity		**b**	balance of opposing processes
3 vaporization		**c**	heat capacity per gram
4 equilibrium		**d**	inward pull at surface
5 heat capacity		**e**	related to density
6 specific heat		**f**	relates heat and temperature
7 cohesion		**g**	attraction to another substance
8 adhesion		**h**	dissolves in water vapor
9 surface tension		**i**	attraction between H_2O molecules
10 meniscus		**j**	reaction of water molecules as single units
11 hydration		**k**	escape of molecules from surface
12 efflorescence		**l**	molecular attraction
13 deliquescence		**m**	curved surface

EXERCISES BY TOPIC

Kinetic theory of liquids and solids (12.1)

14 How do a solid and a liquid differ from a gas at the molecular level?

15 Why does diffusion take place more slowly in a liquid than in a gas?

16 Why is the liquid state ideal for carrying out chemical reactions?

17 Why do solids have a definite shape, but liquids do not?

Structure of water (12.2)

18 Describe the bonding in a water molecule.

19 Explain why water is a polar molecule.

20 What is hydrogen bonding?

21 How does the strength of a hydrogen bond compare with that of a covalent bond?

Density of water and specific gravity (12.3)

22 Describe how the density of water varies with its temperature.

23 Why does ice float?

24 How is specific gravity related to density?

25 Alcohol has a density of 0.789 g/cm³ at 20°C. What is its specific gravity?

Vaporization (12.4)

26 Describe **a** vaporization   **b** evaporation

27 How does vaporization in a closed container reach a state of equilibrium?

28 Explain the effect of temperature on vaporization.

29 Why does sweating cause your body to cool?

Heat capacity and specific heat (12.5)

30 What is the difference between heat capacity and specific heat?

31 Why does water have a high specific heat?

32 A certain metal requires 45 cal to increase the temperature of 51 g of the metal by 10.°C. Find its specific heat.

33 How many calories are needed to raise the temperature of 1.00 kg of water by 100.°C?

34 How many kilojoules are needed to heat 425 g of water from 15.1°C to 79.2°C?

Changes of state (12.6)

35 Why is boiling related to pressure?

36 Why does water have a high heat of vaporization?

37 Why must a substance's heat of condensation be equal and opposite to its heat of vaporization?

38 Why is ice used in cold packs?

39 Calculate the number of calories needed to **a** boil away 1.00 kg of water at its boiling point   **b** melt 1.00 kg of ice at its melting point

40 Find the total amount of energy needed to boil away completely 350. g of water that is initially at 18°C.

Cohesion and adhesion (12.7)

41 What is the difference between cohesion and adhesion?

42 What causes surface tension?

43 Describe how to read **a** a mercury meniscus   **b** a water meniscus

44 What does a surfactant do?

Water of hydration (12.8)

45 Define **a** hydration   **b** water of hydration

46 Compare deliquescence and efflorescence.

47 Why are hygroscopic substances good drying agents?

48 Why is plaster of paris used to make casts?

49 Name the following compounds: **a** $NaS \cdot 9H_2O$ **b** $Cd(C_2H_3O_2)_2 \cdot 2H_2O$ **c** $K_3PO_4 \cdot H_2O$

50 Write the formula for: **a** nickel(II) nitrate hexahydrate **b** anhydrous copper(II) sulfate **c** magnesium sulfate 7-water

Water pollution (12.9)

51 Give examples of five different types of water pollutants.

52 Describe the sources and effects of the pollutants given in Exercise 51.

53 What is the importance of dissolved oxygen? What are oxygen-demanding wastes?

Water treatment (12.10)

54 What takes place during primary treatment of waste water?

55 Describe secondary treatment.

56 What is tertiary wastewater treatment?

Hard water (12.11)

57 What makes water hard?

58 How can **a** permanent **b** temporary hardness be removed?

59 How does an ion exchanger work?

60 How does distillation produce purified water?

ADDITIONAL EXERCISES

61 Describe the properties of water that make it different from most other liquids.

62 Does specific gravity change with temperature?

63 Why do different liquids have different vapor pressures?

64 Why do you feel cool when you step out of a swimming pool?

65 Why are cooking times longer at high altitudes?

66 Describe the basis for capillary action.

67 In the equilibrium between molecules leaving a liquid to form vapor and molecules returning to the liquid, what changes and what stays the same?

68 How much energy in calories is needed to raise the temperature of 25 g of each substance by 40.°C? **a** water **b** alcohol **c** gold

69 How many kilojoules are needed to melt 396 g of ice at its melting point?

70 How many calories are released when 100. moles of steam condense and then cool to 30.°C?

71 Name the following: **a** $Ba(OH)_2 \cdot 8H_2O$ **b** $CrPO_4 \cdot 2H_2O$ **c** $NaC_2H_3O_2 \cdot 3H_2O$

72 Write the formula of: **a** sodium bromide dihydrate   **b** platinum(IV) oxide trihydrate   **c** cobalt(II) sulfate 7-water

73 Why does ice float in water but sink in an alcoholic drink?

74 Calcium chloride forms a hexahydrate. How many moles of water can be absorbed by 50.0 g of $CaCl_2$?

75 How much energy in joules is needed to convert 1.00 kg of snow at $-10.°C$ to steam at $100.°C$? The specific heat of snow is 2.1 J/g°C.

76 Ether has a boiling point of 34.5°C. Why does your hand feel cold after you pour ether on it?

77 Which would evaporate faster, 100 mL of water in a tall glass or 100 mL of water in a flat bowl? Explain.

78 Write a balanced equation for the removal of phosphates (as sodium phosphate) from water, using aluminum sulfate.

79 Why is it that a steel razor blade can be made to float on water?

80 As plaster sets, it becomes hot. How do you explain this observation?

81 Why do eyeglasses fog up when you come inside on a cold winter day?

82 Anhydrous copper(II) sulfate can be used to test for water because it changes color when hydrated. Write an equation for the conversion of greenish-white anhydrous copper(II) sulfate to blue copper(II) sulfate pentahydrate.

PROGRESS CHART   Check off each item when completed.

_____   Previewed chapter

Studied each section:

_____   **12.1**   Kinetic theory of liquids and solids

_____   **12.2**   Structure of water

_____   **12.3**   Density of water and specific gravity

_____   **12.4**   Vaporization

_____   **12.5**   Heat capacity and specific heat

_____   **12.6**   Changes of state

_____   **12.7**   Cohesion and adhesion

_____   **12.8**   Water of hydration

_____   **12.9**   Water pollution

_____   **12.10** Water treatment

_____   **12.11** Hard water

_____   Reviewed chapter

_____   Answered assigned exercises:

#'s _____

Solutions

Chapter Thirteen is about a type of mixture called a solution. Most solutions consist of a solid dissolved in water. Solutions are important because they provide an ideal medium for chemical reactions. Types of solutions and properties of solutions are described in this chapter. Various methods for expressing the concentration of a solution are also presented. To demonstrate an understanding of Chapter Thirteen, you should be able to:

1 Define a solution and give examples of different types of solutions.

2 Describe how an ionic compound dissolves in water.

3 Explain the factors that affect solubility, and use a solubility table to predict solubilities of ionic compounds in water.

4 Describe the relationship between saturation and solubility.

5 Describe the solubility of liquids in liquids, of gases in gases, and of gases in liquids.

6 Give examples of common solvents and important applications of solvents.

7 Solve problems based on the percentage composition of solutions.

8 Solve problems involving concentrations expressed in moles.

9 Solve problems involving dilution of solutions.

10 Given a reaction that takes place in solution, write the balanced net ionic equation.

11 Explain the colligative properties of solutions.

12 Compare the properties of colloids and suspensions with the properties of solutions.

**13.1
TYPES OF
SOLUTIONS**
A **solution** is a uniform mixture of the atoms, ions, or molecules of two or more substances. Often, one of the substances in a solution is a liquid. The components of a solution can be combined in various proportions, which is true of all mixtures.

FIGURE 13-1

*Preparing a solution. A
solution is made by dissolving
a solute in a solvent.*

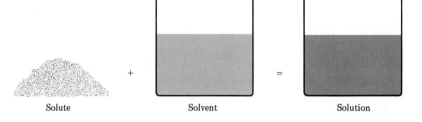

Solute   +   Solvent   =   Solution

The composition and properties of each part of a particular solution are identical to those of every other part. For example, when sugar dissolves in water, the sugar molecules spread out and mix completely with the water. Each drop of the solution contains sugar and water in the same proportions.

Because a solution is a type of mixture, its components can be separated physically. For example, we can separate the components of the sugar solution just mentioned by boiling away the water. However, the components of a solution do not separate by themselves, nor can they be separated by filtration. These properties can be understood if we picture a solution as a random arrangement of very tiny particles: atoms, ions, or small molecules. The particles of a solution are usually too small to be pulled down by gravity and settle, and too small to be trapped by filter paper. In addition, liquid and gaseous solutions are generally transparent. We can see through them because the particles are too small to block the transmission of light through the solution.

A solution consists of two parts. The **solute** is the substance being mixed or dissolved, and the **solvent** is the substance doing the dissolving. We say that a solution is formed by dissolving a solute in a solvent, as shown in Figure 13-1. The solute can also be defined as the substance present in a smaller amount in the solution. The solvent is the substance present in a larger amount; it is the major component of the solution. Many solutions contain water as the solvent and are called *aqueous* solutions (from "aqua," which means water). Two examples of aqueous solutions are syrup (sugar dissolved in water) and household ammonia (gaseous ammonia dissolved in water).

TABLE 13-1 *Examples of solutions*

	Solid solvent	*Liquid solvent*	*Gaseous solvent*
Solid solute	brass (zinc in copper, an alloy)	saline solution (sodium chloride in water)	a
Liquid solute	amalgam (mercury in silver)	liquor (alcohol in water)	a
Gaseous solute	hydrogen in platinum metal	soda water (carbon dioxide in water)	air (oxygen in nitrogen)

[a]To form a solution in a gas, the solute must generally also exist in the gaseous state.

TABLE 13-2 *Important alloys*

Alloy	Composition	Use
amalgam alloy	69% Ag, 26% Sn, 4% Cu, 1% Zn	mixed with mercury for dental restoration
brass	60–85% Cu, 15–40% Zn (or Pb)	plumbing, hardware
bronze	89–92% Cu, 8–11% Sn	artwork, machine parts
gold alloy	70% Au, 15% Ag, 10% Cu, 3% Pd, 1% Pt, 1% Zn	dental casting (inlays, crowns, etc.)
nichrome	80% Ni, 20% Cr	heating coils
pewter	85% Sn, 7% Cu, 6% Bi, 2% Sb	bowls, pots
solder	60% Pb, 40% Sn	joining metal parts
steel	99% Fe, 1% C	building
sterling silver	93% Ag, 7% Cu	cutlery
vinertia alloy	67% Co, 27% Cr, 6% Mo	replacement for body parts (such as hip bone)

As you can see from Table 13-1, solutions can be prepared in many ways. The solute and solvent each may be either an element or a compound, and they can combine as solids, liquids, or gases (with the exceptions noted in the table). The name **alloy** is given to solid solutions made by mixing atoms of different metals. Examples of important alloys and their compositions are listed in Table 13-2. Some other types of solutions also have special names. A saline solution is a solution of

Box 13

Honey, a natural solution

Honey is a natural solution of sugars dissolved in water. It is produced by honeybees from the nectar of flowers. Bees chemically convert the sucrose in flower nectar into two simpler sugars, glucose and fructose. The bees store an aqueous solution of these two sugars in the cells of their hive. As water evaporates, this solution becomes thick and sticky. Evaporation is speeded up by the rapid air circulation caused by the moving wings of worker bees. The resulting concentrated solution is what we call honey.

A typical bee colony needs about 400 to 500 pounds of honey each year to sustain itself. We use the excess of the hive's requirement as food. The flavor and color of the honey depend on the kind of flower from which the bees obtain the nectar. Most bees in the U.S. make honey from the nectar of either alfalfa or clover.

Nutritionally, honey is nearly identical to sucrose, or table sugar. However, it is almost twice as sweet, since it contains fructose, the sweetest-tasting sugar.

Honey is an extremely concentrated solution. It is about 80% sugar by weight. In contrast, jams are approximately 70% sugar, maple syrup is 65% sugar, and molasses is 60% sugar by weight.

sodium chloride in water; a tincture is a solution in which alcohol is the solvent. For example, tincture of iodine, which has been used as an antiseptic, contains iodine dissolved in alcohol.

**13.2
PROCESS OF
DISSOLVING**

When a solute dissolves in a solvent, its atoms, ions, or molecules separate from each other. For this separation to occur, the solute particles must be attracted in some way by the solvent particles. Generally, *a solute will dissolve in a solvent that has a structure similar to its own.* For example, polar solvents tend to attract and dissolve polar or ionic solutes, and nonpolar solvents tend to dissolve nonpolar solutes. The expression "Like dissolves like" can help you remember this fact.

As explained in Chapter Twelve, water molecules are polar. Many ionic compounds, such as sodium chloride, dissolve in water. So do polar covalent compounds like sugar, ammonia, and hydrogen chloride. Most organic molecules, including those of oils and greases, are nonpolar. Therefore they do not dissolve in water; they dissolve in nonpolar organic solvents like gasoline.

Figure 13-2 shows how an ionic compound dissolves in water. The polar water molecules strike the compound and attract its ions, causing

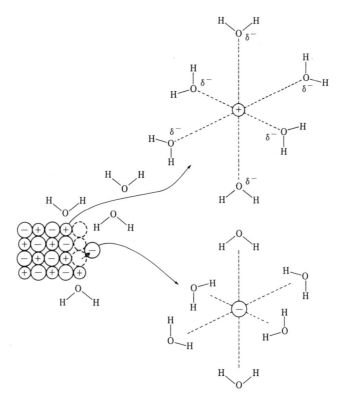

FIGURE 13-2

An ionic solid dissolving in water. The ions dissociate as they become surrounded by water molecules.

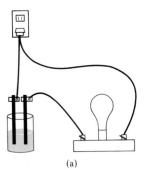

(a)

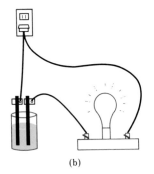

(b)

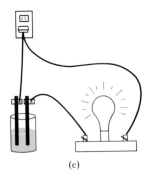

(c)

FIGURE 13-3

Conductivity. A solution of a nonelectrolyte (a) does not conduct electricity because no ions are present. In a solution of weak electrolyte (b) and a solution of a strong electrolyte (c), the presence of ions permits electricity to be conducted.

them to separate, or dissociate. This process is called **dissociation**. Dissociation occurs as the water molecules pull apart the positive and negative ions. Remember that the water molecule is polar; the oxygen atoms are slightly negatively charged and the hydrogen atoms are slightly positively charged. In a solution of an ionic solute in water, the positive cations of the solute become surrounded by water molecules with their negatively charged oxygen atoms facing the cations. The negatively charged anions also become surrounded by water molecules, but here the water molecules have hydrogen atoms facing the anions. This attraction between the ions and the polar water molecules is called **hydration**. Each hydrated ion is surrounded by approximately six water molecules. The general term for the attraction and surrounding of solute particles by solvent molecules is **solvation**.

Polar covalent compounds, like hydrogen chloride, dissolve in water in a process called **ionization**. Ionization is a type of dissociation in which ions are actually *created* when water molecules separate the atoms of solute molecules. For example, hydrogen ions, H^+, and chloride ions, Cl^-, result from the ionization of HCl when it dissolves in water. After ionization these ions are hydrated, or surrounded by water molecules.

Substances that produce ions when they dissolve in water are called **electrolytes**. Because ions move freely through the water, solutions of electrolytes can transmit electrical charge and can therefore conduct electricity. Substances like sodium chloride that separate completely into ions conduct electricity well and are said to be *strong* electrolytes. Substances that produce fewer ions in solution and therefore conduct electricity poorly are called *weak* electrolytes. Substances that do not produce any ions at all when they are dissolved in water are called **nonelectrolytes**. A solution of a nonelectrolyte, like sugar in water, does not readily conduct electricity at all. (See Figure 13-3.) In medicine, the term electrolyte specifically refers to the cations (Na^+, K^+, Ca^{2+}, Mg^{2+}) and anions (Cl^-, HCO_3^-, HPO_4^{2-}, SO_4^{2-}, organic acids, proteins) present in the body fluids.

13.3 SOLUBILITY

The **solubility** of a solute is the mass of the solute that can be dissolved in a certain amount of solvent at a particular temperature. Solubility is usually expressed in grams of solute per 100 mL (100 cm^3) or per 100 g of solvent. For example, the solubility of sodium chloride in water is 36 g per 100 g H_2O at 20°C.

The interaction between a solute and a solvent determines how much solute can dissolve. Thus, in a particular solvent such as water, the nature of the solute is a key factor in determining solubility. Solutes whose molecules are strongly attracted by water molecules tend to dissolve readily in this solvent. For that reason some compounds like table sugar (sucrose) are very soluble in water. Sucrose has a solubility of 204 g per 100 g H_2O at 20°C. Other compounds, like sodium chloride, are also soluble in water, but to a somewhat lesser extent.

Certain compounds are said to be only *slightly soluble*. Calcium hydroxide, Ca(OH)$_2$, for example, has a solubility in water of only 0.17 g per 100 g H_2O at 20°C. Finally, some compounds dissolve to an extremely small degree. For example, the solubility in water of silver chloride, AgCl, is 10^{-4} g per 100 g H_2O at 20°C. This compound is said to be *sparingly soluble* or *insoluble* in water since, for all practical purposes, it hardly dissolves at all in this solvent. The particles of an insoluble compound have a far greater attraction for each other than they do for the molecules of the solvent.

Table 13-3 summarizes the solubilities of common ionic compounds. As indicated, most ionic compounds that contain Na^+, K^+, or NH_4^+ as the cation are soluble. Most compounds that contain $C_2H_3O_2^-$, Cl^-, Br^-, I^-, NO_3^-, or SO_4^{2-} as the anion also are soluble. In contrast, ionic compounds that contain CO_3^{2-}, OH^-, PO_4^{3-}, or S^{2-} are usually sparingly soluble or insoluble (with the exceptions noted in the table). Using these rules, we can predict the solubilities of large numbers of compounds. Appendix F contains a more detailed table of solubilities for ionic compounds.

TABLE 13-3 *Solubility rules for ionic compounds*

A *Soluble in water*	*Example*
1 Na^+, K^+, and NH_4^+ compounds	NaCl
2 acetates (except of Ag^+)	Zn(C$_2$H$_3$O$_2$)$_2$
3 chlorides, bromides, and iodides (except of Ag^+, Hg_2^{2+}, Pb^{2+})	CaCl$_2$
4 nitrates	LiNO$_3$
5 sulfates (except of Ba^{2+}, Sr^{2+}, Pb^{2+}, Hg_2^{2+}, Hg^{2+}, Ag^+)	MgSO$_4$

B *Insoluble or sparingly soluble in water*	*Example*
1 carbonates (except of Group 1A, NH_4^+)	CaCO$_3$
2 hydroxides (except of Group 1A, Ba^{2+}, Sr^{2+})	Fe(OH)$_3$
3 phosphates (except of Group 1A, NH_4^+)	AlPO$_4$
4 sulfides (except of Groups 1A and 2A, NH_4^+)	CdS

Example   Referring to Table 13-3, predict the solubility in water for each of the following compounds.

Compound	Prediction	Rule
NaOH	soluble	A1
$Ca_3(PO_4)_2$	insoluble	B3
$AgNO_3$	soluble	A4
ZnS	insoluble	B4

When a solute dissolves, its molecules or ions must be pulled apart. Therefore, the dissociation or ionization process requires energy and absorbs heat. Then, as solute molecules or ions become attached to solvent molecules, heat is released. In general, the solvation process releases less heat than is absorbed during dissociation. Thus the dissolving of solute is usually endothermic because overall the process absorbs heat.

Because the dissolving of solute usually requires heat, the solubility of most solids in liquids increases with increasing temperature. More solute dissolves as the solvent is heated because raising the temperature favors the process that absorbs heat. The added thermal energy increases the kinetic energy of the solute molecules, causing them to separate more readily.

Table 13-4 presents several examples of changes in solubility with changes in temperature. The sucrose column shows something familiar: a greater amount of sugar dissolves in hot water than in cold water. The increase in solubility as temperature increases is very large; an additional 308 g of sugar dissolve in water as the temperature rises from 0° to 100°C. Sodium chloride increases in solubility by only 4 g over this same temperature range. The solubility of calcium hydroxide, on the other hand, actually *decreases* as the temperature increases.

TABLE 13-4   *Solubilities of solids in water at various temperatures*

Temperature (°C)	Solubility of sucrose (g per 100 g H_2O)	Solubility of sodium chloride (g per 100 g H_2O)	Solubility of calcium hydroxide (g per 100 g H_2O)
0	179	35.7	0.185
20	204	36.0	0.165
40	238	36.6	0.141
60	287	37.3	0.116
80	362	38.4	0.094
100	487	39.8	0.077

(In contrast to the other examples, the dissolving of the solute calcium hydroxide is exothermic.) Appendix C.4 shows how change in solubility due to temperature change can be expressed by using a graph.

In addition to the *amount* of solute that can dissolve, we are also concerned with the *rate* at which the solute dissolves. The rate depends on several factors. A solid dissolves more readily when it is ground up (pulverized), because more of its surface is then exposed to the solvent. Stirring (agitating) increases interaction between solute and solvent particles, and therefore increases the rate at which a solid or liquid dissolves, as you know from mixing sugar with coffee or tea. In fact, any factor that increases the contact between solute particles and solvent molecules will increase the rate of dissolving. Heating makes the molecules move faster and has a similar effect. Note that heating affects the amount of solute that dissolves as well as the rate of dissolving.

13.4 SATURATION

Most solutes are soluble only to a certain extent. In other words, there is a limit to their solubility. For example, when we gradually add sugar to water, we eventually reach a point when no more sugar appears to dissolve. At this point a balance or equilibrium exists. The rate at which the sugar dissolves in the water exactly equals the rate at which it returns from solution to form solid sugar. (This dynamic situation is similar to the liquid-vapor equilibrium described in Section 12.4.) The solution then is said to be **saturated**. (See Figure 13-4.) A saturated solution contains the maximum amount of dissolved solute possible under normal conditions at a particular temperature. The amount of solute needed to make a saturated solution depends on the solubility of the solute in the solvent.

A saturated solution does not necessarily contain a large quantity of solute. If a solute is only slightly soluble, not much of it will dissolve, and we can therefore prepare a saturated solution with relatively little solute. As shown in Table 13-4, a saturated solution of calcium hydroxide at 20°C contains only 0.165 g of $Ca(OH)_2$ per 100 g of water.

When a solution contains less solute than required for equilibrium, it is said to be **unsaturated**. More solute can therefore be dissolved in an unsaturated solution. For example, a solution prepared with 20 g of sodium chloride in 100 g of water at 20°C is unsaturated. As shown in Table 13-4, the solution becomes saturated only when the mass of sodium chloride reaches 36 g per 100 g water at this temperature. A solution can contain a large amount of solute and still be unsaturated if the solubility of the solute is great.

Under certain conditions, we can make a solution containing *more* solute than is required to form a saturated solution at a particular temperature. Such a solution is said to be **supersaturated**. A supersaturated solution can be prepared by first making a saturated solution and then slowly lowering its temperature (assuming that the solubility of the solute decreases upon cooling). If the temperature is lowered slowly enough, excess solute will remain dissolved. Another method

FIGURE 13-4

A saturated solution. The rate at which the solid solute dissolves equals the rate of crystallization of solute from solution.

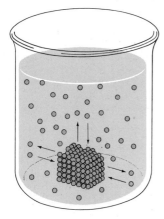

of making a supersaturated solution is to prepare a saturated solution and then allow some of the solvent to evaporate. Honey, for example, is a supersaturated solution of sugars (fructose and glucose) in water, which forms in a beehive by evaporation of water from a saturated solution. A supersaturated solution is unstable. The excess solute can crystallize out of the supersaturated solution, leaving behind a saturated solution. You may have noticed solid sugar that has crystallized in a jar of honey.

13.5 SOLUBILITIES OF LIQUIDS AND GASES

Liquids that mix completely with each other to form a solution in any proportion are said to be **miscible** (pronounced miss' uh-bull). Miscible liquids generally have similar molecular structures and their molecules may interact. Water and alcohol molecules, for example, are polar and can form hydrogen bonds between each other. Because the molecules of these two liquids attract each other, the liquids mix readily and are miscible. Water and gasoline have very different structures. Gasoline is nonpolar and therefore does not mix with water, which is polar. Water and gasoline are said to be **immiscible**. Immiscible liquids arrange themselves in layers according to their densities. For example, when water and gasoline are mixed, the less dense gasoline forms a separate layer on the denser water. This process takes place after an oil spill (see Figure 13-5).

FIGURE 13-5

Oil spill. As shown here, oil and water are immiscible liquids. (Photo courtesy of the U.S. Coast Guard.)

Smithsonian Institution: Photo #49981

In the brewery next door to his home in Leeds, England, Priestley studied the gas produced by fermenting grain. We now call this gas carbon dioxide. Priestley dissolved it in water to make the first soda water. Using the method of collecting gases by displacement of mercury instead of water, he also isolated the water-soluble gases ammonia, hydrogen chloride, and sulfur dioxide.

Priestley decomposed the red compound now known as mercuric oxide by heating it with sunlight through a magnifying glass. It released a gas as it decomposed. Priestley found that this gas made objects burn more brightly and made mice frisky. He tested the ''goodness'' of his gas by measuring the length of time a mouse could live breathing it! Priestley didn't fully realize the importance of this gas, which Lavoisier later called oxygen. Oxygen had actually been discovered several years earlier by the Swedish chemist Carl Wilhelm Scheele, but Priestley published first and is generally given credit for the discovery.

A Unitarian minister, Priestley was a religious and political liberal. He supported the American and French revolutions and opposed slavery and religious persecution. Because of his views, an angry mob destroyed his home in 1791, forcing him to flee to America. His final home in Northumberland, Pennsylvania, has been preserved as a memorial. The American Chemical Society was founded there in 1876, 100 years after Priestley's discovery of oxygen.

Gases can mix with each other in any proportions to form a solution. Air is a solution containing the solute oxygen (along with many other solutes in smaller amounts) dissolved in nitrogen. The normal composition of air is about 21% oxygen and 78% nitrogen. The percentage of oxygen in air is sometimes increased, however, for medical use.

The solubility of gases in liquids varies considerably and depends on several factors. One factor is the interaction between molecules of the gas and molecules of the liquid. A strong attraction between these molecules increases solubility. For example, ammonia is polar and therefore very soluble in water. Solubility of gases in liquids also depends on pressure and temperature. Table 13-5 compares the solubilities of several common gases in water. At constant temperature, the solubility of a gas is directly proportional to the pressure of that gas above the liquid. This relationship is known as Henry's law. Soda, or carbonated water, is made by dissolving carbon dioxide in water under high pressure. It was first produced by Joseph Priestley, whose biography is presented in this chapter. Soda eventually goes "flat" after the bottle or can is opened. It loses carbonation because the carbon dioxide that was forced into solution under high pressure can escape into the atmosphere (where its partial pressure is only about 10^{-4} atm).

Decompression sickness, or "the bends," is caused by the too rapid escape of gases from solution in the blood. The gas forms bubbles as it leaves the blood, and these bubbles collect in the joints and capillaries, causing pain and damage. Decompression sickness is a hazard of underwater diving. As a diver descends, pressure increases by 1 atm for every 10 m of depth. A diver at a depth of 50 m, for example, would experience a pressure 5 atm greater than the pressure at the surface of the water. At such high pressures, more of the gases that the diver breathes dissolve in the blood. (Except for oxygen, none of the gases in a diver's breathing mixture can be used or broken down by the body. Therefore, these gases remain in the blood.) If the diver surfaces too

TABLE 13-5 *Solubilities of gases in water*

Gas	Solubility (g) in 100 g H_2O at 0°C and 1 atm
NH_3	89.5
HCl	82.3
SO_2	22.8
Cl_2	1.46
CO_2	0.335
O_2	6.95×10^{-3}
N_2	2.94×10^{-3}

quickly, the sudden decrease in pressure causes these gases to "bubble out" of solution in the blood, just as CO_2 bubbles rise to the surface of a soft drink when the bottle is opened. The gas bubbles can cause pain in the diver's joints, muscle cramps, paralysis, and eventual death. Fortunately, decompression sickness can be treated by placing the victim in a hyperbaric chamber, in which the pressure is increased above 1 atm to redissolve the gas bubbles. The pressure in the chamber is then gradually decreased until normal atmospheric pressure is again reached. The gradual decrease in pressure allows the gas to escape from the blood slowly, without forming dangerous bubbles.

Unlike most solids, gases are less soluble in liquid at higher temperatures. They are less soluble because increasing the temperature increases the number of gas molecules that have enough kinetic energy to escape from the surface of the liquid. Thus, as pointed out in the last chapter, one of the problems caused by thermal pollution of natural waters is a decrease in the amount of dissolved oxygen in the water. Cooling a liquid, on the other hand, increases the solubility of a gas. When the temperature is decreased, fewer gas molecules have enough kinetic energy to escape from the liquid. Therefore, an open bottle of soda, for example, retains more CO_2 if it is kept in a refrigerator than if it is left out at room temperature.

**13.6
SOLVENTS**

In a chemical laboratory, the term solvent almost always refers to a liquid. In the laboratory, *a solvent often serves as the medium or environment in which reactants mix and form products.* Atoms, molecules, or ions of the reactants can move about rapidly in a solvent, but are still close enough to collide frequently. Thus a common procedure in the laboratory is to mix solutions of appropriate reactants, stir them together for a certain length of time while they react, and then isolate the products from the solution.

The choice of solvent depends on the solubilities of the reactants and products, the temperature of the reaction, and other factors. Table 13-6 compares the properties of some common solvents. Figure 13-6 shows how some of this information is given on the label of a solvent bottle or can.

Liquid solvents serve many other important purposes. In paints, inks, and lacquers they are used as a means for depositing a uniform coating on a surface. In this case, the solvent must be volatile, that is, able to evaporate readily. The solution is spread on the surface, and the solvent is then allowed to evaporate, leaving behind the desired deposit of solute. Solvents are also used in manufacturing. Fibers, for example, can be made by dissolving appropriate compounds in a solvent and then forcing the resulting solution through small holes at high pressure. Solvents also are used in dry cleaning to dissolve grease and oil.

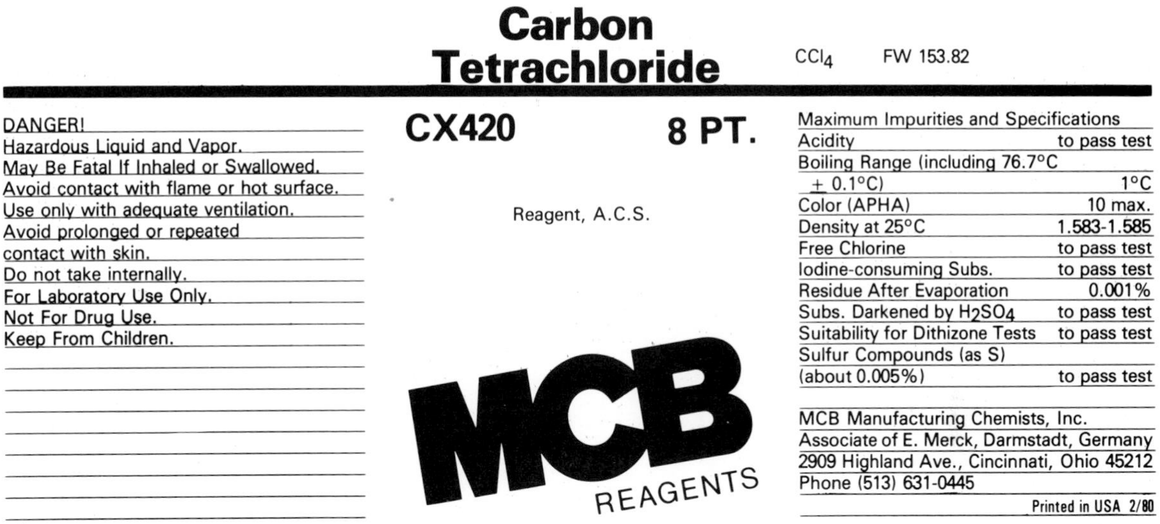
Carbon
Tetrachloride
CCl4 FW 153.82
DANGER!
Hazardous Liquid and Vapor.
May Be Fatal If Inhaled or Swallowed.
Avoid contact with flame or hot surface.
Use only with adequate ventilation.
Avoid prolonged or repeated
contact with skin.
Do not take internally.
For Laboratory Use Only.
Not For Drug Use.
Keep From Children.
CX420
8 PT.
Reagent, A.C.S.
MCB
REAGENTS
Maximum Impurities and Specifications
Acidity to pass test
Boiling Range (including 76.7°C
± 0.1°C) 1°C
Color (APHA) 10 max.
Density at 25°C 1.583-1.585
Free Chlorine to pass test
Iodine-consuming Subs. to pass test
Residue After Evaporation 0.001%
Subs. Darkened by H2SO4 to pass test
Suitability for Dithizone Tests to pass test
Sulfur Compounds (as S)
(about 0.005%) to pass test
MCB Manufacturing Chemists, Inc.
Associate of E. Merck, Darmstadt, Germany
2909 Highland Ave., Cincinnati, Ohio 45212
Phone (513) 631-0445
Printed in USA 2/80

FIGURE 13-6

A solvent label. This label indicates the most important properties of the solvent carbon tetrachloride. (Photo courtesy of MCB Manufacturing Chemists, Inc.)

TABLE 13-6 *Properties of common solvents*

Solvent	*Formula*	*Density (20°C), g/ml*	*Boiling point, °C*	*Water solubility*	*Hazards*
acetone	$(CH_3)_2CO$	0.792	56	miscible	flammable
acetonitrile	CH_3CN	0.783	82	miscible	toxic, flammable
benzene	C_6H_6	0.879	80	poor	toxic, flammable
carbon disulfide	CS_2	1.263	46	poor	irritant, flammable
carbon tetrachloride	CCl_4	1.595	77	poor	toxic
chloroform	$CHCl_3$	1.489	61	poor	irritant
cyclohexane	C_6H_{12}	0.778	81	insoluble	flammable
dimethyl sulfoxide	$(CH_3)_2SO$	1.100	189	soluble	absorbed through skin
dioxane	$C_4H_8O_2$	1.033	101	moderate	toxic, flammable
ethyl ether	$C_2H_5OC_2H_5$	0.719	35	soluble	flammable
ethyl alcohol	C_2H_5OH	0.789	78	miscible	flammable
hexane	C_6H_{14}	0.660	69	insoluble	flammable
methyl alcohol	CH_3OH	0.792	65	miscible	flammable, toxic
pyridine	C_5H_5N	0.982	115	miscible	irritant, flammable
tetrahydrofuran	C_4H_8O	0.888	66	soluble	irritant, flammable
toluene	$C_6H_5CH_3$	0.866	111	poor	irritant, flammable
water	H_2O	0.998	100	miscible	none

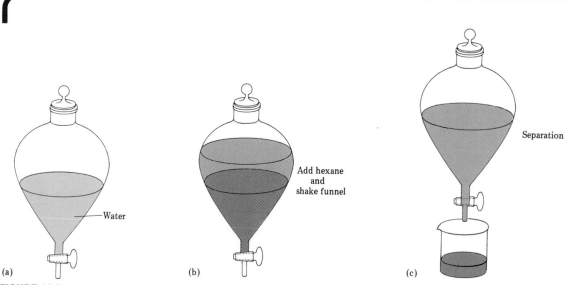

FIGURE 13-7

Solvent extraction. In this method of separation, a solute is distributed between two immiscible solvents. For example, if the water (a) contains a nonpolar solute, some of the solute is transferred to the hexane (b). The two solvent layers can be separated (c) and the process repeated until most of the solute is removed from the water.

Another important application of solvents is the separation of mixtures into their components. In a process called solvent extraction, one solvent containing the mixture to be separated is shaken in a separatory funnel with a second immiscible solvent. (See Figure 13-7.) Some components of the mixture dissolve more readily in the second solvent while others remain behind in the first solvent. The components of the mixture are separated when the denser solvent is allowed to run out of the funnel, leaving the less dense solvent behind.

Solvents are also used to purify compounds. In a process called recrystallization, an impure compound is dissolved in an appropriate solvent by heating. Upon cooling and partial evaporation of the solvent, the solid crystallizes in a purer form, leaving impurities behind in the solvent.

**13.7
PERCENTAGE
CONCENTRATION
OF SOLUTIONS**

The **concentration** of a solution expresses the amount of solute present in a given quantity of solvent or solution. Concentrations can vary greatly. A *concentrated* solution contains a relatively large amount of solute, while a *dilute* solution contains only a small amount. The terms "concentrated" and "dilute" are vague, however, and we must use more precise terms to state the concentration of a solution.

We can express concentration as the percentage of solute in the solution. For example, the concentrations of commercially available

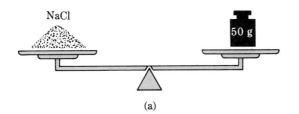

(a)

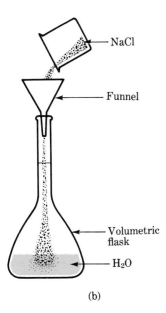

(b)

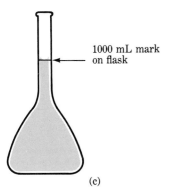

(c)

FIGURE 13-8

Preparation of 1000 mL of 5% (w/v) saline solution. (a) 50 g of NaCl are measured. (b) The solid NaCl is added to a 1000 mL volumetric flask partially filled with water. (c) After the solid dissolves, water is added up to the 1000 mL mark.

solutions, such as the dextrose (glucose) in water used for intravenous feeding, are often given in percent. There are several ways to express concentration in percent. For solutions of solids in liquids, a common unit is **weight-volume** (w/v) **percent**.

$$\text{weight-volume (w/v) percent} = \frac{\text{grams of solute}}{100 \text{ mL of solution}} \times 100\%$$

This unit gives the mass (weight) in grams of solute that is present in 100 mL of the solution. (Note that weight-volume percentage is *not* a true percentage because it is based on a ratio of two different kinds of quantities, weight and volume.)

A 5% (w/v) solution of sodium chloride, for example, contains 5 g of NaCl per 100 mL of solution.

$$5\% \text{ (w/v) NaCl} = \frac{5 \text{ g NaCl}}{100 \text{ mL solution}} \times 100\%$$

Since each 100 mL of solution contains 5 g of solute, we can find the amount present in any other volume of the same solution by using this ratio as a conversion factor. For example, one liter (1000 mL) of this solution contains:

$$1000 \text{ mL solution} \times \frac{5 \text{ g NaCl}}{100 \text{ mL solution}} = 50 \text{ g NaCl}$$

Thus 50 g of sodium chloride are needed to make one liter of a 5% (w/v) solution. Figure 13-8 shows how such a solution could be prepared in the laboratory.

Example   What mass of silver nitrate, $AgNO_3$, is needed to make 500. mL of a 0.40% (w/v) solution?

UNKNOWN	GIVEN	CONNECTION
grams $AgNO_3$	500. mL solution	$\dfrac{0.40 \text{ g } AgNO_3}{100. \text{ mL solution}}$ or $\dfrac{100. \text{ mL solution}}{0.40 \text{ g } AgNO_3}$

$$\text{grams } AgNO_3 = 500. \text{ mL solution} \times \frac{0.40 \text{ g } AgNO_3}{100. \text{ mL solution}}$$

ESTIMATE $5 \times 10^2 \text{ mL solution} \times \dfrac{4 \times 10^{-1} \text{ g } AgNO_3}{1 \times 10^2 \text{ mL solution}} = 2 \text{ g } AgNO_3$

CALCULATION $2.0 \text{ g } AgNO_3$

The concentrations of commercial aqueous solutions are frequently given in **weight-weight** (w/w) **percent**, or simply **weight percent**.

$$\text{weight (w/w) percent} = \frac{\text{grams of solute}}{.00 \text{ g of solution}} \times 100\%$$

Weight percent indicates the mass (weight) in grams of solute that is present in 100 g of solution. A 70% (w/w) nitric acid solution, for example, contains 70 g of HNO_3 in each 100 g of solution.

$$70\% \text{ (w/w) } HNO_3 = \frac{70 \text{ g } HNO_3}{100 \text{ g solution}} \times 100\%$$

Keep in mind that 100 g of solution is not the same as 100 g of solvent. The mass of the solution includes both the mass of the solute and the mass of the solvent. Also, except in very dilute aqueous solutions, 100 g of solution is different from 100 mL of solution. They are the same only if the solution has a density of 1.00 g/mL. Because mass is independent of temperature, the weight percent of a solution remains constant when its temperature changes. In contrast, weight-volume percent changes with temperature because volume changes with temperature.

Example   What mass of 85.0% (w/w) aqueous phosphoric acid contains 20.0 g of H_3PO_4?

UNKNOWN	GIVEN	CONNECTION
grams of solution	20.0 g H_3PO_4	$\dfrac{85.0 \text{ g } H_3PO_4}{100. \text{ g solution}}$ or $\dfrac{100. \text{ g solution}}{85.0 \text{ g } H_3PO_4}$

$$\text{grams of solution} = 20.0 \text{ g } H_3PO_4 \times \frac{100. \text{ g solution}}{85.0 \text{ g } H_3PO_4}$$

ESTIMATE $\quad 2 \times 10^1 \text{ g } H_3PO_4 \times \dfrac{1 \times 10^2 \text{ g solution}}{9 \times 10^1 \text{ g } H_3PO_4} = 20 \text{ g solution}$

CALCULATION $\quad$ 23.5 g solution

If the density (or specific gravity) of a solution is known, its mass can be readily converted to a volume. In the example just worked out, phosphoric acid has a density of 1.69 g/mL. The 23.5 g of solution therefore must have a volume of

$$23.5 \text{ g solution} \times \frac{1 \text{ mL solution}}{1.69 \text{ g solution}} = 13.9 \text{ mL solution}$$

Density is used as a conversion factor in this calculation, as described in Section 3.8.

Volume-volume (v/v) **percent** or **volume percent** is used to give the concentration of a solution that consists only of liquids.

$$\text{volume-volume percent (v/v)} = \frac{\text{mL of solute}}{100 \text{ mL of solution}} \times 100\%$$

For example, a 70% (v/v) alcohol solution consists of 70 mL of alcohol in a solution whose total volume is 100 mL.

$$70\% \text{ (v/v) alcohol} = \frac{70 \text{ mL alcohol}}{100 \text{ mL solution}} \times 100\%$$

You might expect this solution to contain 30 mL of water, but it actually has slightly more. Alcohol and water molecules are similar and can form hydrogen bonds, so they attract each other weakly. Because of this attraction, the combined volume of alcohol and water is actually less than the sum of the two volumes. Thus a volume-volume percent solution must be prepared by measuring the volume of solute and then adding enough solvent to reach the final desired volume.

The "proof" of a solution of alcohol in water is twice its volume-volume percent. Thus a 70% (v/v) alcohol solution is 140 proof. Similarly, a bottle of 80 proof vodka contains 40% (v/v) alcohol in water.

Example   What volume of acetone is needed to prepare 25 mL of a 6.40% (v/v) solution?

UNKNOWN	GIVEN	CONNECTION
milliliters acetone	25 mL solution	$\dfrac{6.40 \text{ mL acetone}}{100. \text{ mL solution}}$
		or $\dfrac{100. \text{ mL solution}}{6.40 \text{ mL acetone}}$

$$\text{milliliters acetone} = 25 \text{ mL solution} \times \frac{6.40 \text{ mL acetone}}{100. \text{ mL solution}}$$

ESTIMATE $3 \times 10^1 \text{ mL solution} \times \dfrac{6 \text{ mL acetone}}{1 \times 10^2 \text{ mL solution}}$

= 2 mL acetone

CALCULATION   1.6 mL acetone

Percent means parts per hundred, whether the parts are given in terms of mass or volume. Sometimes the concentrations that we wish to express are so small that we need a unit that is smaller than percent. In such cases the unit **parts per million** (ppm) is commonly used. Just as 5% means 5 parts out of 100 (or 10^2) parts, 5 ppm means 5 parts out of 1 million (or 10^6) parts.

To help you understand just how small a ppm is, consider that one part per million is equivalent to: one minute in two years, one inch in 16 miles, or one penny in $10,000.

Parts per million are very useful in environmental monitoring, where extremely small concentrations of pollutants may be significant. For example, if the amount of mercury in a 1 kg water sample is 5 mg, its concentration is

$$\frac{5 \text{ mg}}{1 \text{ kg}} = \frac{5 \text{ mg}}{10^6 \text{ mg}} = 5 \text{ ppm}$$

ASIDE . . .

One ppm is equivalent to one large mouthful of food compared to all the food you eat in a lifetime.

For very dilute aqueous solutions such as this one, we can safely assume that 1 kg of solution has a volume of 1 L, based on the density of water (1 g/mL = 1 kg/L). Thus, for practical purposes, the unit ppm is based on the number of milligrams of solute per liter sample.

$$\text{parts per million (ppm)} = \frac{\text{parts solute}}{10^6 \text{ parts solution}} = \frac{\text{mg of solute}}{\text{L of solution}}$$

For still smaller concentrations the unit **parts per billion** (ppb) can be used.

$$\text{ppb} = \frac{\text{parts solute}}{10^9 \text{ parts solution}} = \frac{\mu\text{g of solute}}{\text{L of solution}}$$

Note that 10^9 ppb $= 10^6$ ppm $= 100\% = 1$.

Example The amount of lead in a 500. mL water sample is 3.6 mg. Express this concentration in ppm.

UNKNOWN	GIVEN	CONNECTION
ppm	$\dfrac{3.6 \text{ mg Pb}}{500. \text{ mL sample}}$	$\text{ppm} = \dfrac{\text{mg of solute}}{\text{L of solution}}$

$$\text{ppm} = \frac{3.6 \text{ mg Pb}}{500. \cancel{\text{mL}}} \times \frac{10^3 \cancel{\text{mL}}}{1 \text{ L}}$$

ESTIMATE $\quad \dfrac{4 \text{ mg Pb}}{5 \times 10^2 \cancel{\text{mL}}} \times \dfrac{10^3 \cancel{\text{mL}}}{1 \text{ L}} = 10 \text{ mg Pb/L} = 10 \text{ ppm Pb}$

CALCULATION $\quad$ 7.2 ppm Pb

Practice problem A sodium nitrate solution contains 0.50 g $NaNO_3$ in 200. mL of solution. Express its concentration in **a** weight-volume percent and **b** ppm.

ANSWERS $\quad$ **a** 0.25% (w/v) $\quad$ **b** 2.5×10^3 or 2500 ppm

Practice problem What mass of solute is needed to make 10. g of a 28% (w/w) solution of sodium hydroxide?

ANSWER $\quad$ 2.8 g NaOH

13.8 CONCENTRATION EXPRESSED IN MOLES

Although it is easy to express the concentration of solutions in percent, it is not always useful to do so. For example, a 5% (w/v) solution of sodium chloride and a 5% (w/v) solution of silver nitrate both contain the same *mass* of solute (5 g per 100 mL solution). However, they do not contain the same *number of moles* of solute, because these two compounds have different formula weights. Therefore, because moles are very important for carrying out calculations involving chemical reactions in solution, chemists often use a unit of concentration that is based on the number of moles of solute in a solution. This unit, called **molarity** and abbreviated capital M, is defined as the number of moles of solute per liter (or cubic decimeter) of solution.

$$\text{molarity (M)} = \frac{\text{moles of solute}}{\text{liter (dm}^3\text{) of solution}}$$

Thus, a 5.00 molar (or 5.00 M) solution of sodium chloride contains 5.00 moles of NaCl per liter of solution. The mass of sodium chloride in this solution can be found from its molar mass, 58.4 g/mole. Thus a 5.00 M solution contains

$$\frac{\text{grams NaCl}}{\text{liter of solution}} = \frac{5.00 \text{ moles NaCl}}{\text{liter of solution}} \times \frac{58.4 \text{ g NaCl}}{1 \text{ mole NaCl}}$$

$$= \frac{292 \text{ g NaCl}}{\text{liter of solution}}$$

Note that in calculations, we always write out M in full as moles/liter.

This concentration can be expressed as a weight-volume percent for comparison.

$$\frac{\text{g}}{100 \text{ ml}} = \frac{292 \text{ g NaCl}}{\text{liter of solution}} \times \frac{0.1 \text{ liter}}{100 \text{ mL}}$$

$$= \frac{29.2 \text{ g NaCl}}{100 \text{ mL solution}}$$

Thus the concentration is equal to

$$\text{w/v \%} = \frac{\text{grams of solute}}{100 \text{ mL of solution}} \times 100\%$$

$$= \frac{29.2 \text{ g NaCl}}{100 \text{ mL of solution}} \times 100\%$$

$$= 29.2\%$$

A 5.00 M solution of any other solute, like silver nitrate for example, would also contain 5.00 moles of solute per liter. However, the *mass* of solute would be different, since the formula weight of $AgNO_3$, which is 169.9 amu, is different from that of NaCl.

$$\frac{\text{grams AgNO}_3}{\text{liter of solution}} = \frac{5.00 \text{ moles AgNO}_3}{\text{liter of solution}} \times \frac{169.9 \text{ g AgNO}_3}{1 \text{ mole AgNO}_3}$$

$$= \frac{850. \text{ g AgNO}_3}{\text{liter of solution}}$$

Thus a 5.00 M solution of silver nitrate has a concentration that can be expressed in percent as 85.0% (w/v). Because the formula weight of $AgNO_3$ is so much larger than that of NaCl, a greater mass is required to make a solution of equal molarity. Each solution, however, contains the same number of moles, and therefore the same number of formula units of solute.

To find the molarity of a solution, we simply divide the number of moles of solute by the solution volume in liters. For example, the molarity of a solution that contains 0.600 mole of solute in 2.00 liters of solution is

$$\frac{0.600 \text{ mole solute}}{2.00 \text{ liters solution}} = \frac{0.300 \text{ mole solute}}{1 \text{ liter solution}} = 0.300 \text{ M}$$

If the mass of solute is given in a problem, we must first convert it to moles, as illustrated in the following example.

Example   A 16 oz (or 0.473 L) bottle of cola contains 36.9 g of sucrose, $C_{12}H_{22}O_{11}$. Find the molarity of this solution.

UNKNOWN	GIVEN	CONNECTION
molarity	36.9 g sucrose 0.473 L solu- tion	$\text{molarity} = \dfrac{\text{moles solute}}{\text{liters solution}}$ $\dfrac{1 \text{ mole sucrose}}{342.3 \text{ g sucrose}}$ or $\dfrac{342.3 \text{ g sucrose}}{1 \text{ mole sucrose}}$

$$\text{molarity} = \frac{36.9 \text{ g sucrose}}{0.473 \text{ L solution}} \times \frac{1 \text{ mole sucrose}}{342.3 \text{ g sucrose}}$$

ESTIMATE $$\frac{4 \times 10^1 \text{ g sucrose}}{5 \times 10^{-1} \text{ L solution}} \times \frac{1 \text{ mole sucrose}}{3 \times 10^2 \text{ g sucrose}}$$

$$= \frac{0.3 \text{ mole sucrose}}{1 \text{ liter solution}} = 0.3 \text{ M sucrose}$$

CALCULATION $$\frac{0.228 \text{ moles sucrose}}{1 \text{ liter solution}} = 0.228 \text{ M sucrose}$$

Notice that molarity, the unit for our answer, is a ratio of two different units, moles/liter.

To prepare a solution with a volume that is not exactly one liter, the molarity can be used as a conversion factor. For example, to make 2.00 liters of 5.00 M sodium chloride, we need

$$2.00 \text{ liters of solution} \times \frac{5.00 \text{ moles NaCl}}{1 \text{ liter of solution}} = 10.0 \text{ moles NaCl}$$

Once again the required mass is found by using the number of grams in one mole, based on the formula weight.

$$10.0 \text{ moles NaCl} \times \frac{58.4 \text{ g NaCl}}{1 \text{ mole NaCl}} = 584 \text{ g NaCl}$$

As shown in the following examples, these two steps can be combined into one.

Example   What mass of copper(II) sulfate is needed to make 1.50 L of a 0.690 M solution?

UNKNOWN	GIVEN	CONNECTION	
g CuSO$_4$	1.50 L solution	$\dfrac{0.690 \text{ mole CuSO}_4}{1 \text{ liter solution}}$ or	$\dfrac{1 \text{ liter solution}}{0.690 \text{ mole CuSO}_4}$
		$\dfrac{1 \text{ mole CuSO}_4}{159.6 \text{ g CuSO}_4}$ or	$\dfrac{159.6 \text{ g CuSO}_4}{1 \text{ mole CuSO}_4}$

$$g \text{ CuSO}_4 = 1.50 \text{ L solution} \times \frac{0.690 \text{ mole CuSO}_4}{1 \text{ L solution}} \times \frac{159.6 \text{ g CuSO}_4}{1 \text{ mole CuSO}_4}$$

ESTIMATE $2 \text{ L solution} \times \dfrac{1 \text{ mole CuSO}_4}{1 \text{ L solution}} \times \dfrac{2 \times 10^2 \text{ g CuSO}_4}{1 \text{ mole CuSO}_4}$

$= 4 \times 10^2 \text{ g CuSO}_4$

CALCULATION   165 g CuSO$_4$

Example   How many grams of potassium dihydrogenphosphate are required to make 250. mL of a 0.110 M KH$_2$PO$_4$ solution?

UNKNOWN	GIVEN	CONNECTION
g KH_2PO_4	250. mL solution	$\dfrac{0.110 \text{ mole } KH_2PO_4}{1 \text{ L solution}}$ or $\dfrac{1 \text{ L solution}}{0.110 \text{ mole } KH_2PO_4}$
		$\dfrac{1 \text{ mole } KH_2PO_4}{136.1 \text{ g } KH_2PO_4}$ or $\dfrac{136.1 \text{ g } KH_2PO_4}{1 \text{ mole } KH_2PO_4}$
		$\dfrac{1 \text{ mL}}{10^{-3} \text{ L}}$ or $\dfrac{10^{-3} \text{ L}}{1 \text{ mL}}$

$$g\ KH_2PO_4 = 250.\ \cancel{mL\ solution} \times \frac{10^{-3}\ \cancel{L}}{1\ \cancel{mL}} \times \frac{0.110\ \cancel{\text{mole } KH_2PO_4}}{1\ \cancel{L\ solution}} \times$$
$$\frac{136.1\ g\ KH_2PO_4}{1\ \cancel{\text{mole } KH_2PO_4}}$$

ESTIMATE $\quad 3 \times 10^2\ \cancel{mL\ solution} \times \dfrac{10^{-3}\ \cancel{L}}{1\ \cancel{mL}} \times$
$$\frac{1 \times 10^{-1}\ \cancel{\text{mole } KH_2PO_4}}{1\ \cancel{L\ solution}} \times \frac{1 \times 10^2\ g\ KH_2PO_4}{1\ \cancel{\text{mole } KH_2PO_4}}$$

$$= 3\ g\ KH_2PO_4$$

CALCULATION $\quad$ 3.74 g KH_2PO_4

(Note: If you recognize immediately that 250 mL is equal to 0.250 L, you can make this substitution directly and eliminate the conversion factor for milliliters to liters.)

Molarity is also used as a conversion factor for finding the volume of solution that contains a required number of moles or grams of solute. This kind of problem may arise when we carry out a reaction in which one or more of the reactants is available as a solution. For example, the amount of chloride ion present when sodium chloride is placed in a solution can be determined by adding enough silver ion (as silver nitrate, $AgNO_3$) to precipitate all of the Cl^- as AgCl.

$$AgNO_3(aq) + NaCl(aq) \rightarrow AgCl(s) + NaNO_3(aq)$$

Suppose that the approximate amount of chloride (as NaCl) present in a sample is 0.0068 mole. We would like to know what volume of 0.300 M silver nitrate solution to add in order to react completely with the chloride ions.

UNKNOWN	GIVEN	CONNECTION	
volume $AgNO_3$	0.0068 mole NaCl	$\dfrac{1\text{ mole }AgNO_3}{1\text{ mole NaCl}}$ or $\dfrac{1\text{ mole NaCl}}{1\text{ mole }AgNO_3}$	
		$\dfrac{0.300\text{ mole }AgNO_3}{1\text{ liter of solution}}$ or $\dfrac{1\text{ liter of solution}}{0.300\text{ moles }AgNO_3\text{n}}$	

$$\text{volume } AgNO_3 = 0.0068 \text{ mole NaCl} \times \frac{1\text{ mole }AgNO_3}{1\text{ mole NaCl}} \times \frac{1\text{ liter solution}}{0.300\text{ mole }AgNO_3}$$

ESTIMATE

$$7 \times 10^{-3} \text{ mole NaCl} \times \frac{1\text{ mole }AgNO_3}{1\text{ mole NaCl}} \times \frac{1\text{ liter solution}}{3 \times 10^{-1}\text{ mole }AgNO_3}$$
$$= 2 \times 10^{-2} \text{ L solution}$$

CALCULATION   0.023 L (or 23 mL) solution

Thus 23 mL of 0.300 M silver nitrate solution must be added to provide enough moles of solute to react with the chloride ions of the sample.

Molality (m) is a concentration unit that is closely related to molarity. It too deals with moles of solute in a solution. Molality is defined as the number of moles of solute *per kilogram (or 1000 g) of solvent* (rather than per liter of solution, as in molarity).

$$\text{molality } (m) = \frac{\text{number of moles of solute}}{\text{kilogram of solvent}}$$

Note that we represent molality with a small m, while molarity is represented by a capital M.

For example, a 5.00 m aqueous solution of sodium chloride contains 5.00 moles of NaCl per kilogram of water. For dilute aqueous solutions, the concentrations expressed as molarity and as molality are approximately the same, since 1 kg of water has a volume of about 1 liter. For more concentrated solutions, and for solutions in solvents whose density differs from 1 g/mL, molality and molarity are *not* the same. A 5.00 M NaCl solution, for instance, contains only 892 mL (0.892 L) of water. In this concentrated solution, the sodium chloride makes a significant contribution to the volume of the solution. Using the density of water as a conversion factor, the molality of this solution is:

$$\frac{\text{moles of solute}}{\text{kg of solvent}} = \frac{5.00 \text{ moles NaCl}}{0.892 \text{ L } H_2O} \times \frac{1 \text{ L } H_2O}{1 \text{ kg } H_2O}$$

$$= \frac{5.61 \text{ moles NaCl}}{1 \text{ kg } H_2O} = 5.61 \text{ } m$$

In this case, molality and molarity are clearly not the same.

Example   An antifreeze solution contains 40. grams of ethylene glycol ($C_2H_6O_2$) mixed with 60. g of water. Find the molality of this solution.

UNKNOWN	GIVEN	CONNECTION
molality, m	40. g $C_2H_6O_2$ 60. g H_2O	$\text{molality} = \dfrac{\text{moles solute}}{\text{kilogram of solvent}}$ $\dfrac{1 \text{ mole } C_2H_6O_2}{62.1 \text{ g } C_2H_6O_2}$ or $\dfrac{62.1 \text{ g } C_2H_6O_2}{1 \text{ mole } C_2H_6O_2}$ $\dfrac{1 \text{ kg}}{10^3 \text{ g}}$ or $\dfrac{10^3 \text{ g}}{1 \text{ kg}}$

$$m = \frac{40. \text{ g } C_2H_6O_2}{60. \text{ g solvent}} \times \frac{1 \text{ mole } C_2H_6O_2}{62.1 \text{ g } C_2H_6O_2} \times \frac{10^3 \text{ g}}{1 \text{ kg}}$$

ESTIMATE $\dfrac{4 \times 10^1 \text{ g } C_2H_6O_2}{6 \times 10^1 \text{ g solvent}} \times \dfrac{1 \text{ mole } C_2H_6O_2}{6 \times 10^1 \text{ g } C_2H_6O_2} \times \dfrac{10^3 \text{ g}}{1 \text{ kg}}$

$$= 10 \frac{\text{moles } C_2H_6O_2}{\text{kg solvent}} = 10 \text{ } m$$

CALCULATION $10.7 \dfrac{\text{moles } C_2H_6O_2}{\text{kg solvent}} = 10.7 \text{ } m$

    Molality is useful for describing concentration in solutions whose temperature may vary. Unlike molarity, molality does not involve volume, which varies with temperature. Therefore, concentration expressed in molality is independent of the temperature.

Practice problem   Normal saline solution is an aqueous sodium chloride solution that is used in medicine because its composition matches that of body fluids. Its concentration is 0.90% (w/v). Find the molarity of this solution.

ANSWER   0.15 M NaCl

What mass of ammonium chloride, NH_4Cl, is needed to make 750. ml of a 0.333 M solution?

ANSWER   13.4 g NH_4Cl

13.9
DILUTION
OF SOLUTIONS

Solutions are often prepared by diluting an existing solution of known concentration, called a stock solution. Using a pipet, a portion of this solution (called an aliquot) is transferred from the stock bottle to another flask. Then additional solvent is added to the flask to reduce the initial concentration to the desired concentration. It is often easier to prepare solutions in this way than to start "from scratch." When we dilute an existing solution, we avoid having to weigh out and dissolve the necessary amount of solute.

Dilution calculations are based on the fact that the number of moles of solute taken from the stock solution and the number of moles of solute in the diluted solution are the same. More solvent is present in the final solution, but the diluted solution still contains the exact amount of solute that was transferred from the stock solution. Thus,

$$\text{moles of the solute (initial)} = \text{moles of solute (final)}$$
$$\quad\text{stock solution} \qquad\qquad \text{diluted solution}$$

Since many concentrations are expressed in molarity, this equation can also be written as

$$\frac{\text{moles of solute}}{\text{liter of solution}} \text{ (initial)} \times \text{liters of solution (initial)}$$

$$= \frac{\text{moles of solute}}{\text{liter of solution}} \text{ (final)} \times \text{liters of solution (final)}$$

In abbreviated form, this relationship is

$$M_1 \times V_1 = M_2 \times V_2$$

where M_1 and V_1 are the initial values of the molarity and volume, and M_2 and V_2 are the final values. (The units for V_1 and V_2 must be the same, but need not be expressed in liters.) A similar equation can be written for concentrations expressed as weight-volume percent, since the number of grams of solute must also stay the same during dilution.

Consider the dilution of hydrochloric acid, which is available in bottles as an aqueous 12 M solution. To find the volume of this stock solution needed to make 2.00 L of 0.50 M HCl, we set up the problem as follows.

UNKNOWN	GIVEN	CONNECTION
V_1, volume of 12 M HCl	$12\ M = M_1$ $0.50\ M = M_2$ $2.00\ L = V_2$	$M_1 \times V_1 = M_2 \times V_2$

The molarity of the concentrated stock solution is M_1. The needed volume of this solution, V_1, is unknown. The molarity, M_2, and volume, V_2, of the diluted solution are both given. We substitute the three known variables in the equation and solve for V_1.

$$M_1 \times V_1 = M_2 \times V_2$$

$$12\ M \times V_1 = 0.50\ M \times 2.00\ L$$

$$V_1 = \frac{0.50\ \cancel{M} \times 2.00\ L}{12\ \cancel{M}}$$

Notice that this problem can also be set up by using a conversion factor based on the original and final molarity.

$$V_1 = V_2 \times \frac{M_2}{M_1} = 2.00\ L \times \frac{0.50\ \cancel{M}}{12\ \cancel{M}}$$

The equation can also be written out in full, with notation indicating which solution is concentrated (conc) and which is dilute (dil):

$$V_1 = 2.00\ \cancel{L\ (dil)} \times \frac{0.50\ \cancel{mole\ HCl}}{1\ \cancel{L\ (dil)}} \times \frac{1\ L\ (conc)}{12\ \cancel{moles\ HCl}}$$

If we use this method, we eliminate the problem of having the same unit in the numerator and denominator of the conversion factor.

To be sure that the problem is set up correctly with either method, check that the result makes sense. The volume of the diluted solution (V_2) must be greater than the volume of the concentrated stock solution (V_1) because we added solvent to V_1 to make V_2. Therefore, the ratio of molarities (0.50 M)/(12 M) must be correct, since it decreases the size of V_2. We can now obtain the answer.

ESTIMATE $\quad \dfrac{5 \times 10^{-1}\ \cancel{M} \times 2\ L}{1 \times 10^{1}\ \cancel{M}} = 0.1\ L = 100\ mL$

CALCULATION $\quad 0.083\ L = 83\ mL$

Thus, we can prepare a 0.50 M solution by diluting 83 mL of a 12 M HCl solution to a volume of 2.00 L.

Example A 10.0 mL aliquot of 15 M NH_3 is diluted to 250. mL with water. What is the molarity of the diluted solution?

UNKNOWN	GIVEN	CONNECTION
M_2	15 M $= M_1$ 10.0 mL $= V_1$ 250. mL $= V_2$	$M_1 \times V_1 = M_2 \times V_2$

$$M_1 \times V_1 = M_2 \times V_2$$

$$(15 \text{ M}) \times (10.0 \text{ mL}) = M_2 \times (250. \text{ mL})$$

$$\frac{(15 \text{ M}) \times (10.0 \text{ mL})}{250. \text{ mL}} = M_2$$

Or by the conversion factor method:

$$M_2 = 15 \text{ M} \times \frac{10.0 \text{ mL}}{250. \text{ mL}}$$

ESTIMATE $\quad \dfrac{2 \times 10^1 \text{ M} \times 1 \times 10^1 \text{ mL}}{3 \times 10^2 \text{ mL}} = 1 \text{ M } NH_3$

CALCULATION $\quad$ 0.60 M NH_3

Example What must the final volume be if 25.0 mL of 1.50 M potassium carbonate, K_2CO_3, is diluted to make a 0.560 M solution?

UNKNOWN	GIVEN	CONNECTION
V_2	1.50 M $= M_1$ 25.0 mL $= V_1$ 0.560 M $= M_2$	$M_1 \times V_1 = M_2 \times V_2$

$$M_1 \times V_1 = M_2 \times V_2$$

$$(1.50 \text{ M}) \times (25.0 \text{ mL}) = (0.560 \text{ M}) \times V_2$$

$$\frac{(1.50 \text{ M}) \times (25.0 \text{ mL})}{(0.560 \text{ M})} = V_2$$

Or by the conversion factor method:

$$V_2 = 25.0 \text{ mL} \times \frac{1.50 \cancel{M}}{0.560 \cancel{M}}$$

ESTIMATE $\quad \dfrac{2\cancel{M} \times 3 \times 10^1 \text{ mL}}{6 \times 10^{-1} \cancel{M}} = 1 \times 10^2 \text{ mL}$

CALCULATION $\quad$ 67.0 mL

Note that in order to obtain this final volume, 67.0 mL − 25.0 mL = 42.0 mL of solvent must be added to the original portion.

Practice problem $\quad$ What volume of 18 M H_2SO_4 is needed to make 500. mL of a 0.150 M solution?

ANSWER $\quad$ 4.2 mL

Practice problem $\quad$ If 25.0 mL of 3.60 M NaH_2PO_4 is diluted to 1.00 L, what is the molarity of the new solution?

ANSWER $\quad$ 0.0900 M NaH_2PO_4

**13.10
REACTIONS
IN SOLUTION:
IONIC
EQUATIONS**
When a reaction takes place in aqueous solution, some of the reactants and products may exist as ions. Ions can be formed by the dissociation of soluble ionic compounds or by the ionization of soluble polar covalent compounds, as discussed earlier in this chapter. Once we have determined which reactants and products will exist as ions in aqueous solution, we can write an ionic equation for the reaction.

Consider again the reaction of silver nitrate with sodium chloride in water to form silver chloride and sodium nitrate.

$$AgNO_3(aq) + NaCl(aq) \rightarrow AgCl(s) + NaNO_3(aq)$$
silver $\qquad$ sodium $\qquad$ silver $\qquad$ sodium
nitrate $\qquad$ chloride $\qquad$ chloride $\qquad$ nitrate

According to Table 13-3, all nitrates are soluble. Therefore both silver nitrate and sodium nitrate, which are ionic compounds, dissociate into ions. Silver nitrate is present as Ag^+ and NO_3^- ions, sodium nitrate as Na^+ and NO_3^- ions. Sodium chloride is also soluble and exists as Na^+ and Cl^- ions. Silver chloride, on the other hand, is insoluble in water according to the solubility table. Therefore, it must exist mainly as solid, undissociated AgCl. The ionic equation for this reaction is therefore written as follows:

$$Ag^+ + NO_3^- + Na^+ + Cl^- \rightarrow AgCl + Na^+ + NO_3^-$$

This equation reflects the actual composition of the solution because both of the reactants and one of the products are shown in ionic form.

Notice that two ions, NO_3^- and Na^+, are present on both sides of the equation. These ions are called **spectator ions**, and are unchanged by the reaction because they do not take part in the reaction. We can simplify an ionic equation by removing the spectator ions.

$$Ag^+ + \cancel{NO_3^-} + \cancel{Na^+} + Cl^- \rightarrow AgCl + \cancel{Na^+} + \cancel{NO_3^-}$$

The resulting equation, called the **net ionic equation**, focuses attention on the reacting ions.

$$Ag^+ + Cl^- \rightarrow AgCl$$

To emphasize the formation of a precipitate we can also write this equation as

$$Ag^+ + Cl^- \rightarrow AgCl\downarrow \quad \text{or}$$

$$Ag^+(aq) + Cl^-(aq) \rightarrow AgCl(s)$$

Ionic equations, like all equations, must be balanced by mass as described in Section 9.6. In addition, ionic equations must be balanced by *charge*. The sum of the charges on one side of the equation must equal the sum of the charges on the other side. In the above ionic equation, the sum is zero on both sides of the equation. In general, the sum of the charges need not equal zero, but must have the same value on both sides of the equation.

The procedure for writing net ionic equations can be summarized as follows:

1 Write the complete equation, including all reactants and products in full.

2 Identify the water-soluble reactants and products that dissociate (ionic compounds) or ionize (polar covalent compounds). Use a solubility table if necessary.

3 Separate these substances into the resulting ions, making sure that their charges are correct.

4 Remove all spectator ions, those ions present and unchanged on both sides of the equation.

5 Balance the net ionic equation for both mass and charge.

The following examples illustrate these rules.

Example   Zinc metal reacts in aqueous hydrogen chloride to form zinc chloride and hydrogen gas. Write the balanced net ionic equation.

1 $Zn(s)$ + $HCl(aq)$ $\rightarrow$ $ZnCl_2(aq)$ + $H_2(g)$
 zinc   hydrochloric   zinc   hydrogen
              acid   chloride

2 HCl and $ZnCl_2$ are water soluble and produce ions in solution.

3 Zn + H^+ + $\cancel{Cl^-}$ $\rightarrow$ Zn^{2+} + $\cancel{Cl^-}$ + H_2

Note: In writing the ionic equation, it is necessary only to write what ions are formed, not the actual number produced by each formula unit or molecule. Thus, in **3** we wrote Zn^{2+} + Cl^- rather than Zn^{2+} + $2Cl^-$.

4 Zn + H^+ $\rightarrow$ Zn^{2+} + H_2

5 Zn + $2H^+$ $\rightarrow$ Zn^{2+} + $H_2\uparrow$

or $Zn(s)$ + $2H^+(aq)$ $\rightarrow$ $Zn^{2+}(aq)$ + $H_2(g)$

Example   Reacting sodium sulfate with barium chloride in water produces a precipitate of barium sulfate. Write the balanced net ionic equation.

1 $Na_2SO_4(aq)$ + $BaCl_2(aq)$ $\rightarrow$ $BaSO_4(s)$ + $NaCl(aq)$
 sodium   barium   barium   sodium
 sulfate   chloride   sulfate   chloride

2 Na_2SO_4, $BaCl_2$ and NaCl are water-soluble ionic compounds.

3 $\cancel{Na^+}$ + SO_4^{2-} + Ba^{2+} + $\cancel{Cl^-}$ $\rightarrow$ $BaSO_4$ + $\cancel{Na^+}$ + $\cancel{Cl^-}$

4 and **5** Ba^{2+} + SO_4^{2-} $\rightarrow$ $BaSO_4\downarrow$

or $Ba^{2+}(aq)$ + $SO_4^{2-}(aq)$ $\rightarrow$ $BaSO_4(s)$

Example   Potassium sulfide in a reaction with sulfuric acid in water produces hydrogen sulfide gas. Write the balanced net ionic equation.

1 $K_2S(aq)$ + $H_2SO_4(aq)$ $\rightarrow$ $H_2S(g)$ + $K_2SO_4(aq)$
 potassium   sulfuric   hydrogen   potassium
 sulfide   acid   sulfide   sulfate

2 K_2S, H_2SO_4 and K_2SO_4 are water-soluble and produce ions.

3 $\cancel{K^+}$ + S^{2-} + H^+ + $\cancel{SO_4^{2-}}$ $\rightarrow$ H_2S + $\cancel{K^+}$ + $\cancel{SO_4^{2-}}$

4 H^+ + S^{2-} $\rightarrow$ H_2S

5 $2H^+$ + S^{2-} $\rightarrow$ $H_2S\uparrow$

or $2H^+(aq)$ + $S^{2-}(aq)$ $\rightarrow$ $H_2S(g)$

Example   Lithium hydroxide reacts with aqueous phosphoric acid producing water along with lithium phosphate. Write the balanced net ionic equation.

1 $LiOH(aq) + H_3PO_4(aq) \rightarrow H_2O(l) + Li_3PO_4(aq)$
      lithium   phosphoric   water   lithium
      hydroxide   acid   phosphate

2 $LiOH$, H_3PO_4, and Li_3PO_4 are water soluble and produce ions.

3 $\cancel{Li^+} + OH^- + H^+ + \cancel{PO_4^{3-}} \rightarrow H_2O + \cancel{Li^+} + \cancel{PO_4^{3-}}$

4 and **5** $H^+ + OH^- \rightarrow H_2O$

or $H^+(aq) + OH^-(aq) \rightarrow H_2O(l)$

Practice problem   Write a balanced net ionic equation for each of the following reactions.

a $Al(s) + H_2SO_4(aq) \rightarrow H_2(g) + Al_2(SO_4)_3(aq)$

b $Zn(s) + CuSO_4(aq) \rightarrow ZnSO_4(aq) + Cu(s)$

ANSWERS   **a** $2Al(s) + 6H^+(aq) \rightarrow 3H_2(g) + 2Al^{3+}(aq)$
           b $Zn(s) + Cu^{2+}(aq) \rightarrow Zn^{2+}(aq) + Cu(s)$

13.11
COLLIGATIVE
PROPERTIES
OF SOLUTIONS

Colligative properties are properties that depend only on the *number* of solute particles in a solution. For example, the presence of solute particles (atoms, ions, or molecules) in a solution makes the vapor pressure of the solvent lower than if the solvent contained no solute. The solute in effect dilutes the solvent, reducing the number of solvent molecules that can escape from the surface. However, the number of vapor molecules *returning* to the liquid is unchanged. For this reason, concentrated solutions evaporate much more slowly than pure solvent. Thus vapor pressure depression is an example of a colligative property.

A solvent that has its vapor pressure lowered by the presence of solute will need a higher temperature to make the vapor pressure equal to the atmospheric pressure. This effect results in an elevation (increase) of the boiling point of the solvent in the solution. This boiling point elevation is a colligative property, and is directly related to the molality of the solution, as given by the following expression:

$$\Delta T_b = K_b m$$

ΔT_b is the change in the boiling temperature, m is the molality, and K_b is the molal boiling point elevation constant. The constant K_b is a property of the solvent. For water, the value of K_b is

0.512 (°C)(kg H_2O)/mole.

(Note that the term "mole" in K_b refers to moles of solute.) For example, we can raise the boiling point of water by adding sugar. If we add enough sugar to make a 1.00 m sugar solution we will raise the boiling point of the water by

$$\Delta T_b = 0.512 \frac{(°C)(kg\ H_2O)}{mole} \times 1.00 \frac{mole}{kg\ H_2O} = 0.512°C$$

Since the boiling point of pure water is 100°C, this solution boils at 100.512°C (at 1 atm).

When a solvent freezes, there is an equilibrium between the liquid and solid forms, just as there is an equilibrium between the liquid and gaseous states during boiling. The vapor pressure of the liquid solvent must equal the vapor pressure of the pure solid solvent. But because solute lowers the vapor pressure of a solution, it also depresses (lowers) the freezing point. The lowered vapor pressure means that the solvent is less likely to leave the liquid state to become a solid (or a gas). Therefore, a lower temperature is required for the solid solvent to be in equilibrium with the liquid solvent in the solution.

The relationship for calculating freezing point depression is similar to the one used to calculate boiling point elevation.

$$\Delta T_f = K_f m$$

Here ΔT_f is the change in the freezing temperature, m is the molality, and K_f is the molal freezing point depression constant. Like K_b, K_f is also a property of the solvent. For water, the value of K_f is

1.86 (°C)(kg H_2O)/mole.

(Here again, the term "mole" in K_f refers to moles of solute.)

By adding sugar to water, we also lower the freezing point. If we add enough sugar to make a 1.00 m sugar solution, we lower the freezing point of water by

$$\Delta T_f = 1.86 \frac{(°C)\ (kg\ H_2O)}{mole} \times 1.00 \frac{mole}{kg\ H_2O} = 1.86°C$$

Because pure water freezes at 0°C, this solution will freeze at $-1.86°C$ (at 1 atm). The antifreeze in automobile radiators works by lowering the freezing point of water. A 40% (w/w) solution of ethylene glycol ($C_2H_6O_2$) depresses the freezing point of the water to $-20°C$. Similarly, salt that is spread on icy roads in winter melts the ice by lowering the freezing point of water to a temperature that is colder than the air temperature.

Osmotic pressure is another colligative property of solutions. **Osmosis** is the diffusion of solvent molecules through a membrane through which solute particles cannot pass. Such a membrane is called semipermeable because it is permeable, or capable of being penetrated, by some but not by all types of molecules. When solutions having different concentrations are placed on opposite sides of such a membrane, solvent flows through the membrane. Solvent moves from the more dilute

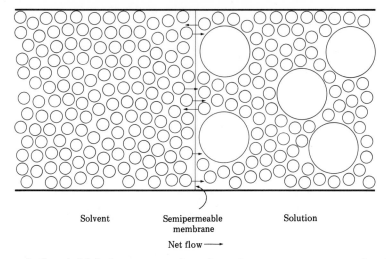

FIGURE 13-9

Osmosis. Solvent molecules tend to diffuse through the membrane toward the more concentrated solution.

Solvent | Semipermeable membrane | Solution

Net flow ⟶

solution (which has more solvent) to the more concentrated solution (which has less solvent) as shown in Figure 13-9. Osmosis continues until the more concentrated solution becomes dilute enough so that solvent molecules pass through the membrane in opposite directions at equal rates.

Osmotic pressure is the pressure that is needed to stop the flow of solvent from the more concentrated solution during osmosis. It is a colligative property because it depends only on the number of solute particles (at constant temperature and volume). Osmotic pressure is important in living systems. It is the driving force in the movement of water through membranes. At the roots of plants and trees, for example, osmotic pressures may reach 50 atm or more.

Since colligative properties depend on the number of particles in solution, the effects are even greater for electrolytes. For example, in a strong electrolyte like NaCl, each formula unit produces two particles, a sodium ion and a chloride ion. Therefore, the vapor pressure decrease, boiling point elevation, freezing point depression, and osmotic pressure are all approximately twice as great as they are in a nonelectrolyte solution that has the same concentration.

13.12
COLLOIDS AND
SUSPENSIONS

We have seen that solutions are mixtures of small molecules, atoms, or ions. **Colloids**, on the other hand, are mixtures of larger particles that range in size from 1 to 1000 nm in diameter. One substance in a colloid is said to be dispersed or distributed in another substance, called the dispersing medium. The colloidal particles are analogous to the solute, and the dispersing medium is analogous to the solvent. Table 13-7 presents common examples of colloids. Many of the fluids in our bodies are colloids.

TABLE 13-7 *Examples of colloids*

Substance being dispersed	In solid medium	In liquid medium	In gaseous medium
solid	concrete (sol)	jelly (gel) india ink (sol)	dust (aerosol)
liquid	cheese (emulsion)	mayonnaise (emulsion)	fog (aerosol)
gas	foam rubber (foam)	whipped cream (foam)	none[a]

[a]Gas mixtures form solutions.

Colloids are given various names, depending on the type of particles involved and on the dispersing medium. For example, a **sol** is a dispersion of a solid in a liquid or solid. Blood plasma, which contains large biological molecules dispersed in an aqueous medium, is an example of a sol.

A **gel** is a type of sol in which the colloidal particles are joined in a random way to form an open, rigid structure. Examples are gelatin desserts and jellies.

An **emulsion** consists of a liquid dispersed in another liquid or in a solid. Homogenized milk is an emulsion of butterfat droplets that have been broken down to colloidal size and dispersed in water. An emulsifying agent is sometimes needed to keep the components from separating. In milk, the protein casein is the emulsifying agent.

Do-it-yourself 13

An edible emulsion

Oil and water do not mix. If you vigorously stir an oil into water, the oil breaks up into tiny droplets. After a while, the oil droplets join to form a layer of oil that floats on the water.

If you add an emulsifying agent, these two immiscible liquids can form an emulsion. The emulsifying agent consists of molecules having both polar and nonpolar regions. Its polar regions bind to water molecules, and its nonpolar regions bind to oil molecules. The emulsifying agent keeps the oil dispersed in water and prevents the oil droplets from coming together. Egg yolk is a good emulsifying agent.

To do: You can make an edible emulsion by putting one egg, two tablespoons of vinegar or lemon juice, and ¼ teaspoon of dry mustard into a blender. Start blending at a very slow speed. With the blender going, slowly add one cup of salad oil. After about a minute, you should have a creamy, stable emulsion called mayonnaise.

FIGURE 13-10

The Tyndall effect. The beam of light is visible in the colloid (right), but not in the solution (left), because only colloidal particles are large enough to scatter light. (Photo by Al Green.)

A **foam** is a gas dispersed in a liquid or solid. Whipped cream, which consists of air in cream, is a foam.

An **aerosol** is a liquid or solid dispersed in a gas. Fog, which consists of fine water droplets in air, is an aerosol. Commercial products that are sold in spray cans produce aerosols.

Although colloidal particles are larger than the particles in a solution, they are still too small to be seen. However, they are big enough to reflect and scatter light. This scattering is called the Tyndall effect, and is shown in Figure 13-10. A beam of light thus becomes visible when it passes through a colloid. You probably have seen this effect in a movie theater, when the projected beam passes through air that has dust or smoke in it. Because of the Tyndall effect, colloids appear either translucent (like frosted glass) or cloudy. Solutions, on the other hand, contain particles that are too small to scatter light and are therefore transparent.

When we observe a colloid under a microscope, we see rapidly moving pinpoints of scattered light. This random, zigzag movement, called Brownian motion, results from collisions of the colloidal particles with molecules of the dispersing medium. This effect is similar to the random motion of molecules in a gas.

A colloidal particle can attract or adsorb other substances. The colloidal particle carries smaller particles with it, much as a dog might carry fleas. This ability to adsorb other substances helps prevent colloidal particles from settling. For example, one type of colloid called a *lyophobic* colloid, contains inorganic particles like sulfur or metals. These colloidal particles become electrically charged by adsorbing ions

TABLE 13-8 *Properties of solutions, colloids, and suspensions*

Property	Solution	Colloid	Suspension
nature of particles	atoms, ions, or small molecules	large molecules or groups of molecules	very large, visible particles
effect of light	transparent	translucent or opaque; shows Tyndall effect	translucent or opaque; shows Tyndall effect
effect of gravity	does not settle	may slowly settle	settles quickly
uniformity	homogeneous	less homogeneous than solution	heterogeneous
separability	cannot separate by filtration	can be separated only with special membranes	can be separated with filter paper

onto their surfaces. The repulsion between like charges then prevents these particles from coming together to form larger particles that would settle due to gravity. Another type of colloid, called a *lyophilic* colloid, contains large organic molecules like proteins or starch. These colloidal particles adsorb a film of molecules (generally water), forming a protective layer that helps prevent the organic molecules from joining together and settling.

Small molecules and ions can be separated from a colloid by the process of **dialysis**. In contrast to osmosis, which involves the movement of liquid through a membrane, dialysis involves diffusion of small *particles* through a dialyzing membrane. Colloidal particles are too large to pass through the small openings in the dialyzing membrane, and therefore they remain on one side. Ions and small molecules, however, can move from the more concentrated side to the more dilute side of the membrane. Dialysis is used to purify the blood of patients with kidney failure. This process is called hemodialysis. As the blood pumps through the dialyzing membrane, small molecules of toxic waste products move into the surrounding solution and are thus removed from the blood.

Suspensions are mixtures of particles even larger than the particles of colloids. The particles in a suspension are big enough to be seen with the naked eye. Suspensions are heterogeneous (their composition is not uniform), they settle when left standing, and they can be separated by using filter paper. A mixture of clay and water is an example of a suspension. Table 13-8 compares some properties of solutions, colloids, and suspensions.

Summary   **13.1** A solution is a homogeneous mixture. The solute is the substance that is mixed or dissolved, and the solvent is the substance that does the dissolving. Solutions can be prepared from solids, liquids, or gases.

13.2 When a solute dissolves in a solvent, its particles separate from each other. Ionic compounds undergo dissociation in water, and polar covalent compounds ionize. Substances that produce ions in water are called electrolytes.

13.3 The solubility of a solute is the mass that can be dissolved in a certain amount of solute at a particular temperature. The nature of the solute itself is a key factor in determining its solubility. The solubility of most solids in water increases with increasing temperature.

13.4 A solution is saturated when it contains the maximum possible amount of dissolved solute under normal conditions at a particular temperature. At this point, the rate at which solute dissolves exactly equals the rate at which solute returns from the solution.

13.5 Liquids that mix completely in any proportion are said to be miscible. Immiscible liquids do not mix. All gases can mix to form solutions. The solubility of a gas in a liquid increases with increasing pressure but decreases with increasing temperature.

13.6 Solvents serve as a medium in which substances react and form products. Solvents are also used as cleaners, to deposit coatings, and in industrial processing.

13.7 The concentration of a solution describes the amount of solute present in any given quantity of solvent or solution. Concentration can be expressed as a percentage based on the ratios of solute and solution in terms of weight to volume, weight to weight, or volume to volume.

13.8 Molarity is defined as the number of moles of solute per liter of solution. Two solutions with the same molarity contain equal numbers of moles of solute per liter. However, they generally contain different masses of solute.

13.9 Solutions are often prepared by diluting a portion of an existing stock solution. Dilution calculations can be made using the relationship $M_1 \times V_1 = M_2 \times V_2$, or by using the conversion factor method.

13.10 When a reaction takes place in aqueous solution, some of the reactants and products are often present as ions. The net ionic equation expresses the reaction in simplified form. Spectator ions, those that do not participate in the reaction, do not appear in a net ionic equation.

13.11 Colligative properties are properties that depend only on the number of solute particles in a solution. Colligative properties include lowering of vapor pressure, freezing point depression, boiling point elevation, and osmotic pressure.

13.12 Colloids are mixtures of particles that are larger than those found in solutions. Colloids are translucent or opaque, and exhibit Brownian motion and the Tyndall effect. Suspensions consist of even larger particles. These particles can settle or be filtered.

Exercises

KEY-WORD MATCHING EXERCISE

For each term on the left, choose a phrase from the right-hand column that most closely matches its meaning.

1 solution	**a** separation of ions
2 solute	**b** mass that can be dissolved at a given temperature
3 solvent	**c** equilibrium exists
4 dissociation	**d** soluble liquids
5 ionization	**e** g solute per 100 mL solution
6 electrolyte	**f** substance dissolved
7 solubility	**g** mL solute per 100 mL solution
8 saturated	**h** moles per liter of solution
9 unsaturated	**i** homogeneous mixture
10 supersaturated	**j** mg/L
11 miscible	**k** no spectator ions
12 immiscible	**l** more can be dissolved
13 concentration	**m** flow of solvent through membrane
14 weight-volume percent	**n** does the dissolving
15 weight percent	**o** settles with time
16 volume-volume percent	**p** dispersion
17 parts per million	**q** unstable state
18 molarity	**r** depends on number of particles
19 molality	**s** formation of ions
20 net ionic equation	**t** moles per kg of solvent
21 colligative property	**u** insoluble liquids
22 osmosis	**v** g solute per 100 g solution
23 colloid	**w** its solution conducts electricity
24 suspension	**x** "strength" of solution

Types of solution (13.1)

25 What is a solution? Describe its properties.

26 How do you know which is the solute and which is the solvent in a solution?

27 Give an example of a solution (other than one given in Table 13-1) of: **a** a solid in a liquid **b** a liquid in a liquid **c** a solid in a solid

28 What is an alloy?

Process of dissolving (13.2)

29 Why is water a good solvent for many solutes?

30 Describe how KBr might dissolve in water.

31 What is an electrolyte? a strong electrolyte? a weak electrolyte? a nonelectrolyte?

Solubility (13.3)

32 What factors determine the solubility of a solid in a liquid?

33 Predict the solubility in water of **a** $NaC_2H_3O_2$ **b** LiOH **c** $MgSO_4$ **d** PbS **e** $BaCO_3$

34 Find the solubility of a compound if 10. g dissolves in 55 g of water.

35 The solubility of sodium nitrate is 92.1 g per 100 mL H_2O at 25°C. How many grams will dissolve in 2.5 L?

36 What factors influence the rate at which a substance dissolves?

Saturation (13.4)

37 What is a saturated solution?

38 Can an unsaturated solution contain a large amount of solute? Explain.

39 How can you make a supersaturated solution?

40 Using Table 13-4, determine whether each of the following solutions is unsaturated, saturated, or supersaturated: **a** 38 g NaCl per 100 g H_2O at 80°C **b** 102 g sucrose per 50 g H_2O at 20°C **c** 1.5 g $Ca(OH)_2$ per kg H_2O at 40°C

41 How can you test a solution to see if it is saturated?

Solubilities of liquids and gases (13.5)

42 What are miscible liquids?

43 Why is air considered a solution?

44 Why does soda "fizz" when you open the bottle?

45 Why does less gas dissolve at higher temperatures?

Solvents (13.6)

46 Why are solvents important?

47 Using Table 13-6, pick a solvent **a** with a high boiling point that is water soluble   **b** with a high density   **c** that is both nontoxic and nonflammable

48 What is meant by the volatility of a solvent?

Percentage concentration of solutions (13.7)

49 How many grams of potassium chloride are needed to make 500. mL of **a** a 0.25% (w/v) solution?   **b** a 1.60% (w/v) solution?

50 What mass of ammonia is present in 1.00 kg of a 29.0% (w/w) solution?

51 What mass of 37.3% (w/w) hydrochloric acid solution is needed to provide 25.0 g of HCl?

52 What volume of alcohol do you need to make 750. mL of a 2.10% (v/v) aqueous solution?

53 A 5.0 kg water sample contains 7.2×10^{-3} g of lead. Express this concentration in **a** ppm   **b** ppb   **c** w/w percent

54 Two liters of solution contain 12.5 ppm of mercury. How many grams of Hg are present?

55 Calculate the number of grams of $KMnO_4$ needed to make **a** 5.0 L of a 0.12% (w/v) solution   **b** 25.0 g of a 1.5% (w/w) solution   **c** 125 mL of a 10. ppm solution

Concentration expressed in moles (13.8)

56 A potassium iodide solution contains 62.5 g of KI in 500. mL. Find the molarity of this solution.

57 What is the molarity of a 1.2% (w/v) solution of sodium sulfate?

58 How many grams of lithium nitrate are needed to make **a** 1.75 L of a 0.020 M solution?   **b** 750. mL of a 1.65 M solution?

59 What mass of cesium chloride do you need to make **a** 25 mL of a 3.3 M solution?   **b** 1.10 L of a 6.7×10^{-3} M solution?

60 What volume of a 0.10 M $KMnO_4$ solution contains 5.0 g of $KMnO_4$?

61 What is the difference in mass of solute between a 5.0 M solution and a 5.0 m solution of ammonium chloride?

Dilution of solutions (13.9)

62 What volume of a 2.0 M solution of copper(II) sulfate is needed to make 500. mL of a 0.100 M solution?

63 If 25 mL of water is added to 100. mL of an aqueous 1.2 M solution, what is the new concentration?

64 How many milliliters of 16 M HNO_3 do you need to make 1.00 L of a 1.0 M solution?

65 How much water must you add to 100. mL of a 0.55 M sodium acetate solution to make it a 0.45 M solution?

66 How can you prepare 400. mL of a 0.95 M solution of silver nitrate starting with a 3.0 M stock solution?

Reactions in solution: ionic equations (13.10)

67 Write the net ionic equation for
$$Al(NO_3)_3(aq) + NaOH(aq) \rightarrow Al(OH)_3(s) + NaNO_3(aq)$$

68 Sodium carbonate reacts with barium chloride to form sodium chloride and barium carbonate. Write the net ionic equation.

69 Nickel reacts with lead chloride to form nickel(II) chloride and lead. Write a net ionic equation.

70 Iron(III) bromide can be reduced to iron(II) bromide by hydrogen gas, forming hydrogen bromide as the other product. Write the net ionic equation.

71 Write the net ionic equation for
$$NaHCO_3(aq) + H_2SO_4(aq) \rightarrow CO_2(g) + Na_2SO_4(aq) + H_2O(l)$$

Colligative properties of solutions (13.11)

72 What are colligative properties?

73 What are the effects of solute on the boiling and freezing points of a solvent?

74 Calculate the freezing point of a 0.20 m aqueous alcohol solution.

75 Find the freezing point of a 0.33 m sucrose solution.

76 What is the boiling point of the solution in Exercise 75?

77 What happens during osmosis?

Colloids and suspensions (13.12)

78 How does a colloid differ from a solution?

79 What is a gel? a sol? an emulsion? a foam? an aerosol?

80 Describe **a** the Tyndall effect **b** Brownian motion

81 Why don't colloidal particles settle?

82 How does a suspension differ from a colloid?

ADDITIONAL EXERCISES

83 If someone hands you a mixture, what can you do to find out if it is a solution, a colloid, or a suspension?

84 How does ionization differ from dissociation?

85 Using Table 13-3, predict the solubility in water of **a** $(NH_4)_2CO_3$ **b** $MgCl_2$ **c** PbS **d** $Al(C_2H_3O_2)_3$ **e** $AgNO_3$

86 How is saturation related to solubility?

87 Why does honey sometimes crystallize in its jar? What can you do to prevent this?

88 If oil and vinegar are immiscible, how can they be combined into a salad dressing?

89 From Table 13-6, pick a solvent that has the **a** lowest boiling point **b** least number of hazards **c** lowest density

90 Describe the cause of and treatment for decompression sickness ("the bends").

91 How many grams of $MgSO_4$ are needed to prepare **a** 125 mL of a 1.3% (w/v) solution? **b** 4.0 g of a 0.95% (w/w) solution? **c** 1.5 L of a 20. ppm solution?

92 What mass of NH_4Cl is needed to make **a** 2.3 L of a 0.111 M solution? **b** 50. mL of a 2.0 M solution? **c** 3.0 L of a 7.5×10^{-2} M solution?

93 What volume of a 0.75 M solution of HCl is needed to make 50.0 g of H_2 in the reaction

$$Zn(s) + 2HCl\ (aq) \rightarrow H_2(g) + ZnCl_2(s)?$$

94 What volume of a 50.0% (w/v) solution is needed to make 125 mL of a 0.15 M solution of NH_4Cl?

95 A 1 mM (millimolar) solution contains one millimole of solute per liter of solution. How many grams of $FeSO_4$ are needed to make 500. mL of a 0.82 mM solution?

96 Write the balanced net ionic equation for each reaction:
a $Cu(s) + AgNO_3(aq) \rightarrow Ag(s) + Cu(NO_3)_2(aq)$
b $FeCl_2(aq) + Na_2S(aq) \rightarrow FeS(s) + NaCl(aq)$
c $Mg(OH)_2(aq) + HNO_3(aq) \rightarrow Mg(NO_3)_2(aq) + H_2O(l)$

97 The boiling point elevation constant for ethyl alcohol is 1.22 (°C)(kg alcohol)/(mole solute). Find the boiling point elevation of a solution containing 1.5 g of I_2 in 250. g of alcohol.

98 How does dialysis differ from osmosis?

99 2.0 L of methyl alcohol are mixed with 3.0 L of ethyl alcohol. Find the **a** volume percent and **b** weight percent of methyl alcohol in this solution. (Note: density of methyl alcohol = 0.793 g/mL; density of ethyl alcohol = 0.714 g/mL; assume final volume = 5.0 L.)

100 Using a solubility table, predict whether a precipitate forms in these reactions:
a $BaCl_2 + K_2SO_4 \rightarrow$?
b $Li_2SO_4 + Mg(NO_3)_2 \rightarrow$?
c $NaOH + FeCl_3 \rightarrow$?

101 Why is pure HCl a nonelectrolyte in the liquid state but a good conductor in aqueous solution?

102 Which solution has a higher Cl^- concentration, 0.05 M HCl or 15% (w/v) NaCl?

103 How would you prepare 250. mL of 0.874 M NaOH using 4.37 M NaOH?

104 A liter of seawater contains 1.9×10^4 mg of Cl^- ions. Express this concentration in molarity.

105 Which of the following concentration units are temperature dependent? molarity   molality   weight percent   volume percent

106 Write a balanced net ionic equation for each of these reactions:
a $Pb(NO_3)_2(aq) + (NH_4)_2SO_4(aq) \rightarrow PbSO_4(s) + NH_4NO_3(aq)$
b $NaBr(aq) + NH_4Cl(aq) \rightarrow NaCl(aq) + NH_4Br(aq)$
c $CO_2(g) + Ca(OH)_2(aq) \rightarrow CaCO_3(s) + H_2O(l)$

107 The density of a 32% by weight solution of HNO_3 is 1.19 g/mL. What is the molarity of this solution?

PROGRESS CHART Check off each item when completed.

——————— Previewed chapter

Studied each section:

——————— **13.1** Types of solutions

——————— **13.2** Process of dissolving

——————— **13.3** Solubility

——————— **13.4** Saturation

——————— **13.5** Solubilities of liquids and gases

——————— **13.6** Solvents

——————— **13.7** Percentage composition of solutions

——————— **13.8** Concentration expressed in moles

——————— **13.9** Dilution of solutions

——————— **13.10** Reactions in solution: ionic equations

——————— **13.11** Colligative properties of solutions

——————— **13.12** Colloids and suspensions

——————— Reviewed chapter

——————— Answered assigned exercises:

#'s ——————————————————————————

Acids, bases, and salts

Chapter Fourteen is about special kinds of solutions, solutions of acids and bases. In this chapter, you will learn the important characteristics and properties of these solutions. The relationship between acids and bases, as well as their reactions, will also be presented. To demonstrate an understanding of Chapter Fourteen, you should be able to:

1 Define an acid and describe its ionization in water.

2 Define the strength of an acid and describe common properties of acids.

3 Define a base and describe its relationship to an acid.

4 Describe the common properties of bases.

5 Convert between pH and hydronium ion concentration.

6 Explain the process of acid-base neutralization, and give the name and formula of the salt formed.

7 Describe the process of titration and solve titration problems.

8 Use the concept of hydrolysis to explain why solutions of some salts are acidic while others are basic.

9 Explain how a buffer solution maintains the pH of a solution.

**14.1
ACIDS**

Although water is written as a molecule, H_2O, it does ionize, but only very slightly. A water molecule can ionize to produce one hydrogen ion, H^+, and one hydroxide ion, OH^-.

$$H_2O(l) \rightarrow H^+(aq) + OH^-(aq) \qquad \text{self-ionization of water}$$
water   hydrogen   hydroxide
           ion   ion

In pure water at room temperature, both the hydrogen ion concentration and the hydroxide ion concentration are 1.0×10^{-7} M. Thus,

Smithsonian Institution: Photo #56566

While he was still a student, the Swedish chemist Arrhenius had a revolutionary idea. He proposed in his doctoral thesis that electrolytes dissociate into ions in water. With this concept, he was able to explain why a solution of an electrolyte has a lower freezing point, greater conductivity, and higher osmotic pressure than a nonelectrolyte. Yet his professors found this notion hard to accept. As a result, Arrhenius was awarded the lowest possible passing grade for his thesis.

Fortunately, two other young chemists, Jacobus Henricus Van't Hoff and Friedrich Wilhelm Ostwald, were interested in Arrhenius' ideas, and promoted them. The discovery of the electron by J. J. Thomson provided additional support for Arrhenius by showing that atoms contained charged particles. In 1903, Arrhenius received the Nobel Prize for the same thesis that had nearly been rejected 19 years earlier.

Arrhenius had broad interests. He proposed the concept of the greenhouse effect—the process in which carbon dioxide in the atmosphere traps heat. Arrhenius noted that the Earth cools down at night by giving off infrared radiation. He predicted that the more carbon dioxide in the atmosphere, the less heat could be given off at night, and the higher the Earth's temperature would eventually become. A significant increase in the Earth's temperature could seriously disrupt food production, and cause massive flooding by melting the polar ice caps. Today, environmentalists are concerned about the greenhouse effect because fossil fuels like coal and oil produce carbon dioxide when burned.

there is only one H^+ and one OH^- ion for each 554 million H_2O molecules. However, when certain substances called acids and bases dissolve in water, the number of hydrogen and hydroxide ions changes.

The Swedish chemist Svante Arrhenius studied the dissociation of electrolytes. In 1884, he defined an **acid** as an electrolyte that furnishes H^+ ions. Acid molecules may have one or more hydrogen atoms that are released by ionization in water. For example, hydrogen chloride ionizes in water according to the following equation:

$$HCl(g) \xrightarrow{\text{in water}} H^+(aq) + Cl^-(aq)$$

hydrogen   hydrogen   chloride
chloride   ion   ion

Hydrogen chloride is an acid because it adds H^+ ions to the water. In this way it changes the balance between the H^+ and OH^- ions that exists in pure water. The solution now contains an excess of hydrogen ions. The number of H^+ ions becomes larger than the number of OH^- ions in solution.

The hydrogen ion, H^+, is simply a proton. It is the positive nucleus of a hydrogen atom that has lost its one electron. Because of its positive charge and small mass, this reactive subatomic particle cannot exist by itself in aqueous solution. It becomes hydrated, that is, surrounded by the partially negative oxygen atoms of water molecules. A hydrated hydrogen ion is called a **hydronium ion**, and its formula is written as H_3O^+. This formula represents the combination H^+ and H_2O. The properties of acids in aqueous solutions result from the presence of this ion, whether it is represented as H^+ or H_3O^+.

We can rewrite the equation for the ionization of hydrogen chloride in water to show the formation of the hydronium ion:

$$HCl(g) + H_2O(l) \rightarrow H_3O^+(aq) + Cl^-(aq)$$

hydrogen   water   hydronium   chloride
chloride   ion   ion

We can see more clearly what happens if we use Lewis structures:

$$H{\overset{..}{\underset{..}{\text{Cl}}}}: + H{\overset{..}{\text{O}}}: \rightarrow \left[H{\overset{H}{\overset{..}{\text{O}}}}: \right]^+ + [{\overset{..}{\underset{..}{\text{Cl}}}}:]^-$$

We see that acid ionization is really a reaction between an acid and water. A proton (shown in color) is transferred from the acid molecule to a water molecule, producing a hydrated proton, the hydronium ion. The rest of the acid molecule is now a negatively charged ion (Cl^-) since it carries the electron that was left behind by the hydrogen atom.

This process is the basis for the definition of acids developed in 1923 by the Danish chemist J. N. Brönsted and the British chemist

J. M. Lowry. According to the Brönsted-Lowry definition, an acid is a **proton donor**, a substance that can transfer a proton to another substance. The Brönsted-Lowry (or simply Brönsted) definition of an acid is broader than the Arrhenius definition because it can apply to reactions involving solvents other than water.

A **polyprotic acid** is an acid whose molecules can donate more than one hydrogen atom. Polyprotic acids ionize in several steps, donating one H^+ ion at a time. For example, sulfuric acid, H_2SO_4, is a polyprotic acid. It is called *di*protic because it can donate two protons. It ionizes in two steps:

$$H_2SO_4(aq) + H_2O(l) \rightarrow H_3O^+(aq) + HSO_4^-(aq)$$

sulfuric acid   water   hydronium ion   hydrogensulfate ion

$$HSO_4^-(aq) + H_2O(l) \rightarrow H_3O^+(aq) + SO_4^{2-}(aq)$$

hydrogensulfate ion   water   hydronium ion   sulfate ion

In the first step, one proton is transferred to water, producing a hydronium ion and the hydrogensulfate ion, HSO_4^-. In the second step, hydrogensulfate ion donates its remaining proton to water, forming a second hydronium ion and the sulfate ion. Phosphoric acid, H_3PO_4, is a *tri*protic acid, which ionizes in three steps.

Section 7.5 describes the names of acids. You may want to review that section before continuing.

14.2
PROPERTIES AND PREPARATION OF ACIDS

Because all solutions of acids in water contain excess H_3O^+ ions, they have certain properties in common. For example, these solutions taste sour like vinegar (which contains acetic acid) or lemons (which contain citric acid). Acid solutions also change the colors of certain dyes. For example, tea contains dyes that become lighter in color when we add lemon juice. In concentrated form, acids are corrosive. This term means that they dissolve substances they touch. They therefore can cause serious damage if spilled on your skin.

The strength of an acid depends on its tendency to donate a proton. A *strong* acid readily reacts with water. That is, nearly all the acid molecules transfer their protons to water molecules. In dilute hydrochloric acid, for example, over 90% of the HCl molecules exist as H_3O^+ and Cl^- ions. In a *weak* acid, on the other hand, only a portion of the molecules donate their protons. In dilute acetic acid, for example, about 99% of the molecules remain in the un-ionized form, $HC_2H_3O_2$. Only 1% donate their protons to water to form H_3O^+ and $C_2H_3O_2^-$ ions. The stronger an acid at a given concentration, the greater the concentration of hydronium ions in solution. (Section 14.5 will explain how to express this concentration.) The last column of Table 14-1 gives the percent ionization for several common acids in dilute form.

TABLE 14-1 *Properties of some common acids*

Name	Formula	Molecular weight (amu)	% (w/w)[a]	Specific gravity[a]	Molarity[a]	Percent ionization in dilute solutions[b]
acetic acid	$HC_2H_3O_2$	60.1	99.8%	1.05	17.4	about 1%
hydrochloric acid	HCl	36.5	37%	1.19	12.0	over 90%
nitric acid	HNO_3	63.0	70%	1.42	15.9	over 90%
phosphoric acid	H_3PO_4	98.0	85%	1.70	4.9	about 30%
sulfuric acid	H_2SO_4	98.0	96%	1.84	18.0	about 60%

[a]These values apply to commercial solutions.
[b]Based on the ratio of the total concentration of all ionized forms of the acid divided by the initial concentration.

Acids typically react with certain metals, oxidizing the metal, and liberating hydrogen gas.

$$Zn(s) + 2H_3O^+(aq) \rightarrow Zn^{2+}(aq) + H_2(g) + 2H_2O(l)$$

zinc   hydronium   zinc   hydrogen   water
ion   ion

Those metals listed to the left of hydrogen in the activity series given in Section 9.6 react with acids by this reaction or by a similar reaction. Even relatively unreactive gold reacts with, and dissolves in, a certain mixture of acids. This mixture of HCl and HNO_3 is called *aqua regia*, which in Latin means "royal water." Acids also react with the oxides and hydroxides of metals. An example is the reaction of iron(II) oxide with acid:

$$FeO(s) + 2H_3O^+(aq) \rightarrow Fe^{2+}(aq) + 3H_2O(l)$$

iron(II)   hydronium   iron(II)   water
oxide   ion   ion

Table 14-1 also contains information about the concentrated forms of commercially available acids. Concentrated acetic acid, $HC_2H_3O_2$, is called glacial acetic acid because it solidifies to an icelike solid at 17°C. Unlike the other acids in Table 14-1, acetic acid contains carbon atoms and is called an organic acid. Although its molecule contains four hydrogen atoms, only one can be ionized. The other three remain firmly bonded to a carbon atom. Vinegar is composed primarily of dilute (5%) acetic acid.

The remaining acids in Table 14-1, called mineral acids, are inorganic. Hydrochloric acid, HCl, consists of the very soluble gas hydrogen chloride dissolved in water. Your stomach secretes this acid at a concentration of 0.5% (0.1 M) to aid in the digestion of food. Nitric acid, HNO_3, is a strong corrosive acid that gives off fumes. Solutions of nitric acid may appear yellow because they contain decomposition products like NO_2. If you acidentally spill any on yourself, your skin

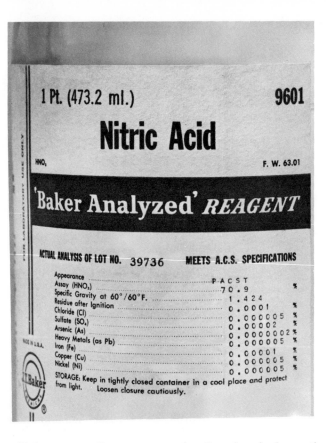

FIGURE 14-1

Nitric acid label. This label lists the concentration and important properties of nitric acid. (Photo courtesy of J.T. Baker Chemical Company.)

will become yellow as a result of a chemical reaction with protein molecules in the skin. (If you do spill nitric acid on yourself, immediately rinse the area with water.) Phosphoric acid, H_3PO_4, also called orthophosphoric acid, is a thick, syrupy liquid in concentrated form. In solution, it ionizes to produce four ions: H_3O^+, $H_2PO_4^-$, HPO_4^{2-} and PO_4^{3-}. Sulfuric acid, H_2SO_4, is viscous (resists flow) and very corrosive. As mentioned in Section 7.6, it is one of the most useful compounds, and is known as the workhorse of the chemical industry. Figure 14-1 shows the label from a bottle of a commercial acid.

There are several general methods for preparing acids. Commercial processes for preparing sulfuric acid and nitric acid have already been described in Section 9.10. The reaction of water with a nonmetallic oxide, such as sulfur trioxide, forms an acid.

$$\underset{\substack{\text{sulfur} \\ \text{trioxide}}}{SO_3(s)} + \underset{\text{water}}{H_2O(l)} \rightarrow \underset{\substack{\text{sulfuric} \\ \text{acid}}}{H_2SO_4(aq)}$$

These oxides, which lack only water to make the acid, are called acid anhydrides ("acid without water").

Certain acids can also be produced by double replacement reactions between an ionic compound that contains the desired anion and another acid having a higher boiling point. For example, hydrochloric acid can be made from NaCl, a compound containing the necessary anion, Cl^-, and sulfuric acid, which has a higher boiling point than HCl.

$$NaCl(s) + H_2SO_4(aq) \rightarrow HCl(g) + NaHSO_4(s)$$

sodium   sulfuric   hydrogen   sodium
chloride   acid   chloride   hydrogensulfate

The $NaHSO_4$ remains behind as HCl is boiled off and collected.

Another method for preparing acids is the direct combination of the elements that make up the acid. For example, HCl can be produced by combining elemental hydrogen and chlorine:

$$H_2(g) + Cl_2(g) \rightarrow 2HCl(g)$$

hydrogen   chlorine   hydrogen
chloride

14.3 BASES AND THEIR RELATION TO ACIDS

Arrhenius defined a **base**, also called an **alkali**, as an electrolyte that furnishes OH^- ions. Thus a water-soluble ionic compound whose anion is OH^- is a base. Sodium hydroxide, for example, dissociates in water as follows:

$$NaOH(s) \xrightarrow{\text{in water}} Na^+(aq) + OH^-(aq)$$

sodium   sodium   hydroxide
hydroxide   ion   ion

Sodium hydroxide is therefore a base. It dissociates into hydroxide ions that make the resulting solution basic or alkaline. The nature of the cation makes relatively little difference.

Other types of compounds, like ammonia, NH_3, are bases even though their molecules do not contain hydroxide ions. Such compounds raise the OH^- concentration of water because they react with water to produce hydroxide ions. For example, ammonia reacts with water as follows:

$$NH_3(g) + H_2O(l) \rightarrow NH_4^+(aq) + OH^-(aq)$$

ammonia   water   ammonium   hydroxide
ion   ion

Thus, the resulting aqueous solution contains a greater number of hydroxide ions than hydrogen ions.

According to the Brönsted-Lowry definition, a base is a **proton acceptor**. An acid can transfer its proton to a base. For example, hydroxide ion is a base by the Brönsted-Lowry definition because it accepts a proton to form water.

$$OH^-(aq) + H^+(aq) \rightarrow H_2O(l)$$

hydroxide   hydrogen   water
ion   ion

$$\left[\ddot{\underset{xx}{\overset{xx}{O}}}^x H\right]^- + H^+ \rightarrow \overset{xx}{\underset{xx}{O}}^x H$$
$$\hspace{6cm} H$$

Ammonia is also a base according to the Brönsted-Lowry definition, since it can accept a proton to form ammonium ion.

$$NH_3(aq) + H^+(aq) \rightarrow NH_4^+(aq)$$

ammonia   hydrogen   ammonium
          ion   ion

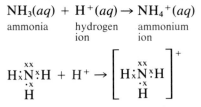

In order to accept a proton, a base must have a lone pair of electrons available to share. Both OH^- and NH_3 have lone pairs of electrons, as we can see from their Lewis structures. (NH_3 has one lone pair, and OH^- has two lone pairs of electrons.)

The strength of a base depends on its tendency to accept a proton, as well as on its solubility. As shown in the last column of Table 14-2, both potassium hydroxide, KOH, and sodium hydroxide, NaOH, produce large numbers of hydroxide ions in water. These two bases are very soluble and dissociate almost completely in dilute solutions. In addition, the hydroxide ion has a strong tendency to accept a proton. Therefore, solutions of potassium hydroxide and sodium hydroxide are strongly basic.

Calcium hydroxide, $Ca(OH)_2$, and magnesium hydroxide, $Mg(OH)_2$, also dissociate in water almost completely to produce hydroxide ions. However, the solubility of these two compounds is low. Thus, only a relatively small number of formula units are able to dissolve and dissociate in water. As a result, the number of hydroxide ions in solution is small, and their solutions are only weakly basic.

Ammonia, NH_3, on the other hand, is quite soluble in water. However, ammonia molecules have relatively little tendency to accept protons. Therefore, when ammonia is dissolved in water, very few hydroxide ions (and ammonium ions) are produced. Most of the molecules remain as NH_3. For this reason, ammonia solutions are weakly basic.

TABLE 14-2 *Properties of some common bases*

Name	Formula	Formula weight (amu)	% (w/w)[a]	Specific gravity[a]	Molarity[a]	Percent OH^- in dilute solutions
ammonia	NH_3	17.0	29	0.90	14.8	about 1%
potassium hydroxide	KOH	56.1	45	1.46	11.7	over 90%
sodium hydroxide	NaOH	40.0	50	1.53	19.1	over 90%

[a]These values apply to commercial solutions.

The relationship between acids and bases is made clear by the Brönsted-Lowry definitions: acids donate protons and bases accept them. By giving up its proton, an acid becomes a base because it can now gain back or accept another proton. The base that is formed when an acid loses a proton is called the **conjugate base** of that acid.

acid → H$^+$ + conjugate base

Acetate ion, $C_2H_3O_2^-$, is the conjugate base of acetic acid, for example.

Similarly, a base becomes an acid by gaining a proton, which it can then donate. The acid that is formed when a base gains a proton is called the **conjugate acid** of that base.

base + H$^+$ → conjugate acid

For example, the base ammonia picks up a proton to form ammonium ion, NH_4^+, which in turn can act as an acid by donating its "extra" proton.

In general, an acid and a base that differ from each other only by the presence or absence of a proton are known as a **conjugate acid-base pair**. (The word "conjugate" means coupled or joined as a pair.) Examples of conjugate acid-base pairs are presented in Table 14-3. Note

TABLE 14-3 *Conjugate acid-base pairs*

| Conjugate acid | | Conjugate Base | |
Name	Formula	Formula	Name
perchloric acid	$HClO_4$	ClO_4^-	perchlorate ion
sulfuric acid	H_2SO_4	HSO_4^-	hydrogensulfate ion
hydrogen chloride	HCl	Cl$^-$	chloride ion
nitric acid	HNO_3	NO_3^-	nitrate ion
hydronium ion	H_3O^+	H_2O	water
hydrogensulfate ion	HSO_4^-	SO_4^{2-}	sulfate ion
phosphoric acid	H_3PO_4	$H_2PO_4^-$	dihydrogenphosphate ion
acetic acid	$HC_2H_3O_2$	$C_2H_3O_2^-$	acetate ion
carbonic acid	H_2CO_3	HCO_3^-	hydrogencarbonate ion
hydrogen sulfide	H_2S	HS$^-$	hydrosulfide ion
ammonium ion	NH_4^+	NH_3	ammonia
hydrogen cyanide	HCN	CN$^-$	cyanide ion
hydrogencarbonate ion	HCO_3^-	CO_3^{2-}	carbonate ion
water	H_2O	OH$^-$	hydroxide ion
ethyl alcohol	C_2H_5OH	$C_2H_5O^-$	ethoxide ion
ammonia	NH_3	NH_2^-	amide ion
methylamine	CH_3NH_2	CH_3NH^-	methylamide ion
hydrogen	H_2	H$^-$	hydride ion

Increasing acid strength (left column, upward arrow)

Increasing base strength (right column, downward arrow)

that the stronger an acid, the weaker its conjugate base. For example, hydrogen chloride (hydrochloric acid) is a strong acid, but its conjugate base Cl^- is weak. Because HCl readily transfers its proton, Cl^- has little tendency to gain one. By the same reasoning, a strong base has a weak conjugate acid.

Certain molecules can act either as acids or as a bases. These molecules are called **amphoteric** (based on a Greek work meaning "both"). The term **amphiprotic** is sometimes used instead, to emphasize that such a molecule can either donate or accept a proton. Water molecules are amphoteric. In the presence of a base, a water molecule will donate a proton and thus act as an acid.

$$NH_3 + H_2O \rightarrow NH_4{}^+ + OH^-$$
$$\underset{\substack{\text{proton donor}\\ \text{(acid)}}}{}$$

In the presence of an acid, a water molecule will accept a proton and thus act as a base.

$$HCl + H_2O \rightarrow H_3O^+ + Cl^-$$
$$\underset{\substack{\text{proton acceptor}\\ \text{(base)}}}{}$$

HSO_4, $HCO_3{}^-$, and other hydrogen-containing anions produced by polyprotic acids are also amphoteric. They too can either act like an acid and donate a proton, or act like a base and accept a proton.

14.4 PROPERTIES OF BASES

Solutions of bases taste bitter and feel slippery. Strong bases are caustic. This term means that they are capable of corroding or otherwise destroying living tissue. The ability of bases to react with fats and oils makes them useful as cleaning agents. For example, Drano, which is used to clear clogged sink drains, contains the base sodium hydroxide. The sodium hydroxide converts grease into a soluble, soapy material that washes away. Cleaning products that contain bases are dangerous if spilled on your skin or splashed in your eyes. Additional properties of commercially available solutions of bases are given in Table 14-2.

Aqueous ammonia, $NH_3(aq)$, a solution of ammonia gas dissolved in water, is often called "ammonium hydroxide." However, although a small number of ammonium and hydroxide ions are present in aqueous ammonia, no undissociated NH_4OH molecules actually exist. (Keep in mind the differences between gaseous ammonia, $NH_3(g)$, aqueous ammonia, $NH_3(aq)$, and the ammonium ion, $NH_4{}^+(aq)$.) Ammonia is the second most produced industrial chemical. The bases sodium hydroxide, NaOH, and potassium hydroxide, KOH, are both water-soluble, hygroscopic solids. Concentrated solutions of these compounds should be stored in plastic containers because they slowly react with glass. Magnesium hydroxide, $Mg(OH)_2$, and calcium hydroxide, $Ca(OH)_2$, are much less soluble, and form weakly basic solutions. Milk of magnesia, an antacid and laxative, is the name given to $Mg(OH)_2$ in water. Limewater is the name for a $Ca(OH)_2$ solution.

Bases can be prepared by the reaction of Group 1A and 2A metals with water. For example, sodium reacts with water to produce sodium hydroxide:

$$2Na(s) + 2H_2O(l) \rightarrow 2Na^+(aq) + 2OH^-(aq) + H_2(g)$$

    sodium   water   sodium   hydroxide   hydrogen
                                ion   ion

The oxides of Group 1A and 2A elements also produce bases by reacting with water. For example, sodium oxide reacts with water as follows:

$$Na_2O(s) + H_2O(l) \rightarrow 2Na^+(aq) + 2OH^-(aq)$$

    sodium   water   sodium   hydroxide
    oxide   ion   ion

Bases can also be produced by using replacement reactions to exchange cations of hydroxide bases. For example, we can make sodium hydroxide by adding sodium carbonate to a calcium hydroxide solution. Calcium carbonate is precipitated, leaving sodium hydroxide in solution.

$$Ca(OH)_2(aq) + Na_2CO_3(aq) \rightarrow CaCO_3(s) + 2NaOH(aq)$$

    calcium   sodium   calcium   sodium
    hydroxide   carbonate   carbonate   hydroxide

Ammonia can be synthesized from N_2 and H_2 by the Haber process, which is described in Chapter Nine.

**14.5
pH OF ACIDS
AND BASES**

The acidity or basicity (also called alkalinity) of an aqueous solution depends on the relative numbers of hydronium ions and hydroxide ions present. Pure water contains molecular H_2O, and equal but small numbers of hydronium and hydroxide ions.

When an acid or a base is added to water, it produces additional hydronium or hydroxide ions. The stronger an acid at a given concentration, the greater the resulting excess of hydronium ions in solution. The stronger a base at a given concentration, the greater the resulting excess of hydroxide ions in solution. In water, the product of the hydronium ion concentration and the hydroxide ion concentration is always equal to 1.00×10^{-14} (at 25°C). This number is called the **ion-product constant** for water. (We will discuss this constant further in Chapter Sixteen.) We can then write:

$$[H_3O^+] \times [OH^-] = 1.00 \times 10^{-14}$$

Fluorosulfonic acid, HSO_3F, is called a superacid. It is over a billion times stronger than a very strong acid like nitric acid. A superacid can even donate protons to acetic acid, making the acetic acid act as a base!

The square brackets stand for the concentration expressed in moles/liter. Note that in pure water, since the hydronium and hydroxide ion concentrations are equal, the concentrations of each ion must be

$$[H_3O^+] = [OH^-] = 1.00 \times 10^{-7}$$

since $(1.00 \times 10^{-7}) \times (1.00 \times 10^{-7}) = 1.00 \times 10^{-14}$.

When the hydronium and hydroxide ion concentrations are *not* equal, the solution is either acidic or basic. If either the hydronium or hydroxide ion concentration is known, the other concentration can be found by using the fact that their product equals 1.00×10^{-14}. Suppose that we know the H_3O^+ concentration is 10^{-4} moles/liter, and that we want to know the OH^- ion concentration. We set up the problem as follows:

UNKNOWN	GIVEN	CONNECTION
$[OH^-]$	$[H_3O^+] = 10^{-4}$	$[H_3O^+] \times [OH^-] = 10^{-14}$

$$[H_3O^+] \times [OH^-] = 10^{-14}$$

$$10^{-4} \times [OH^-] = 10^{-14}$$

$$[OH^-] = \frac{10^{-14}}{10^{-4}}$$

$$[OH^-] = 10^{-10}$$

The usual concentration of either hydronium ion or hydroxide ion in an aqueous solution can range from one (or 10^0) to 10^{-14} moles/liter. However, we can avoid using numbers with negative exponents because chemists have developed a way to express the concentrations of hydronium ions as small positive numbers. We use the unit **pH**, which is always written with a small p and capital H. *The pH of a solution is a measure of its hydronium ion concentration,* as shown in Table 14-4. (Note that on each line in the table the product $[H_3O^+] \times [OH^-]$ is always equal to 1.00×10^{-14}.)

The usual pH scale ranges from 0 to 14. The midpoint of the scale, pH 7, represents pure water, which is *neutral*. It is neither acidic nor basic because the concentrations of hydronium ion and hydroxide ion are equal (each is 10^{-7} moles/liter). An acid solution has a greater hydronium ion concentration and a lower hydroxide ion concentration than pure water. Its pH is less than 7. *The more hydronium ions present, the more acidic is the solution and the lower is its pH.*

In a basic solution, the reverse holds true. A basic solution contains a greater hydroxide ion concentration and a smaller hydronium ion

TABLE 14-4 *The pH scale*

pH	H_3O^+ ion concentration (moles/L)	OH$^-$ ion concentration (moles/L)		General description
0	10^0 = 1.	10^{-14} = 0.000 000 000 000 01		strongly acidic
1	10^{-1} = 0.1	10^{-13} = 0.000 000 000 000 1	acid (excess	
2	10^{-2} = 0.01	10^{-12} = 0.000 000 000 001	H$^+$ ions)	
3	10^{-3} = 0.001	10^{-11} = 0.000 000 000 01		mildly acidic
4	10^{-4} = 0.000 1	10^{-10} = 0.000 000 000 1		
5	10^{-5} = 0.000 01	10^{-9} = 0.000 000 001		
6	10^{-6} = 0.000 001	10^{-8} = 0.000 000 01		
7	10^{-7} = 0.000 000 1	10^{-7} = 0.000 000 1		neutral
8	10^{-8} = 0.000 000 01	10^{-6} = 0.000 001		
9	10^{-9} = 0.000 000 001	10^{-5} = 0.000 01	base (excess	
10	10^{-10} = 0.000 000 000 1	10^{-4} = 0.000 1	OH$^-$ ions)	
11	10^{-11} = 0.000 000 000 01	10^{-3} = 0.001		mildly basic
12	10^{-12} = 0.000 000 000 001	10^{-2} = 0.01		
13	10^{-13} = 0.000 000 000 000 1	10^{-1} = 0.1		
14	10^{-14} = 0.000 000 000 000 01	10^0 = 1.		strongly basic

concentration than pure water. The pH of a basic solution is greater than 7. *The more hydroxide ions present, the more basic is the solution, and the higher is its pH.* Table 14-5 gives the pH of some common solutions.

The pH of a solution is given by the following formula:

$$pH = -\log [H_3O^+]$$

The pH is the negative logarithm of the hydronium ion concentration in moles per liter. The logarithm (log) of a number is the power to which 10 must be raised in order to equal the given number. The following list gives some examples of numbers and their logarithms.

Number	Logarithm of number
10^{-2} = 0.01	$\log(10^{-2})$ = -2
10^{-1} = 0.1	$\log(10^{-1})$ = -1
10^0 = 1	$\log(10^0)$ = 0
10^1 = 10	$\log(10^1)$ = 1
10^2 = 100	$\log(10^2)$ = 2

For example, suppose that the concentration of H_3O^+ ion is 0.001 M. In order to equal 0.001, we must raise 10 to the -3 power. The log of 0.001 therefore is -3.

TABLE 14-5   *The pH of some common solutions*

Solution	pH
hydrochloric acid (1.0 M)	1.1
sulfuric acid (0.50 M)	1.2
phosphoric acid (0.33 M)	1.5
citric acid (lemon juice)	2.2
acetic acid (vinegar)	2.9
carbonic acid (soda water, saturated)	3.8
tomato juice	4.2
coffee, black	5.0
urine	6.0
rainwater	6.2
milk	6.5
water, pure	7.0
saliva	7.2
blood	7.4
sodium bicarbonate (1.0 M)	8.4
magnesium hydroxide (milk of magnesia, saturated)	10.5
ammonia (1.0 M)	11.1
calcium hydroxide (saturated)	12.4
sodium hydroxide (1.0 M)	13.0

The hydronium ion concentration is almost always smaller than 1 (or 10^0) moles/L. The logarithm of a concentration given by a negative power of ten is negative. To make pH values positive, pH has been defined as the *negative* logarithm of the concentration. When we place a minus sign in front of a logarithm, we change a negative value into a positive one, since $(-) \times (-) = (+)$. Therefore, pH values are almost always positive. For example, the pH of neutral water, where $[H_3O^+] = 10^{-7}$, is equal to $-\log(10^{-7}) = -(-7) = 7$.

Example   A solution has a hydronium ion concentration of 10^{-10} moles per liter. Find its pH.

UNKNOWN	GIVEN	CONNECTION
pH	$[H_3O^+] = 10^{-10}$	$pH = -\log[H_3O^+]$

$$pH = -\log[H_3O^+] = -\log(10^{-10}) = -(-10) = 10$$

We can also find the hydronium ion concentration if we are given the pH. To do this, we rewrite the definition of pH in terms of the hydronium ion concentration.

$$[H_3O^+] = 10^{-pH}$$

Using this equation, we can see that a solution with a pH of 3, for example, must have a hydronium ion concentration of 10^{-3} moles/liter.

Example   Find the hydronium ion concentration for a solution whose pH is 8.

UNKNOWN	GIVEN	CONNECTION
$[H_3O^+]$	pH = 8	$[H_3O^+] = 10^{-pH}$

$[H_3O^+] = 10^{-pH}$

$[H_3O^+] = 10^{-8}$ moles/liters

Notice that for every change in pH by a factor of one, the concentration changes by a factor of 10. For example, a solution with a pH of 2 has a hydronium ion concentration that is 10 times greater than that of a solution with a pH of 3.

As shown in Table 14-5, pH values need not be whole numbers. For example, a pH of 1.5 means that the hydronium ion concentration

Do-it-yourself 14

An indicator

A pH indicator is a compound whose color depends on the pH of the surrounding solution. Generally, indicators are complex organic molecules. Litmus is probably the best known example of an indicator. It is a dye obtained from a form of plant called a lichen. (A lichen is a combination of a fungus and an alga living together. Lichens are found growing on rocks and tree trunks.)

By following the procedure given here, you can prepare a pH indicator at home from red cabbage.

To do: Cut up a portion of red cabbage into small pieces. Put the pieces in a pot and add boiling water. Stirring occasionally, let this mixture stand for about $\frac{1}{2}$ hour. Then pour off the liquid into a jar.

This liquid is your indicator. It is red in acid solutions and blue to green in basic solutions. Add a few drops of the indicator to a few milliliters of each of the following: ammonia, tea, fruit juice, soft drink, shampoo. Observe the color of the indicator in each case. You should now be able to say whether each of these common household solutions is acidic or basic.

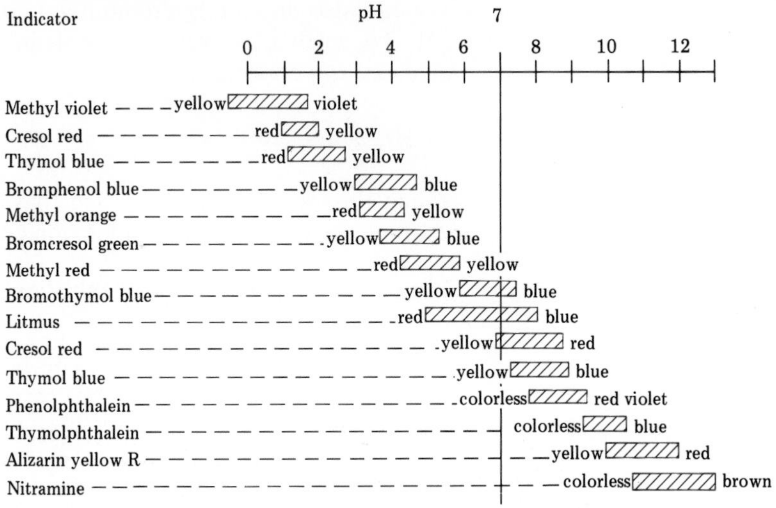

Indicator   pH   7

	0	2	4	6	8	10	12

Methyl violet — — —yellow▨▨▨ violet
Cresol red — — — — — — red▨▨ yellow
Thymol blue — — — — — — —red▨▨ yellow
Bromphenol blue— — — — — — —yellow▨▨ blue
Methyl orange — — — — — — — —red▨▨ yellow
Bromcresol green — — — — — — —yellow▨▨ blue
Methyl red — — — — — — — — —red▨▨ yellow
Bromothymol blue— — — — — — — — —yellow▨▨ blue
Litmus — — — — — — — — — — —red▨▨▨ blue
Cresol red — — — — — — — — — — — yellow▨▨ red
Thymol blue — — — — — — — — — — — yellow▨▨ blue
Phenolphthalein — — — — — — — — — — —colorless▨▨ red violet
Thymolphthalein — — — — — — — — — — — — colorless▨▨ blue
Alizarin yellow R — — — — — — — — — — — — —yellow▨▨ red
Nitramine — — — — — — — — — — — — — — —colorless▨▨ brown

FIGURE 14-2

pH indicators. The indicators change color over the pH ranges shown. For example, phenolphthalein changes from colorless to red-violet as the pH increases from 8 to 10.

is between 10^{-1} moles/liter (or pH 1) and 10^{-2} moles/liter (or pH 2). (Appendix H explains how to determine the exact value.)

In the laboratory, we can find the pH of a solution in several ways. Certain dyes, called **indicators**, change color over a specific pH range, usually 2 pH units, as shown in Figure 14-2. For example, litmus paper contains a dye that changes from red to blue in the region of pH 4.4 to 8.3. This indicator is red in acid solutions having a pH less than 4.4, and blue in base solutions having a pH greater than 8.3. Several indicators can be mixed in such a way that the combined color changes gradually with pH. By comparing the color of the indicator with a printed color scale, the pH can be estimated to within several tenths of a pH unit.

A pH meter, shown in Figure 14-3, measures the pH of a solution more exactly. It makes use of a special sensing device called a glass electrode, which responds to the hydronium ion concentration. The glass electrode is placed in the solution (along with a reference electrode) and the meter measures the response. The meter must always be adjusted first, or standardized, with a solution of known pH.

14.6
NEUTRALIZATION AND SALTS

Neutralization is the name given to the reaction of an acid with a base. In water, neutralization consists of the reaction between a hydronium ion and a hydroxide ion.

$$H_3O^+(aq) + OH^-(aq) \rightarrow 2H_2O(l)$$
hydronium   hydroxide   water
ion   ion

Two neutral water molecules form from the loss of one hydronium ion and one hydroxide ion from the solution. We can say that each hy-

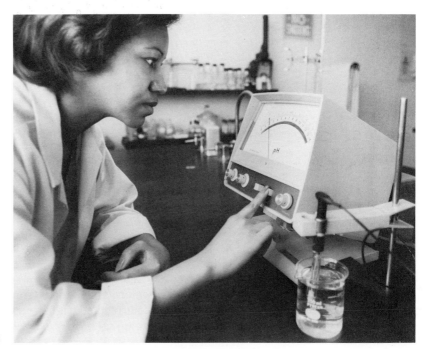

FIGURE 14-3

A pH meter. The scale of the meter is calibrated to read the hydrogen ion concentration in pH units. (Photo by Al Green.)

droxide ion exactly cancels or neutralizes the effect of one hydronium ion. For this reason, certain bases, such as magnesium hydroxide (as milk of magnesia), are used as antacids. They neutralize excess hydrochloric acid in the stomach.

In terms of the Brönsted-Lowry definitions of acids and bases, neutralization in water is the transfer of a proton from a hydronium ion to a hydroxide ion.

$$\left[H \overset{xx}{\underset{\underset{H}{\cdot x}}{\overset{x}{\cdot}}} O \overset{x}{\underset{x}{\cdot}} H \right]^{+} + \left[\overset{xx}{\underset{xx}{\cdot}} O \overset{x}{\underset{x}{\cdot}} H \right]^{-} \rightarrow H \overset{xx}{\underset{\underset{H}{\cdot x}}{\overset{x}{\cdot}}} O \overset{x}{\underset{x}{\cdot}} + \overset{xx}{\underset{\underset{H}{xx}}{\cdot}} O \overset{x}{\underset{x}{\cdot}} H$$

In this reaction, the hydronium ion forms its conjugate base, H_2O, and the hydroxide ion forms its conjugate acid, also H_2O. The neutralization reaction between a Brönsted acid and a Brönsted base can be represented in the following way.

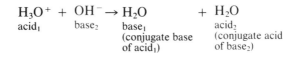

$$\underset{\text{acid}_1}{H_3O^+} + \underset{\text{base}_2}{OH^-} \rightarrow \underset{\substack{\text{base}_1 \\ \text{(conjugate base} \\ \text{of acid}_1)}}{H_2O} + \underset{\substack{\text{acid}_2 \\ \text{(conjugate acid} \\ \text{of base}_2)}}{H_2O}$$

In this example, the reaction of a hydronium ion with a hydroxide ion, both of the products are water molecules. Because water molecules

TABLE 14-6 *Formation of salts*

Base	Acid	Formula of salt	Name of salt
NaOH	H_2SO_4	Na_2SO_4	sodium sulfate
KOH	H_3PO_4	K_3PO_4	potassium phosphate
$Mg(OH)_2$	$HC_2H_3O_2$	$Mg(C_2H_3O_2)_2$	magnesium acetate
$Ca(OH)_2$	HCl	$CaCl_2$	calcium chloride
$Al(OH)_3$	HNO_3	$Al(NO_3)_3$	aluminum nitrate

dissociate so slightly, the reaction effectively goes to completion. (Notice that the neutralization reaction is the reverse of the self-ionization of water.) Once a hydronium ion and hydroxide ion react to form a molecule of water, they lose their separate identities and can no longer act as acid and base.

When an acid and a base react, they generally form another product in addition to water. This product is called a **salt**. For example, if sodium hydroxide is reacted with an equivalent amount of hydrochloric acid and the resulting solution is evaporated, the salt, sodium chloride, remains.

$$NaOH + HCl \rightarrow NaCl + H_2O$$
$$\text{base} \qquad \text{acid} \qquad \text{salt} \qquad \text{water}$$

A salt is an ionic compound that contains a cation other than hydrogen ion and an anion other than hydroxide ion. The salt produced by neutralization consists of the cation from the base and the anion from the acid. In the salt sodium chloride, the cation, Na^+, comes from the base, NaOH, and the anion, Cl^-, comes from the acid, HCl.

$$Na|Cl$$
cation from base (NaOH) | anion from acid (HCl)

Table 14-6 shows examples of other salts formed by neutralization.

14.7 TITRATION

The neutralization reaction is the basis for a procedure called **titration**. Titration is the process of gradually reacting an acid with a base in order to find the concentration of one of them. For example, if the solution of unknown concentration is an acid, we add a base of known concentration slowly until all the acid has been neutralized. The base is usually added by using a buret as shown in Figure 14-4. An indicator signals the point at which all of the hydronium ions have been neutralized by the added hydroxide ions. The pH rises suddenly at this

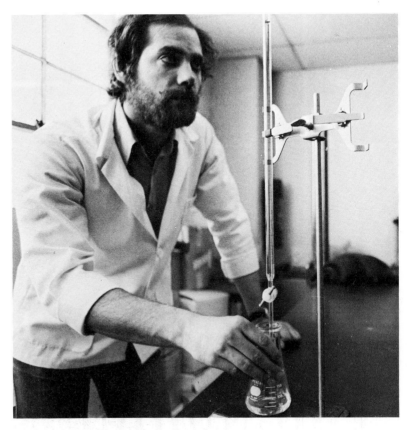

FIGURE 14-4

Titration. In this process, an unknown concentration of acid is determined by adding a base of known concentration until the acid is neutralized. Similarly, an unknown concentration of base can be determined by adding (or titrating with) an acid of known concentration. (Photo by Al Green.)

point, as the added base rapidly converts the acidic solution to a basic one. If the solution of unknown concentration is a base, we follow the reverse procedure, adding acid until the base is neutralized.

We can then calculate the unknown concentration from the volume of acid or base that was used to neutralize the solution. For example, suppose that we want to determine the concentration of acetic acid in a sample of vinegar by performing a titration. We fill a buret with a base solution of known concentration, say 0.500 M NaOH. The buret allows us to accurately deliver variable volumes of solution. We put a measured volume of vinegar, say 10.0 mL, into a flask, and add several drops of an appropriate indicator solution (such as phenolphthalein). Then we slowly add base from the buret to the flask of vinegar until the indicator changes color. At this point all the acetic acid in the vinegar has been neutralized. We can then read the buret to determine what volume of base was added. The calculation of acid concentration is based on the mole ratio of acid to base in the balanced neutralization equation.

$$HC_2H_3O_2(aq) + NaOH(aq) \rightarrow H_2O(l) + NaC_2H_3O_2(aq)$$

acetic acid   sodium   water   sodium
                 hydroxide   acetate

In this case, one mole of acid reacts with one mole of base. Using this mole ratio, we can find the number of moles of acid present in the vinegar, as well as the molarity of the vinegar (or acetic acid solution), based on the volume of base we used.

Suppose that we needed 17.1 mL of NaOH solution to neutralize the acetic acid in the vinegar. We outline the problem as shown:

UNKNOWN	GIVEN	CONNECTION
molarity of $HC_2H_3O_2$	0.500 M NaOH 17.1 mL NaOH solution 10.0 mL $HC_2H_3O_2$ solution	$\dfrac{1 \text{ L}}{10^3 \text{ mL}}$ or $\dfrac{10^3 \text{ mL}}{1 \text{ L}}$ $\dfrac{1 \text{ mole } HC_2H_3O_2}{1 \text{ mole NaOH}}$ or $\dfrac{1 \text{ mole NaOH}}{1 \text{ mole } HC_2H_3O_2}$ $\text{molarity} = \dfrac{\text{moles of solute}}{\text{liter of solution}}$

First we find the number of moles of acetic acid that were neutralized by the base.

$$\text{moles } HC_2H_3O_2 = 17.1 \text{ mL NaOH solution} \times \frac{1 \text{ L}}{10^3 \text{ mL}} \times$$

$$\frac{0.500 \text{ mole NaOH}}{\text{liter of NaOH solution}} \times \frac{1 \text{ mole } HC_2H_3O_2}{1 \text{ mole NaOH}}$$

ESTIMATE $2 \times 10^1 \text{ mL NaOH solution} \times \dfrac{1 \text{ L}}{10^3 \text{ mL}} \times$

$$\frac{5 \times 10^{-1} \text{ mole NaOH}}{\text{liter of NaOH solution}} \times \frac{1 \text{ mole } HC_2H_3O_2}{1 \text{ mole NaOH}}$$

$$= 1 \times 10^{-2} \text{ mole } HC_2H_3O_2$$

CALCULATION 8.55×10^{-3} mole $HC_2H_3O_2$

The molarity of the acetic acid solution is then easily found, because we know that the 8.55×10^{-3} mole of acetic acid was present in 10.0 mL of vinegar.

$$\text{molarity} = \frac{8.55 \times 10^{-3} \text{ mole HC}_2\text{H}_3\text{O}_2}{10.0 \text{ mL solution}} \times \frac{10^3 \text{ mL}}{1 \text{ L}}$$

$$= \frac{0.855 \text{ mole HC}_2\text{H}_3\text{O}_2}{\text{liter of solution}}$$

$$= 0.855 \text{ M HC}_2\text{H}_3\text{O}_2$$

Therefore, in this titration, 10.0 mL of 0.855 M $HC_2H_3O_2$ were neutralized with 17.1 mL of 0.500 M NaOH.

Example   Milk of magnesia contains the base magnesium hydroxide. Find its molarity if 45.1 mL of 1.60 M HCl are needed to titrate 25.0 mL of milk of magnesia:

$$2HCl(aq) + Mg(OH)_2(aq) \rightarrow 2H_2O(l) + MgCl_2(aq)$$

UNKNOWN	GIVEN	CONNECTION
molarity of $Mg(OH)_2$	1.60 M HCl 45.1 mL HCl 25.0 mL $Mg(OH)_2$	$\dfrac{1 \text{ L}}{10^3 \text{ mL}}$ or $\dfrac{10^3 \text{ mL}}{1 \text{ L}}$ $\dfrac{2 \text{ moles HCl}}{1 \text{ mole Mg(OH)}_2}$ or $\dfrac{1 \text{ mole Mg(OH)}_2}{2 \text{ moles HCl}}$ $\text{molarity} = \dfrac{\text{moles of solute}}{\text{liter of solution}}$

moles of $Mg(OH)_2$ = 45.1 m̶L̶ H̶C̶l̶ ̶s̶o̶l̶u̶t̶i̶o̶n̶ $\times \dfrac{1 \text{ L̶}}{10^3 \text{ m̶L̶}} \times$

$$\frac{1.60 \text{ m̶o̶l̶e̶s̶ ̶H̶C̶l̶}}{\text{l̶i̶t̶e̶r̶ ̶o̶f̶ ̶H̶C̶l̶ ̶s̶o̶l̶u̶t̶i̶o̶n̶}} \times \frac{1 \text{ mole Mg(OH)}_2}{2 \text{ m̶o̶l̶e̶s̶ ̶H̶C̶l̶}}$$

ESTIMATE 5×10^1 m̶L̶ H̶C̶l̶ ̶s̶o̶l̶u̶t̶i̶o̶n̶ $\times \dfrac{1 \text{ L̶}}{10^3 \text{ m̶L̶}} \times$

$$\frac{2 \text{ m̶o̶l̶e̶s̶ ̶H̶C̶l̶}}{\text{l̶i̶t̶e̶r̶ ̶o̶f̶ ̶H̶C̶l̶ ̶s̶o̶l̶u̶t̶i̶o̶n̶}} \times \frac{1 \text{ mole Mg(OH)}_2}{2 \text{ m̶o̶l̶e̶s̶ ̶H̶C̶l̶}}$$

$$= 5 \times 10^{-2} \text{ mole Mg(OH)}_2$$

CALCULATION   0.0361 mole $Mg(OH)_2$

$$\text{molarity} = \frac{0.0361 \text{ mole Mg(OH)}_2}{25.0 \text{ mL solution}} \times \frac{10^3 \text{ mL}}{1 \text{ L}}$$

$$= \frac{1.44 \text{ moles Mg(OH)}_2}{\text{liter of solution}}$$

$$= 1.44 \text{ M Mg(OH)}_2$$

Practice problem   If 36.8 mL of 0.333 M NaOH are needed to neutralize 50.0 mL of a sulfuric acid solution, find the molarity of the acid solution.

ANSWER   0.123 M H_2SO_4

**14.8
HYDROLYSIS**   Although the reaction of an acid and a base is called neutralization, the resulting solution is not necessarily neutral. A solution of a salt produced by neutralization can have a pH other than 7 if any of the ions in the salt can react with water. Consider sodium acetate, $NaC_2H_3O_2$, the salt from the neutralization reaction of sodium hydroxide and acetic acid. When it dissolves in water, sodium acetate dissociates completely, since it is a soluble ionic compound. The solution consists of sodium ions, Na^+, and acetate ions, $C_2H_3O_2^-$. Acetate ions can react with water to form slightly ionized acetic acid.

$$C_2H_3O_2^-(aq) + H_2O(l) \rightarrow HC_2H_3O_2(aq) + OH^-(aq)$$
acetate ion   water   acetic acid   hydroxide ion

This reaction is called **hydrolysis** ("hydro" means "water" and "lysis" means "to separate") because a water molecule is broken up. A proton from a water molecule becomes attached to the acetate ion, leaving a free hydroxide ion. The solution is thus basic and has a pH above 7. Although the solution also contains $HC_2H_3O_2$, the solution is *not* acidic because acetic acid is a weak acid and does not affect the pH as much as does the strongly basic hydroxide ion.

On the other hand, consider ammonium chloride, NH_4Cl, the salt from the neutralization reaction of ammonia and hydrochloric acid. When this salt dissolves in water, it dissociates into ammonium ions and chloride ions. Ammonium ions react with water to make ammonia.

$$NH_4^+(aq) + H_2O(l) \rightarrow NH_3(aq) + H_3O^+(aq)$$
ammonium ion   water   ammonia   hydronium ion

Since a hydronium ion also forms, the solution becomes acidic, with a pH below 7. Even though ammonia is present, it is so weakly basic that it does not affect the pH as much as the strongly acidic hydronium ion. Hydrolysis occurs in salt solutions whenever one of the ions of the salt can react with water to produce a weak acid (like acetic acid)

TABLE 14-7 *Effect of some ions on pH*

Form acidic solutions	Form neutral solutions	Form basic solutions
HSO_4^-	Cl^-	CO_3^{2-}
$H_2PO_4^-$	Br^-	HCO_3^-
NH_4^+	I^-	PO_4^{3-}
metal cations $(Zn^{2+}, Cu^{2+}, etc.)$	SO_4^{2-}	$C_2H_3O_2^-$

or a weak base (like ammonia). Table 14-7 lists some common ions and their effect on pH.

Because of hydrolysis, an acid-base titration does not necessarily produce a neutral solution. For example, the titration of acetic acid, a weak acid, with sodium hydroxide can be represented by the following equation:

$$\underset{\text{acetic acid}}{HC_2H_3O_2(aq)} + \underset{\substack{\text{sodium} \\ \text{hydroxide}}}{NaOH(aq)} \rightarrow \underset{\substack{\text{sodium} \\ \text{acetate}}}{NaC_2H_3O_2(aq)} + \underset{\text{water}}{H_2O(l)}$$

At the point of neutralization, the solution no longer contains the original acetic acid or sodium hydroxide. It contains only sodium acetate. But because the acetate ion undergoes hydrolysis, the solution is actually basic at neutralization.

14.9 BUFFERS

For the human body to function properly, the pH of its fluids must be maintained within certain narrow ranges. For example, the pH of blood must stay between 7.35 and 7.45. Similarly, the pH of the solutions involved in industrial reactions and laboratory procedures must often be kept constant.

The pH of a solution can be maintained close to a certain value by a **buffer**. A buffer is a solution whose pH changes only very slightly when moderate amounts of either acid or base are added. The following table compares the effects on pH of adding 0.01 mole of HCl and 0.01 mole of NaOH to one liter each of pure water and of a typical buffer. Both pure water and the buffer are initially at pH 7.0.

Solution tested	Initial pH	pH after adding 0.01 mole HCl	pH after adding 0.01 mole NaOH
pure water	7.0	2.0	12.0
buffer	7.0	6.8	7.2

As we can see, there is quite a difference! Upon adding 0.01 mole of H_3O^+ ion (from the 0.01 M HCl) to the water and to the buffer, the water pH drops 5.0 units, while the pH of the buffer decreases by only 0.2 unit. When 0.01 mole of OH⁻ ion (from the 0.01 M NaOH) is added, the pH of a liter of pure water rises by 5.0 units, but the pH of the buffer increases by only 0.2 unit.

Box 14

Acidosis and alkalosis

In order for the human body to function normally, the blood pH must stay within the range 7.35 to 7.45. If the pH of a person's blood drops below 7.35, he or she is said to have acidosis, or low blood pH. Acidosis can cause disorientation, coma, and ultimately death. If the pH of a person's blood rises above 7.45, he or she has alkalosis, or high blood pH. Alkalosis can cause weak, irregular breathing, muscle cramps, and convulsions.

The blood's buffer system contains carbonic acid, H_2CO_3, and hydrogencarbonate ions, HCO_3^-. Changes in blood pH result from changes in the relative amounts of these two components. Respiration is a major factor in controling the blood's buffer system because the H_2CO_3 concentration in the blood depends on the partial pressure of CO_2 (p_{CO_2}) in the lungs.

$$CO_2(g) + H_2O(l) \rightleftarrows H_2CO_3(aq)$$
carbon   water   carbonic
dioxide   acid

For example, by holding your breath you increase p_{CO_2} in your lungs. As a result, more carbonic acid enters the blood. If you hold your breath for 5 minutes (don't try it!), the blood pH will drop to 6.3. This condition, which is caused by hypoventilation or reduced rate of breathing, is called respiratory acidosis. Respiratory acidosis sometimes occurs in persons with pneumonia, emphysema, poliomyelitis, or in patients who are anesthetized, because these conditions can interfere with breathing. The nervous system responds to acidosis by attempting to increase the rate and depth of breathing in order to decrease p_{CO_2}.

Respiratory alkalosis results from the other extreme—hyperventilation, or increased respiration. In this case, too much CO_2 is lost from the lungs, p_{CO^2} falls, and the H_2CO_3 concentration is reduced below normal.

An anxiety attack or hysteria can sometimes cause a person to breathe very rapidly. Under these conditions, the blood pH may rise to 7.6 or 7.7 within minutes. The nervous system responds to alkalosis by lowering the rate of breathing to increase p_{CO_2} and the H_2CO_3 concentration.

Acidosis and alkalosis can also result from changes in the HCO_3^- concentration. Metabolic acidosis, or low HCO_3^- concentration, can result from severe dehydration or untreated diabetes mellitus. Metabolic alkalosis, or high HCO_3^- concentration, can result from vomiting or an overdose of antacids.

A buffer maintains an almost constant pH by neutralizing small amounts of acid or base that may be added to the solution. The buffer may consist of a weak acid and its conjugate base, such as acetic acid and acetate ion. Or it may consist of a weak base and its conjugate acid, such as ammonia and ammonium ion. In your blood, the pH is kept at around 7.4 by several buffers. These buffers prevent acidosis, a pH below 7.35, as well as alkalosis, a pH about 7.45. The main blood buffer consists of carbonic acid, H_2CO_3 (this formula stands for carbon dioxide in water, $H_2O + CO_2$) and hydrogencarbonate ion, HCO_3^-. If a base enters the blood, OH^- ions react with the acid part of the buffer.

$$OH^-(aq) + H_2CO_3(aq) \rightarrow H_2O(l) + HCO_3^-(aq)$$

hydroxide   carbonic   water   hydrogencarbonate
ion   acid   ion

The base is thus neutralized as it forms more hydrogencarbonate ion and water. If acid enters the blood, H_3O^+ ions react with the hydrogencarbonate ion, converting it to carbonic acid.

$$H_3O^+(aq) + HCO_3^-(aq) \rightarrow H_2O(l) + H_2CO_3(aq)$$

hydronium   hydrogencarbonate   water   carbonic
ion   ion   acid

Since carbonic acid is so slightly ionized, the hydronium ions are effectively removed from the solution.

A buffer must contain two parts so that it can neutralize either an acid or a base, as we have just seen. The free hydronium ion (or hydroxide ion) concentration of the solution stays nearly constant, and therefore, so does the pH. Other examples of common buffers and their approximate pH values are:

Buffer	pH
$HC_2H_3O_2$; $C_2H_3O_2^-$	3.5
$H_2PO_4^-$; HPO_4^{2-}	7
NH_4^+; NH_3	9

NORMALITY: SUPPLEMENT TO CHAPTER FOURTEEN

In neutralization reactions, the numbers of moles of acid and base that react are not necessarily equal. Polyprotic acids have more than one hydrogen atom per molecule that can react. Some bases have several hydroxide ions in their formula units. For example, one mole of sulfuric acid, H_2SO_4, can provide two moles of hydrogen ions when reacted with a base. Similarly, one mole of calcium hydroxide, $Ca(OH)_2$, can provide two moles of hydroxide ions.

To take this factor into account, an alternate method of expressing concentration is sometimes used. Instead of dealing with moles of the compound, we consider the number of moles of H^+ (or H_3O^+) or OH^-

that are available to react. We refer to the number of moles of H^+ (or H_3O^+) or OH^- that are supplied by the acid or base as the number of **equivalents** (or **equiv**). The equivalent weight of a compound is equal to the molar mass of the compound multiplied by a conversion factor based on the number of equivalents that it produces per mole. For an acid, the equivalent weight is the mass in grams that will provide one mole of H^+ (or H_3O^+) ions. For sulfuric acid, the equivalent weight equals the molar mass, 98.08 g/mole, multiplied by the conversion factor relating the number of equivalents (or number of replaceable hydrogens) per mole. In this case, there are two protons per molecule and therefore two equivalents per mole.

$$\text{equivalent weight of } H_2SO_4 = \frac{98.08 \text{ g } H_2SO_4}{1 \text{ mole } H_2SO_4} \times \frac{1 \text{ mole } H_2SO_4}{2 \text{ equivalents}}$$

$$= 49.04 \text{ g } H_2SO_4 \text{ per equivalent}$$

Therefore 49.04 g of H_2SO_4 provides one mole of H^+ (or H_3O^+) ions. One mole of sulfuric acid, 98.08 g, can produce two moles of hydrogen ions, or two equivalents.

For bases, the procedure is similar. The equivalent weight of a base equals the mass in grams that provides one mole of OH^- ions. We find the equivalent weight by multiplying the molar mass by a conversion factor based on the number of hydroxide ions that the base produces. In calcium hydroxide, $Ca(OH)_2$, for example, there are two OH^- ions and therefore two equivalents of OH^- per mole of formula unit.

$$\text{equivalent weight of } Ca(OH)_2 = \frac{74.10 \text{ g } Ca(OH)_2}{1 \text{ mole } Ca(OH)_2} \times \frac{1 \text{ mole } Ca(OH)_2}{2 \text{ equivalents}}$$

$$= 37.05 \text{ g } Ca(OH)_2 \text{ per equivalent}$$

Once we have found the equivalent weight of a solution, concentration can then be expressed as **normality**, the number of equivalents of acid or base in one liter of solution.

$$\text{normality (N)} = \frac{\text{number of equivalents of solute}}{\text{one liter (dm}^3\text{) of solution}}$$

Normality is similar to molarity except that equivalents are used instead of moles. A one-normal (1 N) solution contains one equivalent per liter; a 2 N solution contains two equivalents per liter, and so on. *The normality of an acid or base solution is always greater than or equal to its molarity,* since one mole of the compound will provide at least one equivalent. A 1 M solution of HCl is also a 1 N solution because one mole of this acid provides one equivalent of H_3O^+ ion. But a 1 M solution of H_2SO_4 is a 2 N solution because each mole of this acid provides two equivalents of H_3O^+ ion.

Example What mass of magnesium hydroxide is needed to prepare 1.50 L of a 0.260 N solution?

UNKNOWN	GIVEN	CONNECTION
g $Mg(OH)_2$	1.50 L solution	$\dfrac{1 \text{ mole } Mg(OH)_2}{58.3 \text{ g } Mg(OH)_2}$
	$\dfrac{0.260 \text{ equiv}}{1 \text{ L solution}}$	or $\dfrac{58.3 \text{ g } Mg(OH)_2}{1 \text{ mole } Mg(OH)_2}$
		$\dfrac{2 \text{ equiv}}{1 \text{ mole } Mg(OH)_2}$
		or $\dfrac{1 \text{ mole } Mg(OH)_2}{2 \text{ equiv}}$

$$\text{g } Mg(OH)_2 = 1.50 \text{ L solution} \times \frac{0.260 \text{ equiv}}{1 \text{ L solution}} \times \frac{1 \text{ mole } Mg(OH)_2}{2 \text{ equiv}} \times \frac{58.3 \text{ g } Mg(OH)_2}{1 \text{ mole } Mg(OH)_2}$$

ESTIMATE $\quad 2 \text{ L solution} \times \dfrac{3 \times 10^{-1} \text{ equiv}}{1 \text{ L solution}} \times \dfrac{1 \text{ mole } Mg(OH)_2}{2 \text{ equiv}} \times \dfrac{6 \times 10^{1} \text{ g } Mg(OH)_2}{1 \text{ mole } Mg(OH)_2}$

$$= 2 \times 10^{1} \text{ g } Mg(OH)_2$$

CALCULATION $\quad$ 11.4 g $Mg(OH)_2$

Normality can be used in titration calculations. By the end of a titration, *the number of equivalents of acid must equal the number of equivalents of base,* because the number of moles of H_3O^+ and OH^- are the same in a neutral solution. Therefore, the relationship between the normality of the acid and the normality of the base is the following:

$$\text{equivalents of acid} = \text{equivalents of base}$$

$$\left(\begin{array}{c} \text{normality} \\ \text{of acid} \end{array} \right) \times \left(\begin{array}{c} \text{volume} \\ \text{of acid} \end{array} \right) = \left(\begin{array}{c} \text{normality} \\ \text{of base} \end{array} \right) \times \left(\begin{array}{c} \text{volume} \\ \text{of base} \end{array} \right)$$

$$N_A \times V_A = N_B \times V_B$$

The following example is the same as the one worked out in Section 14.7, except that it is now solved using normality.

Example A sample of vinegar is titrated to determine the concentration of acetic acid. If 17.1 mL of 0.500 N NaOH are needed to completely neutralize 10.0 mL of the vinegar, what is the normality of the vinegar?

UNKNOWN	GIVEN	CONNECTION
N_A	$10.0 \text{ mL} = V_A$ $0.500 \text{ equiv/L} = N_B$ $17.1 \text{ mL} = V_B$	$N_A \times V_A = N_B \times V_B$

$$N_A \times V_A = N_B \times V_B$$

$$N_A \times 10.0 \text{ mL} = 0.500 \text{ equiv/L} \times 17.1 \text{ mL}$$

$$N_A = \frac{0.500 \text{ equiv/L} \times 17.1 \cancel{\text{ mL}}}{10.0 \cancel{\text{ mL}}}$$

ESTIMATE $\quad \dfrac{5 \times 10^{-1} \text{ equiv/L} \times 2 \times 10^{1} \cancel{\text{ mL}}}{1 \times 10^{1} \cancel{\text{ mL}}} = 1 \text{ equiv/L} = 1 \text{ N}$

CALCULATION $\quad 0.855 \text{ equiv/L} = 0.855 \text{ N}$

Note that the volumes need not be converted from mL to L since these units cancel in the calculation.

Summary

14.1 An acid is a proton donor—it can transfer a proton to another substance. Because the proton is a hydrogen ion, an acid raises the hydrogen ion concentration of water. The hydrated hydrogen ion is called the hydronium ion.

14.2 Solutions of acids in water have similar properties because they all contain excess H_3O^+ ions. For example, acid solutions taste sour, change the colors of dyes, and react with certain metals. The strength of an acid depends on its tendency to donate protons.

14.3 A base is a proton acceptor. It is the substance to which an acid can transfer its proton. In water, bases raise the concentration of hydroxide ions, OH^-. Two forms of the same compound that differ only by the presence or absence of one proton are called a conjugate acid-base pair.

14.4 Solutions of bases taste bitter and feel slippery. The most important bases are the Group 1A and Group 2A hydroxides, and ammonia.

14.5 The pH of a solution is defined as the negative logarithm of its hydronium ion concentration. A neutral solution has a pH of 7. In water solution, acids have a pH below 7 and bases have a pH above

7. The stronger an acid, the lower its pH. The stronger a base, the higher its pH. Indicators show the pH by undergoing a change in color.

14.6 Neutralization is the reaction of an acid with a base. In water, neutralization consists of the reaction between hydronium ions and hydroxide ions to form neutral water. A salt is an ionic compound formed by neutralization. It contains a cation from the base and an anion from the acid.

14.7 Titration is the process of reacting an acid and base to determine the unknown concentration of one of them. For example, if we wish to find the concentration of an acid, we add a base of known concentration. We can calculate the concentration of the acid from the volume of acid used and the known concentration and volume of base added.

14.8 Hydrolysis is a reaction of an ion with a water molecule in which the water molecule is broken apart. Hydrolysis can produce free hydrogen (hydronium) ions, making the solution acidic, or free hydroxide ions, making the solution basic.

14.9 A buffer solution consists of either a weak acid and its conjugate base, or a weak base and its conjugate acid. The pH of a buffer changes only slightly upon addition of small amounts of acid or base.

Exercises

KEY-WORD MATCHING EXERCISE For each term on the left, choose a phrase from the right-hand column that most closely matches its meaning.

1 acid	**a** differ only by a proton
2 hydronium ion	**b** a measure of H_3O^+ ion concentration
3 base	**c** either acidic or basic
4 conjugate acid-base pair	**d** equivalents per liter
5 amphoteric	**e** proton acceptor
6 pH	**f** acid-base reaction
7 indicator	**g** used to find a concentration
8 neutralization	**h** formed during neutralization
9 salt	**i** changes pH from 7
10 titration	**j** keeps pH relatively constant
11 hydrolysis	**k** proton donor
12 buffer	**l** changes color with pH
13 normality	**m** hydrated proton

Acids (14.1)

14 Compare the Arrhenius and Brönsted-Lowry definitions of an acid.

15 What makes the hydrogen ion so special?

16 In what form does the hydrogen ion exist in aqueous solution?

17 Write equations for the ionization of **a** nitric acid   **b** phosphoric acid

18 What is meant by the self-ionization of water?

Properties and preparation of acids (14.2)

19 Describe three ways used to prepare acids.

20 Describe a physical and a chemical property of acids.

21 What is an acid anhydride?

22 On what factors does the strength of an acid depend?

23 Name the acid that **a** is in vinegar   **b** is the workhorse of the chemical industry   **c** is stomach acid   **d** turns skin yellow

Bases and their relation to acids (14.3)

24 Compare the Arrhenius and Brönsted definitions of a base.

25 Why is ammonia a base?

26 How are acids and bases related by the Brönsted definition?

27 Write the conjugate base of **a** HNO_3 **b** H_2SO_4 **c** HI **d** H_2O

28 What are amphoteric substances? Give an example.

Properties of bases (14.4)

29 Give one physical and one chemical property of bases.

30 Describe three ways in which bases are prepared.

31 Why should concentrated solutions of NaOH be stored in plastic bottles?

pH of acids and bases (14.5)

32 What determines the acidity or basicity of a solution?

33 If the hydronium ion concentration of a solution is 1×10^{-4} M (at 25°C), what is the hydroxide ion concentration?

34 Why is the pH scale useful?

35 Write the pH when the hydronium ion concentration is **a** 10^{-9} **b** 10^{-3} **c** 10^{-7} **d** 10^{-1} **e** 10^{-5}

36 Which of the solutions in Exercise 35 are acidic?

37 Write the hydronium and hydroxide ion concentrations in a solution whose pH is **a** 4   **b** 8   **c** 13

38 What is an indicator?

Neutralization and salts (14.6)

39 What takes place in a neutralization reaction?

40 How does an antacid work?

41 Complete and balance the following equations:
a $HF + LiOH \rightarrow ?$
b $H_2SO_4 + Ca(OH)_2 \rightarrow ?$
c $HNO_3 + KOH \rightarrow ?$

42 Write the acid and base from which the following salts are prepared:
a NH_4Br　**b** $CsNO_3$　**c** $Ca(C_2H_3O_2)_2$

Titration (14.7)

43 Why are titrations performed?

44 Find the molarity of a solution of boric acid, H_3BO_3, if 25.0 mL are neutralized by 15.2 mL of 0.112 M NaOH.

45 In a titration, 50.0 mL of calcium hydroxide are neutralized by 32.5 mL of 0.531 M HCl. Find the molarity of the base.

46 If 10.0 mL of phosphoric acid are neutralized by 28.3 mL of 0.496 M KOH in a titration, what is the molarity of the acid?

Hydrolysis (14.8)

47 What happens in hydrolysis?

48 Predict whether a solution containing the following salts would be acidic, basic, or neutral: **a** $NaClO_4$　**b** KH_2PO_4　**c** $MgCl_2$

49 Write an equation for the hydrolysis of sodium carbonate. Explain why the solution is acidic or basic.

50 Complete and balance the following equations:
a $PO_4^{3-} + H_2O \rightarrow ?$
b $CN^- + H_2O \rightarrow ?$

Buffer solutions (14.9)

51 What are the components of a buffer?

52 How does a buffer work?

53 Which of the following would act as a buffer? Explain.
a $HCl + NaCl$　**b** $H_2SO_4 + NaHSO_4$　**c** $NaOH + HCl$　**d** H_2S and Na_2S

54 Write two equations to show how a solution containing $H_2PO_4^-$ ions and HPO_4^{2-} ions acts as a buffer.

Normality: Supplement to Chapter Fourteen

55 Find the equivalent weight of **a** HNO_3　**b** $Ba(OH)_2$　**c** H_3PO_4

56 What mass of calcium hydroxide is needed to prepare 500. mL of a 0.15 N solution?

57 Find the normality of **a** 0.10 M H_2SO_4　**b** 6.0 M H_3PO_4　**c** 2.1 M KOH

58 A 10.0 mL acid sample is titrated with 21.3 mL of 0.105 N base. Find the acid normality.

59 If 17.9 mL of 0.253 N sulfuric acid is needed to titrate 25.0 mL of a sodium hydroxide solution, what is the normality of the base?

60 Find the molarity of **a** 0.010 N H_2CO_3 **b** 3.1 N H_3BO_3 **c** 5.0×10^{-3} N $Mg(OH)_2$

ADDITIONAL EXERCISES

61 What is the difference between an organic and a mineral acid? Give an example of each type.

62 Write the conjugate acid of **a** $C_2H_3O_2^-$ **b** CO_3^{2-} **c** HPO_4^{2-} **d** H_2O

63 Using Figure 14-3, which indicator would you choose to measure a pH change that occurs around pH 5?

64 Find the pH of a solution whose hydroxide ion concentration is:
a 10^{-2} M **b** 10^{-7} M **c** 10^{-9} M

65 How can you test the strength of an acid?

66 Black coffee has a pH of 5.0. What are the hydronium ion and hydroxide ion concentrations?

67 In the following equation identify the acid, base, conjugate acid and conjugate base.
$H_2PO_4^-(aq) + OH^-(aq) \rightarrow HPO_4^{2-}(aq) + H_2O(l)$

68 If 17.9 mL of a NaOH solution is needed to titrate 2.15 g of benzoic acid, $HC_7H_5O_2$, what is the molarity of the base?

69 Using Table 14-1, how many milliliters of concentrated nitric acid are needed to make 2.0 L of a 0.514 M solution?

70 Write the conjugate base of: **a** HNO_3 **b** HSO_4^- **c** $H_2PO_4^-$

71 Write a complete equation and then a net ionic equation for the neutralization of phosphoric acid with potassium hydroxide.

72 What is the OH^- concentration in 0.010 M HCl?

73 What is the pH of the solution in Exercise 72?

74 Write equations for the complete ionization of carbonic acid, H_2CO_3, in water.

75 Predict whether solutions of the following compounds would be acidic, basic, or neutral: **a** K_2CO_3 **b** $CaCl_2$ **c** $NaHSO_3$ **d** $AgNO_3$ **e** Li_3PO_4

76 Tomato juice has a pH of 4.2 and lemon juice has a pH of 2.2. How many times greater is the H_3O^+ concentration in lemon juice?

77 If 23.65 mL of an HCl solution neutralizes 25.00 mL of a 0.1037 M NaOH solution, what is the molarity of the HCl?

78 Complete the following equations:
a $HSO_3^- + OH^- \rightarrow$?
b $HSO_3^- + H_3O^+ \rightarrow$?

79 Solid $Ca(OH)_2$ is not very soluble in water. Why does the solid readily dissolve in hydrochloric acid?

80 Label the acids and bases, and their corresponding conjugate bases and conjugate acids, in the following equations:

a $NH_3 + HBr \rightarrow NH_4^+ + Br^-$
b $H_3O^+ + PO_4^{3-} \rightarrow HPO_4^{2-} + H_2O$
c $HSO_3^- + CN^- \rightarrow HCN + SO_3^{2-}$

81 Because it contains dissolved CO_2, natural water is not neutral. How does the CO_2 affect the pH?

PROGRESS CHART Check off each item when completed.

_____ Previewed chapter

Studied each section:

_____ **14.1** Acids

_____ **14.2** Properties and preparation of acids

_____ **14.3** Bases and their relation to acids

_____ **14.4** Properties of bases

_____ **14.5** pH of acids and bases

_____ **14.6** Neutralization and salts

_____ **14.7** Titration

_____ **14.8** Hydrolysis

_____ **14.9** Buffers solutions

_____ Reviewed chapter

_____ Answered assigned exercises:

#'s _____

Review Exercises for Chapters Twelve–Fourteen

1 Why can't a liquid be compressed like a gas?

2 Describe how hydrogen bonding affects one physical property of water.

3 What is **a** vaporization equilibrium?   **b** solubility equilibrium?

4 How much energy in kilocalories and in kilojoules is released when 1.50 kg of steam at 100°C condense and cool to form liquid water at 25.1°C?

5 Why is a water meniscus concave (lower in the center)?

6 Why is anhydrous calcium chloride used as a drying agent?

7 Describe what happens when a nonelectrolyte dissolves in water.

8 The solubility of lead(II) bromide in water is 0.844 g per 100 mL H_2O at 20°C. Is it possible to prepare the following? Explain. **a** 9.5% (w/v) solution   **b** 1.53 M solution

9 A solution of lead(II) bromide contains 8.73×10^{-2} g of $PbBr_2$ in 15.0 mL. Is it saturated? (See Exercise 8.)

10 Find how many grams of $NH_4C_2H_3O_2$ are needed to make **a** 50.0 mL of a 0.99 M solution?   **b** 1.7 L of a 5.11% (w/v) solution?

11 A solution contains 1.87 g of mercury(II) nitrate in 200. mL. Express its concentration in **a** moles/liter   **b** weight-volume percent   **c** parts per million

12 If 25.0 mL of water are added to 125 mL of a 0.121 M $KMnO_4$ solution, what is its final concentration?

13 Write a net ionic equation for the reaction of barium hydroxide with acetic acid.

14 Can the same substance act both as an acid and as a base? Explain.

15 Write equations for the step-by-step ionization of boric acid, H_3BO_3, in water.

16 Find the pH of a 0.365% (w/v) solution of HCl.

17 A solution has a hydroxide ion concentration of 1.0×10^{-6} M. What is its pH?

18 Write the name and formula of the salt formed by the reaction of ammonia and nitric acid.

19 If 18.2 mL of 0.106 M KOH is needed to titrate 25.1 mL of perchloric acid, $HClO_4$, what is the molarity of the acid?

20 Predict whether a solution of Na_3PO_4 is acidic, basic, or neutral. Write an equation to explain your answer.

FIFTEEN

Oxidation-reduction reactions and electrochemistry

The acid-base reactions described in the previous chapter involve the transfer of protons. In this chapter we describe a type of reaction that involves the transfer of electrons. Since oxidation, the loss of electrons, and reduction, the gain of electrons, must always take place at the same time, reactions based on electron transfer are called oxidation-reduction, or redox reactions. In Chapter Fifteen we describe how to identify redox reactions and how to balance them. We also discuss ways in which a flow of electrons can either be produced by a redox reaction or result in a redox reaction. To demonstrate an understanding of Chapter Fifteen, you should be able to:

1 Determine the oxidation numbers of the elements in a compound or ion.

2 Identify the reactants that are oxidized and reduced in a redox reaction based on changes in their oxidation numbers.

3 Balance a redox equation by the oxidation number method.

4 Balance a redox equation by the half-reaction method.

5 Describe the operation of a galvanic cell and give several examples.

6 Using a table of reduction potentials, predict whether a given reaction can take place spontaneously.

7 Describe the operation of an electrolytic cell.

15.1 OXIDATION NUMBERS

Oxidation numbers are helpful aids for identifying and studying redox reactions. The **oxidation number** of an atom is based on an artificial but useful way of assigning electrons in a compound to the atoms present. It is *the charge that an atom would have if the electrons in each bond to that atom were assigned to the more electronegative atom.* For example, in HCl, the electron pair of the covalent bond is assigned to chlorine because that element is more electronegative than hydrogen. Therefore, chlorine is assigned an oxidation number of -1, since it has "gained" an electron. Hydrogen is assigned an oxidation number of $+1$ because it has "lost" an electron according to this bookkeeping method. (Note: Oxidation numbers will be written with the $+$ or $-$ sign in front of a number to make them different from the charges of ions, which are written with the sign after a number in this book.) These numbers represent the oxidation states of the elements.

Many elements, particularly the transition metals, can have several different oxidation states, depending on the compound they are in. Oxidation numbers are useful in naming compounds of these elements, as described in Chapter Seven. Oxidation numbers are assigned according to the following rules:

1 *The oxidation number of an atom in the form of its element is zero.* For example, hydrogen has an oxidation number 0 in the molecule H_2, and chlorine has an oxidation number of 0 in Cl_2. Electrons in bonds between atoms of the same element are shared equally and are not lost or gained by either atom.

2 *The oxidation number of a simple ion is equal to the charge of that ion.* For example, the Group 1A metal ions have oxidation numbers of $+1$, and ions of Group 2A have oxidation numbers of $+2$. This rule, used for naming compounds in Chapter Seven, is based on the loss or gain of electrons that occurs when an ion forms.

3 *The sum of the oxidation numbers of all atoms in a molecule or formula unit must equal 0.* For example, in NaCl, the sum of the oxidation numbers, $(+1) + (-1)$, equals 0. This rule is based on the fact that all compounds are neutral.

4 *In a polyatomic ion, the sum of the oxidation numbers of the atoms must equal the charge of the ion.* For example, in the hydroxide ion, OH^-, oxygen has an oxidation number of -2 and hydrogen has an oxidation number of $+1$. The sum, $(-2) + (+1) = -1$, is equal to the charge of the ion.

5 *The most electronegative atoms in a compound have negative oxidation numbers because they have gained electrons.* Oxygen usually has an oxidation number of -2. (In peroxides, however, oxygen's oxidation number is -1.) The halogens (Group 7A)—F, Cl, Br, I—generally have oxidation numbers of -1. Less electronegative elements, especially metals, usually have positive oxidation numbers. Hydrogen usually has an oxidation number of $+1$. However, when hydrogen is bonded to an even less electronegative element such as a metal, forming a hydride, its oxidation number is -1, because hydrogen gains an electron is these compounds.

In simple ionic compounds, the assignment of oxidation numbers is easy if we are familiar with the charges of the ions involved. For example, in potassium oxide, K_2O, potassium has an oxidation number of $+1$ and oxygen, -2. These oxidation numbers are the same as the charges of the ions. To keep track of oxidation numbers, we write them above the formula, as shown here.

oxidation number of potassium $+1$ -2 oxidation number of oxygen

$$K_2O$$

We then check that the sum of the oxidation numbers is zero: $2(+1) + (-2) = 2 - 2 = 0$. Note that the oxidation numbers written above the formula are for only *one* atom (or ion) of each element. Before taking the sum, the oxidation number of potassium must be multiplied by two because there are two potassium ions in the formula unit.

Aluminum oxide, Al_2O_3, is another simple ionic compound:

$$\begin{array}{c|c} +3 & -2 \\ \hline Al_2 & O_3 \end{array}$$

Check: $2(+3) + 3(-2) = +6 - 6 = 0$

If the metal is not from groups 1A, 2A, or 3A, the charge of its ion, and therefore its oxidation number, may not be immediately obvious. For example, manganese is from Group 7B. We can find the oxidation state of manganese in MnO_2 as follows. We know that oxygen has an oxidation number of -2 and that the sum of the oxidation numbers of the compound must be zero:

$$\begin{array}{c|c} ? & -2 \\ \hline Mn & O_2 \end{array}$$

$$1(?) + 2(-2) = 0$$

Therefore, manganese must have an oxidation number of $+4$, since this is the only number that gives a sum of 0 when added to $2(-2)$ or -4.

This same technique can be used to find the oxidation numbers of atoms in covalent compounds. For example, phosphorus trichloride, PCl_3, is a covalent compound. We may not know offhand the oxidation state for phosphorus, but we do know that chlorine, like the other halogens, has an oxidation number of -1. Using this information, and the rule that the sum is 0, we can find the missing oxidation number.

$$\begin{array}{c|c} ? & -1 \\ \hline P & Cl_3 \end{array}$$

$$1(?) + 3(-1) = 0$$

To make the sum equal to 0, phosphorus must have an oxidation number of $+3$.

In a molecule composed of three elements, we need to know the oxidation numbers of two of them in order to find the oxidation number of the third. Consider H_2SO_4, sulfuric acid:

$$\begin{array}{c|c|c} +1 & ? & -2 \\ \hline H_2 & S & O_4 \end{array}$$

$$2(+1) + ? + 4(-2) = 0$$

The sum of the oxidation numbers of the two hydrogens and four oxygens is $(+2) + (-8)$, or -6. Therefore, the oxidation number of sulfur must be $+6$ in this compound. We can also find the missing oxidation number by using an algebraic equation (see Appendix C.1):

$$\begin{array}{ccccc} H_2 & & S & & O_4 \\ 2(+1) & + & x & + & 4(-2) = 0 \\ +2 & + & x & - & 8 \quad = 0 \\ & & x & & = +6 \end{array}$$

For polyatomic ions, the sum of the oxidation numbers equals the charge. Therefore, the sum of the oxidation numbers in ammonium ion, NH_4^+, is $+1$. The oxidation state of nitrogen in this ion can be found as follows:

$$\begin{array}{c|c} ? & +1 \\ \hline N & H_4^+ \end{array}$$

$$1(?) + 4(+1) = +1$$

$$\text{or} \quad x + 4 \quad = +1$$

$$x \quad = -3$$

Check: $-3 + -4 = +1$

Hydrogencarbonate (bicarbonate) ion, HCO_3^-, is a slightly more complex polyatomic ion. We can find the oxidation number of C in HCO_3^- just as we found the oxidation number of N in NH_4^+. The sum of the oxidation numbers in HCO_3^- is equal to -1, its charge.

$$\begin{array}{c|c|c} +1 & ? & -2 \\ \hline H & C & O_3^- \end{array}$$

$$+1 + ? + 3(-2) = -1$$

$$\text{or} \quad +1 + x - 6 \quad = -1$$

$$x \quad = +4$$

Check: $+1 + 4 - 6 = -1$

Thus carbon has an oxidation number of $+4$.

Figure 15-1 shows oxidation states for the elements. The most common states are in color.

Example Find the oxidation state of xenon in XeF_4, xenon tetrafluoride.

$$
\begin{array}{c|c}
x & -1 \\
\hline
Xe & F_4
\end{array}
$$

$$x + 4(-1) = 0$$

$$x - 4 \quad\quad = 0$$

$$x \quad\quad\quad = +4$$

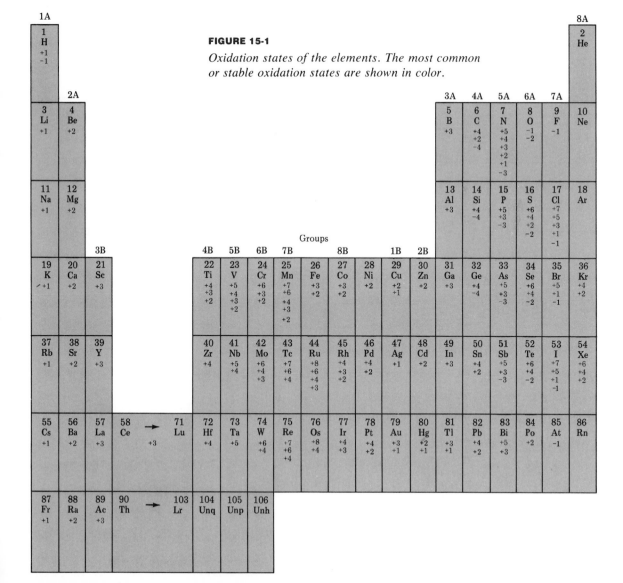

FIGURE 15-1

Oxidation states of the elements. The most common or stable oxidation states are shown in color.

Example   What is the oxidation state of manganese in $MnO_4{}^-$, permanganate ion?

$$\begin{array}{c|c} x & -2 \\ \hline Mn & O_4{}^- \end{array}$$

$$x + 4(-2) = -1$$
$$x - 8 \qquad = -1$$
$$x \qquad\quad = +7$$

Example   What is the oxidation state of nitrogen in KNO_3, potassium nitrate?

$$\begin{array}{c|c|c} +1 & x & -2 \\ \hline K & N & O_3 \end{array}$$

$$+1 + x + 3(-2) = 0$$
$$+1 + x - 6 \qquad = 0$$
$$x \qquad\qquad = +5$$

Since nitrogen is part of the nitrate ion in this compound, its oxidation state can also be found by examining the nitrate ion by itself:

$$\begin{array}{c|c} x & -2 \\ \hline N & O_3{}^- \end{array}$$

$$x + 3(-2) = -1$$
$$x - 6 \qquad = -1$$
$$x \qquad\quad = +5$$

Example   Find the oxidation state of phosphorus in the dihydrogenphosphate ion, $H_2PO_4{}^-$.

$$\begin{array}{c|c|c} +1 & x & -2 \\ \hline H_2 & P & O_4{}^- \end{array}$$

$$2(+1) + x + 4(-2) = -1$$
$$+2 + x - 8 \qquad = -1$$
$$x \qquad\qquad = +5$$

Example Find the oxidation state of chromium in the dichromate ion, $Cr_2O_7^{2-}$.

$$\begin{array}{c|c} x & -2 \\ \hline Cr_2 & O_7^{2-} \end{array}$$

Here we use x to represent the unknown oxidation state of chromium. Then we must use $2x$ to represent the contribution of the two chromium atoms to the sum of oxidation numbers.

$$2x + 7(-2) = -2$$
$$2x + (-14) = -2$$
$$2x - 14 = -2$$
$$2x = +12$$
$$x = +6$$

**15.2
OXIDATION
AND
REDUCTION**

Many of the reactions that we have seen in this book are oxidation-reduction reactions. An easy way to identify these reactions is to look for *changes in the oxidation numbers of the atoms involved*. If an atom is oxidized, it loses electrons and its oxidation number increases (becoming more positive, or less negative) as shown in Figure 15-2 and in the following list.

Before oxidation **After oxidation**

$\overset{0}{Fe}$ $\overset{+3}{Fe^{3+}}$

$\overset{+1}{Cu^+}$ $\overset{+2}{Cu^{2+}}$

$\overset{-2}{S^{2-}}$ $\overset{0}{S}$

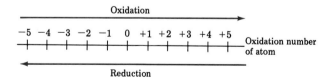

FIGURE 15-2

Change in oxidation number. When an atom undergoes oxidation, its oxidation number increases. When an atom undergoes reduction, its oxidation number decreases.

On the other hand, if an atom is reduced, it gains electrons and its oxidation number decreases (becoming more negative, or less positive).

Before reduction

$\overset{+4}{Sn^{4+}}$

$\overset{+3}{Al^{3+}}$

$\overset{0}{Cl_2}$

After reduction

$\overset{+2}{Sn^{2+}}$

$\overset{0}{Al}$

$\overset{-1}{2Cl^-}$

Using oxidation numbers, we can tell immediately that the reaction of sodium and chlorine (introduced in Sections 6.7 and 6.8) involves oxidation and reduction.

$$\underset{\text{sodium}}{\overset{0}{2Na}} + \underset{\text{chlorine}}{\overset{0}{Cl_2}} \rightarrow \underset{\text{sodium chloride}}{\overset{+1\ -1}{2NaCl}}$$

The sodium atom is oxidized. It loses an electron and its oxidation number increases from 0 in sodium metal to $+1$ in sodium cation. Chlorine atoms are reduced. Each atom gains an electron, and the oxidation number decreases from 0 in chlorine to -1 in chloride ion.

The reaction between zinc and hydrochloric acid is another example of an oxidation-reduction reaction.

$$\underset{\text{zinc}}{\overset{0}{Zn}} + \underset{\substack{\text{hydrochloric} \\ \text{acid}}}{\overset{+1\ -1}{2HCl}} \rightarrow \underset{\substack{\text{zinc} \\ \text{chloride}}}{\overset{+2\ -1}{ZnCl_2}} + \underset{\text{hydrogen}}{\overset{0}{H_2}}$$

Here the zinc is oxidized, increasing in oxidation number from 0 to $+2$ as it loses two electrons. Hydrogen ion is reduced, decreasing in oxidation number from $+1$ to 0. In this reaction, the oxidation state of chlorine (-1) is unchanged, meaning that chlorine is neither oxidized nor reduced.

Even in more complicated reactions, often only two elements undergo changes in oxidation states. For example, consider the formation of chromium(III) chloride from potassium dichromate.

$$\underset{\substack{\text{potassium} \\ \text{dichromate}}}{\overset{+1\ +6\ -2}{K_2Cr_2O_7}} + \underset{\substack{\text{hydrochloric} \\ \text{acid}}}{\overset{+1\ -1}{14HCl}} \rightarrow \underset{\substack{\text{chromium(III)} \\ \text{chloride}}}{\overset{+3\ -1}{2CrCl_3}} + \underset{\text{chlorine}}{\overset{0}{3Cl_2}} + \underset{\text{water}}{\overset{+1\ -2}{7H_2O}} + \underset{\substack{\text{potassium} \\ \text{chloride}}}{\overset{+1\ -1}{2KCl}}$$

In this case, the only atoms that change in oxidation number are

TABLE 15-1 *Examples of redox reactions*

Synthesis (Combination)

$$\overset{0}{2H_2} + \overset{0}{O_2} \rightarrow \overset{+1\ -2}{2H_2O}$$

$$\overset{0}{N_2} + \overset{0}{3H_2} \rightarrow \overset{-3\ +1}{2NH_3}$$

Decomposition

$$\overset{+5\ -2}{2KClO_3} \rightarrow \overset{-1}{2KCl} + \overset{0}{3O_2}$$

$$\overset{+5\ -2}{2Cu(NO_3)} \rightarrow 2CuO + \overset{+4}{4NO_2} + \overset{0}{O_2}$$

Replacement

$$\overset{0}{2Na} + \overset{+1}{2H_2O} \rightarrow \overset{+1}{2NaOH} + \overset{0}{H_2}$$

$$\overset{0}{2Al} + \overset{+3}{Cr_2O_3} \rightarrow \overset{+3}{Al_2O_3} + \overset{0}{2Cr}$$

Other types

$$\overset{0}{C_6H_{12}O_6} + \overset{0}{6O_2} \rightarrow \overset{+4\ -2}{6CO_2} + 6H_2O$$

$$\overset{+3}{Fe_2O_3} + \overset{+2}{3CO} \rightarrow \overset{0}{2Fe} + \overset{+4}{3CO_2}$$

$$\overset{-2}{MnS} + \overset{-1}{4H_2O_2} \rightarrow \overset{+6}{MnSO_4} + \overset{-2}{4H_2O}$$

$$\overset{+3}{2Fe^{3+}} + \overset{-2}{H_2S} \rightarrow \overset{+2}{2Fe^{2+}} + 2H^+ + \overset{0}{S}$$

chromium and chlorine. Chromium is reduced from $+6$ to $+3$, and chlorine is oxidized from -1 to 0. Table 15-1 presents other examples of oxidation-reduction reactions.

When an atom is oxidized, it loses electrons and becomes more positive, or less negative. We can see from the examples given that this change is reflected by an increase in the atom's oxidation number. Because it transfers one or more of its electrons to another atom, thus reducing that atom, the substance being oxidized is called a **reducing agent**. Metals are typical reducing agents, since they have a strong tendency to donate electrons and are therefore easily oxidized.

The substance that accepts the electrons from the reducing agent, causing it to undergo oxidation, is called an **oxidizing agent**. Because of their tendency to attract electrons, electronegative atoms are typical oxidizing agents. Thus, the halogens (F_2, Cl_2, Br_2, I_2), oxygen (O_2), and many oxygen-containing polyatomic ions (such as MnO_4^-, NO_3^-, and $Cr_2O_7^{2-}$) are examples of oxidizing agents. Of course, as an oxidizing agent accepts electrons, it is reduced. It becomes more negative, or less positive. This change is shown by a *decrease* in the oxidation

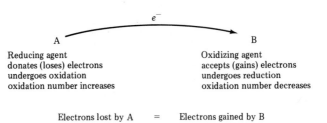

FIGURE 15-3

Summary of the oxidation-reduction process. Here an electron is transferred from substance A to substance B.

number of one of its atoms. These relationships are summarized in Figure 15-3.

It is possible to express separately the processes of oxidation and reduction that take place in a redox reaction. As indicated in Section 9.6, we can write a redox reaction in the form of two **half-reactions**. For example, the redox reaction of sodium and chlorine to form sodium chloride can be written as two half-reactions. The equation for a half-reaction is called a **half-equation**.

$$\text{oxidation half-reaction: } 2Na \rightarrow 2Na^+ + 2e^-$$
$$\text{reduction half-reaction: } Cl_2 + 2e^- \rightarrow 2Cl^-$$
$$\overline{\text{overall redox reaction: } 2Na + Cl_2 \rightarrow 2NaCl}$$

Do-it-yourself 15

Bleach

Bleaches are strong oxidizing agents. They act by oxidizing colored compounds to colorless ones. A material appears colored when it contains a compound in which electrons can be excited by visible radiation. Certain frequencies of light are absorbed and others are reflected, giving the material a particular color. After the coloring compound is oxidized, its electrons can no longer be excited by visible light, and the compound appears colorless.

Household bleach solutions contain approximately 5% sodium hypochlorite, NaClO, in water. During bleaching, the hypochlorite ion is reduced in the following half-reaction:

$$\underset{\substack{\text{hypochlorite} \\ \text{ion}}}{ClO^-} + \underset{\text{water}}{H_2O} + 2e^- \rightarrow \underset{\substack{\text{chloride} \\ \text{ion}}}{Cl^-} + \underset{\substack{\text{hydroxide} \\ \text{ion}}}{2OH^-}$$

Bleach powders containing calcium hypochlorite, $Ca(ClO)_2$, work by the same reaction.

To do: Carefully mix a small amount of household bleach with an equal volume of water. (**CAUTION: Bleach can be dangerous. Read the label carefully before using.**) Examine the effect of the bleach by adding several drops to a small volume of coffee, red wine, ink, grape juice, and food coloring.

Neither half-reaction can take place by itself, since oxidation cannot occur without reduction and reduction cannot occur without oxidation. The overall reaction is the sum of the two half-reactions. Note that the electrons "cancel" because the number of electrons donated must equal the number of electrons accepted during the reaction. For this reason, electrons do not appear in the equation for the overall redox reaction.

15.3 BALANCING REDOX EQUATIONS BY OXIDATION NUMBERS

Many redox equations are too complicated to be balanced easily by the trial-and-error method described in Section 9.4. Therefore, it is helpful to have a systematic method that can be used in these cases. One such method, called the **oxidation number method**, is based on the changes in oxidation numbers that occur in the reaction. We know that the number of electrons lost by the reducing agent must equal the number gained by the oxidizing agent. Thus, *the total increase in oxidation number for one type of atom must equal the total decrease in oxidation number for another*. This principle is the basis for balancing equations using the oxidation number method.

The oxidation number method consists of 4 steps. We can illustrate these steps by using the unbalanced equation for the production of chlorine from manganese(IV) oxide and hydrochloric acid.

$$MnO_2(s) + HCl(aq) \rightarrow Cl_2(g) + MnCl_2(aq) + H_2O(l)$$

manganese(IV) oxide   hydrochloric acid   chlorine   manganese(II) chloride   water

1 In the unbalanced overall equation, identify those atoms that change oxidation number in going from reactants to products.

Until you have enough experience to tell which atoms are oxidized and which atoms are reduced by simply inspecting the equation, you should determine the oxidation number of each atom.

$$\overset{+4\ -2}{Mn\,O_2} + \overset{+1\ -1}{H\,Cl} \rightarrow \overset{0}{Cl_2} + \overset{+2\ -1}{MnCl_2} + \overset{+1\ -2}{H_2O}$$

In our sample reaction, manganese is reduced from $+4$ to $+2$ and chlorine is oxidized from -1 to 0.

2 Write balanced half-equations for the atoms that undergo changes in oxidation number.

change in oxidation number

$$\overset{+4}{Mn} \rightarrow \overset{+2}{Mn} \qquad -2$$

$$\overset{-1}{Cl} \rightarrow \overset{0}{\tfrac{1}{2}Cl_2} \qquad +1$$

491

Note that we have added the coefficient $\frac{1}{2}$ to the product Cl_2. This coefficient balances the single chlorine atom in the reactant.

3 Using coefficients, make the total increase in oxidation number equal to the total decrease in oxidation number.

Since the oxidation number of the manganese atoms increases by two, while the oxidation number of the chlorine atoms decreases by one, the chlorine atoms must be multiplied by two to make the increase equal to the decrease. (Multiplying the coefficient of Cl_2 by 2 changes it from $\frac{1}{2}$ to 1.)

total change in oxidation number

$$\overset{+4}{Mn} \quad \rightarrow \quad \overset{+2}{Mn} \qquad \qquad -2$$
$$\overset{-1}{2Cl} \quad \rightarrow \quad \overset{0}{Cl_2} \qquad 2(+1) = \underline{+2}$$
$$0$$

Using the coefficient of 1 for both forms of manganese, 2 for the reactant Cl^-, and 1 for Cl_2, the equation becomes:

$$MnO_2 + 2HCl \rightarrow Cl_2 + MnCl_2 + H_2O$$

(As usual, the coefficient 1 is omitted in the equation.)

4 Balance the rest of the equation by trial-and-error, leaving hydrogen and oxygen for last.

In our example, chlorine atoms of unchanged oxidation number (-1) are also present as products in the form of $MnCl_2$. They must be accounted for by changing the coefficient of HCl.

$$MnO_2 + 4HCl \rightarrow Cl_2 + MnCl_2 + H_2O$$

Finally, the hydrogen and oxygen atoms can be balanced. Check to see that the same number of each kind of atom now appears on both sides of the equation.

$$MnO_2 + 4HCl \rightarrow Cl_2 + MnCl_2 + 2H_2O$$

reactants	products
$\overset{+4}{1Mn}$	$\overset{+2}{1Mn}$
$\overset{-1}{2O}$	$\overset{-1}{2O}$
$\overset{+1}{4H}$	$\overset{+1}{4H}$
$\overset{-1}{4Cl}$	$\overset{0}{2Cl}, \overset{-1}{2Cl}$

Now let's consider another example, the reaction of copper(II) oxide with ammonia:

$$CuO(s) + NH_3(l) \rightarrow Cu(s) + N_2(g) + H_2O(l)$$
copper(II)   ammonia   copper   nitrogen   water
oxide

Step **1** in the oxidation number method is to identify the atoms that change oxidation numbers.

$$\overset{+2\,-2}{CuO} + \overset{-3\,+1}{NH_3} \rightarrow \overset{0}{Cu} + \overset{0}{N_2} + \overset{+1\,-2}{H_2O}$$

Copper is reduced from $+2$ to 0, and nitrogen is oxidized from -3 to 0.

Step **2** is to write balanced half-equations. We place the coefficient $\frac{1}{2}$ in front of N_2 to balance the single N on the left side.

		change in oxidation number
$\overset{+2}{Cu}$	$\rightarrow \quad \overset{0}{Cu}$	-2
$\overset{-3}{N}$	$\rightarrow \quad \overset{0}{\frac{1}{2}N_2}$	$+3$

The total changes in oxidation number must be balanced in step **3** by multiplying the copper atoms by three and the nitrogen atoms by two, since $3(-2) + 2(+3) = 0$.

		total change in oxidation number
$\overset{+2}{3Cu}$	$\rightarrow \quad \overset{0}{3Cu}$	$3(-2) = -6$
$\overset{-3}{2N}$	$\rightarrow \quad \overset{0}{N_2}$	$2(+3) = \underline{+6}$
		0

These coefficients are placed in the equation.

$$3CuO + 2NH_3 \rightarrow 3Cu + N_2 + H_2O$$

In step **4** the remainder of the equation is balanced.

$$3CuO + 2NH_3 \rightarrow 3Cu + N_2 + 3H_2O$$

reactants	products
$\overset{+2}{3Cu}$	$\overset{0}{3Cu}$
$\overset{-2}{3O}$	$\overset{-2}{3O}$
$\overset{-3}{2N}$	$\overset{0}{2N}$
$\overset{+1}{6H}$	$\overset{+1}{6H}$

Additional equations are balanced by the oxidation number method in the following examples.

Example Phosphorus (P_4) can be produced in the following reaction. Balance this equation by the oxidation number method.

$$Ca_3(PO_4)_2(s) + C(s) + SiO_2(s) \rightarrow P_4(s) + CaSiO_3(s) + CO(g)$$

calcium   carbon   silicon   phosphorus   calcium   carbon
phosphate   dioxide   silicate   monoxide

Step 1 $\overset{+2\ +5-2}{Ca_3(PO_4)_2} + \overset{0}{C} + \overset{+4-2}{SiO_2} \rightarrow \overset{0}{P_4} + \overset{+2\ +4\ -2}{Ca\ Si\ O_3} + \overset{+2-2}{C\ O}$

Step 2   change in oxidation number

$\overset{+5}{P} \rightarrow \overset{0}{\tfrac{1}{4}P_4}$ -5

$\overset{0}{C} \rightarrow \overset{+2}{C}$ $+2$

Step 3   total change in oxidation number

$\overset{+5}{2P} \rightarrow \overset{0}{\tfrac{1}{2}P_4}$ $2(-5) = -10$

$\overset{0}{5C} \rightarrow \overset{+2}{5C}$ $5(+2) = \underline{+10}$
                                       0

Multiply each coefficient by 2 to make the coefficient of P_4 an integer:

$\overset{+5}{4P} \rightarrow \overset{0}{P_4}$

$\overset{0}{10C} \rightarrow \overset{+2}{10C}$

Place these coefficients in the equation:

$$2Ca_3(PO_4)_2 + 10C + SiO_2 \rightarrow P_4 + CaSiO_3 + 10CO$$

Step 4 $2Ca_3(PO_4)_2 + 10C + 6SiO_2 \rightarrow P_4 + 6CaSiO_3 + 10CO$

reactants	products
$\overset{+2}{6Ca}$	$\overset{+2}{6Ca}$
$\overset{+5}{4P}$	$\overset{0}{4P}$
$\overset{-2}{28O}$	$\overset{-2}{28O}$
$\overset{0}{10C}$	$\overset{+2}{10C}$
$\overset{+4}{6Si}$	$\overset{+4}{6Si}$

Example Bromine reacts with sodium hydroxide to form sodium bromate, sodium bromide, and water. Balance this equation by the oxidation number method.

$$Br_2(l) + NaOH(aq) \rightarrow NaBrO_3(aq) + NaBr(aq) + H_2O(l)$$

bromine   sodium   sodium   sodium   water
hydroxide   bromate   bromide

Step 1
$$\overset{0}{Br_2} + \overset{+1 \ -2 +1}{NaOH} \rightarrow \overset{+1 \ +5 \ -2}{NaBrO_3} + \overset{+1 \ -1}{NaBr} + \overset{+1 \ -2}{H_2O}$$

Step 2   change in oxidation number

$$\overset{0}{\tfrac{1}{2}Br_2} \rightarrow \overset{+5}{Br} \quad +5$$

$$\overset{0}{\tfrac{1}{2}Br_2} \rightarrow \overset{-1}{Br} \quad -1$$

Note: Some of the bromine is oxidized to bromate and some is reduced to bromide in this reaction.

Step 3   total change in oxidation number

$$\overset{0}{\tfrac{1}{2}Br_2} \rightarrow \overset{+5}{Br} \quad\quad\quad +5$$

$$\overset{0}{\tfrac{5}{2}Br_2} \rightarrow \overset{-1}{5Br} \quad 5(-1) = \underline{-5}$$

$$0$$

$$3Br_2 + NaOH \rightarrow NaBrO_3 + 5NaBr + H_2O$$

Note: The coefficient 3 for Br_2 is the sum of the coefficients in the partial equations, $\tfrac{1}{2} + \tfrac{5}{2} = \tfrac{6}{2} = 3$.

Step 4 $3Br_2 + 6NaOH \rightarrow NaBrO_3 + 5NaBr + 3H_2O$

reactants	products
$\overset{0}{6Br}$	$\overset{+5}{1Br},\ \overset{-1}{5Br}$
$\overset{+1}{6Na}$	$\overset{+1}{6Na}$
$\overset{-2}{6O}$	$\overset{-2}{6O}$
$\overset{+1}{6H}$	$\overset{+1}{6H}$

In summary, the steps in the oxidation number method are:

1 Identify the atoms that change oxidation number in the unbalanced equation.

2 Write balanced half-equations for the atoms that change oxidation number.

3 Add coefficients to make the total increase in oxidation number equal to the total decrease in oxidation number.

4 Balance the rest of the equation by trial-and-error.

Practice problem Balance the following equations by the oxidation number method.

a $C + HNO_3 \rightarrow CO_2 + NO_2 + H_2O$

b $Na_2Cr_2O_7 + HCl \rightarrow CrCl_3 + NaCl + Cl_2 + H_2O$

ANSWERS **a** $C + 4HNO_3 \rightarrow CO_2 + 4NO_2 + 2H_2O$

b $Na_2Cr_2O_7 + 14HCl \rightarrow 2CrCl_3 + 2NaCl + 3Cl_2 + 7H_2O$

15.4
BALANCING BY
HALF-REACTIONS

Another way to balance redox equations is by the **method of half-reactions**. We divide the redox reaction into its two half-reactions and then make the number of electrons involved in the two half-equations equal. Because we begin with the net ionic equation for the reaction, we eliminate the complication of spectator ions in the intermediate steps. This approach is also called the ion-electron method.

We will illustrate the steps involved in the half-reaction method by using the reaction of copper metal with nitric acid. (This reaction serves as a laboratory synthesis for nitric oxide, NO.) The unbalanced equation is:

$$Cu(s) + HNO_3(aq) \rightarrow NO(g) + Cu(NO_3)_2(aq) + H_2O(l)$$

copper nitric nitric copper(II) water
acid oxide nitrate

1 Write the net ionic equation for the overall reaction.

$$Cu + H^+ + NO_3^- \rightarrow NO + Cu^{2+} + H_2O$$

Both nitric acid and copper(II) nitrate exist primarily as ions. Note that nitrogen atoms appear in two forms on the right side of the original equation, as NO and as NO_3^- in $Cu(NO_3)_2$. The NO_3^- form is eliminated from the right side of the net ionic equation because it is unchanged from the NO_3^- ion on the left side of the equation.

2 Identify those atoms that change oxidation number.

$$\overset{0}{Cu} + \overset{+1}{H^+} + \overset{+5\ -2}{N\ O_3^-} \rightarrow \overset{+2\ -2}{N\ O} + \overset{+2}{Cu^{2+}} + \overset{+1\ -2}{H_2O}$$

The copper is oxidized. Its oxidation number increases from 0 to $+2$. The nitrogen is reduced. Its oxidation number decreases from $+5$ in NO_3^- to $+2$ in NO.

3 Write half-equations for the oxidation and reduction processes, leaving out any substance whose oxidation number does not change.

oxidation: $Cu \rightarrow Cu^{2+}$

reduction: $NO_3^- \rightarrow NO$

4 Balance each half-equation so that the number of atoms of each element is the same on both sides. We can add H^+ and H_2O in acidic solutions, or OH^- and H_2O in basic solutions.

$Cu \rightarrow Cu^{2+}$

$NO_3^- + 4H^+ \rightarrow NO + 2H_2O$

In the copper half-equation, no balancing of atoms is required. In the half-equation involving nitrogen reduction, however, the number of oxygen atoms must be balanced. Since this reaction takes place in acid solution, H^+ and H_2O may be added, H^+ to one side and H_2O to the other. H_2O molecules are added to the side with fewer oxygen atoms to balance the number of oxygens. To make the hydrogens balance in this half-equation, two H^+ ions then must be added to the other side for each H_2O added.

5 Balance the charges in each half-equation by adding electrons to the more positively charged side.

$Cu \rightarrow Cu^{2+} + 2e^-$

$NO_3^- + 4H^+ + 3e^- \rightarrow NO + 2H_2O$

In the first half-equation, we add two electrons to the right side to make the charge the same (zero) on both sides. For the second half-equation, the initial sum of charges on the left side is $+3$, while the right side is neutral. Therefore, we add three electrons to the left side so that both sides have the same charge.

Note that the electrons must always be on opposite sides in the two half-equations, since one half-equation involves oxidation and the other half-equation involves reduction. In this example, both half-equations end up with a charge of zero on each side, but that is not always the case. The charges of a half-equation are balanced as long as the sum of charges is *the same* on both sides. The sum of charges need not be zero.

6 Multiply the half-equations so that the number of electrons lost in one half-equation equals the number of electrons gained in the other.

$3 (Cu \rightarrow Cu^{2+} + 2e^-)$

$2 (NO_3^- + 4H^+ + 3e^- \rightarrow NO + 2H_2O)$

The number of electrons is the same in both half-equations if the first half-equation is multiplied by three ($3 \times 2e^- = 6e^-$), and the second half-equation is multiplied by two ($2 \times 3e^- = 6e^-$). This multiplication results in the following equations:

$$3Cu \rightarrow 3Cu^{2+} + 6e^-$$

$$2NO_3^- + 8H^+ + 6e^- \rightarrow 2NO + 4H_2O$$

7 Add the two half-equations, canceling any like substances that appear on both sides.

$$3Cu \rightarrow 3Cu^{2+} + \cancel{6e^-}$$

$$2NO_3^- + 8H^+ + \cancel{6e^-} \rightarrow 2NO + 4H_2O$$
$$\overline{}$$
$$3Cu + 2NO_3^- + 8H^+ \rightarrow 3Cu^{2+} + 2NO + 4H_2O$$

In this case, only the electrons cancel. The resulting equation is the *balanced net ionic equation for the reaction*. In many cases, this answer is sufficient.

8 If spectator ions were removed from the equation, replace them, making sure that they balance also.

$$3Cu + 8HNO_3 \rightarrow 3Cu(NO_3)_2 + 2NO + 4H_2O$$

Nitrate ion is also a spectator ion in this reaction. Notice that the reactants in the net ionic equation can be regrouped as $(2H^+ + 2NO_3^-) + 6H^+$. Thus, six more nitrate ions can be added to the left side, making a total of $2HNO_3 + 6HNO_3$, or $8HNO_3$. These six added ions are balanced, since they appear on the right side of the equation in $3Cu(NO_3)_2$.

As another example, consider the reaction of potassium permanganate with hydrochloric acid.

$$\underset{\substack{\text{potassium} \\ \text{permanganate}}}{KMnO_4(aq)} + \underset{\substack{\text{hydrochloric} \\ \text{acid}}}{HCl(aq)} \rightarrow \underset{\substack{\text{manganese(II)} \\ \text{chloride}}}{MnCl_2(aq)} + \underset{\text{chlorine}}{Cl_2(g)} + \underset{\substack{\text{potassium} \\ \text{chloride}}}{KCl(aq)} + \underset{\text{water}}{H_2O(l)}$$

The net ionic equation (step **1**) is

$$MnO_4^- + H^+ + Cl^- \rightarrow Mn^{2+} + Cl_2 + H_2O$$

Potassium and chloride ions are spectator ions, although Cl^- is also a reactant, as shown. The oxidation numbers of the atoms (step **2**) are:

$$\overset{+7\ -2}{MnO_4^-} + \overset{+1}{H^+} + \overset{-1}{Cl^-} \rightarrow \overset{+2}{Mn^{2+}} + \overset{0}{Cl_2} + \overset{+1\ -2}{H_2O}$$

and the half-equations (step **3**) are:

oxidation: $Cl^- \rightarrow Cl_2$

reduction: $MnO_4^- \rightarrow Mn^{2+}$

The numbers of atoms in these half-equations are then balanced (step **4**).

$2Cl^- \rightarrow Cl_2$

$MnO_4^- + 8H^+ \rightarrow Mn^{2+} + 4H_2O$

The coefficient 2 is placed in front of Cl^- to balance the two chlorine atoms in Cl_2. Four water molecules are added to the right side of the second half-equation to balance the four oxygen atoms in MnO_4^-. Then $8H^+$ ions are added to the left side to balance the hydrogen atoms in H_2O. The charges are then balanced (step **5**).

$2Cl^- \rightarrow Cl_2 + 2e^-$

$MnO_4^- + 8H^+ + 5e^- \rightarrow Mn^{2+} + 4H_2O$

In the first half-equation, two electrons are added to the right side to balance the -2 charge of the left side. In the second half-equation, five electrons are added to the left side so that now each side has a net charge of $+2$. The number of electrons lost in the first half reaction is made equal to the number of electrons gained in the second (step **6**) by multiplying the first half-equation by five and the second by two, so that each half-equation now has ten electrons.

$5 (2Cl^- \rightarrow Cl_2 + 2e^-)$

$2 (MnO_4^- + 8H^+ + 5e^- \rightarrow Mn^{2+} + 4H_2O)$

The two half-equations are now added (step **7**):

$10Cl^- \qquad\qquad\qquad \rightarrow 5Cl_2 + \cancel{10e^-}$

$2MnO_4^- + 16H^+ + \cancel{10e^-} \rightarrow 2Mn^{2+} + 8H_2O$

$$\overline{2MnO_4^- + 10Cl^- + 16H^+ \rightarrow 2Mn^{2+} + 5Cl_2 + 8H_2O}$$

This equation is the balanced net ionic equation. When we add the spectator ions (step **8**), the equation becomes:

$2KMnO_4 + 16HCl \rightarrow 2MnCl_2 + 5Cl_2 + 8H_2O + 2KCl$

Two K^+ ions are added to the permanganate ion. They are balanced by two KCl formula units on the right side. Six more chloride ions are added to the left side, making $16H^+$ and $16Cl^-$, or 16HCl. These six ions are balanced on the right side of the equation by the four Cl^- in $2MnCl_2$ and the two Cl^- in 2KCl.

Reactions that take place in basic solutions are handled in the same way, except that OH^- and H_2O are used in step **5** to balance the half-equations for mass. The oxidation of sulfite ion to sulfate ion by chlorine serves as an example.

$$Na_2SO_3(aq) + Cl_2(g) + NaOH(aq) \rightarrow$$

sodium   chlorine   sodium
sulfite   hydroxide

$$Na_2SO_4(aq) + NaCl(aq) + H_2O(l)$$

sodium   sodium   water
sulfate   chloride

Step 1 $SO_3^{2-} + Cl_2 + OH^- \rightarrow SO_4^{2-} + Cl^- + H_2O$

Step 2 $\overset{+4\,-2}{SO_3^{2-}} + \overset{0}{Cl_2} + \overset{-2\,+1}{OH^-} \rightarrow \overset{+6\,-2}{SO_4^{2-}} + \overset{-1}{Cl} + \overset{+1\,-2}{H_2O}$

Step 3   oxidation: $SO_3^{2-} \rightarrow SO_4^{2-}$
             reduction: $Cl_2 \rightarrow Cl^-$

Step 4 $SO_3^{2-} + 2OH^- \rightarrow SO_4^{2-} + H_2O$
             $Cl_2 \rightarrow 2Cl^-$

In balancing the oxidation half-equation with hydroxide ion, note that OH^- is added to the side that has fewer oxygens. Two OH^- ions must be added for each H_2O on the other side to keep the hydrogen atoms balanced.

Steps 5, 6, 7
$$SO_3^{2-} + 2OH^- \rightarrow SO_4^{2-} + H_2O + 2e^-$$
$$Cl_2 + 2e^- \rightarrow 2Cl^-$$
$$\overline{SO_3^{2-} + Cl_2 + 2OH^- \rightarrow SO_4^{2} + 2Cl^- + H_2O}$$

Step 8 $Na_2SO_3 + Cl_2 + 2NaOH \rightarrow Na_2SO_4 + 2NaCl + H_2O$

Example   Ferrous ion, iron(II), can be oxidized to ferric ion, iron(III), by dichromate ion in acid solution. Balance this equation by the half-reaction method.

$$Fe^{2+}(aq) + Cr_2O_7^{2-}(aq) + H^+(aq) \rightarrow Fe^{3+}(aq) + Cr^{3+}(aq) + H_2O(l)$$

iron(II)   dichromate   hydrogen   iron(III)   chromium(III) water
ion   ion   ion   ion

Steps 1 and 2 $\overset{+2}{Fe^{2+}} + \overset{+6\ -2}{Cr_2O_7^{2-}} + \overset{+1}{H^+} \rightarrow \overset{+3}{Fe^{3+}} + \overset{+3}{Cr^{3+}} + \overset{+1\ -2}{H_2O}$

Step 3 oxidation: $Fe^{2+} \rightarrow Fe^{3+}$
reduction: $Cr_2O_7^{2-} \rightarrow Cr^{3+}$

Step 4 $Fe^{2+} \rightarrow Fe^{3+}$
$Cr_2O_7^{2-} + 14H^+ \rightarrow 2Cr^{3+} + 7H_2O$

Step 5 $Fe^{2+} \rightarrow Fe^{3+} + e^-$
$Cr_2O_7^{2-} + 14H^+ + 6e^- \rightarrow 2Cr^{3+} + 7H_2O$

Step 6 $6(Fe^{2+} \rightarrow Fe^{3+} + e^-)$
$Cr_2O_7^{2-} + 14H^+ + 6e^- \rightarrow 2Cr^{3+} + 7H_2O$

Step 7

$$6Fe^{2+} \rightarrow 6Fe^{3+} + \cancel{6e^-}$$
$$Cr_2O_7^{2-} + 14H^+ + \cancel{6e^-} \rightarrow 2Cr^{3+} + 7H_2O$$
$$\overline{6Fe^{2+} + Cr_2O_7^{2-} + 14H^+ \rightarrow 6Fe^{3+} + 2Cr^{3+} + 7H_2O}$$

Example Calcium in the blood can be measured as the oxalate ($C_2O_4^{2-}$) salt by the following reaction. Balance the equation by the half-reaction method.

$CaC_2O_4(s) + KMnO_4(aq) + H_2SO_4(aq) \rightarrow CO_2(g) + MnSO_4(aq) +$
calcium   potassium   sulfuric   carbon   manganese(II)
oxalate   permanganate   acid   dioxide   sulfate

$CaSO_4(s) + K_2SO_4(aq) + H_2O(l)$
calcium   potassium   water
sulfate   sulfate

Step 1 $C_2O_4^{2-} + MnO_4^- + H^+ \rightarrow CO_2 + Mn^{2+} + H_2O$

Step 2 $\overset{+3\ -2}{C_2O_4^{2-}} + \overset{+7\ -2}{MnO_4^-} + \overset{+1}{H^+} \rightarrow \overset{+4-2}{CO_2} + \overset{+2}{Mn^{2+}} + \overset{+1\ -2}{H_2O}$

Step 3 oxidation: $C_2O_4^{2-} \rightarrow CO_2$
reduction: $MnO_4^- \rightarrow Mn^{2+}$

Step 4 $C_2O_4^{2-} \rightarrow 2CO_2$
$MnO_4^- + 8H^+ \rightarrow Mn^{2+} + 4H_2O$

Step 5 $C_2O_4^{2-} \rightarrow 2CO_2 + 2e^-$
$MnO_4^- + 8H^+ + 5e^- \rightarrow Mn^{2+} + 4H_2O$

Step 6 $5(C_2O_4^{2-} \rightarrow 2CO_2 + 2e^-)$
$2(MnO_4^- + 8H^+ + 5e^- \rightarrow Mn^{2+} + 4H_2O)$

Step 7

$$5C_2O_4^{2-} \rightarrow 10CO_2 + \cancel{10e^-}$$
$$2MnO_4^- + 16H^+ + \cancel{10e^-} \rightarrow 2Mn^{2+} + 8H_2O$$
$$\overline{2MnO_4^- + 5C_2O_4^{2-} + 16H^+ \rightarrow 10CO_2 + 2Mn^{2+} + 8H_2O}$$

Step 8 $2KMnO_4 + 5CaC_2O_4 + 8H_2SO_4 \rightarrow 10CO_2 + 2MnSO_4 + 5CaSO_4 + K_2SO_4 + 8H_2O$

In summary, the steps in the half-reaction method are:

1 Write a net ionic equation.

2 Identify atoms that change in oxidation number.

3 Write half-equations for the oxidation and reduction steps.

4 Balance the numbers of atoms in each half-equation.

5 Balance the charges in each half-equation.

6 Multiply the half-equations so that the number of electrons lost and the number of electrons gained are equal.

7 Add the half-equations, canceling substances that appear on both sides in equal quantities.

8 Balance the spectator ions.

Both the oxidation number method and the half-reaction method lead to balanced redox equations. The oxidation number method is useful for reactions that do not take place in aqueous solution or do not involve the formation of ions. The half-reaction method is most useful for reactions in aqueous solution, since it starts with the net ionic equation. This method clearly emphasizes the actual oxidation and reduction half-reactions. We will see in the next section that these half-reactions can be physically separated.

Practice problem   Balance the following equations by the half-reaction method.

a $Cr_2O_7^{2-} + H^+ + SO_3^{2-} \rightarrow Cr^{3+} + SO_4^{2-} + H_2O$

b $Zn + HNO_3 \rightarrow Zn(NO_3)_2 + NH_4NO_3 + H_2O$

ANSWERS   a $Cr_2O_7^{2-} + 8H^+ + 3SO_3^{2-} \rightarrow 2Cr^{3+} + 3SO_4^{2-} + 4H_2O$

b $4Zn + 10HNO_3 \rightarrow 4Zn(NO_3)_2 + NH_4NO_3 + 3H_2O$

**15.5
GALVANIC CELLS**   As we will see in this section, a redox reaction may be used to convert energy from the form of chemical bonds into the form of electricity. This type of redox reaction takes place in batteries. On the other hand, electrical energy can be used to make a chemical change take place through a redox reaction. This process is used in industry to manufacture certain materials such as aluminum.

Consider what happens if we put a piece of zinc metal into a blue aqueous solution of copper sulfate. The following redox reaction takes place:

$$Zn(s) + Cu^{2+}(aq) \rightarrow Zn^{2+}(aq) + Cu(s)$$
zinc   copper(II)   zinc ion   copper
          ion

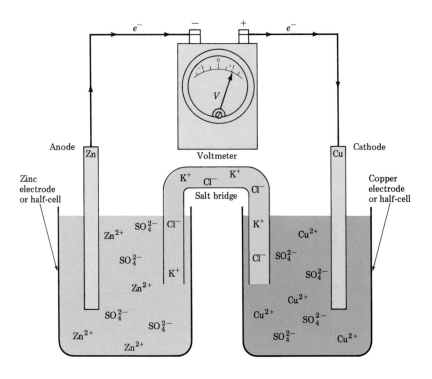

FIGURE 15-4

A galvanic (voltaic) cell. In this cell, an oxidation-reduction reaction generates a flow of electrons.

Electrons are transferred from zinc atoms to copper ions in this reaction. As it reacts, the zinc metal slowly dissolves. It loses electrons, becoming oxidized to zinc ions, which then enter the solution. Copper ions gain electrons, becoming reduced to metallic copper. This reduction removes Cu^{2+} ions from solution, making it appear less blue. The equations for the two half-reactions are:

oxidation: $Zn \rightarrow Zn^{2+} + 2e^-$
reduction: $Cu^{2+} + 2e^- \rightarrow Cu$

These two half-reactions can be physically separated, as shown in Figure 15-4. Electrons can be made to move through an external wire or circuit while passing from the oxidation to the reduction step. The set-up in the Figure is called an **electrochemical cell**. It contains two electrical conductors, called electrodes, each of which are immersed in an electrolyte solution. In an electrochemical cell, a flow of electrons can either be produced by a redox reaction or result in a redox reaction. In the particular cell shown in the Figure, electricity is *produced* by a redox reaction and the cell is called **galvanic**, after Luigi Galvani, or **voltaic**, after Alessandro Volta. (Galvani and Volta were 18th century Italian scientists.) Electrolytic cells, which *require* electrical energy, are described in Section 15.7.

Electrons leave and enter an electrochemical cell through the electrodes. In the galvanic cell in Figure 15-4, the electrodes are pieces of metal—zinc and copper. Oxidation occurs at one of the electrodes (Zn). This electrode is called the **anode**. The electrons leave the cell at the anode and pass through a wire connecting the anode to the other electrode. Reduction occurs at the other electrode (Cu), called the **cathode**. Here the electrons reenter the cell. (It may help you to remember these terms by noting that *a*node and *o*xidation both begin with vowels, and *c*athode and *r*eduction begin with consonants.)

Accompanying the flow of electrons outside the cell is a movement of ions inside the cell. Positively charged cations move towards the cathode, and negatively charged anions move towards the anode. The processes occurring at the electrodes can be summarized as follows:

Anode	**Cathode**
oxidation occurs	reduction occurs
electrons leave cell	electrons enter cell
attracts anions	attracts cations

In every electrochemical cell, oxidation and reduction take place at separate electrodes, electrons pass through an external circuit, and ions flow through the electrolyte solution in the cell.

Each electrode and its accompanying electrolyte solution is called a **half-cell**. For example, zinc metal and aqueous zinc ions make up a half-cell. The half-cells are connected by an external wire between the electrodes, and by a salt bridge. The wire allows electrons to pass from one half-cell to the other. The salt bridge, which generally contains aqueous potassium chloride, permits the movement of ions between the half-cells, but prevents the physical mixing of the electrolyte solutions.

Galvanic cells can be constructed in many ways. Commercial galvanic cells that are used as sources of electrical energy are known as **batteries**. The type of battery used in flashlights, called a dry cell, or Leclanché cell, is shown in Figure 15-5. It consists of a zinc container, which serves as the anode, and a carbon rod, which is the cathode. A moist paste of ammonium chloride, zinc chloride, and manganese dioxide fills the inside of the battery. This paste serves as the electrolyte "solution." The half-cell reactions can be represented as:

oxidation: $Zn \rightarrow Zn^{2+} + 2e^-$
reduction: $MnO_2 + NH_4^- + e^- \rightarrow MnO(OH) + NH_3$

Note that the carbon rod serves only as a conductor in this cell; it is actually manganese atoms that are reduced. No salt bridge is needed because the thick electrolyte paste prevents mixing of the oxidizing agent, MnO_2, and the reducing agent, Zn.

FIGURE 15-5

A dry or Leclanché cell. This common galvanic cell is the battery used in flashlights.

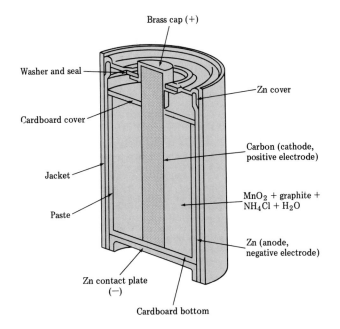

Brass cap (+)

Washer and seal

Zn cover

Cardboard cover

Carbon (cathode, positive electrode)

Jacket

MnO_2 + graphite + NH_4Cl + H_2O

Paste

Zn (anode, negative electrode)

Zn contact plate (−)

Cardboard bottom

FIGURE 15-6

A lead storage battery. This galvanic cell is the rechargeable battery used in automobiles.

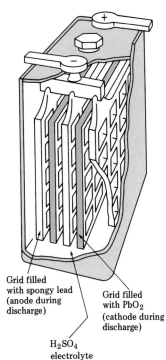

Grid filled with spongy lead (anode during discharge)

Grid filled with PbO_2 (cathode during discharge)

H_2SO_4 electrolyte

The dry cell has been replaced in many applications by the alkaline (manganese oxide-zinc) cell and the compact mercury (Ruben-Mallory) cell. The half-reactions of these cells are as follows:

alkaline cell   oxidation: $Zn + 2OH^- \rightarrow ZnO + H_2O + 2e^-$
reduction: $2MnO_2 + H_2O + 2e^- \rightarrow Mn_2O_3 + 2OH^-$
mercury cell   oxidation: $Zn + 2OH^- \rightarrow ZnO + H_2O + 2e^-$
reduction: $HgO + H_2O + 2e^- \rightarrow Hg + 2OH^-$

These batteries generally last longer than ordinary dry cells.

The automobile lead storage battery is illustrated in Figure 15-6. This cell consists of pairs of lead electrodes, half of which are packed with metallic lead and half with lead(IV) oxide, PbO_2. The half-reactions are:

oxidation: $Pb + SO_4^{2-} \rightarrow PbSO_4 + 2e^-$
reduction: $PbO_2 + 4H^+ + SO_4^{2-} + 2e^- \rightarrow PbSO_4 + 2H_2O$

The overall reaction that takes place in the lead storage battery is:

$$Pb(s) + PbO_2(s) + 2H_2SO_4(aq) \xrightarrow{\text{discharge}} 2PbSO_4(s) + 2H_2O(l)$$

lead   lead(IV)   sulfuric   lead(II)   water
          oxide   acid   sulfate

When the battery produces a flow of electrons by this reaction, the

505

battery is said to *discharge*. However, by applying an electric current from an outside source to the battery, it is possible to force the reaction to take place in reverse (as described in Section 15.7) or to *recharge*.

$$2PbSO_4(s) + 2H_2O(l) \xrightarrow{\text{recharge}} Pb(s) + PbO_2(s) + 2H_2SO_4(aq)$$

In recharging, the products are converted back to reactants. Thus, recharging extends the useful life of a battery by allowing it to discharge over and over again. The lead storage battery is specially constructed so that recharging restores it almost completely to its original condition.

The electrolyte in the lead storage battery is a solution of sulfuric acid. This acid is consumed, along with Pb and PbO_2, as the battery discharges. Since the density of the sulfuric acid solution decreases with use, measuring its specific gravity gives an indication of the battery's condition.

Box 15

Solar cells

Solar cells, or photovoltaic cells, were developed by scientists at Bell Laboratories in the early 1950s, not long after the transistor was introduced. These cells convert solar energy to electricity directly without the use of a redox reaction or any moving parts.

Solar cells are made from semiconductors like silicon. Two distinct regions are built into each cell, one positive and one negative. These two regions are separated by a discontinuity, called a junction. When sunlight strikes the surface of a solar cell it knocks electrons loose from their original locations. Some electrons receive enough energy to cross the junction. They cannot return to their original locations, however, without first passing through an external circuit. Thus the sunlight creates a flow of electrons, which is an electric current.

The main advantage of solar cells is that they produce electricity directly. Solar cells are also quiet, reliable, and easy to operate. Their small size permits them to be used in many different ways and on any scale. Spacecraft use solar cells to obtain energy. Solar cells are also being installed on rooftops to supply homes and apartments with energy, and can be built into large centralized installations.

The main disadvantages of solar cells are high cost and low efficiency. Large amounts of energy are needed to produce these cells. Yet present cells trap only about 10 to 15% of the solar energy available to them. Also, accumulation of dirt on solar cells reduces their power output by about 6 to 12% per year in Nebraska, and 35 to 40% per year in New York City. However, engineering research is now under way to make solar cells less costly and more efficient.

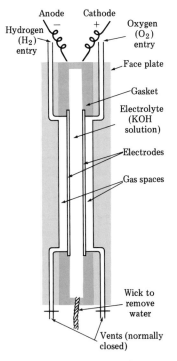

Anode   Cathode
Hydrogen $-$ $+$ Oxygen
(H_2) (O_2)
entry   entry

Face plate

Gasket

Electrolyte
(KOH
solution)

Electrodes

Gas spaces

Wick to
remove
water

Vents (normally
closed)

FIGURE 15-7

*A fuel cell. Electricity is
produced by the reaction of
hydrogen and oxygen in this
cell.*

Another common storage battery, used in electronic calculators, is the nickel-cadmium (or ni-cad) cell. Its half-reactions are:

oxidation: $Cd + 2OH^- \rightarrow Cd(OH)_2 + 2e^-$
reduction: $2NiO(OH) + 2H_2O + 2e^- \rightarrow 2Ni(OH)_2 + 2OH^-$

The fuel cell, illustrated in Figure 15-7, is a special type of galvanic cell that has renewable electrodes. It produces electrical energy by the oxidation of a fuel that is supplied continuously to the cell. In one fuel cell, hydrogen gas, the fuel, is supplied at the anode, and oxygen gas is supplied at the cathode. A concentrated aqueous solution of potassium hydroxide serves as the electrolyte in this fuel cell. The overall reaction is simply the combustion of hydrogen to form water.

oxidation: $2H_2 + 4OH^- \rightarrow 4H_2O + 4e^-$
reduction: $O_2 + 2H_2O + 4e^- \rightarrow 4OH^-$
$$\overline{2H_2(g) + O_2(g) \rightarrow 2H_2O(l)}$$

Compact hydrogen-oxygen fuel cells were used on the Apollo flights to the moon. They not only provided energy, but also produced drinkable water as a waste product.

The fuel cell converts chemical energy directly to electrical energy. This process is far more efficient than normal fuel combustion in which chemical energy must first be converted to thermal energy before it can be converted to electrical energy.

**15.6
REDUCTION
POTENTIALS**

Each half-reaction in an electrochemical cell involves both the oxidized and reduced forms of a substance, and can be written as either oxidation or reduction. For example, consider a half-reaction involving copper metal (Cu) and copper(II) ion, Cu^{2+}.

$$Cu^{2+} + 2e^- \underset{\text{oxidation}}{\overset{\text{reduction}}{\rightleftharpoons}} Cu$$

Copper metal is a reducing agent because it can donate electrons, and copper(II) ion is an oxidizing agent because it can accept electrons. A pair of substances related as oxidizing and reducing agents on opposite sides of a half-equation is called a **redox couple**.

In many respects, a redox couple is similar to the conjugate acid-base pairs discussed in Section 14.3. The members of a redox couple are related in an important way: *the stronger an oxidizing agent, the weaker its corresponding reducing agent; the stronger a reducing agent, the weaker its corresponding oxidizing agent.* For example, if a substance is a good oxidizing agent, it has a strong tendency to gain electrons. Once it has gained the electrons, however, the reduced form of the compound has little tendency to lose them. Therefore, it is a poor reducing agent. We used this same kind of reasoning to explain why a strong acid has a weak conjugate base.

The tendency of the oxidized form of a redox couple to be reduced is given by its **reduction potential**, E, measured in volts (V). Typical values of E are listed in Table 15-2. They are based on comparison

TABLE 15-2 *Standard reduction potentials*

Oxidizing agent (oxidized form of redox couple)		Reducing agent (reduced form of redox couple)	$E°$, volts
$F_2 + 2e^-$	$\rightarrow$	$2F^-$	$+ 2.87$
$H_2O_2 + 2H^+ + 2e^-$	$\rightarrow$	$2H_2O$	$+ 1.77$
$MnO_4^- + 8H^+ + 5e^-$	$\rightarrow$	$Mn^{2+} + 4H_2O$	$+ 1.51$
$Cl_2 + 2e^-$	$\rightarrow$	$2Cl^-$	$+ 1.36$
$Cr_2O_7^{2-} + 14H^+ + 6e^-$	$\rightarrow$	$2Cr^{3+} + 7H_2O$	$+ 1.33$
$O_2 + 4H^+ + 4e^-$	$\rightarrow$	$2H_2O$	$+ 1.23$
$Br_2 + 2e^-$	$\rightarrow$	$2Br^-$	$+ 1.09$
$Ag^+ + e^-$	$\rightarrow$	Ag	$+ 0.80$
$Fe^{3+} + e^-$	$\rightarrow$	Fe^{2+}	$+ 0.77$
$I_2 + 2e^-$	$\rightarrow$	$2I^-$	$+ 0.54$
$Cu^{2+} + 2e^-$	$\rightarrow$	Cu	$+ 0.34$
$2H^+ + 2e^-$	$\rightarrow$	H_2	0.00
$Pb^{2+} + 2e^-$	$\rightarrow$	Pb	$- 0.13$
$Fe^{2+} + 2e^-$	$\rightarrow$	Fe	$- 0.44$
$Zn^{2+} + 2e^-$	$\rightarrow$	Zn	$- 0.76$
$2H_2O + 2e^-$	$\rightarrow$	$H_2 + 2OH^-$	$- 0.83$
$Al^{3+} + 3e^-$	$\rightarrow$	Al	$- 1.66$
$Mg^{2+} + 2e^-$	$\rightarrow$	Mg	$- 2.37$
$Na^+ + e^-$	$\rightarrow$	Na	$- 2.71$
$K^+ + e^-$	$\rightarrow$	K	$- 2.93$
$Li^+ + e^-$	$\rightarrow$	Li	$- 3.05$

increasing oxidizing strength

increasing reducing strength

with the hydrogen half-cell, which is assigned the value of 0.0000 V. The values in this table are referred to as standard reduction potentials, $E°$, because the measurements were made at certain standard conditions: solution concentrations at 1 M, gas pressures at 1 atm, and solids and liquids pure and in their most stable forms, usually at 25°C.

The strongest oxidizing agent shown in Table 15-2 is F_2, and the weakest reducing agent is F^-. Similarly, the strongest reducing agent is Li, and the weakest oxidizing agent is Li^+. In general, the higher the positive value of the reduction potential, the stronger the oxidizing agent in the couple. The more negative the potential, the stronger the reducing agent. Thus couples at the top of this table tend to react in the direction shown by the arrow, as reductions. Couples at the bottom of this table are more likely to react in the opposite direction of the arrow, as oxidations. In some tables of reduction potentials the order of potentials is reversed, with the most negative values appearing at the top. For such tables, the relations described here must be reversed.

The table of reduction potentials can be used to predict chemical reactions. The reactions most likely to occur are between strong oxidizing agents (on the upper left in the table) and strong reducing agents (on the lower right in the table). Thus, fluorine, F_2, will react readily with lithium metal, Li. Similarly, oxygen, O_2, will react with magnesium metal, Mg. In general, the reduced form of a couple tends to react with the oxidized form of the couples above it in Table 15.2. This principle is the basis for the activity series described in Section 9.6. Metals listed below the hydrogen couple (those having negative values of $E°$, like zinc) will react with acids, liberating H_2 (under standard conditions). Also, metals will displace the oxidized form of metals above them in the table for the same reason. Therefore, Zn will react with Cu^{2+}, as we saw in the previous section.

Example Using Table 15-2, predict whether each of the following substances will react under standard conditions.

reactants	reaction?
Fe^{2+} and Cl_2	yes
Na^+ and Ag	no
H_2O_2 and Pb	yes
Cl^- and Br_2	no
O_2 and Mn^{2+}	no
Fe^{3+} and Al	yes

The voltage generated by a galvanic cell depends on the standard reduction potentials of the two half-cells. It is found by adding the potentials as shown:

Smithsonian Institution: Photo # 781831

The son of an English blacksmith, Faraday had little formal schooling. But he was apprenticed to a bookbinder who encouraged him to read. In his spare time, Faraday attended several lectures given by the chemist Humphry Davy. He made his notes into a book and sent it to Davy, who was impressed enough to hire him the following year as a laboratory assistant.

Much to the bitterness of Davy, Faraday soon became a greater scientist than his master. It was said that of all Davy's discoveries, Faraday was the greatest one. Faraday became director of the laboratory after 10 years, and then a professor of chemistry.

Davy had isolated new metals (K, Mg, Cu, Ba, Sr) by passing an electric current through their molten salts. Faraday continued work with his process, which he called electrolysis. Stated in modern-day terms, Faraday discovered that the number of moles of a substance produced at an electrode is directly proportional to the number of moles of electrons transferred at that electrode. This principle is known as Faraday's law, and the term faraday is now used to describe one mole of electrons (96,500 coulombs).

Faraday invented both the electric motor and the electric generator. These inventions eventually made possible the practical production and use of electricity.

Faraday was clearly one of the greatest experimentalists. His experiments yielded some of the most significant principles and inventions in the history of science.

ASIDE . . .

About 5% of all the electricity generated in the United States is used to prepare aluminum by electrolysis.

$$Cu^{2+} + 2e^- \rightarrow Cu \qquad +0.34 \text{ V}$$
$$Zn \rightarrow Zn^{2+} + 2e^- \qquad \underline{+0.76 \text{ V}}$$
$$+1.10 \text{ V}$$

Notice, however, that the zinc half-reaction occurs not as a reduction, but as an oxidation, and therefore the sign of the potential in the table (-0.76) must also be reversed. The resulting value, $+1.10$ V, is a measure of the driving force of the reaction.

The maximum value of the potential of a cell (under conditions where no current flows) is called its electromotive force, or **emf** (pronounced E-M-F). It is determined by the relative oxidizing and reducing strengths of the reactants, which, of course, are reflected in their respective reduction potentials. The greatest emf is produced when a strong oxidizing agent reacts with a strong reducing agent. The emf of a cell also depends on the concentrations of the reactants, and on the temperature.

**15.7
ELECTROLYTIC CELLS**

Reactions that do not normally take place can be forced to occur by providing electrical energy from an external source such as a battery. This process is called **electrolysis**, and is carried out in an **electrolytic cell**. Electrolysis is the opposite of the process that takes place in a galvanic cell:

$$\text{chemical energy} \xrightleftharpoons[\text{electrolytic cell}]{\text{galvanic cell}} \text{electrical energy}$$

In a galvanic cell, a chemical reaction generates a flow of electrons, as we saw in Section 15.5. But in an electrolytic cell, electrical energy is supplied to the cell to bring about a chemical change.

The electrolysis of water is illustrated in Figure 9-4, and in Figure 15-8. Two electrodes of relatively unreactive platinum are placed in water. A small amount of electrolyte (H_2SO_4) is added to the water to increase its ability to conduct electricity. The electrodes are connected to a power supply, causing the following half-reactions:

oxidiation: $2H_2O \rightarrow O_2 + 4H^+ + 4e^-$

reduction: $4H_2O + 4e^- \rightarrow 2H_2 + 4OH^-$

Electrons flow from the source of electric current to one of the electrodes, giving this electrode a negative charge and making it, in effect, a reducing agent. This electrode becomes the cathode, and is the source

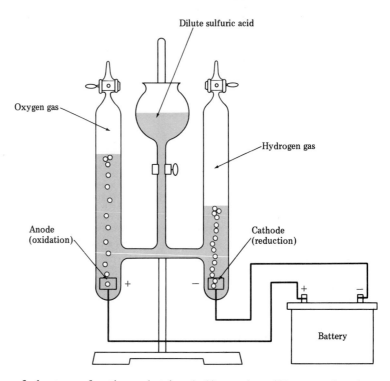

Dilute sulfuric acid

Oxygen gas

Hydrogen gas

Anode
(oxidation)

Cathode
(reduction)

Battery

FIGURE 15-8

*Electrolysis of water. An
electric current is used to
separate water into H_2 and O_2.
Note that the volume of H_2
produced is twice that of O_2.*

of electrons for the reduction half-reaction. Water molecules are reduced at the cathode, forming hydrogen gas. The power supply removes
electrons from the other electrode, giving it a positive charge, and
making it the anode. At the anode, water molecules are oxidized to
form oxygen.

Many important substances are produced by electrolysis. For example, the chlor-alkali industry is based on the electrolysis of salt water
(brine) to make chlorine, sodium hydroxide, and hydrogen. The electrolytic cell used for this process is shown in Figure 15-9 and the half-
reactions are:

oxidation: $2Cl^- \rightarrow Cl_2 + 2e^-$
reduction: $2H_2O + 2e^- \rightarrow H_2 + 2OH^-$

Chlorine gas is produced at the anode and hydrogen gas at the cathode.
Since chloride ions are eliminated from the solution and hydroxide ions
are formed, the original sodium chloride is gradually replaced by sodium hydroxide, which is then removed from the cell as a 10% solution.

Another commerical use of electrolysis is the reduction of aluminum
ore (bauxite), Al_2O_3, to aluminum metal, as shown in Figure 15-10. The
ore is dissolved in a molten (melted) salt, cryolite, Na_3AlF_6, at about
1000°C. The steel container lined with graphite is the cathode at which
the reduction occurs.

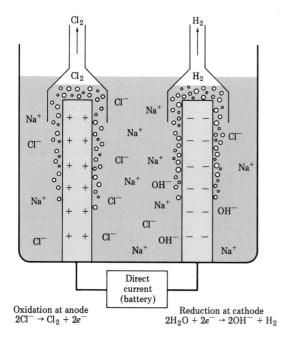

Oxidation at anode
$2Cl^- \rightarrow Cl_2 + 2e^-$

Reduction at cathode
$2H_2O + 2e^- \rightarrow 2OH^- + H_2$

FIGURE 15-9

A chlor-alkali cell. This cell is named after two of its products: chlorine gas and sodium hydroxide solution.

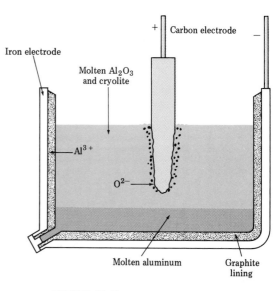

FIGURE 15-10

Electrolysis of alumina, Al_2O_3. The cell is used in industry to make aluminum.

oxidation: $2O^{2-} + C \rightarrow CO_2 + 4e^-$
reduction: $Al^{3+} + 3e^- \rightarrow Al$

The carbon anode gradually gets used up and must be replaced. About 5% of the electrical energy generated in the U.S. is used to prepare aluminum by electrolysis.

Metals can also be purified by electrolysis. For example, Figure 15-11 shows the refining of copper by electrolysis. The impure or "blister" copper serves as the anode, while a thin sheet of very pure copper is used as the cathode. The electrolyte is aqueous copper(II) sulfate. The half-reactions are:

oxidation: $Cu \rightarrow Cu^{2+} + 2e^-$
reduction: $Cu^{2+} + 2e^- \rightarrow Cu$

During electrolysis the impure anode reduces in size as the copper metal is oxidized to Cu^{2+} ions and dissolves. Impurities either dissolve or fall to the bottom of the cell, forming what we call anode sludge.

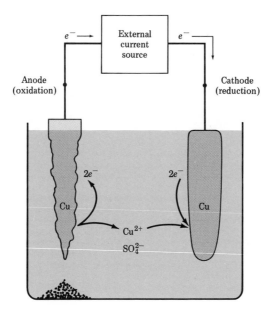

FIGURE 15-11

Electrolytic refining of copper. Impure copper is oxidized at the anode, and pure copper is deposited at the cathode. Note the impurities that have piled up beneath the anode.

At the cathode, only Cu^{2+} ions are reduced, increasing the amount of pure copper on the cathode, which is later removed.

Electroplating, the coating of an object with a metal, also is done by electrolysis. The object that we wish to coat with metal is used as the cathode. The solution contains ions of the metal that will be plated onto the object. For example, automobile bumpers are usually coated with chromium by electrolysis. The bumper is used as the cathode, and the solution contains dichromate ions, $Cr_2O_7^{2-}$, which are reduced and deposited onto the bumper as a thin coating of chromium metal.

Summary

15.1 Oxidation numbers are helpful aids for identifying and following redox reactions. The oxidation number of an element is found by assigning electrons to the atoms of a compound or ion according to certain rules.

15.2 One way to identify redox reactions is to look for changes in oxidation numbers. In oxidation, a reducing agent donates electrons, becoming oxidized and increasing its oxidation number. In reduction, an oxidizing agent accepts electrons, becoming reduced and decreasing its oxidation number.

15.3 Redox equations can be balanced by the oxidation number method. This approach is based on the fact that the total increase in oxidation number for one element must equal the total decrease in oxidation number in another.

15.4 Redox equations can also be balanced by the half-reaction method. This method involves making equal the number of electrons in the oxidation half-equation and the number of electrons in the reduction half-equation.

15.5 An electrochemical cell is an arrangement in which a flow of electrons is either produced by a redox reaction or results in a redox reaction. There are two types of electrochemical cells, galvanic (or voltaic) and electrolytic. In a galvanic, or voltaic, cell, electricity is generated by a chemical reaction. Electrons are produced by oxidation at the anode, pass through a wire, and enter the cell again at the cathode, where reduction occurs.

15.6 A redox couple is a pair of substances that appear on opposite sides of a half-equation. The tendency of the oxidized form of a redox couple to be reduced is given by a reduction potential in volts. Reduction potentials can be used to predict the direction of reactions.

15.7 In electrolysis, electrical energy is used to drive a redox reaction that would not normally take place. The redox reaction occurs in an electrolytic cell, in which one electrode causes reduction while the other causes oxidation. Electrolysis is used industrially to produce many important substances.

Exercises

KEY-WORD MATCHING EXERCISE

For each term on the left, choose a phrase from the right-hand column that most closely matches its meaning.

1 redox reaction	**a** donates electrons
2 oxidation number	**b** either oxidation or reduction
3 reducing agent	**c** oxidation occurs here
4 oxidizing agent	**d** requires electricity
5 half-reaction	**e** galvanic or electrolytic
6 galvanic cell	**f** electron transfer reaction
7 electrochemical cell	**g** members differ by one or more electrons
8 anode	**h** creates electricity
9 cathode	**i** accepts electrons
10 half-cell	**j** reduction occurs here
11 electrolytic cell	**k** charge based on rules
12 redox couple	**l** acts as anode or cathode

Oxidation numbers (15.1)

13 Determine the oxidation state for each atom in the following compounds:
a CO_2 **b** CO **c** SF_6 **d** N_2H_4 **e** P_4O_{10}

14 Write the oxidation number for each atom in the following compounds:
a $CrCl_6$ **b** $MgBr_2$ **c** Al_2O_3 **d** PtI_4 **e** HgS

15 Identify the oxidation number for each atom: **a** NO_2^- **b** CO_3^{2-}
c PO_4^{3-} **d** HS^- **e** Mo^{4+}

16 Find the oxidation number for each atom: **a** Na_2SO_4 **b** NH_4Cl
c $Ca(OH)_2$ **d** $KHCO_3$

Oxidation and reduction (15.2)

17 Why do atoms change oxidation numbers in redox reactions?

18 Identify the following changes as oxidation or reduction:
a $Zn^{2+} \rightarrow Zn$ **b** $2Br^- \rightarrow Br_2$ **c** $Sn^{2+} \rightarrow Sn^{4+}$

19 Identify the atoms that undergo oxidation number changes in these reactions:
a $2MnO_2 + 2K_2CO_3 + O_2 \rightarrow 2KMnO_4 + 2CO_2$
b $Sn + 4HNO_3 \rightarrow SnO_2 + 4NO_2 + 2H_2O$

20 Identify the oxidizing agents and reducing agents in the reactions of Exercise 19.

21 Write the equations for the half-reactions of the following reaction:
$$Cu + Zn^{2+} \rightarrow Cu^{2+} + Zn$$

Balancing redox equations by oxidation numbers (15.3)

22 Burning ammonia in oxygen produces nitrogen by the following reaction:
$$NH_3(g) + O_2(g) \rightarrow N_2(g) + H_2O(l)$$
Balance this equation by the oxidation number method.

23 Silver is extracted as a soluble compound by reacting it with sodium cyanide:
$$Ag(s) + NaCN(aq) + O_2(g) + H_2O(l) \rightarrow NaAg(CN)_2(aq) + NaOH(aq)$$
Balance the equation by the oxidation number method.

24 The following equation represents the action of sulfuric acid on iron:
$$Fe(s) + H_2SO_4(aq) \rightarrow Fe_2(SO_4)_3(aq) + SO_2(g) + H_2O(l)$$
Balance it by the oxidation number method.

25 The roasting of stibnite, an antimony ore, produces the oxide:
$$Sb_2S_3(s) + O_2(g) \rightarrow Sb_4O_6(s) + SO_2(g)$$
Balance this equation by the oxidation number method.

Balancing by half-reactions (15.4)

26 A test for sulfur dioxide involves watching orange dichromate ion change to blue-green chromium(III) ion:
$$Cr_2O_7^{2-}(aq) + SO_2(g) + H^+(aq) \rightarrow Cr^{3+}(aq) + SO_4^{2-}(aq) + H_2O(l)$$
Balance this equation by the half-reaction method.

27 The following equation represents the change of Fe(II) to Fe(III):

$$FeSO_4(aq) + O_2(g) + H_2SO_4(aq) \rightarrow Fe_2(SO_4)_3(aq) + H_2O(l)$$

Balance it by the half-reaction method.

28 Hydrogen peroxide, usually an oxidizing agent, acts as a reducing agent in a reaction with permanganate ion:

$$H_2O_2(l) + MnO_4^-(aq) + H^+(aq) \rightarrow Mn^{2+}(aq) + O_2(g) + H_2O(l)$$

Balance this equation by the half-reaction method.

29 Chromium(III) ion is readily oxidized in basic solution by chlorite ion:

$$Cr(OH)_3(aq) + NaClO(aq) + NaOH(aq) \rightarrow Na_2CrO_4(aq) + NaCl(aq) + H_2O(l)$$

Balance this equation by the half-reaction method.

Galvanic cells (15.5)

30 How is an electrochemical cell related to a redox reaction?

31 In a galvanic cell, describe what happens at the **a** anode **b** cathode

32 How does a galvanic cell differ from an electrolytic cell?

33 Why is the specific gravity of the electrolyte in a car battery a measure of the condition of the battery?

34 Write the overall cell reaction in **a** a dry cell **b** an alkaline battery

Reduction potentials (15.6)

35 What is a redox couple?

36 What is measured by a reduction potential?

37 Predict whether each of the following pairs of substances will react under standard conditions: **a** Zn^{2+} and Mg **b** Cl^- and I_2 **c** Cu^{2+} and Fe **d** Zn and Al^{3+} **e** Ag^+ and Pb

38 What is the emf of a cell? How is it found?

Electrolytic cells (15.7)

39 Why does an electrolytic cell require an outside source of electricity?

40 Describe how an electrolytic cell is used for electroplating.

41 Write the overall cell reaction in **a** a chlor-alkali cell **b** an aluminum electrolysis cell

42 How can metals be purified by electrolysis?

ADDITIONAL EXERCISES   **43** Identify the oxidation number of each atom in **a** $K_2Cr_2O_7$ **b** CaC_2O_4 **c** $Al(H_2PO_4)_3$ **d** $Mg_3(AsO_4)_2$

44 Why are metals good reducing agents?

45 Iron dissolves in nitric acid by the following reaction:

$$Fe(s) + HNO_3(aq) \rightarrow Fe(NO_3)_3(aq) + NO_2(g) + H_2O(l)$$

Balance this equation by the oxidation number method.

46 Dichromate ion can be reduced using hydrogen iodide:

$$K_2Cr_2O_7(aq) + HI(aq) \rightarrow CrI_3(aq) + I_2(s) + KI(aq) + H_2O(l)$$

Balance this equation by the half-reaction method.

47 Silane, SiH_4, reacts with a base in the following way:

$$SiH_4(g) + NaOH(aq) + H_2O(l) \rightarrow Na_2SiO_3(aq) + H_2(g)$$

Balance this equation by the oxidation number method. (Hint: Silane is a hydride.)

48 Nitrite ion can act as a reducing agent as illustrated here:

$$NO_2^-(aq) + MnO_4^-(aq) + H^+(aq) \rightarrow Mn^{2+}(aq) + NO_3^-(aq) + H_2O(l)$$

Balance this equation by the half-reaction method.

49 What is a salt bridge? Why is it necessary in an electrochemical cell?

50 Why do batteries "go dead"?

51 Calculate cell potentials for the reactions given in Exercise 37.

52 Describe the operation of a fuel cell.

53 How is a redox couple similar to a conjugate acid-base pair?

54 Predict whether the following will react: **a** Al^{3+} and Pb **b** Mg and Cu^{2+} **c** K^+ and Fe

55 Balance the following equations either by the oxidation number or half-reaction method:

a $Ag + NO_3^- + H^+ \rightarrow Ag^+ + HNO_2 + H_2O$
b $Cd + Sb_2O_3 + HCl \rightarrow CdCl_2 + Sb + H_2O$
c $FeBr_2 + N_2O_4 + HBr \rightarrow FeBr_3 + NO + H_2O$
d $NiO_2 + S_2O_3^{2-} + OH^- \rightarrow Ni(OH)_2 + SO_3^{2-} + H_2O$
e $Br_2 + Ce^{4+} + H_2O \rightarrow BrO_3^- + Ce^{3+} + H^+$
f $H_2C_2O_4 + HClO_3 \rightarrow 2CO_2 + Cl_2 + H_2O$
g $CrCl_3 + F_2 + KCl + H_2O \rightarrow K_2Cr_2O_7 + HF + HCl$
h $Cu + I^- + H_2SO_3 + H^+ \rightarrow CuI + S + H_2O$
i $Mn^{3+} + I^- + OH^- \rightarrow Mn^{2+} + IO_3^- + H_2O$
j $Ti^{2+} + ClO^- + H_2O \rightarrow Ti^{3+} + Cl^- + OH^-$

56 Find the oxidation state of S in: S S^{2-} H_2S SO_2 SO_3 SO_3^{2-} HSO_4^-

57 In each equation, pick out the oxidizing agent and reducing agent.
a $4Al(s) + 3O_2(g) \rightarrow 2Al_2O_3(s)$
b $Cr_2O_7^{2-}(aq) + 3SO_3^{2-}(aq) + 8H^+(aq) \rightarrow 2Cr^{3+}(aq) + 3SO_4^{2-}(aq) + 4H_2O(l)$

58 Predict whether or not each of the following reactions will occur as written:
a $Mg(s) + Cu^{2+}(aq) \rightarrow Mg^{2+}(aq) + Cu(s)$
b $Pb(s) + 2H^+(aq) \rightarrow H_2(g) + Pb^{2+}(aq)$
c $2Ag^+(aq) + Cu(s) \rightarrow 2Ag(s) + Cu^{2+}(aq)$
d $2Al^{3+}(aq) + 3Zn(s) \rightarrow 3Zn^{2+}(aq) + 2Al(s)$
e $H_2(g) + Zn^{2+}(aq) \rightarrow 2H^+(aq) + Zn(s)$

59 Balance these equations by the oxidation number method:
a $SiO_2(s) + Al(s) \rightarrow Si(s) + Al_2O_3(s)$
b $MnO_2(s) + PbO_2(s) + HNO_3(aq) \rightarrow HMnO_4(aq) + Pb(NO_3)_2(aq) + H_2O(l)$
c $H_2O_2(aq) + HI(aq) \rightarrow I_2(s) + H_2O(l)$

60 Balance these equations by the half-reaction method.

a $H^+(aq) + I^-(aq) + NO_2^-(aq) \rightarrow NO(g) + I_2(s) + H_2O(l)$

b $Al(s) + H^+(aq) + SO_4^{2-}(aq) \rightarrow Al^{3+}(aq) + H_2O(l) + SO_2(g)$

c $Zn(s) + NO_3^-(aq) + OH^-(aq) + H_2O(l) \rightarrow NH_3(aq) + Zn(OH)_4^{2-}(aq)$

61 Describe how you could electroplate a fresh coat of silver onto an old silver spoon.

PROGRESS CHART Check off each item when completed.

———————— Previewed chapter

Studied each section:

———————— **15.1** Oxidation numbers

———————— **15.2** Oxidation and reduction

———————— **15.3** Balancing redox equations by oxidation numbers

———————— **15.4** Balancing by half-reactions

———————— **15.5** Galvanic cells

———————— **15.6** Reduction potentials

———————— **15.7** Electrolytic cells

———————— Reviewed chapter

———————— Answered assigned exercises:

#'s ————————————————————

SIXTEEN

Chemical kinetics and chemical equilibrium

Many different kinds of chemical reactions have been described in previous chapters. Yet certain general questions regarding chemical change have been left unanswered. For example, how fast does a reaction occur? By what process or steps does a reaction occur, and how can the process be changed to make the reaction faster? What is the extent of a reaction—does it go nearly to completion or does only a small amount of product form? How can the relative amounts of reactants and products be changed? These important questions are vital to the production of chemical compounds both in industry and in the laboratory, and they are discussed in the following sections. To demonstrate an understanding of Chapter Sixteen, you should be able to:

PART A
CHEMICAL KINETICS
(16.1–16.4)

1 Explain the meaning of a rate equation and describe the factors that affect reaction rates.

2 Describe the reaction process in terms of activation energy.

3 Explain how a rate equation can be related to the mechanism of a reaction.

4 Explain the effect of catalysis on the rate of a reaction.

PART B
CHEMICAL EQUILIBRIUM
(16.5–16.11)

5 Describe the nature of chemical equilibrium.

6 Use Le Châtelier's principle to explain the effects on a reaction at equilibrium of changes in concentration, pressure, and temperature.

7 Given a balanced chemical equation, write the equilibrium constant expression.

8 Explain the meaning of K_w, K_a, K_b and K_{sp}.

9 Given an equilibrium constant, solve problems involving chemical equilibrium.

10 Solve problems involving acid and base equilibrium.

11 Solve problems involving solubility equilibrium.

Part A   Chemical kinetics

16.1 REACTION RATE   The **rate** of a reaction describes how rapidly a chemical change takes place. It is expressed as a change in concentration of reactant or product with time. For a reaction in solution, the rate may be expressed in moles per liter per second. For example, a reaction product might be formed at the rate of 2.5 moles per liter per second. The rate of a reaction is an experimental quantity found by measuring the disappearance of a reactant or the appearance of a product during a period of time. The changing concentration of reactant or product may be determined by watching the changing color of the solution (if the products and reactants differ in color), by observing the changing pressure of the reaction system if a gas is involved, or by many other methods.

The rate of a reaction depends on the nature of the reactants. After all, the rate at which bonds break or form depends on the nature of the bonds and on the nature of the molecules in which the bonds are located. The rate is also related to the concentrations of the reactants. The relationship between rate and concentrations is given by a **rate equation**, or rate law.

The following are two chemical equations and their experimentally determined rate equations:

Chemical equation	*Rate equation*
$H_2 + I_2 \rightarrow 2HI$	$\text{rate} = k[H_2][I_2]$
$H_2 + Br_2 \rightarrow 2HBr$	$\text{rate} = \dfrac{k[H_2][Br_2]}{1 + \dfrac{k'[HBr]}{[Br_2]}}$

(Remember that the square brackets around a substance stand for its concentration in moles/liter.) The rate equation for the formation of hydrogen iodide shows that the rate depends on the product of the H_2 and I_2 concentrations. The rate equation for the formation of hydrogen bromide, however, shows that the rate depends on the H_2, Br_2, and HBr concentrations, and in a more complicated way. Even though the HBr reaction has exactly the same stoichiometry as the HI reaction, its rate equation is very different because the reaction takes place in a different manner. We see then that a rate equation cannot be written simply by looking at the chemical equation, but must be found experimentally.

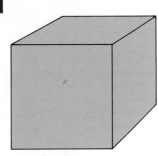

One piece:
small surface
area

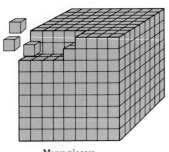

Many pieces:
large surface
area

FIGURE 16-1

Surface area. The conversion of large solid pieces into a powder greatly increases the solid's surface area, allowing it to react more rapidly.

The symbol k in a rate equation is called the **rate constant**. It is a number that relates the concentrations to the measured rate. A large value of k means that the reaction is fast; a small value indicates that the reaction is slow. The value of k depends mainly on the nature of the reactants and products, and usually remains unchanged during a reaction. It is affected by temperature, however. For each 10°C increase in temperature, the value of k, and therefore the rate of reaction, approximately doubles. Similarly, lowering the temperature decreases the rate. We store milk in a refrigerator, for example, to slow down the reactions that cause it to turn sour.

The major factors that influence the rate of a particular reaction are the concentrations of reactants and the temperature. We can speed up a reaction by raising the concentrations of those substances that appear in the rate equation, or by increasing the temperature. (If the reactants are gases, increasing the pressure leads to increased concentration in moles/liter, and a greater rate.) One additional factor that affects the reaction rate is the amount of contact between reactants. When one of the reactants is a solid, the reaction may take place at the solid's surface. As shown in Figure 16-1, converting the solid to a powder increases its surface area. Such an increase in surface area causes an increase in reaction rate because more of the solid's atoms or molecules are now exposed and can react. Substances that burn slowly as lumps, like coal, can burn explosively when powdered and dispersed in air. Reaction rates can be even faster when the reactants are in solution than when they are in the solid state. In a solution, the reactants can be separated by solvent into still smaller particles, thus increasing the surface area even more.

**16.2
ACTIVATION
ENERGY**

We can describe how chemical reactions take place in terms of the collisions between atoms, ions, or molecules. According to the kinetic molecular theory, the particles that make up matter are constantly moving. The degree of motion depends on the state of matter (solid, liquid, or gas) and the temperature. When two or more different substances are mixed together, it is likely that their atoms, molecules, or ions will collide with each other. Some of these collisions will cause bonds to break and new bonds to form in such a way that different combinations of atoms are created. When this happens, we say that a chemical reaction has taken place.

The reaction rate depends on the numbers of atoms, molecules, or ions that react after colliding. These numbers, in turn, depend on the nature of the reactants, on their concentration, and on their energy of collision. For example, some reactants, such as oppositely charged ions, attract each other. Therefore, they react readily and the rate of reaction is high. On the other hand, other types of reactants, such as nonpolar molecules, must overcome repulsive forces and therefore must collide with greater energy in order to react. In this case, the rate of reaction is low.

Since the kinetic energy of atoms, ions, and molecules is related to temperature, raising the temperature increases the number of particles having high kinetic energy. Therefore, the number of collisions that result in a reaction increases. Thus, raising the temperature causes the reaction rate to increase.

The minimum amount of energy needed for colliding molecules to react is called the **activation energy**. If many of the molecules in a reaction have this energy, the reaction will take place rapidly. As

Box 16

Inhibitors

Inhibitors are substances that slow down chemical reactions. In many cases, they do this by interfering with a catalyst. Inhibitors can be either helpful or harmful. Some inhibitors are drugs or food preservatives, while others are poisons.

Sometimes it is very useful to be able to slow down a chemical reaction. For example, sulfa drugs help the body to fight certain bacterial infections by inhibiting an enzyme in the invading bacteria. By inhibiting this enzyme, the sulfa drug slows down a reaction that the bacteria need to produce an essential substance for their growth.

Some food preservatives are inhibitors. If you examine food labels, you may come across the letters EDTA. This compound, *ethylenedi-aminetetraacetic* acid, can bind strongly to metal ions, and this prevents them from entering into other reactions. Certain metal ions, such as Cu^{2+}, catalyze oxidation reactions that spoil food. Thus, the use of EDTA inhibits those reactions, and prevents food from spoiling.

Some inhibitors are poisons because they interfere with important enzymes in our bodies. For example, cyanide ion, CN^-, is one of the most toxic poisons known. It binds to the iron in an enzyme involved in reducing oxygen in the cell. By preventing this enzyme from transporting electrons, cyanide blocks respiration, causing respiratory failure and death.

Mercury, another poisonous inhibitor, binds to sulfur strongly and inhibits sulfur-containing enzymes, which are common in the body. Mercury seems to interfere most with enzymes involved in the brain and nervous system, and mercury poisoning therefore can result in paralysis and mental disturbances.

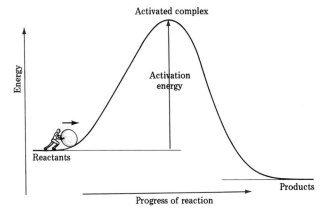

FIGURE 16-2

The energy barrier between reactants and products. Reactants must have sufficient energy, called the activation energy, in order to react and form products.

illustrated in Figure 16-2, the activation energy can be thought of as a hill or barrier that must be crossed before the reactants can be converted to products. Activation energies are usually about 10 to 50 kcal/mole (or 40 to 210 kJ/mole).

Consider a mixture hydrogen, H_2, and iodine, I_2, in a flask. At room temperature the reaction rate is too slow to be measured. The rate is slow because few of the collisions are energetic enough to break the bonds between iodine atoms in the I_2 molecules. The activation energy is too great. When the temperature is raised, however, the molecular collisions become energetic enough to break the bonds, and the liberated iodine atoms can react with the hydrogen molecules to produce hydrogen iodide, HI.

Another way to visualize the reaction process is to imagine the formation of an unstable, short-lived combination of atoms called the **activated complex** (or transition state). This special molecule is intermediate between the reactants and products, and exists only at the top of the energy barrier shown in Figure 16-2. In the activated complex, the new bonds needed to make the products have begun to form, and

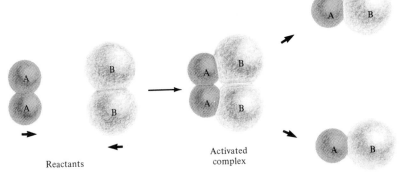

FIGURE 16-3

Formation of an activated complex. An activated complex is a short-lived, high-energy combination that is intermediate between the reactants and products.

the old bonds have weakened. Figure 16-3 shows the formation and breakdown of an activated complex.

16.3 REACTION MECHANISM

The **mechanism** of a reaction is the series of steps by which the reaction takes place. It is the pathway that converts reactant molecules into product molecules. The mechanism may involve a single step or it may involve several steps. For example, the simplified overall equation for a substitution reaction might be written as follows.

$$AX + Y \rightarrow AY + X$$

reactants products

This overall equation shows that Y replaces X in the molecule AX. However, this equation does not indicate how the substitution is made. We know that this reaction must involve the breaking of the A-X bond and the formation of an A-Y bond. Therefore, one possible mechanism for this reaction is:

mechanism I:

$$AX \xrightarrow{\text{slow}} A + X \qquad \text{first step}$$
$$A + Y \xrightarrow{\text{fast}} AY \qquad \text{second step}$$
$$\overline{AX + Y \longrightarrow AY + X} \qquad \text{overall reaction}$$

The two steps of this mechanism are illustrated in Figure 16-4(a). The first step, breaking the A-X bond, is likely to be slow because it probably involves a high activation energy. The slowest step in a reaction is called the **rate determining step** (or rate limiting step) because it limits the speed of the overall reaction. The rate of the faster second step, the formation of the A-Y bond, does not affect the rate of the overall equation. No matter how fast this second step is, the overall reaction can never be faster than the rate determining step. As an analogy, consider a family at dinner in which no one may leave the table until everyone is finished eating. Family members may not get up, no matter how fast they eat, until the slowest one has finished. The slowest eater is the rate-determining step in the meal.

FIGURE 16-4

Possible mechanisms for a substitution reaction. In (a), X dissociates from A in a slow step, followed by the rapid addition of Y. In (b), Y adds to A–X in a slow step, followed by the rapid elimination of X.

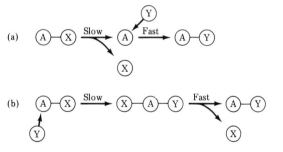

We can write a rate equation for mechanism I since we know the rate determining step. The slow first step, the breakdown of AX molecules, depends only on the concentration of AX molecules. Therefore, the rate equation for the overall reaction is:

rate $= k[AX]$ mechanism I

Another possible reaction mechanism is:

mechanism II: $AX + Y \xrightarrow{\text{slow}} AXY$ first step

$\dfrac{AXY \xrightarrow{\text{fast}} AY + X}{AX + Y \longrightarrow AY + X}$ second step

overall reaction

In this mechanism, shown in Figure 16-4(b), Y forms a bond to AX in a slow step. This step is followed by the rapid loss of X. Again, the slow step is the rate determining step. Both AX and Y are reactants in the rate determining step. Therefore, the rate of the overall reaction depends on the concentrations of AX and Y. The rate equation for this mechanism is:

rate $= k[AX][Y]$ mechanism II

Note that the rate equation obtained for this mechanism is quite different from the rate equation obtained for mechanism I. If we find by experiment the *actual* rate equation for a substitution reaction, we can then rule out one or both of these possible mechanisms.

The study of the relationship between reaction rates and mechanisms is called **chemical kinetics**. The proposed or theoretical mechanism for a reaction is the chemist's way of explaining what is happening at the molecular level to give the observed rate equation. By experimentally determining a rate equation, the chemist eliminates reaction mechanisms that predict rate equations that are different from the one observed. For example, if the measured rate equation for the substitution reaction just described is rate $= k[AX][Y]$, then the reaction cannot occur by mechanism I. Instead, it probably takes place by mechanism II, or some other mechanism which gives the observed rate equation.

16.4
CATALYSIS

A **catalyst** is a substance that increases the rate of a reaction without itself being consumed by the reaction. The catalyst takes part in the reaction but can be recovered intact at the end of the reaction. The reaction is said to be catalyzed or to undergo **catalysis**.

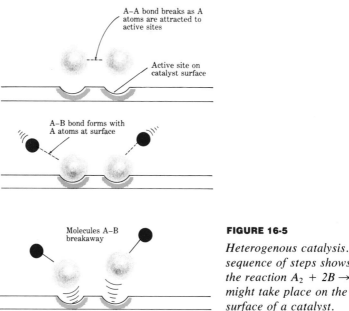

A–A bond breaks as A
atoms are attracted to
active sites

Active site on
catalyst surface

A–B bond forms with
A atoms at surface

Molecules A–B
breakaway

FIGURE 16-5

*Heterogenous catalysis. This
sequence of steps shows how
the reaction $A_2 + 2B \rightarrow 2AB$
might take place on the
surface of a catalyst.*

An example of catalysis is the reaction of hydrogen and oxygen gas in the presence of a small amount of platinum metal.

$$2H_2(g) \ + \ O_2(g) \ \xrightarrow{\text{Pt}} \ 2H_2O(l)$$
$$\text{hydrogen} \quad \text{oxygen} \qquad\qquad \text{water}$$

The uncatalyzed reaction between these two gases at room temperature is very slow, and the gases can remain mixed together for years without producing much water. But the addition of the platinum catalyst makes the reaction so rapid that an explosion occurs. The platinum catalyst can be recovered unchanged after the hydrogen and oxygen have been converted to water.

Catalysts increase the reaction rate by providing an alternate reaction mechanism. In other words, catalysis lowers the energy barrier shown in Figure 16-2. Thus, more molecules have the energy needed to react, and the rate of reaction increases.

The platinum catalyst just mentioned is called a surface catalyst, because it allows the reaction to take place on its surface. See Figure 16-5. The platinum catalyst enables H_2 molecules to dissociate into H atoms, which then react very quickly with O_2 molecules. This catalyst is also called a heterogeneous catalyst, because it is physically distinct from the reaction mixture. The catalytic converter in an automobile (illustrated in Figure 16-6) contains heterogeneous catalysts, which take part in reactions involving the car's exhaust gases.

A homogeneous catalyst is a catalyst that is dissolved in the reaction

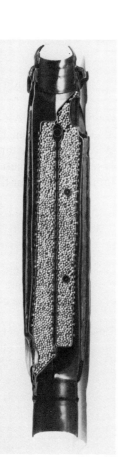

FIGURE 16-6

*A catalytic converter. The
catalyst, which contains
platinum, palladium, and
rhodium, increases the rate of
oxidation of the gases released
by an automobile engine.
(Photo courtesy of General
Motors.)*

Photo courtesy of the New York Public Library

We came across Ostwald's name in Chapter Nine of this book. It was he who developed the Ostwald process for the synthesis of nitric acid. The first step in that synthesis is the reaction of ammonia with oxygen in the presence of a platinum-rhodium catalyst. This reaction was made possible by Ostwald's research in catalysis. In 1909 he received the Nobel Prize in chemistry for that research.

This German chemist proposed that catalysts increase the rate of a reaction without affecting the energy relationship between reactants and products. He was an early supporter of Arrhenius' notion of activation energy. (Arrhenius worked in Ostwald's laboratory for a time, and was always a welcome guest in Ostwald's home. In fact Ostwald's family referred to Arrhenius as "jolly, fat, Uncle Svante.") Ostwald suggested that catalysts work by lowering the activation energy. He also recognized that ions, which were introduced by Arrhenius, can act as catalysts in certain reactions.

Ostwald is considered one of the founders of physical chemistry. This branch of chemistry deals with the theoretical aspects of the properties and reactions of atoms, ions, and molecules. Ostwald and his students carried out a good deal of significant research in this area, studying colligative properties, oxidation-reduction, and electrochemistry, in addition to catalysis.

Ostwald founded the famous journal "Zeitschrift für physikalische Chemie" and published many books and papers. When asked why he was able to accomplish so much, Ostwald replied: "Because my work has always brought me so much pleasure."

mixture. Nitric oxide, NO, is a homogeneous catalyst used in the oxidation of sulfur dioxide to sulfur trioxide.

$$2SO_2(g) + O_2(g) \rightarrow 2SO_3(s)$$

sulfur   oxygen   sulfur
dioxide   trioxide

This reaction is normally very slow. The addition of nitric oxide permits the reaction to take place by the following mechanism:

$2NO + O_2 \rightarrow 2NO_2$	first step
$2SO_2 + 2NO_2 \rightarrow 2SO_3 + 2NO$	second step
$2SO_2 + O_2 \rightarrow 2SO_3$	overall reaction

The nitric oxide is rapidly oxidized to nitrogen dioxide, NO_2, which in turn rapidly oxidizes the sulfur dioxide. This alternate mechanism is very fast. Note that although the catalyst is consumed in the first step, it is formed in the second step. Therefore none of the NO is used up in the overall reaction. In both homogeneous and heterogeneous catalysis, only a small amount of catalyst is needed. This small amount constantly converts reactants to products and is not consumed by the reaction.

Do-it-yourself 16

Enzymes in yeast

Yeasts are living microorganisms. They contain enzymes that are useful to us. One such enzyme is zymase, which catalyzes the conversion of sugar (glucose) to alcohol and carbon dioxide.

$$C_6H_{12}O_6(s) \xrightarrow{\text{zymase}} 2C_2H_5OH(l) + 2CO_2(g)$$

glucose   ethyl   carbon
                    alcohol   dioxide

This type of reaction, known as fermentation, is used in producing alcohol and in bread making.

Bread dough made with yeast rises because alcohol and carbon dioxide gas are produced and driven off during fermentation. Fermentation reactions take place very slowly at room temperature and therefore bread dough rises more quickly when it is heated than it does at room temperature.

You can demonstrate enzyme activity in yeast as follows.

To do: Pour a package of "active dry yeast," or one tablespoon of yeast, into $\frac{1}{2}$ cup of warm water at 100° to 115°F. Add two teaspoons of granulated sugar. Stir well and set the mixture aside. If the yeast is active, small bubbles of carbon dioxide will appear at the surface in a few minutes. The mixture will swell as the gas is given off.

The catalysts found in living systems are called **enzymes**. Enzymes consist of very large molecules called proteins (described in Chapter Eighteen). Without enzymes, the reactions in our bodies would take place too slowly to support life. Enzymes are highly specific for certain reactants. For example, an enzyme called amylase breaks down starch. Amylase cannot digest cellulose ("fiber"), however, even though the molecular structure of cellulose is quite similar to that of starch. Thousands of different enzymes are present in our bodies, and each of these enzymes catalyzes its own specific reaction or reactions.

Part B   Chemical equilibrium

**16.5
INTRODUCTION
TO CHEMICAL
EQUILIBRIUM**

In Part A of this chapter we examined an important property of chemical reactions called the reaction rate. Another important property of chemical reactions is the amount of product formed. In many chemical reactions, the reactants are only partially converted to products. In other words, these reactions do not go to completion. In fact, even in those reactions that are said to go to completion (those that produce a gas, or a sparingly soluble precipitate, or a slightly ionized compound), some reactant still remains at the end of the reaction. For this reason, to varying degrees, all reactions can be considered **reversible**. They occur in the forward direction to form products and in the reverse direction to make the reactants again.

$$\text{reactants} \underset{\text{reverse}}{\overset{\text{forward}}{\rightleftarrows}} \text{products}$$

A double arrow $\rightleftarrows$ is often used to indicate a reversible reaction.

If only reactants are present at the beginning of a reaction, their molecules collide with each other and react to form a product. As the reaction progresses, more and more product molecules will be present. Unless these product molecules are removed, (as a gas or precipitate, for example) they in turn will collide with each other, and some will react to produce the reactants again. Eventually the rate of this reverse reaction will become equal to the rate of the forward reaction, as shown in Figure 16-7. When these two rates are equal, the reaction is said to be in **equilibrium**.

For example, consider the reversible reaction between hydrogen and iodine to form hydrogen iodide:

$$H_2(g) + I_2(g) \rightleftarrows 2HI(g)$$

If we mix together the hydrogen and iodine in a heated container, their molecules react, causing hydrogen iodide to form. As the reaction proceeds, more and more hydrogen iodide molecules form. Because the reaction is reversible, hydrogen iodide molecules can react with each other to form separate hydrogen and iodine molecules again. The rate of this reverse reaction depends on the concentration of hydrogen

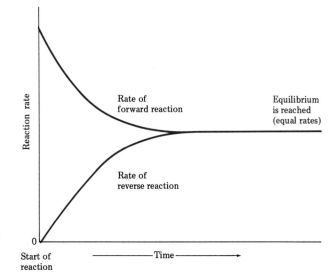

FIGURE 16-7

A reversible reaction. As the reaction progresses, the rate of the forward reaction decreases and the rate of the reverse reaction increases. At equilibrium, these two rates are equal.

iodide. As the concentration of hydrogen iodide builds up, the rate of the reverse reaction increases. Eventually, it exactly balances the rate of the forward reaction. At that point, the rate at which hydrogen iodide forms equals the rate at which it breaks down. The hydrogen iodide reaction is said to have reached equilibrium.

In general, equilibrium is a balance between opposing processes. Figure 16-8 illustrates this basic concept. We have already described

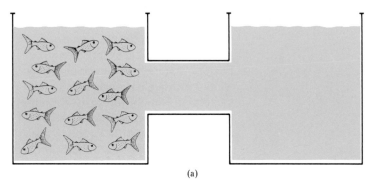

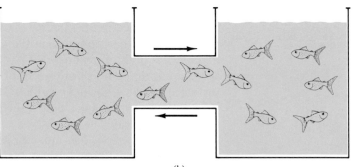

FIGURE 16-8

Equilibrium. If the fish (a) are free to move between the two tanks, equilibrium (b) is soon established. The rate of fish moving to the left exactly equals the rate of fish moving to the right at equilibrium.

531

two examples of equilibrium: vaporization (Section 12.4) and saturation (Section 13.4). In vaporization, equilibrium is reached between molecules leaving a liquid to become vapor and molecules leaving the vapor to return to the liquid state. In saturation, an equilibrium exists between solid dissolving into a solution and precipitating from the solution.

The type of equilibrium that exists between opposing chemical reactions is called **chemical equilibrium**. In chemical equilibrium, the forward and reverse reactions take place at the same rate, and there is no further net product formation. Any new product that forms is balanced by the conversion of the same amount of product back to reactants. At this stage it may seem as if no reaction is taking place, because the concentrations of reactants and products in the equilibrium mixture do not change. But the forward and reverse reactions are still taking place. Because of the continuing activity, chemical equilibrium is referred to as a dynamic equilibrium.

When a reaction reaches equilibrium, the final *amounts* of reactants and products need not be equal. Equilibrium means only that the *rates* of reactant formation and product formation are the same. Equilibrium may exist with only a small amount of product present, as in the ionization of a weak acid. On the other hand, equilibrium may exist with a large amount of product present, as in the neutralization reaction between hydronium and hydroxide ions.

**16.6
LE CHÂTELIER'S
PRINCIPLE**

In 1888 the French chemist Le Châtelier (pronounced luh-sha-tuh-lee-ay') proposed a principle that describes the effects of changes made on a system that is at equilibrium. This principle states: *If a stress is applied to a system at equilibrium, the system will adjust to relieve the stress.* In chemical systems, a stress is either a physical or a chemical change, generally a change in concentration, pressure, or temperature. Thus **Le Châtelier's principle** can be restated in the following way: A system at equilibrium resists attempts to change the concentrations, pressure, or temperature. If we try to change one of these factors, the equilibrium will shift to reduce the effect of that change.

Consider again the equilibrium analogy shown in Figure 16-8. At equilibrium, fish swim from bowl to bowl in opposite directions at the same rate. Now suppose that more fish are added to the bowl on the left, making this bowl crowded. We can think of the addition of fish as a stress applied to the system. Because of this stress, fish no longer move between the two bowls at the same rate. Instead, fish will tend to move out of the more crowded bowl and into the less crowded bowl at a faster rate than fish move out of the less crowded bowl and into the more crowded bowl. Eventually, a *new* equilibrium is established. Now both bowls contain more fish than they did originally, but the system has adjusted to the stress in such a way that the stress is relieved. In this case, fish moved into the less crowded bowl until a new equilibrium was established.

Now let's consider the effect of putting additional amounts of reactant into a chemical reaction mixture that is at equilibrium. Since the rate of the forward reaction depends on the concentrations of reactants, that rate increases when more reactants are added. Thus, more product forms. The presence of more product molecules speeds up the rate of the reverse reaction. Eventually a new equilibrium is established, and the two rates are again equal. But now the equilibrium mixture contains relatively more product than it did originally. Adding the reactant caused a shift to the right, or an increase in the forward reaction. This shift to the right relieves the stress created by the additional reactant molecules, because these molecules are consumed by the forward reaction. For example, in the synthesis of ammonia by the Haber process,

$$\underset{\text{nitrogen}}{N_2(g)} + \underset{\text{hydrogen}}{3H_2(g)} \overset{\text{forward}}{\rightleftharpoons} \underset{\text{ammonia}}{2NH_3(g)}$$

the addition of either nitrogen or hydrogen at equilibrium will cause a shift to the right to relieve the stress, and will increase the amount of ammonia produced.

Adding more *product* to an equilibrium mixture has the opposite effect. It causes a shift to the left, or an increase in the reverse reaction, thus forming more reactant. This shift relieves the stress of added product molecules because some of them are converted into reactant molecules. For example, if ammonia were added to the Haber process reaction at equilibrium, there would be a shift to the left and more of the reactants N_2 and H_2 would be formed.

The *removal* of reactants or products from a reaction mixture at equilibrium also places a stress on the system. In such a case, the reaction shifts in the direction that replaces the substance removed. For example, if ammonia is removed from the Haber process reaction at equilibrium, the reverse reaction becomes slower than the forward reaction. This change takes place because fewer ammonia molecules are now available to collide and form the reactants. The reactants, nitrogen and hydrogen, produce additional ammonia until a new equilibrium is reached. Thus, the removal of ammonia favors the forward reaction, which forms additional ammonia to help make up for the loss. Similarly, the removal of either nitrogen or hydrogen favors the reverse reaction.

If a reaction mixture at equilibrium contains one or more gases, and if the number of molecules of gaseous reactants differs from the number of molecules of gaseous products as given in the balanced chemical equation, then a change in pressure places a stress on the system and affects the equilibrium. According to Le Châtelier's principle, if the pressure is increased, the reaction will shift in the direction that causes a decrease in pressure. In order to decrease the pressure, the shift must reduce the number of molecules of gas. For example, in the ammonia

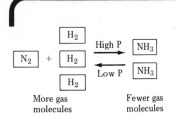

More gas | Fewer gas
molecules | molecules

FIGURE 16-9

The effect of pressure on the synthesis of ammonia, NH$_3$. High pressure favors the production of ammonia because there are fewer gas molecules in the products than there are in the reactants.

synthesis reaction, 4 molecules of gas (N$_2$ + 3H$_2$) are converted to 2 molecules of gas (2NH$_3$) by the forward reaction. Therefore, an increase in pressure on the system will favor the forward reaction because this reaction reduces the number of molecules and causes a decrease in pressure. (See Figure 16-9.) Ammonia synthesis is usually carried out at pressures of 150 atm or greater because these high pressures favor the ammonia-producing forward reaction.

Other reactions that illustrate the effects of pressure changes are the following:

Reaction	Direction favored by increased pressure
$C(s) + H_2O(g) \rightleftarrows CO(g) + H_2(g)$	reverse ($\leftarrow$)
$2NO_2(g) \rightleftarrows N_2O_4(g)$	forward ($\rightarrow$)
$H_2(g) + Cl_2(g) \rightleftarrows 2HCl(g)$	neither
$CaCO_3(s) \rightleftarrows CaO(s) + CO_2(g)$	reverse ($\leftarrow$)
$2SO_2(g) + O_2(g) \rightleftarrows 2SO_3(g)$	forward ($\rightarrow$)
$CaCl_2(aq) + 2AgNO_3(s) \rightleftarrows 2AgCl(s) + Ca(NO_3)_2(aq)$	neither

A change in temperature also causes a change in an equilibrium mixture. An increase in temperature favors the process that absorbs the added thermal energy, and a decrease in temperature favors the process that releases heat. Reversible chemical reactions are exothermic (release heat) in one direction, and endothermic (absorb heat) in the other direction. For example, the formation of ammonia is an exothermic reaction, and the breakdown of ammonia is endothermic. That is, the forward reaction releases heat, and the reverse reaction absorbs heat.

$$N_2 + 3H_2 \underset{\text{absorbs heat}}{\overset{\text{releases heat}}{\rightleftharpoons}} 2NH_3$$

Therefore the production of ammonia is favored by lowering the temperature because this reaction releases heat and counteracts the stress of a temperature decrease. In general, *lowering* the temperature will favor product formation if the forward reaction is *exothermic*. On the other hand, *raising* the temperature will favor product formation if the forward reaction is *endothermic*.

Forward reaction	Reverse reaction	Favored by increased T	Favored by decreased T
exothermic	endothermic	reverse	forward
endothermic	exothermic	forward	reverse

Another way to predict the effect of temperature change on a reaction is to consider heat as a reactant in an endothermic reaction, and as a product in an exothermic reaction. For example, the ammonia reaction could be represented as:

$$N_2 + 3H_2 \underset{endothermic}{\overset{exothermic}{\rightleftharpoons}} 2NH_3 + heat$$

This equation shows that the exothermic forward reaction produces heat, and the endothermic reverse reaction consumes it. Thus, if heat is added to the system, the temperature is increased and the reverse reaction is favored. If heat is removed, the temperature is lowered and the forward reaction is favored.

A catalyst has no effect on a system that is at equilibrium. However, it can *shorten the time it takes for equilibrium to be reached.* A catalyst decreases the activation energy and thus speeds up both the forward and reverse reactions. This effect can be quite useful. In the synthesis of ammonia, for example, an iron catalyst decreases the activation energy and permits the reaction to be run at a relatively low temperature (500°C). Without the catalyst, the reaction would occur very slowly at this temperature. Because the ammonia reaction is exothermic, the lower the reaction temperature the greater the amount of ammonia formed. At 500°C the concentration of ammonia in the reaction mixture is about 15%. If the reaction were run at an even lower temperature, the concentration of ammonia would be greater, but the reaction would take place too slowly to be practical, even with the catalyst.

In the ammonia reaction, and in many other reactions, the practical reaction conditions are a compromise between maximum yield (at a low rate) and maximum rate (with low yield). Both equilibrium and kinetics must be taken into account when designing a reaction system. Since ammonia is such an important industrial compound, chemists are trying to develop either a better catalyst or an alternate reaction, because the present relatively inefficient process requires great amounts of energy.

The following list summarizes the effects on the ammonia synthesis reaction discussed in this section.

Change	*Reaction favored*
add N_2 or H_2	forward
add NH_3	reverse
remove N_2 or H_2	reverse
remove NH_3	forward
increase pressure	forward
decrease pressure	reverse
increase temperature	reverse
decrease temperature	forward
add catalyst	both

Practice problem   For each of the following reactions, predict in which direction the equilibrium will be shifted by   **a** adding CO_2 **b** decreasing the pressure   **c** raising the temperature

1 $C(s) + CO_2(g) \rightleftarrows 2CO(g)$ forward reaction endothermic

2 $CH_4(g) + 2O_2(g) \rightleftarrows CO_2(g) + 2H_2O(g)$ forward reaction exothermic

ANSWERS   **1 a** forward   **b** forward   **c** forward
2 a reverse   **b** no effect   **c** reverse

16.7
WRITING
EQUILIBRIUM
CONSTANT
EXPRESSIONS

An **equilibrium constant**, K, is associated with every reversible reaction that reaches equilibrium. Note that the equilibrium constant is always a capital letter K, in contrast to the rate constant, which is a small letter, k. *The equilibrium constant is equal to the ratio of the equilibrium concentrations of the products over the equilibrium concentrations of the reactants.* This ratio is always the same for a particular reaction at a given temperature. The form of the ratio depends on the balanced equation for the reaction. For the general reaction

$$aA + bB \rightleftarrows cC + dD$$

where A and B are the reactants, C and D are the products, and a, b, c, and d are coefficients in the balanced equation, the equilibrium constant is given by the following expression:

$$K = \frac{[C]^c[D]^d}{[A]^a[B]^b}$$

The coefficients in the balanced equation become the exponents of the concentrations in the equilibrium constant expression. As indicated by the square brackets, concentrations are usually expressed in moles/liter; the units for K are generally omitted. Thus for the synthesis of ammonia, $N_2 + 3H_2 \rightleftarrows 2NH_3$, the equilibrium constant expression is:

$$K = \frac{[NH_3]^2}{[N_2][H_2]^3}$$

It is the ratio of the square of the ammonia concentration over the nitrogen concentration times the cube ($[H_2] \times [H_2] \times [H_2]$) of the hydrogen concentration at equilibrium. More examples of reversible reactions and their equilibrium constant expressions are the following:

Reaction	Equilibrium constant expression
$N_2O_4(g) \rightleftarrows 2NO_2(g)$	$K = \dfrac{[NO_2]^2}{[N_2O_4]}$
$H_2(g) + I_2(g) \rightleftarrows 2HI(g)$	$K = \dfrac{[HI]^2}{[H_2][I_2]}$
$2SO_3(g) \rightleftarrows 2SO_2(g) + O_2(g)$	$K = \dfrac{[SO_2]^2[O_2]}{[SO_3]^2}$
$2Fe^{3+}(aq) + Sn^{2+}(aq) \rightleftarrows 2Fe^{2+}(aq) + Sn^{4+}(aq)$	$K = \dfrac{[Fe^{2+}]^2[Sn^{4+}]}{[Fe^{3+}]^2[Sn^{2+}]}$

Example Write the equilibrium constant expression for the reaction

$$4HCl(g) + O_2(g) \rightleftarrows 2Cl_2(g) + 2H_2O(g)$$

hydrogen oxygen chlorine water
chloride

ANSWER $K = \dfrac{[Cl_2]^2[H_2O]^2}{[HCl]^4[O_2]}$

If any of the reactants or products of a reversible reaction are always present as pure solids or liquids, they are omitted from the equilibrium constant expression. Their concentrations are not variables because they are generally present in large excess compared to the other reacting substances, and their concentration is nearly constant. For example, the equilibrium constant for the reaction

$$C(s) \;\; + \; CO_2(g) \rightleftarrows 2CO(g)$$

carbon carbon carbon
dioxide monoxide

is expressed as:

$$K = \dfrac{[CO]^2}{[CO_2]}$$

The concentration of CO at equilibrium depends only on the concentration of CO_2 (and the temperature), and not on the amount of solid carbon present. Other examples of reactions in which a reactant or product is present as a pure solid or liquid are the following.

Reaction	Equilibrium constant expression
$CaCO_3(s) \rightleftarrows CaO(s) + CO_2(g)$	$K = [CO_2]$
$Zn(s) + Cu^{2+}(aq) \rightleftarrows Zn^{2+}(aq) + Cu(s)$	$K = \dfrac{[Zn^{2+}]}{[Cu^{2+}]}$
$SnO_2(s) + 2H_2(g) \rightleftarrows Sn(s) + 2H_2O(g)$	$K = \dfrac{[H_2O]^2}{[H_2]^2}$
$CuSO_4 \cdot 5H_2O(s) \rightleftarrows CuSO_4(s) + 5H_2O(g)$	$K = [H_2O]^5$
$2H_2O(l) \rightleftarrows H_3O^+(aq) + OH^-(aq)$	$K = [H_3O^+][OH^-]$

The numerical value of the equilibrium constant describes the extent of the reaction, or how far it goes in converting the reactants to products. A very large value of K, such as that obtained in the formation of carbon dioxide,

$$C(s) + O_2(g) \rightleftharpoons CO_2(g)$$

$$K = \frac{[CO_2]}{[O_2]} = 1.0 \times 10^{69}$$

means that the reaction essentially goes to completion. In other words, the equilibrium mixture contains mostly product, CO_2, and only very small amounts of the reactants remain. In general, when K is greater than one, the equilibrium is said to lie on the side of the products. On the other hand, a small value of the equilibrium constant, like that found in the formation of nitrogen dioxide,

$$N_2(g) + 2O_2(g) \rightleftharpoons 2NO_2(g)$$
nitrogen   oxygen   nitrogen dioxide

$$K = \frac{[NO_2]^2}{[N_2][O_2]^2} = 9.5 \times 10^{-10}$$

means that the equilibrium mixture contains mostly the reactants, N_2 and O_2. Only a small amount of the product is present. In general, when K is less than one, the equilibrium lies on the side of the reactants.

If we write a reaction backward, so that the products become the reactants, the equilibrium constant must be inverted. For example, if we rewrite the equation for the formation of NO_2 as the *decomposition* of NO_2,

$$2NO_2(g) \rightleftharpoons N_2(g) + 2O_2(g)$$
nitrogen   nitrogen   oxygen
dioxide

the equilibrium constant expression becomes

$$K = \frac{[N_2][O_2]^2}{[NO_2]^2} = 1.1 \times 10^9$$

Switching the right and left sides of a chemical equation switches the numerator and denominator of the equilibrium constant expression. The numerical value of the equilibrium constant for the decomposition of NO_2 is $1/(9.5 \times 10^{-10})$, or 1.1×10^9. Because this value of K is much larger than one, we know that the equilibrium lies almost entirely on the side of the products. Note that both values of K have the same meaning. Whether we write the equation forward or backward, the corresponding value of K indicates that the equilibrium lies on the side of N_2 and O_2.

16.8
EXAMPLES
OF IONIC
EQUILIBRIUM

Since many reactions take place in water and involve ions, the study of ionic equilibrium is very important to chemists. As mentioned in Chapter Fourteen, water itself ionizes to a small extent:

$$2H_2O(l) \rightleftarrows H_3O^+(aq) + OH^-(aq)$$
water   hydronium   hydroxide
             ion   ion

The equilibrium constant expression for this reaction can be written

$$K = \frac{[H_3O^+][OH^-]}{[H_2O]^2}$$

But the concentration of undissociated water, H_2O, is so much larger than the concentrations of the ions that it essentially remains constant. Therefore we can include it in the equilibrium constant. The resulting new equilibrium constant can be written:

$$K_w = [H_3O^+][OH^-]$$

K_w (w represents water) is called the **ion product constant** for water. This new equilibrium constant equals the product of two other constants, $K \times [H_2O]^2$.

At 25°C, the ion product constant for water has the value

$$K_w = 1.00 \times 10^{-14}$$

This small value shows that only a small number of hydronium and hydroxide ions form compared to the total number of water molecules. Because K_w is equal to $[H_3O^+][OH^-]$ we may write

$$[H_3O^+][OH^-] = 1.00 \times 10^{-14}$$

This expression relates the hydronium and hydroxide ion concentrations in water at 25°C, and was originally presented in Chapter Fourteen. As we saw previously, if the concentration of one of the two ions is known, the other concentration can be readily calculated. For example, if the hydronium ion concentration is 1.0×10^{-3} moles/liter (pH 3.0), the hydroxide ion concentration must be

$$[OH^-] = \frac{1.00 \times 10^{-14}}{[H_3O^+]} = \frac{1.00 \times 10^{-14}}{1.0 \times 10^{-3}} = 1.0 \times 10^{-11}$$

A major type of equilibrium reaction that takes place in water is the ionization of a weak acid. For example, weak hydrofluoric acid ionizes in water as follows:

$$HF(aq) + H_2O(l) \rightleftarrows H_3O^+(aq) + F^-(aq)$$

hydrofluoric   water   hydronium   fluoride
acid   ion   ion

The equilibrium constant expression for this reaction is

$$K = \frac{[H_3O^+][F^-]}{[HF][H_2O]}$$

Once again, the water concentration is so much larger than the concentrations of HF and its ions that it can be considered constant. Therefore we can remove $[H_2O]$ from the equilibrium expression. This simplication gives us the **acid ionization constant K_a**, where a stands for acid.

$$K_a = K \times [H_2O]$$

For the ionization of HF, the acid ionization constant is:

$$K_a = \frac{[H_3O^+][F^-]}{[HF]}$$

K_a is a measure of the strength of an acid. The larger the value of K_a, the more hydronium ions in the equilibrium mixture and the stronger the acid. Strong acids have acid ionization constants of 10^2 or greater, which means that most of their molecules are ionized. Hydrogen fluoride is a weak acid because it has a K_a of only 3.5×10^{-4}. Other K_a values are listed in Table 16-1. As we will see in the following sections, the pH of an acid solution depends on both the ionization constant and on the concentration of the acid.

The equilibrium reaction of a base with water is described in a similar way by a **base ionization constant K_b** (b represents base). For example, ammonia ionizes in water according to the following equation

$$NH_3(g) + H_2O(l) \rightarrow NH_4^+(aq) + OH^-(aq)$$

The base ionization constant is

$$K_b = \frac{[NH_4^+][OH^-]}{[NH_3]}$$

As with K_a, the water concentration is included in K_b. The larger the

TABLE 16-1 *Weak acid ionization constants at 25°C*

Name	Ionization reaction	K_a
phosphoric acid	$H_3PO_4 + H_2O \rightleftarrows H_2PO_4^- + H_3O^+$	7.5×10^{-3}
hydrogen fluoride	$HF + H_2O \rightleftarrows F^- + H_3O^+$	3.5×10^{-4}
acetic acid	$HC_2H_3O_2 + H_2O \rightleftarrows C_2H_3O_2^- + H_3O^+$	1.76×10^{-5}
carbonic acid	$CO_2 + 2H_2O \rightleftarrows HCO_3^- + H_3O^+$	4.3×10^{-7}
hydrogen sulfide	$H_2S + H_2O \rightleftarrows HS^- + H_3O^+$	1.1×10^{-7}
ammonium ion	$NH_4^+ + H_2O \rightleftarrows NH_3 + H_3O^+$	5.6×10^{-10}
hydrogen cyanide	$HCN + H_2O \rightleftarrows CN^- + H_3O^+$	4.9×10^{-10}

value of K_b, the stronger the base. Values of K_b are given in Table 16-2. Ammonia is a weak base because its K_b is only 1.77×10^{-5}. Therefore, the equilibrium mixture contains mostly NH_3 molecules.

The acid and base ionization constants for a conjugate acid-base pair (described in Section 14.3) are closely related. Their product is K_w, the ion product for water.

$$K_a \times K_b = K_w$$

For example, the product of the ionization constants for the acid HF and its conjugate base, F^- is:

$$K_a \times K_b = (3.5 \times 10^{-4})(2.8 \times 10^{-11})$$

$$= 1 \times 10^{-14} = K_w$$

This relationship between K_a and K_b confirms that the stronger an acid or base, the weaker is its conjugate base or conjugate acid.

Another type of ionic equilibrium exists between a slightly soluble solid electrolyte and its ions in solution. The equilibrium constant is

TABLE 16-2 *Weak base ionization constants at 25°C*

Name	Ionization reaction	K_b
phosphate ion	$PO_4^{3-} + H_2O \rightleftarrows HPO_4^{2-} + OH^-$	2.3×10^{-2}
carbonate ion	$CO_3^{2-} + H_2O \rightleftarrows HCO_3^- + OH^-$	1.8×10^{-4}
cyanide ion	$CN^- + H_2O \rightleftarrows HCN + OH^-$	2.0×10^{-5}
ammonia	$NH_3 + H_2O \rightleftarrows NH_4^+ + OH^-$	1.77×10^{-5}
acetate ion	$C_2H_3O_2^- + H_2O \rightleftarrows HC_2H_3O_2 + OH^-$	5.7×10^{-10}
fluoride ion	$F^- + H_2O \rightleftarrows HF + OH^-$	2.8×10^{-11}

called a **solubility product constant**, and is given the symbol K_{sp}. For example, the equilibrium reaction of sparingly soluble silver chloride in water is:

$$AgCl(s) \rightleftarrows Ag^+(aq) + Cl^-(aq)$$

silver   silver   chloride
chloride   ion   ion

and the solubility product expression is written

$$K_{sp} = [Ag^+][Cl^-]$$

Remember that the concentration of a pure solid, silver chloride in this case, does not appear in the equilibrium constant expression. A solubility product constant is a measure of the solubility of an electrolyte. The larger the value of K_{sp} the more solid dissolves, and the greater the concentration of ions in solution. K_{sp} for AgCl is 1.8×10^{-10}. This low value indicates that the equilibrium lies almost entirely to the left, and that the equilibrium mixture contains very few ions. In general, if the K_{sp} is less than 10^{-4}, the compound is considered insoluble. K_{sp} values for other weak electrolytes are given in Table 16-3.

Le Châtelier's principle can be applied to ionic equilibria. Suppose for example that we wish to analyze a solution for chloride ions. We could add Ag^+ ions to the solution to precipitate the Cl^- ions as AgCl, which can then be dried and weighed. It is important to make the

TABLE 16-3 *Solubility product constants of weak electrolytes at 25°C*

Name	Equilibrium equation	K_{sp}
aluminum hydroxide	$Al(OH)_3 \rightleftarrows Al^{3+} + 3OH^-$	1.3×10^{-33}
barium sulfate	$BaSO_4 \rightleftarrows Ba^{2+} + SO_4^{2-}$	1.1×10^{-10}
calcium carbonate	$CaCO_3 \rightleftarrows Ca^{2+} + CO_3^{2-}$	2.8×10^{-9}
calcium fluoride	$CaF_2 \rightleftarrows Ca^{2+} + 2F^-$	2.7×10^{-11}
iron(II) hydroxide	$Fe(OH)_2 \rightleftarrows Fe^{2+} + 2OH^-$	8.0×10^{-16}
iron(III) hydroxide	$Fe(OH)_3 \rightleftarrows Fe^{3+} + 3OH^-$	4×10^{-38}
iron(II) sulfide	$FeS \rightleftarrows Fe^{2+} + S^{2-}$	6.3×10^{-18}
lead chloride	$PbCl_2 \rightleftarrows Pb^{2+} + 2Cl^-$	1.6×10^{-5}
lead sulfide	$PbS \rightleftarrows Pb^{2+} + S^{2-}$	8.0×10^{-28}
magnesium hydroxide	$Mg(OH)_2 \rightleftarrows Mg^{2+} + 2OH^-$	1.8×10^{-11}
silver chloride	$AgCl \rightleftarrows Ag^+ + Cl^-$	1.8×10^{-10}
silver chromate	$Ag_2CrO_4 \rightleftarrows 2Ag^+ + CrO_4^{2-}$	1.1×10^{-12}
zinc sulfide	$ZnS \rightleftarrows Zn^{2+} + S^{2-}$	2.5×10^{-22}

solubility of the AgCl precipitate as small as possible so that nearly all the chloride ions are removed from the solution. We can reduce the solubility of AgCl by adding more Ag^+ ions than are necessary to form the precipitate. The equilibrium will then be shifted to the left, toward solid silver chloride, making the precipitate less soluble.

$$AgCl(s) \rightleftarrows Ag^+(aq) + Cl^-(aq)$$

This shift of an ionic equilibrium by the addition of an ion that is already present (a common ion) is called the **common ion effect.**

16.9 EQUILIBRIUM CALCULATIONS

As shown in the preceding section, equilibrium constants have been experimentally determined for many reactions. These constants are available in tables such as Table 16-1, Table 16-2, and Table 16-3 in this chapter. We saw in that section that the value of K indicates the extent of an equilibrium reaction. A large value of K_a, for example, means that an acid is highly ionized and therefore is a strong acid. In this section, we will use the actual value of the equilibrium constant of a reaction to find the concentrations of reactants and products present when the reaction has reached equilibrium. We will illustrate the steps in the procedure, using the synthesis of hydrogen iodide from hydrogen and iodine. Suppose that we start with 1.00 mole/liter of H_2 and 1.00 mole/liter of I_2 at 450°C. The equilibrium constant K for this reaction at 450°C is 50.0. Let's find the concentrations of reactants and products at equilibrium.

1 Write a balanced chemical equation for the reaction.

The equation for our sample reaction is:

$$H_2(g) + I_2(g) \rightleftarrows 2HI(g)$$

2 Write the equilibrium constant expression for the reaction.

The equilibrium constant expression for the formation of HI from H_2 and I_2 is:

$$K = \frac{[HI]^2}{[H_2][I_2]}$$

3 Identify the unknown in the problem.

In our reaction, an equal number of H_2 and of I_2 molecules will be converted to twice that number of HI molecules. We will let x be the amount by which the concentrations of H_2 and I_2 decrease. Then according to the stoichiometry of the reaction, $2x$ will be the equilibrium concentration of HI molecules.

4 In terms of the unknown, x, set up a table indicating the initial concentration, the change in concentration (the amount formed or lost), and the equilibrium concentration of each substance.

The table for our sample problem has the form:

	H_2	+	I_2 $\rightleftarrows$	$2HI$
initial concentration	1.00		1.00	0
change in concentration	$-x$		$-x$	$+2x$
equilibrium concentration	$1.00 - x$		$1.00 - x$	$2x$

Notice that the equilibrium concentrations are found by adding the change in concentration to the initial concentration.

5 Substitute the equilibrium concentration from the table into the equilibrium constant expression.

The equilibrium constant expression becomes:

$$K = \frac{[HI]^2}{[H_2][I_2]} = \frac{(2x)^2}{(1.00 - x)(1.00 - x)}$$

6 Set the equilibrium constant expression equal to the value of the equilibrium constant and solve for the unknown.

In our example, the value of K (given at the beginning of the problem) is 50.0. We set this value equal to the equilibrium constant expression:

$$\frac{(2x)^2}{(1.00 - x)(1.00 - x)} = 50.0$$

We now solve this algebraic equation for x. Notice that the left side of the equation consists of a fraction whose numerator and denominator are squares.

$$\frac{(2x)^2}{(1.00 - x)^2} = 50.0$$

To solve the equation, we can eliminate the exponents by taking the square root of both sides.

$$\frac{2x}{1.00 - x} = \sqrt{50.0} = 7.07$$

We then eliminate the denominator on the left side of the equation by multiplying both sides by $(1.00 - x)$.

$$\frac{2x}{1.00 - x}(1.00 - x) = 7.07(1.00 - x)$$

$$2x = 7.07(1.00 - x) = 7.07 - 7.07x$$

Finally, we collect the x terms and solve for x.

$$2x + 7.07x = 7.07$$

$$9.07x = 7.07$$

$$x = \frac{7.07}{9.07} = 0.779$$

7 Using the value for the unknown found in the previous step, determine the equilibrium concentrations.

Since the concentration of hydrogen iodide is $2x$, it is equal to 2×0.779 mole/liter $= 1.56$ moles/liter. The concentrations of hydrogen and iodine are each $1.00 - x$, or $1.00 - 0.779 = 0.22$ mole/liter.

8 Check the result.

We check the calculated concentrations by substituting them into the equilibrium expression.

$$\frac{(2x)^2}{(1.00 - x)^2} = \frac{(1.56)^2}{(0.22)^2} = \frac{2.43}{0.048} = 51$$

This substitution gives us the number 51, which is approximately equal to the given value of K. Therefore, we know that our calculations were correct. We must also make sure that the equilibrium concentrations make sense. The relatively large value of K means that the equilibrium lies on the side of product. The concentrations support this fact:

$$H_2 \quad + \quad I_2 \quad \rightleftarrows 2HI$$

$$0.22 \quad 0.22 \quad 1.56 \quad \text{concentration (moles/liter)}$$

**16.10
CALCULATIONS
INVOLVING
ACID AND BASE
EQUILIBRIUM**

We use essentially the same procedure described in the previous section to solve ionic equilibrium problems involving weak acids or bases. However, in these problems, the equilibrium constant is the acid ionization constant, K_a, or the base ionization constant, K_b. The unknown is often the pH, and it is found by determining the concentration of hydronium ions at equilibrium. Suppose that we want to find the pH of 0.0100 M acetic acid, $HC_2H_3O_2$.

1 Write a balanced chemical equation for the reaction.

$$HC_2H_3O_2(aq) + H_2O(l) \rightleftarrows H_3O^+(aq) + C_2H_3O_2^-(aq)$$

2 Write the equilibrium constant expression for the reaction.

$$K_a = \frac{[H_3O^+][C_2H_3O_2{}^-]}{[HC_2H_3O_2]}$$

3 Identify the unknown.

The unknown x is the hydronium ion concentration. Since one acetate ion is formed along with each hydronium ion, its concentration is also x. The equilibrium concentration of acetic acid will be the original concentration minus the amount ionized, or $0.0100 - x$.

4 Set up a table for the reaction.

	$HC_2H_3O_2$	$+$ H_2O $\rightleftharpoons$	H_3O^+	$+$ $C_2H_3O_2{}^-$
initial concentration	0.0100		0	0
change in concentration	$-x$		$+x$	$+x$
equilibrium concentration	$0.0100 - x$		x	x

5 Substitute into the equilibrium expression.

$$\frac{[H_3O^+][C_2H_3O_2{}^-]}{[HC_2H_3O_2]} = \frac{(x)(x)}{0.0100 - x}$$

6 Set the equilibrium expression equal to the ionization constant and solve for the unknown.

We find the value of the acid ionization constant from Table 16-1. It is equal to 1.76×10^{-5} (at 25°C). Therefore, the equation is:

$$\frac{x^2}{0.0100 - x} = 1.76 \times 10^{-5}$$

In problems that involve small equilibrium constants such as this one, we can often make a simplifying approximation. It is likely that x will be quite small since the value of the equilibrium constant shows that relatively little product exists at equilibrium. Therefore the original concentration of un-ionized molecules will stay nearly the same. We can assume that the term in the denominator, $0.0100 - x$, will be about 0.0100 at equilibrium. In other words, we drop the x in the denominator because we assume that x is small. The equation then becomes:

$$\frac{x^2}{0.0100} = 1.76 \times 10^{-5}$$

which can be solved more easily.

$$x^2 = (0.0100)\ 1.76 \times 10^{-5}$$

$$x^2 = 1.76 \times 10^{-7} = 17.6 \times 10^{-8}$$

Note that 1.76×10^{-7} is changed to an even power of ten to simplify taking the square root, which is found by dividing the exponent by 2.

$$x = \sqrt{17.6 \times 10^{-8}} = \sqrt{17.6} \times \sqrt{10^{-8}}$$

$$x = 4.20 \times 10^{-4}$$

Before going further, we must check that the approximation was reasonable. We assumed that x is much less than 0.0100 and, in fact, it is: 4.20×10^{-4} is about 4% of 1.0×10^{-2}. We can make such approximations as long as the error is less than 10%.

7 Using the value of x, find the equilibrium concentrations.

The concentration of hydronium ions and the concentration of acetate ions are both equal to x, or 4.20×10^{-4} mole/liter. Using the method described in Appendix H, the pH of the equilibrium mixture is

$$\begin{aligned}
\text{pH} = -\log[H_3O^+] &= -\log(4.20 \times 10^{-4}) \\
&= -\log(4.20) - \log(10^{-4}) \\
&= -0.623 - (-4) = 4 - 0.623 \\
&= 3.38
\end{aligned}$$

8 Check the result.

Substituting into the equilibrium expression, we get:

$$\frac{x^2}{0.0100 - x} = \frac{(4.20 \times 10^{-4})^2}{0.0100 - (4.20 \times 10^{-4})} = 1.83 \times 10^{-5}$$

This result is close enough to the actual value of K_a, 1.76×10^{-5}, considering that we made a simplifying approximation. As indicated by the small ionization constant, and the results of the calculation, the acid is only slightly ionized. Yet at a concentration of 0.0100 M, enough hydronium ions are formed to bring the pH into the moderately acidic region.

Problems involving weak bases are solved in exactly the same way, as shown in the following example.

Example What is the pH of a 5.0×10^{-3} M solution of ammonia?

Step 1 $NH_3 + H_2O \rightleftharpoons NH_4^+ + OH^-$

Step 2 $K_b = \dfrac{[NH_4^+][OH^-]}{[NH_3]}$

Step 3

UNKNOWN	GIVEN	CONNECTION
$x = [OH^-]$	5.0×10^{-3} M NH_3	$K_b = 1.77 \times 10^{-5}$ (Table 16-2)

Step 4

$$NH_3 + H_2O \rightleftharpoons NH_4^+ + OH^-$$

	NH_3	NH_4^+	OH^-
initial concentration	5.0×10^{-3}	0	0
change in concentration	$-x$	$+x$	$+x$
equilibrium concentration	$5.0 \times 10^{-3} - x$	x	x

Step 5 $\quad K_b = \dfrac{[NH_4^+][OH^-]}{[NH_3]} = \dfrac{(x)(x)}{5.0 \times 10^{-3} - x}$

Step 6 $\quad \dfrac{x^2}{5.0 \times 10^{-3}} = 1.77 \times 10^{-5}$ (assuming x is very small compared to 5×10^{-3})

$$x^2 = (5.0 \times 10^{-3})(1.77 \times 10^{-5}) = 8.9 \times 10^{-8}$$

$$x = \sqrt{8.9 \times 10^{-8}} = \sqrt{8.9} \times \sqrt{10^{-8}}$$

$$x = 3.0 \times 10^{-4} \text{ (approximation is satisfactory)}$$

Step 7 $\quad [OH^-] = [NH_4^+] = x = 3.0 \times 10^{-4} \text{ M}$

$$[NH_3] = 5.0 \times 10^{-3} - x = 4.7 \times 10^{-3} \text{ M}$$

$$[H_3O^+] = \dfrac{K_w}{[OH^-]} = \dfrac{1.0 \times 10^{-14}}{3.0 \times 10^{-4}} = 3.3 \times 10^{-11}$$

$$
\begin{aligned}
pH = -\log[H_3O^+] &= -\log(3.3 \times 10^{-11}) \\
&= -\log(3.3) - \log(10^{-11}) \\
&= -0.52 + 11 \\
&= 10.5
\end{aligned}
$$

Step 8 $\quad \dfrac{x^2}{5.0 \times 10^{-3} - x} = \dfrac{(3.0 \times 10^{-4})^2}{(5.0 \times 10^{-3}) - (3.0 \times 10^{-4})} = 1.9 \times 10^{-5}$

Ammonia ionizes slightly in water, but enough hydroxide ions are produced in a 5.0×10^{-3} M solution to raise the pH to a moderately basic value.

Practice problem Find the pH of 0.010 M HCN.

ANSWER 5.7

16.11 CALCULATIONS INVOLVING SOLUBILITY EQUILIBRIUM We can also use an equilibrium calculation to find the solubility of a weak electrolyte. We follow the same steps that we did in Sections 16.9 and 16.10, except that in this case we use the solubility product. For example, Table 16-3 shows that the K_{sp} for barium sulfate is 1.1×10^{-10}. Let us find the solubility of barium sulfate, by determining the concentration of barium and sulfate ions in solution at equilibrium.

1 Write the balanced equation.

$$BaSO_4(s) \rightleftarrows Ba^{2+}(aq) + SO_4{}^{2-}(aq)$$

2 Write the equilibrium expression.

$$K_{sp} = [Ba^{2+}][SO_4{}^{2-}]$$

3 Identify the unknown.

The unknown x is the Ba^{2+} concentration at equilibrium. Since one $SO_4{}^{2-}$ ion is formed along with each Ba^{2+} ion, its concentration is x also.

4 Set up a table.

	$BaSO_4 \rightleftarrows Ba^{2+}$	$+ \; SO_4{}^{2-}$
initial concentration	0	0
change in concentration	$+x$	$+x$
equilibrium concentration	x	x

5 Substitute into the equilibrium expression.

$$K_{sp} = [Ba^{2+}][SO_4{}^{2-}] = (x)(x) = x^2$$

6 Equate the equilibrium expression with the solubility product constant.

$$x^2 = 1.1 \times 10^{-10}$$
$$x = \sqrt{1.1 \times 10^{-10}} = \sqrt{1.1} \times \sqrt{10^{-10}}$$
$$x = 1.0 \times 10^{-5}$$

7 Find the equilibrium concentrations.

The equilibrium concentrations for Ba^{2+} and $SO_4{}^{2-}$ are each 1.0×10^{-5}.

8 Check the result.

$$(1.0 \times 10^{-5})^2 = 1.0 \times 10^{-10}$$

As expected from the small value of K_{sp}, the solubility of $BaSO_4$ is very slight and thus the concentrations of ions in solution are low.

Example Find the solubility of CaF_2 in water at 25°C.

Step 1 $CaF_2(s) \rightleftarrows Ca^{2+}(aq) + 2F^-(aq)$

Step 2 $K_{sp} = [Ca^{2+}][F^-]^2$

Step 3

UNKNOWN	GIVEN	CONNECTION
$x = [Ca^{2+}]$	$CaF_2(s)$	$K_{sp} = 2.7 \times 10^{-11}$ (Table 16-3)

Step 4

$$CaF_2 \rightleftarrows Ca^{2+} + 2F^-$$

	Ca^{2+}	$2F^-$
initial concentration	0	0
change in concentration	$+x$	$+2x$
equilibrium concentration	x	$2x$

Step 5 $K_{sp} = [Ca^{2+}][F^-]^2 = (x)(2x)^2$

Step 6
$$x(2x)^2 = 2.7 \times 10^{-11}$$
$$x(4x^2) = 2.7 \times 10^{-11}$$
$$4x^3 = 2.7 \times 10^{-11}$$
$$x^3 = 6.8 \times 10^{-12}$$
$$x = \sqrt[3]{6.8 \times 10^{-12}}$$
$$x = \sqrt[3]{6.8} \times \sqrt[3]{10^{-12}}$$
$$x = 1.9 \times 10^{-4}$$

Step 7 $[Ca^{2+}] = x = 1.9 \times 10^{-4}$ M
$[F^-] = 2x = 3.8 \times 10^{-4}$ M

Step 8 $[Ca^{2+}][F^-]^2 = (1.9 \times 10^{-4})(3.8 \times 10^{-4})^2 = 2.7 \times 10^{-11}$

As predicted by the small K_{sp} of CaF_2, the solubility of CaF_2 is slight, and concentrations of Ca^{2+} and F^- are low.

Practice problem What is the solubility of lead sulfide in water at 25°C?

ANSWER 2.8×10^{-14} mole/liter

Summary **16.1** The rate of a reaction describes how rapidly a chemical change takes place. The rate is related to the concentrations of the reactants by an experimentally determined rate equation or rate law. The symbol k in the rate equation is called the rate constant. The rate constant relates the concentrations of the reactants to the measured rate. Its value depends on the temperature.

16.2 Molecules react when they collide with enough energy. The amount of energy needed to cause a reaction is called the activation energy.

16.3 The mechanism of a reaction is the series of steps by which a reaction takes place. Chemical kinetics is the study of the relationship between experimentally determined reaction rates and theoretically determined mechanisms.

16.4 A catalyst is a substance that increases the rate of a reaction. It can be recovered unchanged at the end of the reaction. Enzymes are biological catalysts.

16.5 Most reactions are reversible. They occur in the forward direction to form products and in the reverse direction to make the reactants. Chemical equilibrium occurs when the rates of these two opposing processes become equal.

16.6 Le Châtelier's principle states: If a stress is applied to a system at equilibrium, the system will adjust to relieve the stress. This principle can be used to predict the effects of changing the concentration, pressure, or temperature of a reaction at equilibrium.

16.7 An equilibrium constant, K, is associated with every reversible reaction that reaches equilibrium. The equilibrium constant at a particular temperature is equal to the ratio of the equilibrium concentrations of the products over the equilibrium concentrations of the reactants, each raised to their respective coefficients in the balanced chemical equation.

16.8-16.11 The equilibrium constant for the self-ionization of water, K_w, is known as the ion product constant for water. Other equilibrium constants are K_a, the acid ionization constant, K_b, the base ionization constant, and K_{sp}, the solubility product constant. Equilibrium constants can be used to find the concentrations of reactants and products for reactions that have reached equilibrium.

Exercises

For each term on the left, choose a phrase from the right-hand column that most closely matches its meaning.

1 reaction rate	**a**	slowest step
2 rate equation	**b**	needed for molecules to react
3 rate constant	**c**	series of steps
4 activation energy	**d**	relates rates to mechanisms
5 activated complex	**e**	change in concentration with time

6 reaction mechanism	**f** goes in both directions
7 rate determining step	**g** shows extent of reaction
8 chemical kinetics	**h** relates rate to concentration
9 catalyst	**i** short-lived intermediate
10 reversible reaction	**j** forward and reverse rates equal
11 chemical equilibrium	**k** increases reaction rate
12 equilibrium constant	**l** proportionality factor, k

EXERCISES BY TOPIC

Reaction rate (16.1)

13 What does a reaction rate measure?

14 Explain in words the following rate equation: rate $= k[Cu^{2+}][H_3O^+]^2$.

15 What is the meaning of the rate constant?

16 Discuss the factors that affect that rate of a reaction.

Activation energy (16.2)

17 Describe in a general way how chemical reactions take place at the molecular level.

18 What is activation energy?

19 How does the activation energy affect the rate of a reaction?

20 What is an activated complex? Why is one hard to isolate?

Reaction mechanism (16.3)

21 What is a reaction mechanism?

22 What is meant by the rate determining step?

23 A chemist says her field is chemical kinetics. What does she do?

24 How is a rate equation useful for studying the mechanism of a reaction?

Catalysis (16.4)

25 What is a catalyst? How does it work?

26 What is the difference between a homogeneous and heterogeneous catalyst?

27 What is the role of enzymes in a living organism?

Chemical equilibrium (16.5)

28 What does it mean for a reaction to be reversible?

29 What happens in a chemical equilibrium? What is constant and what is changing?

30 Must the amount of reactants and products be the same at equilibrium? Explain.

Le Châtelier's principle (16.6)

31 State Le Châtelier's principle in your own words.

32 What is the effect of increasing the pressure on these reactions at equilibrium:
a $2KClO_3(s) \rightleftarrows 2KCl(s) + 3O_2(g)$
b $6CO_2(g) + 6H_2O(l) \rightleftarrows C_6H_{12}O_6(s) + 6O_2(g)$

33 What is the effect of removing O_2 in the two reactions in Exercise 32?

34 If the forward reaction in Exercise 32a is exothermic and the forward reaction in Exercise 32b is endothermic, what will be the effects of raising the temperature in these two cases?

35 What would be the effect of adding a catalyst to the reactions of Exercise 32?

36 Hydrogen bromide exists in equilibrium in a closed container with hydrogen and bromine: $2HBr(g) \rightleftarrows H_2(g) + Br_2(g)$. If the forward reaction is endothermic, describe the effect of **a** lowering the temperature **b** decreasing the pressure **c** adding hydrogen **d** removing bromine

Writing equilibrium constant expressions (16.7)

37 What is an equilibrium constant?

38 Write equilibrium constant expressions for the following reactions:
a $N_2(g) + O_2(g) \rightleftarrows 2NO(g)$
b $ZnO(s) + H_2(g) \rightleftarrows Zn(s) + H_2O(g)$
c $2Cr(s) + 3Cu^{2+}(aq) \rightleftarrows 2Cr^{3+}(aq) + 3Cu(s)$
d $NO_2(g) + SO_2(g) \rightleftarrows SO_3(g) + NO(g)$

39 What does a small value for K mean?

40 N_2O_4 and NO_2 are in equilibrium at 135°C:
$$N_2O_4(g) \rightleftarrows 2NO_2(g)$$
If the concentration of N_2O_4 is 4.5 moles/liter and that of NO_2 is 3.0 moles/liter, find the value of K.

41 If the equilibrium constant for the reaction
$$Fe(s) + Cl_2(g) \rightleftarrows FeCl_2(s)$$
is 1×10^{53}, what is the equilibrium constant for the reaction:
$$FeCl_2(s) \rightleftarrows Fe(s) + Cl_2(g)$$

42 Explain the meaning of your answer to Exercise 41.

Examples of ionic equilibrium (16.8)

43 Write the hydroxide ion concentration when the hydronium ion concentration at 25°C is **a** 1.0×10^{-3} **b** 5.6×10^{-8} **c** 9.2×10^{-2}

44 What does the value of K_a tell you about an acid?

45 If nitrous acid, HNO_2, has a K_a of 5×10^{-4}, find K_b for nitrite ion, NO_2^-.

46 What does a solubility product show?

47 Using Table 16-3, which has a greater solubility, PbS or FeS? Explain.

48 What is the common ion effect?

Equilibrium calculations (16.9)

49 Why must an equation be balanced when used for an equilibrium calculation?

50 The air pollutant nitric oxide is produced by the high temperature combination of nitrogen and oxygen:
$$N_2(g) + O_2(g) \rightleftarrows 2NO(g)$$
The equilibrium constant at 1500 K is 1.0×10^{-5}. If the equilibrium concentration of N_2 is 1.0 mole/liter and that of O_2 is 0.25 mole/liter, find the equilibrium concentration of NO.

51 The equilibrium constant for the conversion of the graphite form of carbon to the diamond form, C(*graphite*) $\rightleftarrows$ C(*diamond*) is 0.31 at 25°C. If the initial graphite concentration is 188 mole/dm^3, find the diamond concentration at equilibrium.

Calculations involving acid and base equilibrium (16.10)

52 Find the pH of 0.10 M HF.

53 What is the pH of a 0.025 M solution of NH_3?

Calculations involving solubility equilibrium (16.11)

54 Find the solubility of $Mg(OH)_2$ in water at 25°C.

55 How many grams of iron(II) sulfide dissolve in 500. mL of water at 25°C?

ADDITIONAL EXERCISES

56 State the meaning in words of this expression: rate = $k[Cl^-]^2[OH^-]$.

57 How is it possible for a catalyst to influence a reaction but not appear in the overall equation?

58 If the rate determining step in a reaction is A + B $\rightleftarrows$ C + D, propose a rate equation for the reaction.

59 Why can't a mechanism be *proven* by a rate equation?

60 How is chemical equilibrium similar to and how is it different from vaporization equilibrium (described in Section 12.4)?

61 The following reaction, the combustion of acetylene, is exothermic (in the forward direction):
$$2C_2H_2(g) + 5O_2(g) \rightleftarrows 4CO_2(g) + 2H_2O(g)$$
Describe the effect on the equilibrium of **a** increasing the pressure **b** adding CO_2 **c** lowering the temperature **d** removing O_2 **e** adding a catalyst

62 Write equilibrium constant expressions for the following equations:
a $H_2O(l) \rightleftarrows H_2O(g)$
b $3Fe(s) + 4H_2O(g) \rightleftarrows Fe_3O_4(s) + 4H_2(g)$
c $C_6H_6(l) + Cl_2(g) \rightleftarrows HCl(g) + C_6H_5Cl(l)$
d $Na_2B_4O_7 \cdot 10H_2O(s) \rightleftarrows Na_2B_4O_7(s) + 10H_2O(g)$

63 The value of K for the reaction in Exercise 61 is very large. What does this large value mean?

64 What is K_b for HS^-?

65 Why is barium sulfate less soluble in dilute sulfuric acid than in water?

66 Vinegar is about 5.0% (w/v) acetic acid. Find its pH.

67 The solubility of cobalt(II) sulfide is 3.8×10^{-4} g/100 mL at 18°C. Find its solubility product at this temperature.

68 The following reaction has come to equilibrium:

$$PCl_3(g) + Cl_2(g) \rightleftarrows PCl_5(g)$$

If the equilibrium concentrations of PCl_3, Cl_2, and PCl_5 are 10, 9, and 12 moles/liter respectively, find K.

69 What is the pH of a 0.30 M solution of nitrous acid, HNO_2. (Note: $K_a = 4.5 \times 10^{-4}$)

70 Write an equilibrium constant expression for each of the following reactions:

a $2SO_2 + O_2 \rightleftarrows 2SO_3$
b $4SO_2 + 2O_2 \rightleftarrows 4SO_3$
c $SO_2 + \frac{1}{2}O_2 \rightleftarrows SO_3$
d $2SO_3 \rightleftarrows 2SO_2 + O_2$

71 For the photosynthesis reaction,

$$6CO_2(g) + 6H_2O(l) \rightleftarrows C_6H_{12}O_6(s) + 6O_2(g)$$

predict the effect of **a** increasing the CO_2 concentration **b** removing glucose, $C_6H_{12}O_6$ **c** increasing the total pressure **d** increasing the temperature (note: the forward reaction is endothermic) **e** adding a catalyst

72 The solubility product of Cu_2S is 3×10^{-48}. What is the solubility of Cu_2S in water in grams per liter?

73 Iodide ion is oxidized by hydrogen peroxide in the following reaction:

$$H_2O_2(l) + 2H^+(aq) + 2I^-(aq) \rightarrow I_2(s) + 2H_2O(l)$$

The rate equation for this reaction was found to be rate = $k[H_2O_2][I^-]$. Is this rate equation consistent with the following mechanism?

Step 1 $H_2O_2 + I^- \xrightarrow{\text{slow}} H_2O + OI^-$

Step 2 $H^+ + OI^- \xrightarrow{\text{fast}} HOI$

Step 3 $HOI + H^+ + I^- \xrightarrow{\text{fast}} I_2 + H_2O$

74 Write equilibrium constant expressions for the following reactions:

a $SO_3(g) + H_2(g) \rightleftarrows SO_2(g) + H_2O(g)$
b $Ag_2S(s) \rightleftarrows 2Ag^+(aq) + S^{2-}(aq)$
c $BaCO_3(s) + C(s) \rightleftarrows BaO(s) + 2CO(g)$
d $4NH_3(g) + 5O_2(g) \rightleftarrows 4NO(g) + 6H_2O(g)$
e $CO(g) + 2H_2(g) \rightleftarrows CH_3OH(l)$

75 At 25°C, the equilibrium constant for the reaction

$$N_2O_4(g) \rightleftarrows 2NO_2(g)$$

is $K = 5.85 \times 10^{-3}$. If 10.0 g of N_2O_4 is present initially in a 5.00 L flask at 25°C, find the number of moles of NO_2 present at equilibrium.

76 The following equation represents the ionization of periodic acid:

$$HIO_4(aq) + H_2O(l) \rightleftarrows H_3O^+(aq) + IO_4^-(aq)$$

If a 0.100 M solution of HIO_4 at 25°C contains 0.038 M H_3O^+, what is K_a for this acid?

77 Why do foods stay fresh longer when they are stored in a freezer?

Check off each item when completed.

_____   Previewed chapter

Studied each section:

Part A   Chemical kinetics

_____   **16.1** Reaction rate

_____   **16.2** Activation energy

_____   **16.3** Reaction mechanism

_____   **16.4** Catalysis

Part B   Chemical equilibrium

_____   **16.5**   Introduction to chemical equilibrium

_____   **16.6**   Le Châtelier's principle

_____   **16.7**   Writing equilibrium constant expressions

_____   **16.8**   Examples of ionic equilibrium

_____   **16.9**   Equilibrium calculations

_____   **16.10** Calculations involving acid and base equilibrium

_____   **16.11** Calculations involving solubility equilibrium

_____   Reviewed chapter

_____   Answered assigned exercises:

#'s _____

SEVENTEEN

Nuclear chemistry

In previous chapters we studied chemical properties and reactions, which depend on the number and arrangement of the outermost electrons in atoms. Nuclear reactions, on the other hand, involve the protons and neutrons within the nuclei of atoms. Nuclear chemistry, the study of changes in the nuclei of atoms, is presented in this chapter. We describe radiation, which is produced by nuclear changes, and discuss its uses and effects. To demonstrate an understanding of Chapter Seventeen, you should be able to:

1 Identify the types of radiation emitted by radioactive atoms.

2 Interpret equations for nuclear reactions.

3 Describe a radioactive decay series.

4 Describe how radioisotopes can be produced.

5 Explain the relationship between matter and energy.

6 Describe the significance of nuclear binding energy.

7 Relate the processes of fission and fusion to nuclear stability.

8 Define half-life and describe its application.

9 Describe the major methods for detecting radiation, and units for measuring it.

10 Give general applications of radiation and radioisotopes.

11 Describe the short-term and long-term effects of radiation on human health.

**17.1
RADIOACTIVITY**

The nuclei of certain atoms are unstable. As these nuclei break down, they release particles and energy, which we call **nuclear radiation**. Nuclei that emit radiation are called **radioactive** nuclei. The process of nuclear breakdown or decay in which nuclei emit radiation is known as **radioactivity**.

Radioactivity was first observed by the French scientist Henri Becquerel in 1896 when he placed a uranium salt next to a covered piece of photographic film. Much to his surprise, the film became exposed even though no visible light had reached it. Becquerel concluded that the uranium had emitted an invisible form of energy which exposed the film. This energy was a form of nuclear radiation.

The *stability* of a nucleus, its tendency *not* to break down, depends on its ratio of neutrons to protons. Nuclei with small masses are found to be most stable when the number of neutrons equals the number of protons. Nuclei with large masses, however, require approximately 1.5 neutrons per proton for stability. For example, a stable mercury atom has 80 protons and 120 neutrons in its nuclei. Radioactive nuclei break down until they reach a stable neutron-to-proton ratio. In the process, they emit one or more of the following three major types of radiation.

Alpha radiation is named after and symbolized by the first letter of the Greek alphabet, α. It consists of low-energy positively charged particles, called alpha particles. Each particle contains two neutrons and two protons. Alpha particles are identical to the nuclei of helium atoms. Each has a mass of 4 amu and a charge of $2+$. Because of their relatively large size and low energy, they can be stopped easily even by a thin piece of paper or by the layer of dead cells on your skin.

Beta radiation, named after the second Greek letter, β, also consists of particles. Beta particles are simply electrons, emitted with high speed and high energy by nuclei. (Section 17.3 explains this process.) They have a charge of $1-$ and a very small mass (1/1827 amu). Compared to the mass of a nucleus, the mass of the electron is so small that it is usually ignored. Beta particles penetrate materials deeper than do alpha particles. They can enter the skin, causing burns, but are stopped before they reach internal organs.

Gamma radiation, named after the third Greek letter, γ, differs from alpha radiation and beta radiation in that it does not consist of particles of matter. Gamma rays have no mass or charge but are a form of electromagnetic radiation. We can think of gamma rays as high-energy photons, or "particles" of energy. They are millions of times more energetic than photons of visible radiation. They can pass through most materials, even lead or concrete. Gamma radiation can therefore penetrate and pass through the human body.

Some properties of alpha, beta, and gamma radiation are summarized in Table 17-1.

17.2 NUCLEAR REACTIONS

We can represent a nuclear reaction with a nuclear equation, just as we express a chemical reaction with a chemical equation. In a nuclear equation, each nucleus is represented by the chemical symbol for the element with the mass number and atomic number written on the left

TABLE 17-1 *Properties of radiation*

Name	Symbol	Composition	Mass (amu)	Charge	Penetrating ability
alpha	α	helium nucleus (2 protons and 2 neutrons)	4	2+	low
beta	β	electron	1/1827	1−	moderate
gamma	γ	energy wave	0	0	very high

side. For example, the symbol for a particular radium nucleus is $^{226}_{88}Ra$.

$$\text{mass number} \rightarrow \,^{226}_{88}Ra \leftarrow \text{symbol for radium}$$
$$\text{atomic number} \rightarrow$$

The superscript is the mass number of the nucleus, and the subscript is the atomic number. As we saw in Chapter Four, the atomic number of an element equals the number of protons in the nucleus of an atom of that element. Every nucleus of a given element has the same number of protons. For example, all radium atoms have 88 protons in their nuclei. Once we write the symbol of the element, Ra, we do not have to write the atomic number because it has a fixed value. We often do include it, however, as an aid in following nuclear reactions.

The mass number, the sum of the number of protons and neutrons in a nucleus, must never be omitted. Nuclei of the same element can contain different numbers of neutrons and therefore have different mass numbers. As we saw previously, such nuclei are called isotopes of the element. For example, radium has 16 isotopes. They are:

$$^{213}_{88}Ra \quad ^{214}_{88}Ra \quad ^{215}_{88}Ra \quad ^{216}_{88}Ra \quad ^{217}_{88}Ra \quad ^{219}_{88}Ra \quad ^{220}_{88}Ra \quad ^{221}_{88}Ra$$

$$^{222}_{88}Ra \quad ^{223}_{88}Ra \quad ^{224}_{88}Ra \quad ^{225}_{88}Ra \quad ^{226}_{88}Ra \quad ^{227}_{88}Ra \quad ^{228}_{88}Ra \quad ^{230}_{88}Ra$$

As we just mentioned, the atomic number, 88, may be omitted from these symbols. An isotope is named by its mass number. For example, $^{226}_{88}Ra$ is called radium-226.

The three different forms of nuclear radiation are represented in equations in the following way:

alpha particle	$^{4}_{2}He$	or	$^{4}_{2}\alpha$	
beta particle	$^{0}_{-1}e$	or	$^{0}_{-1}\beta$	
gamma ray	$^{0}_{0}\gamma$	or	γ	

Because the alpha particle is identical to the nucleus of a helium atom, it can be represented by the symbol He. Similarly, because the beta particle is an electron, it can be represented by the symbol e. The atomic number of a nuclear particle is simply its charge. For example, since the charge of an electron is -1 we give the beta particle the subscript -1. Because gamma rays have neither mass nor charge, they are usually not shown at all in nuclear equations. When protons and neutrons take part in nuclear reactions, they are written in the following ways:

proton $\quad {}_1^1H \quad$ or $\quad {}_1^1p$

neutron $\quad {}_0^1n$

Now we are ready to write equations for nuclear reactions. The breakdown of radium is a typical nuclear reaction.

Box 17

How many elementary particles exist?

As our knowledge of the composition of matter has increased, the definition of the term "elementary particle" has changed. First, the atom was thought to be an elementary, or smallest possible, particle. Then we discovered that the electron, the proton, and the neutron are even smaller particles that are part of the atom. We now know that protons and neutrons themselves consist of still smaller units.

According to the current theory of nuclear particles, protons and neutrons are composed of several particles called quarks. This name, given to these elementary particles by the Nobel Prize-winning physicist Murray Gell-Mann, is a nonsense word from James Joyce's novel "Finnegan's Wake."

Eighteen different types of quarks are believed to exist. They are classified in a rather unusual way. Each quark is said to have both a "flavor" and a "color." There are 6 possible flavors, called up, down, strange, charm, top (or truth), and bottom (or beauty). And, a quark must have one of three possible colors, called red, green, and blue.

For example, the proton consists of two up quarks and one down quark. One of these quarks is red, another is green, and the third is blue. The neutron consists of two down quarks and one up quark. These three quarks are also red, green, and blue.

Evidence for the existence of quarks has been gathered with the help of a device called a particle accelerator. Physicists use particle accelerators to shatter nuclei at high energies and then they observe the resulting particles. Only time will tell whether we have at last identified the most fundamental particle, or whether quarks consist of still smaller units.

$$^{226}_{88}\text{Ra} \rightarrow {}^{222}_{86}\text{Rn} + {}^{4}_{2}\text{He}$$
radium   radon   alpha particle

Radium forms another element, radon, by losing an alpha particle. In this reaction, and in all other nuclear reactions, the protons and neutrons are rearranged, just as atoms are rearranged in chemical reactions. But the total number of protons and neutrons remains the same. Therefore, *the sum of the mass numbers must be the same on both sides of the equation.*

$$\frac{\text{sum of the mass}}{\text{numbers of reactants}} = \frac{\text{sum of the mass}}{\text{numbers of products}}$$

$$226 \text{ amu} = 222 \text{ amu} + 4 \text{ amu}$$

The sum of the charges (or atomic numbers) also must be the same on both sides of the equation.

$$\text{sum of the charges of reactants} = \text{sum of the charges of products}$$

$$88 = 86 + 2$$

Keeping in mind these rules about the sum of the masses and the sum of the charges, we can find a reactant or product that is missing in a nuclear equation. For example, if we know that uranium-238 decays by emitting an alpha particle, the product nucleus can readily be found.

$$^{238}_{92}\text{U} \rightarrow ? + {}^{4}_{2}\text{He}$$

Since the mass of the reactant equals 238, the mass of the products must also equal 238. Because the alpha particle has a mass of 4, the nucleus formed must have a mass number of $238 - 4 = 234$. Similarly, since the charge of the reactant is 92, the charges of the products must add up to 92. Because the alpha particle has a charge of $2+$, the missing nucleus must have a charge (or atomic number) of $92 - 2 = 90$. Looking at the periodic table (see inside front cover), we find that the element thorium has a nucleus with atomic number 90. Therefore, the product must be thorium, and the complete equation is:

$$^{238}_{92}\text{U} \rightarrow {}^{234}_{90}\text{Th} + {}^{4}_{2}\text{He}$$
uranium-238   thorium-234   alpha particle

Example Complete each of the equations by identifying the missing reactant or product.

	Missing mass	Missing charge	Missing particle
a $^{14}_{6}C \rightarrow ? + ^{0}_{-1}e$	14	7+	$^{14}_{7}N$
b $^{14}_{7}N + ^{4}_{2}He \rightarrow ? + ^{1}_{1}H$	17	8+	$^{17}_{8}O$
c $^{207}_{84}Po \rightarrow ^{203}_{82}Pb + ?$	4	2+	$^{4}_{2}He$
d $? + ^{2}_{1}H \rightarrow ^{55}_{26}Fe + 2^{1}_{0}n$	55	25+	$^{55}_{25}Mn$

Practice problem Find the missing product in the following nuclear equation:

$$^{235}_{92}U + ^{1}_{0}n \rightarrow ^{139}_{56}Ba + ? + 3^{1}_{0}n$$

ANSWER $^{94}_{36}Kr$

**17.3
NATURAL
RADIOACTIVITY**

Many heavy elements, and in fact *all known isotopes of elements heavier than bismuth* ($^{209}_{83}Bi$), are radioactive. The presence of 84 or more protons in a nucleus tends to make the nucleus unstable regardless of the number of neutrons present. Those elements beyond uranium, as well as technetium, are so radioactive that any amounts present when the Earth formed have long since decayed. Even stable elements like potassium have naturally occurring radioactive isotopes. Of the nearly 330 different nuclei found in nature, over 50 are unstable.

An unstable, or radioactive nucleus decays in steps until it forms a stable nucleus. Such a series of nuclear reactions is known as a **radioactive disintegration series**. Figure 17-1 illustrates the series that begins with uranium-238. Other series begin with uranium-235 and thorium-232. In each of these series, a "parent" nucleus spontaneously decays to form unstable "daughter" nuclei, which also decay, until finally a stable nucleus forms.

Each decay in a disintegration series is accompanied by the emission of an alpha or beta particle, and sometimes also by gamma radiation.

FIGURE 17-1

Uranium disintegration series. The symbols above the arrows indicate the types of radiation emitted in each step.

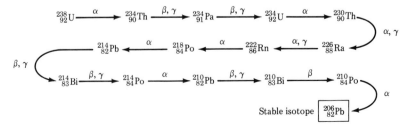

Since nuclei contain protons and neutrons, it is relatively easy to understand how an unstable nucleus can give off an alpha particle consisting of two protons and two neutrons. This process, known as alpha decay, was illustrated in the preceding section with the disintegration of radium to form the lighter element radon. However, since nuclei contain no electrons, it is more difficult to understand how a beta particle (an electron) can be emitted by a nucleus. It turns out that a neutron is converted to a proton and an electron during beta decay. In this way, an electron is released and the atomic number of the nucleus increases by one. An example of beta decay is the conversion of lead-214 to bismuth-214.

$$\underset{\text{lead-214}}{^{214}_{82}\text{Pb}} \rightarrow \underset{\text{bismuth-214}}{^{214}_{83}\text{Bi}} + \underset{\text{electron}}{^{0}_{-1}\text{e}}$$

In the uranium-238 disintegration series, a total of eight alpha particles and six beta particles are released, finally resulting in the formation of stable lead-206:

$$\underset{\text{uranium-238}}{^{238}_{92}\text{U}} \rightarrow \underset{\text{lead-206}}{^{206}_{82}\text{Pb}} + \underset{\substack{\text{alpha}\\\text{particle}}}{8\,^{4}_{2}\text{He}} + \underset{\text{electron}}{6\,^{0}_{-1}\text{e}}$$

**17.4
ARTIFICIAL
RADIOACTIVITY**

Over 1500 radioactive isotopes, called **radioisotopes**, have been produced artificially. One of the most common methods for making radioisotopes is bombardment, the hitting of a nucleus with a high energy particle. If a stable nucleus is bombarded with neutrons, a radioisotope may form (see Figure 17-2). Such a reaction, known as

FIGURE 17-2

(Industrial Research, *October 1973*.)

"I Thought Thorium Gave Off A Gamma Ray When It Absorbed A Neutron"

Before collision

a

X
Target nucleus

At collision

Compound nucleus
(unstable)

After collision

Y   b

$$a + X \rightarrow Y + b$$

FIGURE 17-3

*A nuclear reaction: a + X →
Y + b. In this reaction,
particle a strikes nucleus X,
forming nucleus Y and
particle b.*

neutron capture or neutron activation, is used to make radioactive
cobalt-60.

$$\underset{\text{cobalt-59}}{^{59}_{27}\text{Co}} + \underset{\text{neutron}}{^{1}_{0}\text{n}} \rightarrow \underset{\text{cobalt-60}}{^{60}_{27}\text{Co}}$$

Not all of the cobalt-59 is converted to cobalt-60 in this reaction. Therefore, after the reaction we are left with a mixture of stable cobalt-59
and radioactive cobalt-60. It is impossible to separate the two forms
of cobalt because they are chemically alike. Therefore, the stable form
is called the carrier because it carries, or contains, the radioisotope
after neutron capture.

By bombarding a nucleus with subatomic particles, or with smaller
nuclei, it is also possible to change a nucleus into the nucleus of another
element. This kind of reaction, called **transmutation**, is illustrated in
Figure 17-3. Transmutation became practical after the development of
nuclear reactors and accelerators. Nuclear reactors can produce neutrons and other particles, and accelerators are huge machines that can
give charged particles great energy. An example of transmutation is
the production of radioactive phosphorus-32 from stable sulfur-32.

$$\underset{\text{sulfur-32}}{^{32}_{16}\text{S}} + \underset{\text{neutron}}{^{1}_{0}\text{n}} \rightarrow \underset{\text{phosphorus-31}}{^{32}_{15}\text{P}} + \underset{\text{hydrogen}}{^{1}_{1}\text{H}}$$

Again, not all of the sulfur nuclei are converted to radioactive phos-
phorus-32 in this reaction. However, in transmutation reactions, the
product is a different element than the starting element and can therefore be separated. Such a radioisotope is said to be carrier-free.

As shown in Table 17-2, the elements beyond uranium, called the
transuranium elements, can be produced by bombardment with a variety of particles. These particles range from neutrons to nuclei as large
as that of oxygen-18. Many transuranium elements were first made at
the Lawrence Radiation Laboratory of the University of California.

**17.5
MATTER
AND ENERGY**

In previous chapters, we came across the law of conservation of energy
and the law of conservation of mass. These two separate conservation
laws were unchallenged until this century, and are still satisfactory for
chemical changes. When dealing with nuclear changes, however, mass
and energy must be considered together.

As proposed by Albert Einstein, mass can be converted into energy
or energy can be converted into mass in a nuclear process. Thus, we
can consider matter and energy to be *equivalent*. They are different
forms of the same quantity. The conservation law can now be stated
in its most general terms: *In any process, the total amount of mass*

TABLE 17-2 *The transuranium elements*

Atomic number	Name	How element is made
93	neptunium	$^{238}_{92}U + ^{1}_{0}n \rightarrow ^{239}_{93}Np + ^{0}_{-1}e$
94	plutonium	$^{238}_{92}U + ^{2}_{1}H \rightarrow ^{238}_{93}Np + 2^{1}_{0}n$
		$^{238}_{93}Np \rightarrow ^{238}_{94}Pu + ^{0}_{-1}e$
95	americium	$^{239}_{94}Pu + ^{1}_{0}n \rightarrow ^{240}_{95}Am + ^{0}_{-1}e$
96	curium	$^{239}_{94}Pu + ^{4}_{2}He \rightarrow ^{242}_{96}Cm + ^{1}_{0}n$
97	berkelium	$^{241}_{95}Am + ^{4}_{2}He \rightarrow ^{243}_{97}Bk + 2^{1}_{0}n$
98	californium	$^{242}_{96}Cm + ^{4}_{2}He \rightarrow ^{245}_{98}Cf + ^{1}_{0}n$
99	einsteinium	$^{238}_{92}U + 15^{1}_{0}n \rightarrow ^{253}_{99}Es + 7^{0}_{-1}e$
100	fermium	$^{238}_{92}U + 17^{1}_{0}n \rightarrow ^{255}_{100}Fm + 8^{0}_{-1}e$
101	mendelevium	$^{253}_{99}Es + ^{4}_{2}He \rightarrow ^{256}_{101}Md + ^{1}_{0}n$
102	nobelium	$^{246}_{96}Cm + ^{12}_{6}C \rightarrow ^{254}_{102}No + 4^{1}_{0}n$
103	lawrencium	$^{252}_{98}Cf + ^{10}_{5}B \rightarrow ^{257}_{103}Lr + 5^{1}_{0}n$
104	unnilquadium	$^{249}_{98}Cf + ^{12}_{6}C \rightarrow ^{257}_{104}Unq + 4^{1}_{0}n$
105	unnilpentium	$^{249}_{98}Cf + ^{15}_{7}N \rightarrow ^{260}_{105}Unp + 4^{1}_{0}n$
106	unnilhexium	$^{249}_{98}Cf + ^{18}_{8}O \rightarrow ^{263}_{106}Unh + 4^{1}_{0}n$

plus energy remains constant. For example, if the mass decreases, an equivalent amount of energy must be formed.

The mathematical relationship between mass and energy is given by the famous Einstein equation:

$$E = mc^2$$

In this equation, E stands for energy in joules, m for the mass in kilograms, and c for the speed of light, 3.0×10^8 m/sec. For example, consider the complete conversion of 1.0 g of matter into energy.

$$E = mc^2$$
$$= (1.0 \times 10^{-3} \text{ kg})(3.0 \times 10^8 \text{ m/sec})^2$$
$$= (1.0 \times 10^{-3} \text{ kg})(9.0 \times 10^{16} \text{ m}^2/\text{sec}^2)$$
$$= 9.0 \times 10^{13} \text{ (kg)(m}^2)/\text{sec}^2$$
$$= 9.0 \times 10^{13} \text{ J}$$

From this calculation, we see that a huge amount of energy can be released from a very small quantity of matter. Thus, nuclear reactions can convert small amounts of matter into large amounts of energy.

A nucleus is lighter than the sum of its parts. For example, we might expect the mass of a helium nucleus to be the sum of the masses of two protons and two neutrons:

mass of 2 protons + 2 neutrons

$$= 2\,(1.00728 \text{ amu}) + 2\,(1.00867 \text{ amu})$$

$$= 4.03190 \text{ amu}$$

But the experimentally determined mass of a helium nucleus is only 4.0015 amu. This value is 0.0304 amu smaller than the sum of the masses of the protons and neutrons. The missing mass has been converted to energy, the **binding energy** of the helium nucleus. This is the energy that would be needed to break apart the helium nucleus into the separate protons and neutrons.

The binding energy of a nucleus is found by measuring the difference between the sum of the masses of protons and neutrons that make up the nucleus, and the actual mass of the nucleus. For example, when we subtract 4.0015 amu (the actual mass of a helium nucleus) from 4.03190 amu (the sum of the masses of protons and neutrons in the nucleus), we get 0.0304 amu, the binding energy of helium. If we divide the binding energy of a nucleus by the mass number (sum of protons and neutrons), we obtain the binding energy per nuclear particle. The binding energies per nuclear particle for different nuclei are shown on the graph in Figure 17-4. Notice that iron-56 has the highest binding energy per nuclear particle. Therefore, it is the most stable nucleus. As we will see in the next section, when either light or heavy elements are converted to medium-weight elements like iron, energy is given off because the nucleus produced is more stable.

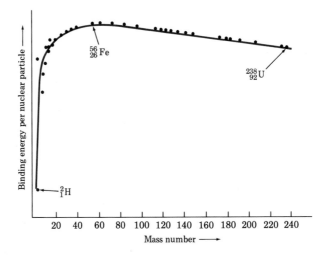

FIGURE 17-4

Binding energy per nuclear particle for the elements. Iron-56 is the most stable nucleus because it has the highest binding energy.

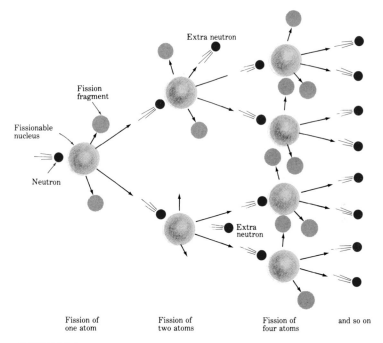

Extra neutron

Fission
fragment

Fissionable
nucleus

Neutron

Extra
neutron

| Fission of | Fission of | Fission of | and so on |
| one atom | two atoms | four atoms | |

FIGURE 17-5

A chain reaction. Each fission reaction releases neutrons, which set off more fission reactions.

17.7
FISSION
AND FUSION

One method for releasing the binding energy of a nucleus is **fission**, the splitting of a heavy nucleus into two or more smaller pieces by a neutron. Uranium-235, for example, is fissionable. It first absorbs a bombarding neutron to form an unstable nucleus, uranium-236. This unstable nucleus then breaks into two smaller nuclei, plus 2 or 3 neutrons. For example, uranium-236 can break down into barium and krypton nuclei:

$$\underset{\text{uranium-235}}{^{235}_{92}\text{U}} + \underset{\text{neutron}}{^{1}_{0}\text{n}} \rightarrow \underset{\text{uranium-236}}{\left[^{236}_{92}\text{U}\right]} \rightarrow \underset{\text{krypton-90}}{^{90}_{36}\text{Kr}} + \underset{\text{barium-144}}{^{144}_{56}\text{Ba}} + \underset{\text{neutrons}}{2^{1}_{0}\text{n}}$$

The brackets in this equation indicate that uranium-236 is unstable.

Only one neutron starts the reaction, but more than one neutron is produced. (The precise number depends on the fission products that are formed.) These additional neutrons can react with other uranium-235 atoms, which in turn form more neutrons that react with still other nuclei, and so on. This process, shown in Figure 17-5, is called a **chain reaction.** The smallest mass of fissionable material that allows the chain reaction to continue is called the critical mass. If fission is uncontrolled, it can produce a huge amount of energy in a very short time, causing a nuclear explosion.

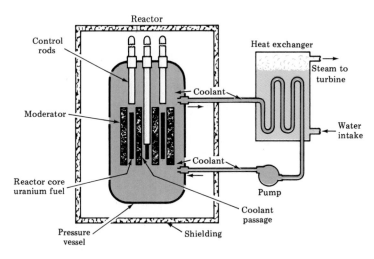

Reactor

Control
rods

Heat exchanger

Steam to
turbine

Coolant

Moderator

Water
intake

Reactor core
uranium fuel

Coolant

Pump

Coolant
passage

Pressure
vessel

Shielding

FIGURE 17-6

A nuclear reactor. The coolant is heated by the nuclear fuel in the reactor. In a heat exchanger, the heated coolant converts water into steam. The steam drives a turbine that is used to generate electricity.

In a nuclear reactor, such as the one shown in Figure 17-6, fission is controlled by substances that absorb neutrons. Therefore, energy is released *slowly* as heat, and the heat converts water to steam that is used to generate electricity. Fission of about one-half gram of uranium-235 provides the same amount of energy as the burning of one ton of coal. Fission products from nuclear reactors are often useful radioisotopes. However, other radioactive products have no practical use, and their disposal can be a complex problem.

The neutrons that are released by fission can be channeled into reactions that form other radioisotopes. Thus not only does fission produce energy, it produces more nuclear fuel. One such reaction is the formation of plutonium-239 from abundant uranium-238 in a special reactor called a "breeder." The starting material in a breeder reactor is a mixture of plutonium-239 and uranium-238. The plutonium undergoes fission, releasing energy and neutrons. The neutrons strike the uranium-238 nuclei, converting them to plutonium-239 nuclei.

$$\underset{\text{uranium-238}}{^{238}_{92}\text{U}} + \underset{\text{neutron}}{^{1}_{0}\text{n}} \rightarrow \underset{\text{plutonium-239}}{^{239}_{94}\text{Pu}} + \underset{\text{electron}}{2\ ^{0}_{-1}\text{e}}$$

Thus, more plutonium is produced, or "bred," at the same time that plutonium is consumed in the reactor.

Nuclear **fusion** is a second method of releasing the large amounts of energy present in the nuclei of atoms. In fusion, smaller nuclei are combined to form a larger, more stable nucleus. The formation of this more stable nucleus releases energy. To overcome the repulsions between the colliding nuclei, extremely high temperatures (millions of kelvins) are required. Therefore, fusion is a thermonuclear reaction.

The sun's energy is generated by fusion. The nuclei of hydrogen atoms in the sun combine to form helium nuclei and release energy. Scientists have not yet found a way to harness fusion in a nuclear reactor. However, fusion remains a potential source of vast energy.

The reaction most likely to be used will be a reaction between two isotopes of hydrogen: deuterium, $^2_1$H, and tritium, $^3_1$H.

$$\underset{\text{deuterium}}{^2_1\text{H}} + \underset{\text{tritium}}{^3_1\text{H}} \rightarrow \underset{\text{helium}}{^4_2\text{He}} + \underset{\text{neutron}}{^1_0\text{n}}$$

This reaction is the same one that takes place when a hydrogen bomb explodes. In an H-bomb, a fission reaction is the source of the high starting temperature that is necessary for the fusion reaction. A different approach must obviously be used if we are to harness fusion in a nuclear reactor. Research is now under way to generate the required high temperature under controlled conditions.

17.8 HALF-LIFE

One of the most important properties of every radioactive element, whether that element is used in medicine or is present in nuclear waste, is its rate of decay. The breakdown of a radioactive element follows the rate equation

rate $= kN$

where N is the number of radioactive nuclei and k is the rate constant. Thus the number of nuclei that decay per unit time is directly proportional to the number N of nuclei present. Initially, a radioactive element will decay rapidly. But as the breakdown proceeds, the rate slows down because fewer and fewer radioactive nuclei remain.

The time required for one-half of a given number of nuclei of a radioactive element to decay is called the **half-life** of that element. The half-life, abbreviated $t_{1/2}$, can vary from a fraction of a second for a very unstable radioisotope, to millions of years for a relatively stable isotope. Iodine-131, for example, has a relatively short half-life of 8 days. If 100 g are present originally, then after 8 days, only 50 g of iodine-131 remain (along with the decay products). After another 8 days, only 25 g of iodine-131 are left. In the following 8 days, decay continues at the same rate, leaving 12.5 g of iodine-131. This process continues until all of the radioactive isotope has decayed. With the passing of each half-life, as shown in Figure 17-7, one-half of the remaining unstable

FIGURE 17-7

The decay of a radioisotope. After each half-life, the radioactivity decreases by one-half.

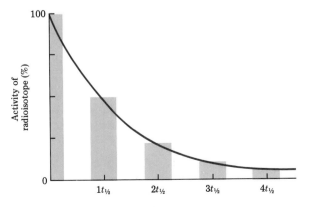

TABLE 17-3 *Properties of radioisotopes*

Name	Half-life	Radiation emitted	Decay product	Applications
carbon-14	5770 years	β	^{14}N	dating
cobalt-60	5.3 years	β, γ	^{60}Ni	cancer treatment
hydrogen-3 (tritium)	12.3 years	β	^{3}He	tracer studies
iodine-131	8 days	β, γ	^{131}Xe	thyroid studies
phosphorus-32	14 days	β	^{32}S	tracer studies, leukemia therapy
plutonium-239	24,000 years	α, γ	^{235}U	nuclear fuel
potassium-40	1.3×10^9 years	β, γ	$^{40}Ar, ^{40}Ca$	source of background radiation; dating
radium-226	1600 years	α, γ	decay series	cancer treatment
strontium-90	27.7 years	β	^{90}Y	radioactive fallout
technetium-99	6 hours	γ	^{99}Ru	organ scans
uranium-235	7×10^8 years	α, γ	decay series	nuclear fuel

nuclei break down, forming the products of decay. Table 17-3 presents half-lives for the nuclei of several radioactive elements.

An interesting application of the half-life of a radioisotope is the technique of radiocarbon dating. This technique is used in archaeology for dating plant and animal remains that are 100 to 50,000 years old. Living plants normally take in carbon dioxide molecules as part of the process of photosynthesis. Most of these molecules contain carbon-12, but some contain radioactive carbon-14. After a plant dies, the amount of carbon-14 it contains decreases slowly, due to radioactive decay, while the amount of stable carbon-12 stays the same. The age of the plant's remains can be found by comparing the relative amounts of these two isotopes: $^{14}C/^{12}C$. Because the half-life of carbon-14 is 5770 years, the ratio of C-14 to C-12 will decrease by one-half every 5770 years. Therefore, the older the remains, the smaller the $^{14}C/^{12}C$ ratio. Remains of animals that consume plants can also be dated by using this method.

Radioisotopes are sometimes taken internally for medical diagnosis (as described in Section 17.10). However, only radioisotopes with short half-lives are used internally because the less time that radioactivity is present in the body, the less harm it can do. The health effects of an isotope depend not only on the length of its half-life, but on the rate at which it is expelled from the body. The radioisotope strontium-90, for example, which is released in nuclear explosions, can replace calcium in the bones and is therefore retained in the body. Since the radioactive half-life of strontium-90 is 21 years, this isotope will remain radioactive for a very long time. Because it stays in the body and has a long half-life, strontium-90 has considerable potential for causing damage.

**17.9
RADIATION
DETECTION
AND UNITS OF
RADIATION**

Radiation can be detected and measured in various ways. A common instrument used to survey an area for radiation is the Geiger-Müller counter, or Geiger counter, shown in Figure 17-8. It contains a tube filled with gas molecules at low pressure, which is connected to a battery. Radiation that enters the window in front of the tube ionizes the gas. The ionized gas conducts electricity produced by the battery, and the resulting current is amplified and indicated by the movement of a meter or by a clicking sound. Beta particles are easily detected with a Geiger counter. The number of particles detected is expressed in counts per minute (cpm).

Photographic film also can be used to detect radiation. As Becquerel accidentally discovered, radiation exposes photographic film. The radiation causes the grains of silver bromide to be reduced to silver metal when the film is developed. The darker the developed film appears, the more radiation it received. Another radiation detector is the scintillation (pronounced sin-tih-lay'-shun) counter, which contains a crystal (NaI) that gives off flashes of light when it is hit by radiation. Semiconductors can also be used as radiation detectors.

The radioactivity of an isotope is commonly expressed in **curies (Ci)**, millicuries (mCi), or microcuries (μCi), in honor of Marie Curie. One curie equals 3.7×10^{10} disintegrations per second, which corresponds to the activity of one gram of radium. The SI unit for radioactivity is the **becquerel (Bq)**, which stands for one disintegration per second. Thus 1 Ci = 3.7×10^{10} Bq.

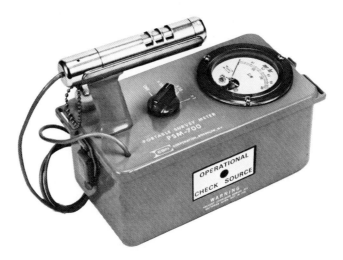

FIGURE 17-8

Geiger counter. This instrument is used to detect the presence of radiation. (Photo courtesy of Eon Corporation.)

Smithsonian Institution: Photo # 57430

Poor and often hungry, Marie Curie nevertheless graduated at the top of her class at the Sorbonne in Paris. When she learned about Henri Becquerel's discovery of radiation in uranium, she began studying its properties herself. She created the term "radioactivity." Marie Curie found that certain uranium ores (called pitchblende) were even more radioactive than pure uranium. She realized that other elements must be present. With the assistance of her husband, Pierre, she isolated the element polonium (Po), which she named after Poland, her homeland. The Curies then found a trace of a still more radioactive element, which they named radium.

To obtain more radium, the Curies used their savings to buy waste ores from mines in Bohemia (present-day Czechoslovakia). They worked for four years in a wooden shed purifying the ore into smaller samples of increasing radioactivity. They finally obtained one gram of radium from eight tons of ore!

The Curies shared with Becquerel the 1903 Nobel Prize in physics for their initial studies of radioactivity. Three years later, Pierre Curie was killed tragically when he was struck by a horse-drawn carriage. Marie Curie took over his professorship, and became the first woman to teach at the Sorbonne. Because she was a woman, she was rejected for membership in the French Academy. But in 1911 she was awarded a Nobel Prize in chemistry for her discovery of two new elements. She was the first person to win two Nobel Prizes in science.

Marie Curie supported the use of radium for cancer treatment, but she rejected financial rewards for her discovery of the element, saying: "Radium is not to enrich anyone. It is an element; it is for all people." She died in 1934 of leukemia, which was probably caused by excessive exposure to radiation.

Exposure to gamma rays, as well as to X rays, is measured in **roentgens (R)**. One roentgen of radiation produces 2.1 billion pairs of ions in one cubic centimeter of dry air at STP. This amount of ionization is equivalent to a charge of 2.58×10^{-4} coulombs per kilogram of air. (The coulomb is a unit of electrical charge.) Geiger counters are often calibrated to read in milliroentgens per hour (mR/hr).

The dose of radiation actually absorbed by living things can be expressed in **rads**, which stand for *r*adiation *a*bsorbed *d*ose. One rad corresponds to the absorption of a certain small amount of energy (100 ergs, equal to 2.4×10^{-6} cal) per gram of tissue receiving the radiation. The SI unit is the **gray (Gy)**, given in joules per kilogram; thus 1 rad $= 10^{-2}$ Gy. A dose of 600 rads (6 Gy) is deadly for most people.

Dose equivalent is a measure of the quantity of absorbed radiation on a common scale for different types of radiation. Each kind of radiation is assigned a "quality factor" based on its biological effect in

---

Do-it-yourself 17

Detecting fires with radiation

Fire detectors are devices that sound an alarm during the initial stages of a fire. They save many lives each year by detecting fires early enough to allow people to escape. One type of detector contains a tiny amount of the radioactive element americium-241. This element breaks down in the following way:

$$^{241}_{95}\text{Am} \rightarrow ^{237}_{93}\text{Np} + ^{4}_{2}\text{He}$$
americium-241　　neptunium-237　　alpha particle

The resulting radiation ionizes the air in and near the fire detector, making the air a conductor of electricity. A small battery inside the fire detector serves as a source of electricity. When smoke particles enter the detector, they block the flow of electric current. This reduction in current sets off the alarm.

Because such a small amount of the radioisotope is present in a fire detector, there is little radiation hazard. If a person stood 10 inches from the fire detector for 8 hours a day, every day of the year, he or she would receive only 0.5 millirem annually. This amount of radiation is about 1/10 the amount that an airline passenger receives from cosmic rays during a roundtrip between New York and Los Angeles. The americium used in detectors is a waste product of nuclear processes carried out at Oak Ridge National Laboratory in Tennessee.

To do: Test the fire detector in your home. Hold a lighted candle under the detector. How close must the candle be to the detector in order to set off the alarm? Must the smoke be visible to activate the detector? Experiment with the smoke from a snuffed out candle.

tissue. For example, an alpha particle has a factor of 10, compared to a factor of 1 for a beta particle, because the larger mass and charge of an alpha particle make its effect on living tissue greater. Dose equivalent equals the absorbed dose in rads multiplied by the quality factor and any other modifying factors which tend to increase or decrease the destruction of tissue. It is expressed in **rems**, *r*oentgen *e*quivalents for *m*an. One rem is the quantity of absorbed radiation that has the same biological effect as the absorption of 1 R or 1 rad. The SI unit of dose equivalent is the **sievert (sv)**; 1 rem = 10^{-2} sv. A recommended dose limit for the general public is no more than 0.5 rem (or 0.005 sv) per year. Table 17-4 presents estimated dose rates in millirems per year from various sources.

Table 17-5 summarizes the radiation measurements described in this section.

17.10 APPLICATIONS OF RADIATION

A radioisotope behaves chemically like the stable form of the same element, but its presence and location can be detected by a radiation detector. Therefore, it is possible to follow, or trace, a radioisotope as it takes part in normal chemical reactions or biological processes. For example, the radioisotope iodine-131 can be used as a tracer to study the thyroid gland in the neck. Iodine normally accumulates in a healthy thyroid in a certain amount and at a certain rate. When iodine-131 is administered to a patient, this radioisotope will also accumulate in the thyroid. The amount of iodine-131 that accumulates and the rate at which it enters the thyroid can be determined by placing a radiation detector at the patient's neck, and counting the gamma rays that are emitted. If the iodine-131 accumulates in an unusual amount, at an abnormal rate, the patient may have a thyroid disorder. For example, in hyperthyroidism, the thyroid gland is overactive. As a result, it takes up iodine more quickly than a normal thyroid. Thus in a 24-hour test using iodine-131, more of the administered dose will accumulate than would in a healthy gland.

Chemical compounds can be prepared or purchased with radioactive atoms substituted for stable ones. For example, radioactive tritium can replace a hydrogen atom, and radioactive phosphorus-32 can replace a stable phosphorus-31 atom in a molecule. A molecule containing a radioactive atom is said to be *labeled* because we can distinguish it from other molecules by using a radiation detector. Labeled molecules can be used to investigate complicated reactions, or series of reactions. For example, if labeled molecules are used as reactants in photosynthesis (a series of reactions through which plants manufacture their own food), their progress through the reactions can be followed. Often a reaction pathway can be determined by discovering which product molecules contain the radioactive atoms. Radioisotope tracers are also used to locate leaks in water pipes, and to solve other engineering problems.

TABLE 17-4 *Estimated dose rates in the United States*[a]

Source	Average dose rate (mrem/year)
environmental	
natural[b]	102
fallout	4
nuclear power	0.003
medical	
diagnostic	72
radiopharmaceuticals	1
occupational	0.8
miscellaneous	2
total	182

[a]From "The Effects on Populations of Exposure to Low Levels of Ionizing Radiation." National Academy of Sciences, National Research Council, Washington, D.C., 1972.
[b]Primarily cosmic rays and radioisotopes in minerals.

TABLE 17-5 *Units for measurement of radiation*

Quantity measured	Customary unit	SI unit
radioactivity	curie (Ci)	becquerel (Bq)
exposure (gamma or X rays)	roentgen (R)	coulomb per kilogram (C/kg)
absorbed dose	rad	gray (Gy)
dose equivalent	rem	sievert (sv)

Radiation can be used to take pictures of the interiors of various objects and of the human body. In an organ scan, for example, a radioisotope is administered to a patient and then pictures are taken with a scintillation detector. A picture of an abnormal thyroid gland might reveal a "cold" (nonradioactive) spot where no radioisotope has penetrated because that region is diseased. See Figure 17-9(a). Using a brain scan, shown in Figure 17-9(b), a tumor can be located as a "hot" (radioactive) spot because more of the radioisotope enters the damaged tissue. Gamma rays, neutrons, and X rays from external sources are also used to study the internal composition of both inan-

FIGURE 17-9

In the thyroid scan (a), the site of cancer (indicated by the arrow) shows up as a "cold" spot because there is no radioactive isotope in that region. In the brain scan (b), a tumor shows up as a "hot" area after the radioisotope is administered. [Photo (a) courtesy of Oak Ridge National Laboratories; photo (b) from "Clinical Scintillation Imaging," L. M. Freeman and P. M. Johnson, Eds. (New York: Grune & Stratton, 1975); Courtesy of Dr. G. V. Taplin.]

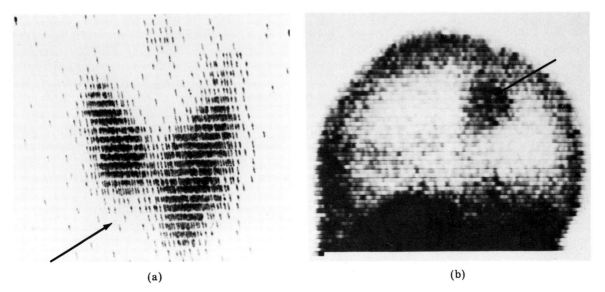

(a)   (b)

imate objects and the human body. This application is known as radiography.

Radiation is used in a type of cancer treatment called teletherapy. Gamma rays from an external source, such as cobalt-60, can destroy cancer cells. Radiation is also used to sterilize male insects in order to control insect populations. New strains of plants can be created by exposing seeds to radiation.

17.11
EFFECTS OF
RADIATION
ON HUMAN HEALTH

Alpha particles, beta particles, and gamma rays are known as **ionizing radiation**. As ionizing radiation passes through matter, it interacts with molecules, creating ions and high-energy molecular fragments called free radicals. Free radicals are very reactive, and cause further chemical changes that are harmful to living cells. The cells most sensitive to ionizing radiation include those in the lymphatic tissue, bone marrow, intestinal mucous membranes, gonads, and lens of the eye.

Overexposure to ionizing radiation, in one large dose or in many small doses, can cause radiation sickness. Exposure to 25 to 100 rems causes a drop in the white blood cell count. A dose of 100 to 200 rems causes nausea and fatigue. Doses of 200 to 400 rems cause nausea and vomiting on the first day after exposure. Exposure to 450 rems is fatal within 30 days to about one out of every two people exposed. A 600-rem dose is fatal for everyone.

Large doses of radiation are fatal because they cause failure of the blood-forming system, the gastrointestinal system, and the central nervous system. Smaller doses have effects that may not be observed until years later, such as impaired fertility, shortened life span, or cancer. People who survived the explosion of the atomic bomb at Hiroshima in World War II later had a high incidence of the blood cancer leukemia. A survivor's chances of developing leukemia were found to be directly related to that person's nearness to the site of the atomic blast.

The most sensitive target of radiation in the body is the hereditary material DNA (discussed further in Section 18.14). Radiation can cause changes called mutations in DNA molecules. Because these molecules carry information from one generation to the next, the effects of mutations are long-range, and generally harmful or even lethal to future offspring.

It is impossible to avoid exposure to radiation, because small amounts of radiation, called background radiation, are constantly present around us. High energy radiation from outer space, called cosmic rays, contribute to the background radiation. Background radiation is also produced by the naturally occurring radioisotopes in minerals and in the construction materials that contain these minerals. As shown in Table 17-4, background radiation is the major source of radiation exposure for most people. Radiation used in medical diagnosis is the next major source. We do not yet know what effect, if any, long-term exposure to small amounts of radiation has on human beings.

Summary　**17.1** The nuclei of certain atoms are unstable. They break down, and emit some combination of alpha particles, beta particles and gamma rays. This breakdown of unstable nuclei is called radioactivity.

17.2 Nuclear reactions can be represented by nuclear equations. The sum of the mass numbers and the sum of the charges (atomic numbers) must be the same on both sides of a nuclear equation.

17.3 All known isotopes of elements heavier than bismuth ($^{209}_{83}$Bi) are radioactive. Radioactive nuclei take part in series of reactions known as radioactive disintegration series. In these series, each unstable nucleus spontaneously decays to form another, until a stable nucleus forms.

17.4 Over 1500 radioactive isotopes, called radioisotopes for short, have been produced artificially. A common method for making radioisotopes is to bombard nuclei with other particles, such as neutrons. When bombardment results in the formation of a radioisotope of the stable starting element, the process is called neutron capture. When bombardment converts atoms of an element into a radioisotope of a different element, the process is called transmutation.

17.5 Matter and energy are equivalent. They are related by the equation $E = mc^2$. In any process, the total amount of mass plus energy remains constant.

17.6 Binding energy is the energy that is needed to break apart a nucleus into separate protons and neutrons. The greater the binding energy per nuclear particle, the more stable the nucleus. Iron-56 is the most stable nucleus.

17.7 Nuclear reactions can release large amounts of energy. One type of nuclear reaction, called fission, involves splitting heavy nuclei into smaller pieces by bombarding them with neutrons. In fusion, smaller nuclei are combined to form larger nuclei.

17.8 The half-life, $t_{1/2}$, of a radioactive element is the time required for half of a given number of nuclei to decay. Half-lives vary from less than a second for some radioactive isotopes to millions of years for more stable isotopes.

17.9 Radiation can be measured with a Geiger counter, photographic film, or a scintillation detector. Common units of radiation (and their SI equivalents) are the curie (or becquerel), roentgen (or coulomb per kilogram), rad (or gray), and rem (or sievert).

17.10 Radioisotopes play an important role as tracers in medicine and chemistry. Radiation can be used to scan or take pictures inside various materials and the human body.

17.11 Alpha particles, beta particles, and gamma rays are known as ionizing radiation. They can interact with molecules in the body, and cause harmful chemical changes.

Exercises

For each term on the left, choose a phrase from the right-hand column that most closely matches its meaning.

1 radioactivity	**a**	forms larger nuclei
2 radioisotope	**b**	decay time
3 half-life	**c**	radioactive isotope
4 fission	**d**	emission of radiation
5 fusion	**e**	splitting of atoms

EXERCISES BY TOPIC

Radioactivity (17.1)

6 What is radioactivity? What forms does it take?

7 Describe the major differences between alpha, beta, and gamma radiation.

8 Explain how a uranium salt can expose a covered piece of photographic film.

9 What causes radioactivity?

Nuclear reactions (17.2)

10 Complete the following nuclear equations:

a $^{11}_{6}C \rightarrow {}^{7}_{4}Be + ?$

b $^{4}_{2}He + ? \rightarrow {}^{30}_{15}P + {}^{1}_{0}n$

c $^{9}_{4}Be + {}^{4}_{2}He \rightarrow ? + {}^{1}_{0}n$

d $^{87}_{36}Kr \rightarrow {}^{86}_{36}Kr + ?$

e $^{2}_{1}H + {}^{3}_{1}H \rightarrow ? + {}^{1}_{0}n$

f $^{235}_{92}U + {}^{1}_{0}n \rightarrow {}^{146}_{57}La + ? + 3\,{}^{0}_{-1}e + 3\,{}^{1}_{0}n$

g $^{7}_{3}Li + ? \rightarrow 2\,{}^{4}_{2}He$

11 How is a nuclear reaction different from a chemical reaction?

Natural radioactivity (17.3)

12 Which elements have only radioactive isotopes?

13 What is a radioactive disintegration series?

14 Write equations for the following conversions:
a $^{230}_{90}$Th to $^{236}_{88}$Ra **b** $^{214}_{83}$Bi to $^{214}_{84}$Po **c** $^{210}_{84}$Po to $^{206}_{82}$Pb

Artificial radioactivity (17.4)

15 How are radioisotopes prepared?

16 What is transmutation?

17 Describe the process of neutron capture.

Matter and energy (17.5)

18 How are mass and energy related?

19 Why were separate conservation laws for mass and for energy satisfactory until the 20th century?

20 Calculate the amount of **a** energy that is equivalent to 3.2 kg of matter **b** mass that is equivalent to 9.2×10^6 J. Note: J = (kg)(m^2)/(sec^2)

Nuclear binding energy (17.6)

21 Why is a nucleus lighter than the sum of its parts?

22 Which nucleus is most stable?

Fission and fusion (17.7)

23 How does fusion differ from fission?

24 What is a chain reaction?

25 Describe how fission can be used to generate electricity.

Half-life (17.8)

26 What does the half-life of a nucleus measure?

27 What does a long $t_{1/2}$ mean?

28 If 80 μg of phosphorus-32 are present initially, what mass of ^{32}P remains after 56 days?

29 A patient receives 20 μg of technetium-99 at 8:00 AM. What mass of this substance remains by 8:00 PM that night?

Detection and units of radiation (17.9)

30 How does a Geiger counter work?

31 Describe two other means for detecting radiation in addition to the Geiger counter.

32 What do each of these units measure? **a** rad **b** curie **c** gray **d** roentgen **e** rem

Applications of radiation (17.10)

33 How are radioisotopes used as tracers?

34 How is radiation used to take pictures?

35 Describe two medical applications of radiation.

Effects of radiation on human health (17.11)

36 What is ionizing radiation?

37 What are the short-term and long-term effects of overexposure to radiation?

38 What are the sources of background radiation?

ADDITIONAL EXERCISES

39 Which type of radiation **a** has the greatest mass? **b** has a negative charge? **c** has the most energy? **d** has a positive charge?

40 Describe the basis for radiocarbon dating.

41 What happens in a breeder reactor?

42 Why is it important for a radioisotope that is used medically to have a short half-life?

43 Doesn't the fact that nucleus is lighter than the sum of its parts violate the law of conservation of mass? Explain.

44 What source of radiation is responsible for the highest dose rate to which human beings are commonly exposed?

45 If uranium-235 is radioactive, why is it still present on the Earth?

46 Describe these equations in words:

a $^{246}_{96}Cm + ^{12}_{6}C \rightarrow ^{254}_{102}No + 4^{1}_{0}n$

b $^{238}_{92}U + 15^{1}_{0}n \rightarrow ^{253}_{99}Es + 7^{0}_{-1}e$

47 What is a carrier-free radioisotope?

48 Complete the following equations:

a $^{27}_{13}Al + ^{4}_{2}He \rightarrow ? + ^{1}_{0}n$

b $^{240}_{92}U \rightarrow ? + ^{0}_{-1}\beta$

49 How does the sun obtain its tremendous energy?

50 Protactinium-234 has a half-life of only 1.18 minutes. How is it possible that this isotope is found naturally?

51 Write a balanced nuclear equation for each of the following:
a capture of a neutron by $^{14}_{7}N$ to form $^{3}_{1}H$
b emission of eight alpha particles and six beta particles by $^{238}_{92}U$

52 The actual atomic mass of $^{40}_{20}Ca$ is 39.96259 amu. Using 1.008665 amu for the mass of a neutron and 1.007825 amu for the mass of a proton, calculate the binding energy for this nucleus. Note: 1 amu = $1.6605655 \times 10^{-27}$ kg; J = $(kg)(m)^2/(sec)^2$

53 Why is it possible to produce energy from the fission of uranium but not from the fission of oxygen?

54 The last step in the uranium decay series is the formation of $^{206}_{82}\text{Pb}$ and the emission of an alpha particle. Write the nuclear equation for this step.

55 If alpha particles do not penetrate the outer layers of the skin, why can nuclei that emit alpha particles pose a health hazard?

56 Why are many nuclear reactions carried out with particle accelerators?

PROGRESS CHART Check off each item when completed.

_____ Previewed chapter

Studied each section:

_____ **17.1** Radioactivity

_____ **17.2** Nuclear reactions

_____ **17.3** Natural radioactivity

_____ **17.4** Artificial radioactivity

_____ **17.5** Matter and energy

_____ **17.6** Nuclear binding energy

_____ **17.7** Fission and fusion

_____ **17.8** Half-life

_____ **17.9** Radiation detection and units of radiation

_____ **17.10** Applications of radiation

_____ **17.11** Effects of radiation on human health

_____ Reviewed chapter

_____ Answered assigned exercises:

#'s _____

Organic chemistry and biological chemistry

Chapter Eighteen deals specifically with compounds that contain the element carbon. These compounds are called organic compounds. In contrast, most of the compounds that we have studied so far are inorganic ("mineral") compounds, which generally do not contain carbon atoms. The first part of this chapter, Part A, is an introduction to organic chemistry. In it, we describe the major types of organic compounds. In Part B we extend our discussion to the organic compounds that are the basis of life. The study of these particular compounds is called biological chemistry or biochemistry. To demonstrate an understanding of Chapter Eighteen, you should be able to:

PART A
ORGANIC CHEMISTRY
(18.1–18.9)

1 Explain the relationship between the electronic structure of a carbon atom and the number and type of bonds it can form.

2 Describe the structure and bonding in methane.

3 Identify the common properties of alkanes.

4 Explain the difference between an alkyl group and an alkane.

5 Draw and recognize different isomers of an alkane.

6 Compare alkenes and alkynes to alkanes.

7 Describe the special features of an aromatic molecule.

8 Recognize the structural formulas of alcohols, ethers, aldehydes, ketones, carboxylic acids, and esters.

9 Recognize the structural formulas of halogenated hydrocarbons, amines, amides, nitrogen heterocycles, and polymers.

**PART B
BIOLOGICAL CHEMISTRY
(18.10–18.14)**

10 Describe the major types of carbohydrates.

11 Describe the most important lipids.

12 Describe the structure and function of proteins.

13 Describe the two major aspects of metabolism—catabolism and anabolism.

14 Describe how the structure of DNA allows it to control protein synthesis and heredity.

Part A   Organic chemistry

**18.1
CARBON ATOM**

Until the 19th century, chemists thought that the compounds found in living things contained a mysterious "vital force," and that these compounds could be made only in living tissue. The compounds found in living things were called organic compounds. But in 1828 the German chemist Friedrich Wöhler (pronounced vuh'ler) made a discovery that radically changed the way chemists thought about organic compounds. Wöhler accidentally prepared urea, a biological waste product present in urine, by heating ammonium cyanate, an inorganic compound that is not found in living things. Wöhler showed that organic compounds can be prepared synthetically, in the laboratory, and need not come from living things. The synthesis of urea showed that the idea that organic compounds could be made only in living tissue was incorrect. Later experiments revealed that urea and most other compounds in living systems contain carbon atoms to which atoms of other elements are covalently bonded.

Organic chemistry is now simply defined as the study of compounds of carbon. Carbon compounds are called organic compounds. The study of compounds of all the other elements is called **inorganic chemistry**.

The electronic structure of a carbon atom can be represented in any of the following ways:

$$\left(\begin{matrix} 6p \\ 6n \end{matrix}\right) \; 2e \; 4e \qquad \text{or} \qquad \cdot \overset{\cdot}{\underset{\cdot}{C}} \cdot \qquad \text{or} \qquad \cdot \overset{\cdot}{C} \colon \qquad \text{or} \qquad 1s^2 2s^2 2p^2$$

Each of these representations shows that carbon has four outermost electrons. To fill its outermost level, carbon must gain another four electrons through covalent bonds. *Thus carbon forms compounds in which it has four covalent bonds.* Carbon can form these bonds in any of the combinations shown in Table 18-1. A long dash (—) indicates a bond consisting of a shared pair of electrons, one from the carbon atom and one from another atom. With four outer electrons of its own $(2s^2 2p^2)$ plus a shared electron from each of the four bonds, carbon in a compound has a complete outer level of eight electrons $(2s^2 2p^6)$. If you remember that each carbon atom must always have four covalent

TABLE 18-1 *The possible bonding arrangements of a carbon atom*

Arrangement	Description	Total number of bonds
$-\overset{\textstyle\mid}{\underset{\textstyle\mid}{C}}-$	4 single bonds	4
$-\overset{\textstyle\|}{C}-$ or $\overset{\textstyle\|}{\underset{\diagup\ \diagdown}{C}}$	2 single bonds, 1 double bond	4
$=C=$	2 double bonds	4
$-C\equiv$	1 single bond, 1 triple bond	4

bonds around it, you will find it easy to draw the structures of organic compounds.

Carbon forms several million different compounds. This number is greater than the number of compounds formed by any other element. There are about 100 times more organic compounds than there are inorganic compounds. One reason for the large number of organic compounds is that carbon atoms can form strong bonds with other carbon atoms, as well as with atoms of other nonmetals. Carbon's ability to form bonds with itself makes possible the formation of compounds that consist of chains or rings of carbon atoms that are bonded together. Additional properties of organic compounds are presented in Table 18-2, and are contrasted with the properties of inorganic compounds.

TABLE 18-2 *Comparison between typical organic and inorganic compounds*

Organic compounds	Inorganic compounds
based on carbon	based on elements other than carbon
largely covalent bonding	largely ionic bonding
relatively large number of atoms in molecule	relatively few atoms in formula unit
low boiling and melting points	high boiling and melting points
low solubility in water	high solubility in water
nonelectrolytes	electrolytes
flammable	nonflammable

18.2 METHANE

Hydrocarbons are the simplest organic compounds. They consist of only carbon and hydrogen atoms. The simplest hydrocarbon is methane, CH_4. In this molecule, a carbon atom is covalently bonded to four hydrogen atoms. The **structural formula** of methane is a representation of the molecule based on its Lewis structure.

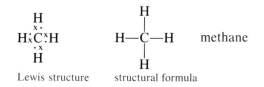

Lewis structure   structural formula   methane

Each dash in the structural formula corresponds to an electron pair in the Lewis structure. The structural formula shows how the atoms in the molecule are attached. It does not show the actual three-dimensional shape of the molecule, however.

Figure 18-1 does show the actual shape of a methane molecule. The ball-and-stick model (a) illustrates the bonding in a three-dimensional way. The space-filling model (b) represents more accurately the relative sizes and distances between the atoms. If we draw lines connecting the four atoms surrounding carbon, we form a tetrahedron (c). A tetra-

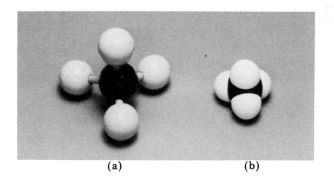

(a)   (b)

FIGURE 18-1

Methane, CH_4. A ball-and-stick model (a) and a space-filling model (b) of this molecule are shown. If the hydrogen atoms are connected as in (c), a tetrahedron is formed. (Photos by Al Green.)

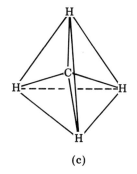

(c)

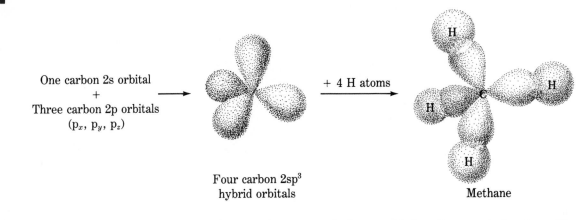

One carbon 2s orbital
+
Three carbon 2p orbitals
(p_x, p_y, p_z)

$+$ 4 H atoms

Four carbon 2sp³
hybrid orbitals

Methane

FIGURE 18-2

Formation of methane. In methane, hydrogen 1s orbitals overlap with sp³ hybrid orbitals of carbon.

hedron is a regular geometric figure with four faces and six edges. The bonds formed by carbon in methane and related molecules are said to be tetrahedral. The angles between the bonds in a tetrahedral molecule are each 109.5°. Even though flat structural formulas do not represent the actual shapes of molecules, they are commonly used in organic chemistry because of their simplicity.

We might expect the carbon 2s and 2p electrons to form different types of bonds, yet all four bonds in methane are alike. To explain this fact, we say that the one 2s and three 2p orbitals *mix* to create a new set of four equivalent orbitals. These new orbitals, called sp³ (read S-P-three) **hybrid orbitals**, are shown in Figure 18-2. Each of the four resulting hybrid orbitals contains one of the four outermost electrons. A methane molecule forms when each of the carbon sp³ orbitals overlaps with a 1s orbital of a hydrogen atom, as illustrated in Figure 18-2. The covalent bonds that result from this overlap contain shared electrons located directly between the centers of the atomic nuclei. They are called sigma bonds, and are symbolized by the Greek letter σ.

18.3 ALKANES

The **alkanes** (or paraffins) are a family of hydrocarbons. The first member of this family is methane, which we just described. The second member is ethane, C_2H_6.

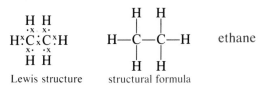

Lewis structure   structural formula   ethane

Ethane consists of two carbon atoms joined by a single bond. Each carbon atom in an ethane molecule shares one outer electron with the other carbon atom. In this way, the carbon atoms each gain one elec-

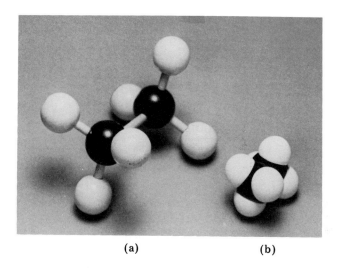

FIGURE 18-3

Ethane, CH_3CH_3. A ball-and-stick model (a) and space-filling model (b) are shown. (Photos by Al Green.)

(a)   (b)

tron, giving each atom five outer electrons. The two carbon atoms complete their outer levels by combining with three hydrogen atoms each. Figure 18-3 presents two models of the ethane molecule. The bonding in terms of hybrid orbitals is illustrated in Figure 18-4.

The structural formula of ethane can be simplified by writing a condensed structural formula:

$$CH_3\text{---}CH_3 \quad \text{or} \quad CH_3CH_3 \qquad \text{ethane (condensed structural formula)}$$

In a condensed formula, the symbols of the atoms bonded to each carbon atom are written on the right side of the symbol for carbon. For example, the formula CH_3CH_3 means that the first carbon has three hydrogen atoms attached and is bonded to another carbon atom that also has three hydrogen atoms attached. If no dash appears between

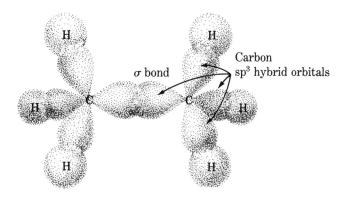

FIGURE 18-4

The bonding in ethane. A sigma (σ) bond is formed by the overlap of carbon sp^3 orbitals.

(a)

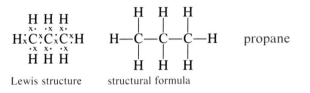

(b)

FIGURE 18-5

Propane, CH₃CH₂CH₃. A ball-and-stick model (a) and a space-filling model (b) are shown. (Photos by Al Green.)

atoms, a single bond is understood to be present. Condensed formulas are the most common way of representing organic molecules because they are clear and easy to write.

Propane, C_3H_8, is the third member of the alkane family. It contains three carbon atoms joined by single bonds.

$$\begin{array}{ccc}
\text{H H H} & \text{H} \quad \text{H} \quad \text{H} & \\
\overset{\cdot}{\underset{\cdot}{\text{H}}}\overset{\times}{\underset{\times}{\text{C}}}\overset{\cdot}{\underset{\times}{\text{C}}}\overset{\times}{\underset{\cdot}{\text{C}}}\text{H} & \text{H}\text{—}\text{C}\text{—}\text{C}\text{—}\text{C}\text{—}\text{H} & \text{propane} \\
\text{H H H} & \text{H} \quad \text{H} \quad \text{H} & \\
\text{Lewis structure} & \text{structural formula} &
\end{array}$$

Notice that the middle carbon forms covalent bonds to only two hydrogen atoms because it already has two covalent bonds, one to each of the other two carbon atoms. The condensed formula for propane is:

$$CH_3\text{—}CH_2\text{—}CH_3 \quad \text{or} \quad CH_3CH_2CH_3 \qquad \text{propane (condensed formula)}$$

Models of this molecule are shown in Figure 18-5.

There are many more compounds in the alkane family. The first ten alkanes are listed in Table 18-3. Alkane molecules are said to be **saturated**. In a saturated molecule, each carbon atom is bonded to four other atoms by single covalent bonds; no double or triple bonds are present between C atoms. The alkanes form a **homologous series**, a sequence of molecules in which each molecule differs from the next by a constant number of atoms. Each member of the alkane series contains one more —CH₂— than the molecule that comes before it in the series. For example, propane, $CH_3CH_2CH_3$, contains one more —CH₂— than ethane, CH_3CH_3.

The general formula for the alkanes is C_nH_{2n+2}. According to this formula, the number of hydrogen atoms in an alkane is always two more than twice the number of carbon atoms. Thus an alkane with 50 carbons (n = 50) has $(2 \times 50) + 2 = 100 + 2 = 102$ hydrogen atoms. Its molecular formula is $C_{50}H_{102}$. As shown in Table 18-3, the names of all the alkanes end in the letters "-ane."

Mixtures of alkanes are found in petroleum and natural gas. Both petroleum and natural gas are the products of prolonged pressure and heat on buried living matter. Natural gas is 60 to 95% methane and contains smaller amounts of ethane, propane, butane, and pentane. Petroleum is a more complex mixture of alkanes. The alkanes in petroleum can be separated by distillation into various mixtures, called fractions, that have different boiling ranges. Figure 18-6 shows a fractionating tower, the device used to separate the fractions by distillation.

TABLE 18-3 *Simple alkanes*

Name	Molecular formula	Condensed structural formula
methane	CH_4	CH_4
ethane	C_2H_6	CH_3CH_3
propane	C_3H_8	$CH_3CH_2CH_3$
butane	C_4H_{10}	$CH_3CH_2CH_2CH_3$
pentane	C_5H_{12}	$CH_3CH_2CH_2CH_2CH_3$
hexane	C_6H_{14}	$CH_3CH_2CH_2CH_2CH_2CH_3$
heptane	C_7H_{16}	$CH_3CH_2CH_2CH_2CH_2CH_2CH_3$
octane	C_8H_{18}	$CH_3CH_2CH_2CH_2CH_2CH_2CH_2CH_3$
nonane	C_9H_{20}	$CH_3CH_2CH_2CH_2CH_2CH_2CH_2CH_2CH_3$
decane	$C_{10}H_{22}$	$CH_3CH_2CH_2CH_2CH_2CH_2CH_2CH_2CH_2CH_3$

FIGURE 18-6

A fractionating tower. Petroleum is separated into various components on the basis of their boiling points. (Photo courtesy of the Mobil Oil Corporation.)

(a)

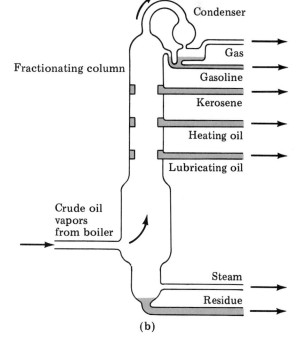

Condenser

Gas

Fractionating column

Gasoline

Kerosene

Heating oil

Lubricating oil

Crude oil vapors from boiler

Steam

Residue

(b)

TABLE 18-4 *Typical petroleum fractions*

Number of carbon atoms	Classification	Boiling point (°C)
1–5	natural gases	less than 40
6–10	gasoline	40–180
11–12	kerosene	180–230
13–17	light gas oil	230–305
18–25	heavy gas oil	305–405
26–38	lubricants	405–515
over 39	asphalts	over 515

Table 18-4 lists some typical fractions of petroleum. Each consists of alkanes with similar numbers of carbon atoms. The properties of an alkane depend on the number of carbon atoms in the molecule. As the number of carbon atoms increases, the molecular weight increases, and the properties of the alkane change. Thus, alkanes with similar numbers of carbon atoms have similar properties. The low molecular weight alkanes, like methane, are gases; middle range alkanes like octane are liquids; higher molecular weight alkanes are solids.

Although alkanes are relatively unreactive, they undergo combustion readily. Alkanes burn in oxygen to form carbon dioxide and water. For this reason, petroleum and natural gas are used mainly as fuels. However, they also serve as raw materials for other compounds. In fact, over 80% of the organic products in the United States are manufactured either directly or indirectly from petroleum and natural gas.

**18.4
ALKYL GROUPS**

An **alkyl group** is an alkane that is missing one hydrogen atom, including its electron. An alkyl group is a *molecular fragment,* not a real molecule, and it cannot exist by itself. The simplest alkyl group is the methyl group. It is a methane molecule that is missing one hydrogen atom.

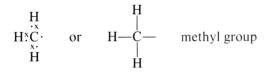

Notice that the carbon atom in the methyl group has only seven outer electrons. It must form another covalent bond to fill the outer level. The open dash in the structural formula indicates that another atom must bond here in order for a molecule to form. For example, the methyl group can be attached to a chlorine atom to form a molecule of methyl chloride:

TABLE 18-5 *Alkyl groups*[a]

Name	Formula
methyl	CH_3-
ethyl	CH_3CH_2-
propyl	$CH_3CH_2CH_2-$
butyl	$CH_3CH_2CH_2CH_2-$
pentyl (amyl)	$CH_3CH_2CH_2CH_2CH_2-$
hexyl	$CH_3CH_2CH_2CH_2CH_2CH_2-$

[a]Only normal (straight-chain) groups are shown.

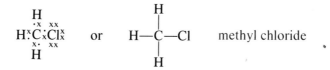

methyl chloride

Table 18-5 lists the first six simple alkyl groups. The names of alkyl groups all end in "-yl." The first part of each name comes from the name of the alkane from which the alkyl group is derived. For example, the group derived from the two carbon alkane, called ethane, is called the ethyl group.

eth	yl
based on	ending
ethane	for alkyl
(2 carbons)	groups

Most of the important organic compounds contain one or more alkyl groups as part of their molecules.

18.5 ISOMERS

The hydrocarbons described so far consist of straight chains of carbon atoms. However, not all hydrocarbons are straight chains. Some are branched chains, in which a carbon atom is bonded to more than two other carbon atoms. The following structural formulas represent four different branched chain alkanes.

A $CH_3-CH-CH_2-CH_2-CH_3$
         $|$
         CH_3

B $CH_3-CH_2-CH-CH_2-CH_3$
              $|$
              CH_3

         CH_3
         $|$
C $CH_3-C-CH_2-CH_3$
         $|$
         CH_3

    CH_3 CH_3
    $|$ $|$
D $CH-CH$
    $|$ $|$
    CH_3 CH_3

Notice that when a chain is branched, you cannot draw a line through all of the carbon atoms in the structural formula without either lifting your pencil off the page or retracing part of the line. Each of these four branched chain alkanes contains 6 carbon atoms and 14 hydrogen atoms. Therefore, they each have the molecular formula C_6H_{14}. However, the atoms in the four molecules are not connected to each other in the same way. Therefore, the molecules have different properties. They are **isomers**, molecules that have the same molecular formula, but different structures.

These four isomers have the same molecular formula as the straight chain alkane called hexane:

E $CH_3CH_2CH_2CH_2CH_2CH_3$

However, the five isomers (**A** through **E**) cannot all be named hexane because each has a different structure and different properties. The nomenclature recommended by the International Union of Pure and Applied Chemistry, IUPAC (called yoo'pak), is to name isomers after the longest unbranched carbon chain in the molecule. For example, isomer **E** is a hexane since it contains six carbon atoms in a straight chain. (It is called normal hexane, or *n*-hexane, because there are no branches in the carbon chain.) Isomers **A** and **B** are pentanes because the longest chain in each has five carbon atoms. The carbon atoms in the longest chain are numbered in sequence. Carbon atom branches are named as alkyl groups, and are given the number of the carbon atom to which they are attached. For example, isomer **A** is called 2-methylpentane, because a methyl group is attached to the second carbon atom of a five carbon chain.

$$\textbf{A} \quad \overset{1}{C}H_3 - \overset{2}{C}H - \overset{3}{C}H_2 - \overset{4}{C}H_2 - \overset{5}{C}H_3 \leftarrow \text{five carbon chain}$$
$$\qquad\qquad \underset{\displaystyle CH_3}{|} \leftarrow \text{methyl group on second carbon}$$

Note that this structure for isomer **A** is identical to the following structure:

$$\overset{5}{C}H_3 - \overset{4}{C}H_2 - \overset{3}{C}H_2 - \overset{2}{C}H - \overset{1}{C}H_3$$
$$\qquad\qquad\qquad \underset{\displaystyle CH_3}{|}$$

This structure is also called 2-methylpentane, but notice that in this case we have numbered the carbon atoms in the longest chain from right to left, rather than from left to right as before. According to IUPAC rules, a carbon chain must always be numbered so that the carbon atoms that have branches attached receive the *lowest* possible numbers. It would therefore be *incorrect* to number this molecule from left to right, and name it 4-methylpentane. Isomer **B** is named 3-methylpentane.

$$
\begin{array}{cccccc}
& \overset{1}{} & \overset{2}{} & \overset{3}{} & \overset{4}{} & \overset{5}{}
\end{array}
$$

B CH_3—CH_2—CH—CH_2—CH_3 ← five carbon chain

CH_3 ← methyl group on third carbon

The methyl group is attached to the third carbon atom on a five carbon chain.

Isomer **C** is named as a butane (see Table 18-3) because the longest unbroken chain has four carbon atoms. Two methyl groups are attached to the second carbon atom.

$$
CH_3
$$

C $\overset{1}{CH_3}$—$\overset{2}{C}$—$\overset{3}{CH_2}$—$\overset{4}{CH_3}$ ← four carbon chain

CH_3 ← two methyl groups on second carbon

The name of this isomer is 2,2-dimethylbutane; the prefix "di" means that both alkyl groups on carbon number 2 are the same. Note that the numbers are separated by a comma and followed by a hyphen in front of the name.

The name for isomer **D** is slightly more complicated. If you trace the longest possible carbon chain (without backtracking or lifting your pencil off the page), you will see that it is four carbons long.

$$
\overset{1}{CH_3} \quad CH_3
$$
$$
\overset{2}{CH}—\overset{3}{CH}
$$
$$
CH_3 \quad \overset{4}{CH_3}
$$

The structural formula of isomer **D** can be redrawn so that the four carbon chain is more obvious.

$$
\begin{array}{cccc}
\overset{1}{} & \overset{2}{} & \overset{3}{} & \overset{4}{}
\end{array}
$$

D CH_3—CH—CH—CH_3 ← four carbon chain

$CH_3 \quad CH_3$ ← methyl groups on second and third carbons

Note that the connections between atoms have not been changed, only the way in which the structure is written. The longest chain has four carbon atoms and a methyl group is bonded to carbon 2 and to carbon 3. Isomer **D** is named 2,3-dimethylbutane.

As the number of atoms in a molecule increases, the number of

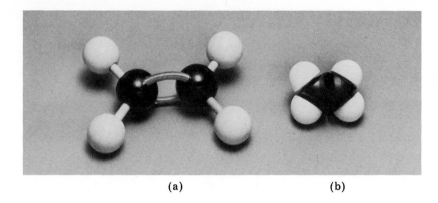

FIGURE 18-7

Ethene (ethylene),
$CH_2{=}CH_2$. *A ball-and-stick*
model (a) and a space-filling
model (b) are shown. (Photos
by Al Green.)

(a)   (b)

possible isomers also increases. For example, the 15 carbon alkane $C_{15}H_{32}$ has nearly 5000 isomers. The existence of isomers contributes to the tremendous variety of organic compounds.

**18.6
ALKENES
AND ALKYNES**

Alkanes are hydrocarbons in which the carbon atoms are all connected by single bonds. Alkenes and alkynes, on the other hand, are hydrocarbons that contain multiple bonds. An **alkene** (or olefin) contains a double bond between two carbon atoms. The simplest alkene is ethene, C_2H_4, which is commonly called ethylene. Figure 18-7 shows two models of ethene.

$$\begin{array}{cc} H & H \\ \overset{x\cdot}{C}\!\!:\!\!\overset{\cdot x}{C} \\ \cdot x & x\cdot \\ H & H \end{array} \quad \text{or} \quad CH_2{=}CH_2 \quad \text{ethene (ethylene)}$$

Each carbon atom receives two electrons from the double bond to the other carbon atom, plus two electrons from single bonds to two hydrogen atoms, and thus obtains the four electrons needed to complete its outer level.

In contrast to ethane, ethene molecules are flat, and the carbon atoms are bonded in a different way. The 2s orbital on each carbon atom in ethene mixes with only two of the 2p orbitals to form sp^2 (read S-P-two) hybrid orbitals. As shown in Figure 18-8, the three resulting orbitals are 120° apart. Each of the three orbitals is involved in a sigma bond, one with the other carbon atom and two with the hydrogen atoms. The remaining p orbitals on both carbon atoms overlap side-to-side, forming a pi (π) bond, located above and below the line connecting the centers of the carbon nuclei. The double bond between the carbon atoms thus consists of a sigma bond plus a pi bond. To better represent the actual geometry of this molecule, the following structural formula is often used:

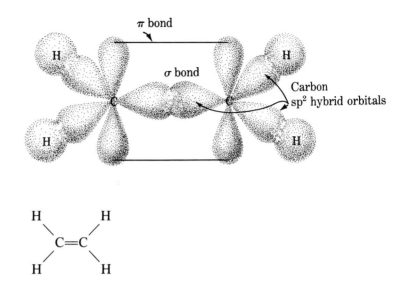

FIGURE 18-8

The bonding of ethene (ethylene). The double bond consists of a sigma (σ) bond formed by the overlapping of sp^2 orbitals and a pi (π) bond formed by the overlapping of p orbitals.

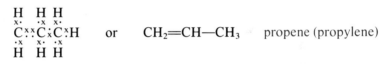

The second member of the alkene family is propene (or propylene), C_3H_6.

$$\begin{array}{ccc} H & H & H \\ \overset{x\cdot}{\underset{\cdot x}{C}} \overset{\cdot x}{\underset{x\cdot}{{}^{x}_{x}}} \overset{x\cdot}{\underset{\cdot x}{C}} {}^{x}_{x} \overset{}{C} {}^{x}H \\ H & H & H \end{array} \quad \text{or} \quad CH_2{=}CH{-}CH_3 \quad \text{propene (propylene)}$$

Other members of the alkenes are listed in Table 18-6. They all have one double bond between two carbon atoms. All the names of the alkenes end in "-ene." Some of these molecules have common names (given in parentheses) in addition to their formal IUPAC names. The general formula of an alkene is C_nH_{2n}; the number of hydrogen atoms is twice the number of carbon atoms in the molecule.

Alkenes offer additional possibilities for isomer formation. Starting

TABLE 18-6 *Simple alkenes*

Molecular formula	Name[a]	Structural formula
C_2H_4	ethene (ethylene)	$CH_2{=}CH_2$
C_3H_6	propene (propylene)	$CH_2{=}CHCH_3$
C_4H_8	butene (butylene)	$CH_2{=}CHCH_2CH_3$[b]
C_5H_{10}	pentene	$CH_2{=}CHCH_2CH_2CH_3$[b]
C_6H_{12}	hexene	$CH_2{=}CHCH_2CH_2CH_2CH_3$[b]

[a]Common name in parentheses.
[b]One of several possible isomers.

with four carbon butene, the double bond can appear in different positions in the molecule as shown:

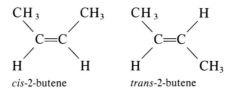

$$\overset{1}{C}H_2\!\!=\!\!\overset{2}{C}H\!-\!\overset{3}{C}H_2\!-\!\overset{4}{C}H_3 \qquad \overset{1}{C}H_3\!-\!\overset{2}{C}H\!\!=\!\!\overset{3}{C}H\!-\!\overset{4}{C}H_3$$

1-butene   2-butene

The number of the first carbon atom involved in the double bond is always placed at the beginning of an alkene's name. For example, the butene isomer with a double bond at the end of the chain is called 1-butene, and the isomer with a double bond in the middle of the chain is called 2-butene.

Another type of isomerism depends on the relative positions of atoms on opposite ends of the double bond:

$$\begin{array}{cc} CH_3 \quad CH_3 & CH_3 \quad\quad H \\ \diagdown \quad\diagup & \diagdown \quad\diagup \\ C\!=\!C & C\!=\!C \\ \diagup \quad\diagdown & \diagup \quad\diagdown \\ H \quad\quad H & H \quad\quad CH_3 \end{array}$$

cis-2-butene   *trans*-2-butene

The molecule with alkyl groups on the same side is called the *cis* isomer. The molecule with the alkyl groups diagonally across from each other is called the *trans* isomer. *Cis-trans* isomerism occurs because the double bond is rigid. Atoms cannot rotate around a double bond as they can around a single bond.

Alkynes are hydrocarbons that contain a triple bond between two carbon atoms. The simplest alkyne is ethyne, C_2H_2, which is commonly called acetylene. Models of ethyne are shown in Figure 18-9.

$$H\!:\!C\!\overset{x}{\underset{x}{\cdot}}\!\overset{x}{\underset{x}{\cdot}}\!C\!:\!H \quad\quad \text{or} \quad\quad CH\!\equiv\!CH \quad \text{or} \quad HC\!\equiv\!CH \quad\quad \text{ethyne (acetylene)}$$

As illustrated in Figure 18-10, the 2s and the 2p orbital on each carbon atom are mixed to form sp (read S-P) hybrid orbitals. One end of the sp orbital forms a sigma bond to the second carbon atom, and the opposite end of the same sp orbital forms a bond to a hydrogen atom. The two remaining 2p orbitals on each carbon atom form two sets of

FIGURE 18-9

Ethyne (acetylene), CH≡CH. A ball-and-stick model (a) and a space-filling model (b) are shown. (Photos by Al Green.)

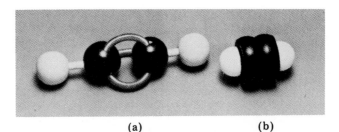

(a)   (b)

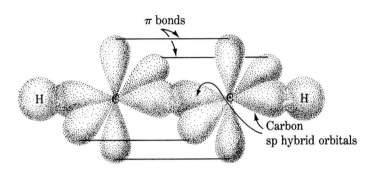

FIGURE 18-10

The bonding in ethyne (acetylene). The triple bond consists of a sigma (σ) bond formed by the overlapping of sp orbitals and two pi (π) bonds formed by the overlapping of p orbitals.

pi bonds with the other carbon atom. The triple bond therefore consists of one sigma bond and two pi bonds. The formal names for all alkynes end in "-yne." The prefixes, based on the number of carbon atoms, are the same as the prefixes used for alkenes and alkanes. The general formula for the alkyne family is C_nH_{2n-2}. This formula shows that the number of hydrogen atoms is two less than twice the number of carbon atoms.

Alkenes and alkynes are called **unsaturated** hydrocarbons because they contain double or triple bonds. Their double or triple bonds allow other molecules to be added to the hydrocarbon chain through a chemical reaction. For example, both an alkene and an alkyne can be converted to the corresponding alkane by an addition (combination) reaction with H_2. This reaction, called hydrogenation, is illustrated for ethene (ethylene) and ethyne (acetylene):

$$CH_2{=}CH_2(g) + H_2(g) \rightarrow CH_3CH_3(g)$$
ethene   hydrogen   ethane

$$CH{\equiv}CH(g) + 2H_2(g) \rightarrow CH_3CH_3(g)$$
ethyne   hydrogen   ethane

The possibility of undergoing hydrogenation and other reactions makes alkenes and alkynes much more chemically active than alkanes. Due in part to this reactivity, ethene (ethylene) is the most produced industrial organic compound. Over 25 billion pounds are manufactured each year. One important use of ethene is in the production of polyethylene, a compound described later in this chapter.

18.7
CYCLIC AND
AROMATIC
HYDROCARBONS

If the two ends of a straight-chain hydrocarbon are connected, a ring or cyclic molecule results. For example, if the two ends of a propane molecule are connected, a cyclic alkane called cyclopropane results. Cyclopropane, C_3H_6, is used as an anesthetic and in organic synthesis.

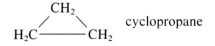
          cyclopropane

Smithsonian Institution: Photo #55587

Kekulé did some of his best work while he was sleeping. His insight into the nature of carbon atoms came to him while he was dozing on a bus. As he recalled, "I watched the carbon atoms link up and form pairs; smaller atoms were embraced; three, four, and more carbon atoms were dancing in rows."

Kekulé became the first to realize that carbon atoms form four bonds and link together in chain-like compounds. Representing the bonds with dashes, he drew "Kekulé structures," which we now call structural formulas. Using these formulas, he showed that structural isomers could exist.

In 1861 this German chemist published a textbook in which he defined organic chemistry simply as the study of carbon compounds. In the past, organic chemistry had been defined as the study of compounds found in living things.

Kekulé is probably best known for proposing the ring structure of benzene, C_6H_6. This compound was discovered by Faraday in 1825, but its structure had remained a mystery. This time, Kekulé's insight came while he was dozing at his desk. He imagined chains of carbon atoms twisting like snakes before his eyes. Suddenly, one seized its own tail to form a ring. As a result of this dream, Kekulé proposed a ring structure for benzene. However, he tested his idea before publishing it, saying: "Let us learn to dream, but let us beware of publishing our dreams before they have been put to the proof of the waking understanding." Kekulé's work has been called "the most brilliant piece of scientific prediction to be found in the whole range of organic chemistry."

Aromatic hydrocarbons are a special type of cyclic compound. The most important aromatic hydrocarbon is benzene, C_6H_6, which can be drawn in either of two ways:

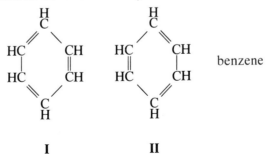

benzene

Each of these structural formulas shows three double bonds and three single bonds in alternating positions around a six carbon ring. Sometimes the following shorthand notation is used for the benzene ring.

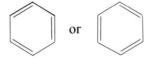

In this notation, each corner (or point at which two lines meet) represents a carbon atom to which a hydrogen atom is attached. Models of benzene are shown in Figure 18-11.

Although benzene looks like an alkene, it does not undergo the typical addition reactions of alkenes. Benzene does not react like an alkene because it does not really contain three distinct double bonds. Instead, all of the carbon-carbon bonds in the molecule are chemically alike, even though structures I and II show distinct double and single bonds. The actual structure of benzene is a combination of forms I and II. The electrons of the three double bonds can be thought of as being spread out, or *delocalized*, over all six carbon atoms in the ring. (See Biography 18 for the origin of this idea.) Each bond is more than a single bond but less than a double bond, and it is said to have partial double bond character. This property, called **resonance**, is characteristic of all aromatic compounds and can be represented by a circle in the center of the aromatic ring.

FIGURE 18-11

Benzene, C_6H_6. A ball-and-stick model (a) and a space filling model (b) are shown. (Photos by Al Green.)

(a)   (b)

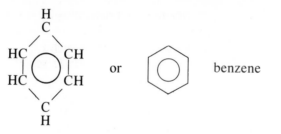 benzene

To summarize, aromatic hydrocarbons are unsaturated molecules that have electrons evenly distributed in a ring system like benzene. (In contrast, hydrocarbons that are *not* aromatic, like those described in previous sections, are called aliphatic.)

Aromatic hydrocarbons undergo substitution reactions. In a substitution reaction, hydrogen atoms are replaced by other atoms. For example, toluene (the solvent in airplane glue) results from the substitution of a methyl group for one hydrogen atom in benzene.

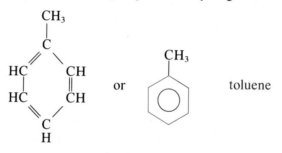 toluene

The atom or group that is substituted for hydrogen is called a **substituent**. When only one hydrogen atom is replaced in a benzene ring, it makes no difference at which position around the ring we draw the substituent. However, when two hydrogen atoms are replaced, three different isomers are possible. For example, the aromatic hydrocarbon xylene (pronounced zy'leen) is formed when two methyl groups replace two hydrogen atoms in benzene. Three different isomers of xylene can form, depending on the relative positions of the methyl groups.

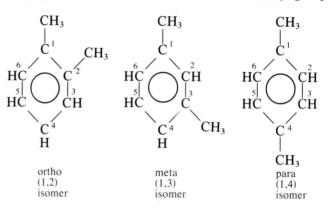

ortho (1,2) isomer   meta (1,3) isomer   para (1,4) isomer

The first isomer, called ortho-xylene, abbreviated *o*-xylene, is 1,2-dimethylbenzene. The two methyl group substituents are on carbon atoms directly next to each other. The second isomer, meta-xylene, abbreviated *m*-xylene, is 1,3-dimethylbenzene; the methyl groups are attached to carbons separated by a single carbon atom. The third possible isomer, para-xylene, abbreviated *p*-xylene, is 1,4-dimethylbenzene; it has the two substituents directly opposite each other across the ring. It does not matter on which carbon atoms the groups are placed as long as they have the correct *relative* positions. For example, each of the following structures represents the same isomer, *p*-xylene.

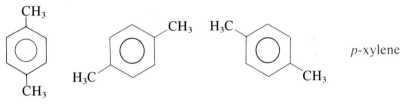

p-xylene

More complicated aromatic hydrocarbons often consist of benzene rings that are "fused" together. These rings have carbon atoms in common. For example, naphthalene, $C_{10}H_8$, formerly used in mothballs, has the following structure:

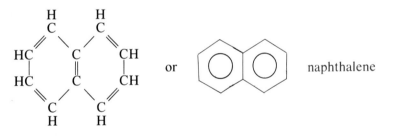

naphthalene

Such molecules are called polynuclear or polycyclic aromatic hydrocarbons. In these hydrocarbons, electrons from the double bonds are spread out over the entire ring system. Some polynuclear aromatic hydrocarbons, like benzopyrene (or benzpyrene), are known to be potent carcinogens (cancer-causing agents).

benzopyrene

Benzopyrene is formed by the incomplete combustion of tobacco, coal, and oil. Its presence in cigarette "tar" may be a factor in the relationship between cigarette smoking and cancer.

The major types of hydrocarbons are summarized in Figure 18-12.

FIGURE 18-12

*Summary of hydrocarbons.
Compounds in parentheses are
examples of the classes listed.*

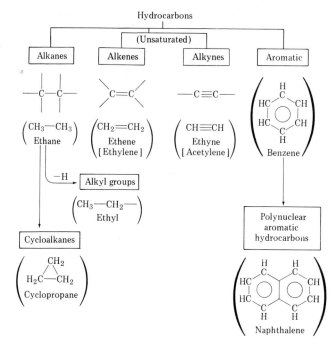

A **functional group** is an atom or group of atoms in a molecule that
gives the molecule its characteristic chemical properties. For example,
the double bond of an alkene can be considered a functional group.
Several important oxygen-containing functional groups are shown in
Table 18-7. These groups cannot exist by themselves, but are bonded
to alkyl groups or to other atoms. The long dashes in the structural
formulas of the functional groups show where these bonds exist. The
functional groups of some compounds, like those of alcohols, aldehydes, and carboxylic acids, form bonds to only one hydrocarbon
group. Functional groups of other compounds, like those of ethers,
ketones, and esters, form bonds with two other hydrocarbon groups.

   Alcohols are compounds that contain the *hydroxyl* or alcohol group,
—OH, covalently bonded to a hydrocarbon.

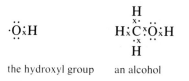

the hydroxyl group   an alcohol

Unlike the hydroxide ion, OH^-, in an ionic compound, the hydroxyl
group in an alcohol does not dissociate; therefore, alcohols are not
basic. The alcohol shown on the right in the preceding diagram consists
of a methyl group, CH_3—, attached to a hydroxyl group. Its common

TABLE 18-7

Major oxygen-containing functional groups

Type of compound	Structural formula of functional group
alcohol	—OH
ether	—O—
aldehyde	$-\overset{\overset{\text{O}}{\|}}{\text{C}}\text{H}$
ketone	$-\overset{\overset{\text{O}}{\|}}{\text{C}}-$
carboxylic acid	$-\overset{\overset{\text{O}}{\|}}{\text{C}}-\text{OH}$
ester	$-\overset{\overset{\text{O}}{\|}}{\text{C}}-\text{O}-$

name is methyl alcohol, and the IUPAC name is methanol. The IUPAC system for naming alcohols is to add the ending "-ol" to a root that indicates the number of carbons. These roots consist of the name of the corresponding alkane without the final "-e." They are listed in Table 18-8. As in branched alkanes, a number in front of the name indicates to which carbon atom the hydroxyl group is bonded. Thus a three carbon alcohol with the hydroxyl group at the middle carbon atom is called 2-propanol.

$$\overset{1}{\text{CH}_3}-\overset{\overset{2}{}}{\underset{\underset{\text{OH}}{\|}}{\text{CH}}}-\overset{3}{\text{CH}_3} \quad \text{2-propanol}$$

The common name of this liquid is isopropyl alcohol. Isopropyl alcohol is used for sponge baths, and is also known as rubbing alcohol.

When people use the word alcohol, they are usually referring to ethyl alcohol or ethanol.

$$\text{CH}_3\text{CH}_2\text{OH} \quad \text{ethanol (ethyl alcohol)}$$

This alcohol, present in beer (3–6%), wine (10–15%) and liquor (40–50%), is the only alcohol that is relatively safe to drink in limited quantities. Since 100% ethanol ("absolute" ethanol) is defined as 200 proof, a bottle marked 80 proof is actually 40% ethanol in water. Today, gasohol is being considered as an alternate fuel for automobiles. It generally consists of 10% ethanol, made from grain, and 90% unleaded gasoline. Denatured alcohol is ethanol that contains methanol, or other poisons. Phenol is an aromatic alcohol used as a disinfectant.

TABLE 18-8 *Roots for formal names of organic compounds*

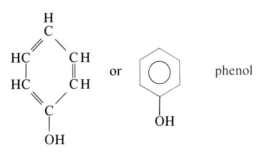

phenol

Number of carbons	Root
1	methan-
2	ethan-
3	propan-
4	butan-
5	pentan-
6	hexan-

Ethers contain an oxygen atom covalently bonded to two hydrocarbon groups. (In this case a single atom, O, is the functional group.) Diethyl ether, an anesthetic, if often simply referred to as ether.

$$\text{H}\overset{\cdot\text{x}}{\underset{\text{x}\cdot}{\text{C}}}\overset{\text{H}}{\underset{\text{H}}{\overset{\text{x}\cdot}{\underset{\cdot\text{x}}{\text{C}}}}}\overset{\cdot\cdot}{\underset{\cdot\cdot}{\text{O}}}\overset{\text{H}}{\underset{\text{H}}{\overset{\cdot\text{x}}{\underset{\text{x}\cdot}{\text{C}}}}}\overset{\text{H}}{\underset{\text{H}}{\overset{\cdot\text{x}}{\underset{\text{x}\cdot}{\text{C}}}}}\text{H} \qquad \text{CH}_3\text{CH}_2-\text{O}-\text{CH}_2\text{CH}_3 \qquad \text{diethyl ether}$$

Two important cyclic ethers are tetrahydrofuran (THF) and dioxane. They are both organic solvents.

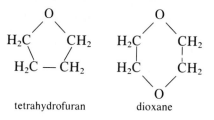

tetrahydrofuran   dioxane

The remaining oxygen-containing functional groups all contain a carbon atom double bonded to an oxygen atom. The structural formula of the *carbonyl* group for example, is:

:O:
  ×:
  ×:
 ×C×   the carbonyl group

$$\overset{O}{\underset{|}{\overset{||}{-C-}}}$$

Aldehydes contain a hydrogen atom bonded to one side of a carbonyl group. For example, ethanal, commonly called acetaldehyde, is an aldehyde.

$$\underset{CH_3CH}{\overset{O}{\overset{||}{}}}$$ ethanal (acetaldehyde)

Just as for alcohols, systematic names of the aldehydes are based on roots indicating the number of carbons. The ending for aldehydes is "-al." The roots for the common names are:

Number of carbon atoms	Common name root
1	form-
2	acet-
3	propion-
4	butyr-

and the common name ending is "-aldehyde." For example, the common name for the one carbon aldehyde is formaldehyde, and the IUPAC name is methanal.

$$\underset{HCH}{\overset{O}{\overset{||}{}}}$$ methanal (formaldehyde)

A 40% aqueous solution of formaldehyde, called formalin, is used as a preservative for biological samples and as a disinfectant.

Ketones also contain a carbonyl group, but they differ from aldehydes in that the carbonyl group has hydrocarbon groups bonded on both sides. The simplest and most important ketone is acetone.

$$\overset{\displaystyle O}{\underset{\displaystyle \|}{}}$$
CH₃CCH₃ propanone (acetone)

The systematic names of ketones end with "-one." Therefore, the systematic name of this three carbon ketone is propanone. Acetone is an excellent solvent and is used frequently in industry and the laboratory.

The *carboxyl* group is a combination of the carbonyl group with a hydroxyl group.

:O:
x·
x·
×C×O×H   —C—OH   the carboxyl group

The bond between the hydrogen and oxygen atoms is weakened by the carbonyl group. Compounds containing the carboxyl group are therefore proton donors and are called **carboxylic acids**. We have already seen one example of a carboxylic acid, acetic acid, which is present in vinegar.

$$CH_3\overset{O}{\overset{\|}{C}}-OH(l) + H_2O(l) \rightleftarrows CH_3\overset{O}{\overset{\|}{C}}-O^-(aq) + H_3O^+(aq)$$
acetic acid   water   acetate ion   hydronium ion

We can now see why one hydrogen atom was written separately in the formula $HC_2H_3O_2$. Only the hydrogen atom attached to oxygen can ionize. The common names of acids are written by adding the ending "-ic acid" to the same common roots that are used for aldehydes. Systematic names consist of the formal root followed by "-oic acid." For example, the systematic name for acetic acid is ethanoic acid. The one carbon acid, called formic or methanoic acid, is produced by certain insects and is responsible for the irritation caused by some insect bites.

$$HC\overset{O}{\overset{\|}{}}-OH \quad \text{methanoic (formic) acid}$$

The functional group that results from the ionization of a carboxylic acid is called a carboxylate group.

$$-C\overset{O}{\overset{\|}{}}-O^- \quad \text{the carboxylate group}$$

The resulting anion is named by dropping the ending "-ic acid" and replacing it by "-ate ion." Thus the anion formed by acetic acid is called acetate ion (or ethanoate ion in the IUPAC system).

605

An **ester** results from the reaction of a carboxylic acid with an alcohol. It contains an ester group.

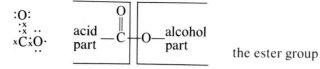

the ester group

The carbon end of the ester group comes from the acid, and the oxygen end comes from the alcohol. An example of esterification is the formation of methyl acetate from methyl alcohol and acetic acid.

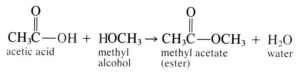

Note that the name of an ester consists of two words. The first word is the name of the hydrocarbon group of the alcohol; in our example, it is methyl from methyl alcohol. The second word is the name of the carboxylate group of the acid; in our example, it is acetate from acetic acid. The IUPAC name is methyl ethanoate. Esters are used as flavorings and fragrances. Ethyl formate, produced by the reaction of formic acid with ethyl alcohol, gives a rum-like flavor to candies.

$$\overset{\displaystyle O}{\underset{}{H\overset{\|}{C}}}{-}OH + HOCH_2CH_3 \rightarrow H\overset{O}{\overset{\|}{C}}{-}OCH_2CH_3 + H_2O$$

formic acid   ethyl alcohol   ethyl formate (ester)   water

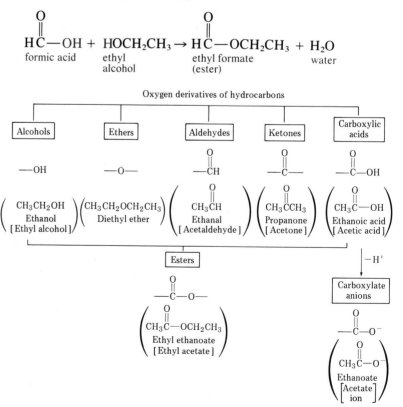

FIGURE 18-13

Summary of oxygen derivatives of hydrocarbons. Compounds shown in parentheses are examples of the classes listed.

Of all the oxygen-containing organic compounds, ethers are the least reactive. Alcohols, on the other hand, can be oxidized to aldehydes or ketones, which in turn may be oxidized further to carboxylic acids. Carboxylic acids undergo typical reactions of acids, such as neutralization. Esters can be separated into their acid and alcohol components by a reaction with water called hydrolysis.

Figure 18-13 summarizes the types of oxygen-containing organic compounds described in this section.

18.9 OTHER ORGANIC DERIVATIVES AND POLYMERS

Other elements besides carbon, hydrogen, and oxygen are often found in organic compounds. These elements can be substituted into a hydrocarbon in various ways. These substitutions result in classes of compounds containing functional groups that give them characteristic properties. For example, atoms of the halogens can replace the hydrogen atoms in a hydrocarbon to form **alkyl halides**. Chlorine derivatives of methane, illustrated in Table 18-9, are examples of this class of compounds. The pesticide DDT (*di*chloro*di*phenyl*t*richloroethane) is a more complicated chlorinated hydrocarbon.

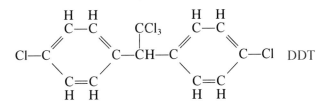

TABLE 18-9 *Chlorine derivatives of methane*

Formula	Lewis structure	Name[a]	Typical use
CH_3Cl	H H×C×Cl× H	chloromethane (methyl chloride)	silicone production
CH_2Cl_2	Cl H×C×Cl× H	dichloromethane (methylene chloride)	grease and paint remover, solvent
$CHCl_3$	Cl H×C×Cl× Cl	trichloromethane (chloroform)	solvent, formerly an anesthetic
CCl_4	Cl Cl×C×Cl× Cl	tetrachloromethane (carbon tetrachloride)	fluorochlorocarbon synthesis, solvent

[a]Common name in parentheses.

Amines are nitrogen-containing organic compounds. They are produced by replacing one, two or all three hydrogen atoms of ammonia, NH_3, by hydrocarbon groups. For example, methyl group substitution results in the following amines:

CH_3NH_2 $(CH_3)_2NH$ $(CH_3)_3N$
methylamine   dimethylamine   trimethylamine

The functional group in amines is called the **amino group**; it often appears as $-NH_2$. An important class of compounds, the **amino acids**, contain both an amino group and a carboxylic acid group. A simple example is glycine.

$$H_2NCH_2\overset{\displaystyle O}{\overset{\|}{C}}-OH$$ glycine (an amino acid)

Box 18

Drugs

Drugs are chemical substances that affect the body. In medicine, they are used for diagnosis, preventing or curing disease, treating the symptoms of disease, and birth control. Most drugs act by stimulating or depressing certain activities in the body cells, by replacing a deficient substance, or by killing or weakening foreign organisms.

Aspirin is probably the most common nonprescription drug used for medical purposes. Its chemical name is acetylsalicylic acid.

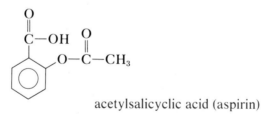

acetylsalicyclic acid (aspirin)

Aspirin is an analgesic, a drug that relieves pain. It is also antipyretic, meaning that it lowers elevated body temperature. An aspirin substitute that also has these properties is acetaminophen, which is sold under such names as Tylenol.

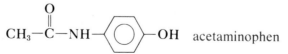

Aspirin, but not acetaminophen, is antiinflammatory, reducing swelling and redness.

One of the most common prescription drugs is diazepam (Valium), an antianxiety drug.

Like ammonia, amines are basic because the nitrogen atom has an unshared pair of electrons (called a lone pair). The nitrogen atom can accept a proton to form a substituted ammonium ion.

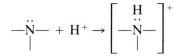

Thus methylamine can gain a proton to form the methylammonium ion.

$$CH_3NH_2(aq) + H^+(aq) \rightarrow CH_3NH_3^+(aq)$$

methylamine   hydrogen   methylammonium
                     ion   ion

diazepam (Valium)

Diazepam has a calming effect on the human nervous system, but causes relatively little sleepiness or interference with motor activities.

Antibiotics are a class of prescription drugs that are used to treat infectious diseases. Antibiotics are produced by living plant or animal cells, and act by selective toxicity. In other words, although they are toxic to most cells, they are more harmful to the invading organisms than they are to the patient. The penicillins are the most potent anti-bacterial drugs. Penicillin G (benzyl penicillin) is the most commonly used form of penicillin.

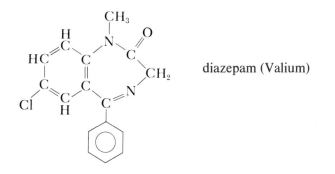

penicillin G

Many drugs are administered in their ammonium form because they are much more soluble in aqueous solution. A substituted ammonium ion with four organic groups around the nitrogen atom is called a quaternary ammonium ion, or "quat" for short.

The reaction of an amine with a carboxylic acid produces an **amide**, which contains the following functional group:

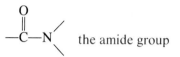
the amide group

For example, if acetic acid reacts with ammonia, the product is acetamide; its IUPAC name is ethanamide.

$$CH_3\overset{\displaystyle O}{\overset{\|}{C}}-OH(aq) + NH_3(aq) \rightarrow CH_3\overset{\displaystyle O}{\overset{\|}{C}}-NH_2(aq) + H_2O(l)$$

acetic acid   ammonia   acetamide   water

One of the most important amides is urea, an end product of the breakdown of nitrogen-containing molecules in the body.

$$H_2N-\overset{\displaystyle O}{\overset{\|}{C}}-NH_2 \quad \text{urea}$$

Like oxygen, nitrogen can replace carbon in cyclic molecules. The resulting molecule, called a **heterocycle**, contains two or more different kinds of atoms arranged in a ring. Pyrrole and pyridine (both used for making drugs) are examples of nitrogen heterocycles.

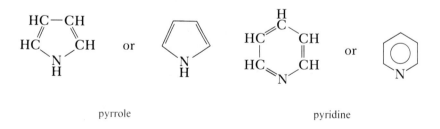

pyrrole   pyridine

Tetrahydrofuran and dioxane, mentioned previously, are examples of oxygen heterocycles.

Alkaloids are nitrogen compounds of vegetable origin that are physiologically active. Alkaloids usually contain a nitrogen heterocycle. A familiar example of an alkaloid is caffeine, which is found in coffee, tea and cocoa. Cocaine, codeine, morphine, and nicotine are also alkaloids.

FIGURE 18-14

Polymer. A polymer is a long molecule made up of repeating units called monomers.

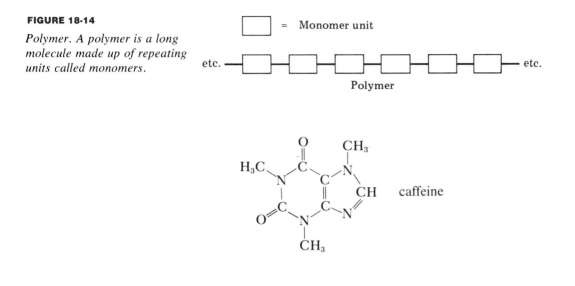

□ = Monomer unit

etc. Polymer

caffeine

Organic **polymers** ("poly" means many) are very large molecules produced by bonding together many small molecules called monomers, as shown in Figure 18-14. Polymers can be made by addition reactions, in which all the monomer atoms become part of the polymer, or by condensation reactions, in which some of the monomer atoms are removed upon bonding. Plastics contain solid polymers that have the ability to flow into a desired shape when heat and pressure are applied. They keep that shape when the heat and pressure are withdrawn. The most widely used plastic is polyethylene, an addition polymer of ethylene.

$$-CH_2-CH_2-CH_2-CH_2-CH_2-CH_2-CH_2-CH_2- \text{ polyethylene}$$

(Color is used here to show the original ethylene monomer units in the long chain polymer.) Polyethylene is used in making squeeze bottles, film packaging, tubing, and electrical insulation. Polyvinyl chloride, abbreviated PVC, is similar to polyethylene. Its monomer, vinyl chloride, contains a chlorine atom in place of one of the hydrogen atoms. PVC is a clear, flexible polymer that is used to make plumbing, raincoats, toys, containers, and flooring.

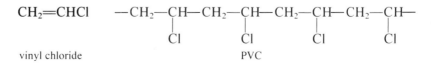

$CH_2=CHCl$

vinyl chloride

PVC

Polystyrene, a foam-like plastic, is based on a styrene monomer that contains an aromatic ring.

TABLE 18-10 *Summary of major types of organic compounds*

Class	Functional group (identifying characteristic)	Example Formula	Name[a]
alkane	saturated hydrocarbon, $-\overset{\mid}{\underset{\mid}{C}}-\overset{\mid}{\underset{\mid}{C}}-$	CH_3CH_3	ethane
alkene	hydrocarbon (double bond), $C{=}C$	$CH_2{=}CH_2$	ethene (ethylene)
alkyne	hydrocarbon (triple bond), $-C{\equiv}C-$	$CH{\equiv}CH$	ethyne (acetylene)
cycloalkane	cyclic saturated hydrocarbon	(ring structure)	cyclohexane
aromatic hydrocarbon	based on a benzene type of ring system	(ring structure)	benzene
alcohol	hydroxyl group, $-OH$	CH_3CH_2OH	ethanol (ethyl alcohol)
ether	oxygen atom, $-O-$	$CH_3CH_2OCH_2CH_3$	diethyl ether
aldehyde	carbonyl group with one alkyl group, $-\overset{O}{\overset{\|}{C}}H$	$CH_3\overset{O}{\overset{\|}{C}}H$	ethanal (acetaldehyde)

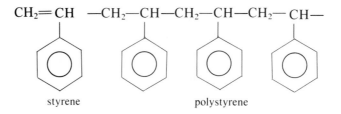

styrene   polystyrene

Fibers are long, slender polymers. Nylon was the first fiber to be synthesized in the laboratory. It is a polymer made by a condensation reaction such as the following:

$$HO-\overset{O}{\overset{\|}{C}}CH_2CH_2CH_2CH_2\overset{O}{\overset{\|}{C}}-OH \ + \ H_2NCH_2CH_2CH_2CH_2CH_2CH_2NH_2 \rightarrow$$

adipic acid   hexamethylenediamine

TABLE 18-10 *Summary of major types of organic compounds*

Class	Functional group (identifying characteristic)	Example Formula	Name[a]
ketone	carbonyl group with two alkyl groups, $-\overset{\overset{\displaystyle O}{\|}}{C}-$	$CH_3\overset{\overset{\displaystyle O}{\|}}{C}CH_3$	propanone (acetone)
carboxylic acid	carboxyl group, $-\overset{\overset{\displaystyle O}{\|}}{C}-OH$	$CH_3\overset{\overset{\displaystyle O}{\|}}{C}-OH$	ethanoic acid (acetic acid)
ester	$-\overset{\overset{\displaystyle O}{\|}}{C}-O-$ group	$CH_3\overset{\overset{\displaystyle O}{\|}}{C}-OCH_3$	methyl ethanoate (methyl acetate)
alkyl halide	halogen atom	CH_3CH_2Cl	chloroethane (ethylene chloride)
amine	nitrogen with one, two, or three alkyl groups	$CH_3CH_2NH_2$	ethylamine
amide	$-\overset{\overset{\displaystyle O}{\|}}{C}-N\diagup$ group	$CH_3\overset{\overset{\displaystyle O}{\|}}{C}-NH_2$	ethanamide (acetamide)
nitrogen heterocycle	N atom replacing carbon in ring	(pyridine ring structure)	pyridine
polymer	large number of repeating units, or monomers	$-[CH_2-CH_2]_n$	polyethylene

[a]Common name in parentheses.

$$-\overset{\overset{\displaystyle O}{\|}}{C}CH_2CH_2CH_2CH_2\overset{\overset{\displaystyle O}{\|}}{C}-NHCH_2CH_2CH_2CH_2CH_2CH_2NH- + H_2O$$

nylon monomer

The most popular synthetic fibers are polyesters. Polyesters are polymers containing esters. Dacron is a well-known polyester, which is made by the following reaction:

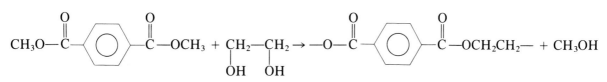

dimethylterephthalate   ethylene glycol   Dacron monomer   methanol

Elastomers are polymers that have the properties of rubber. They can be stretched, but then return rapidly to their original size. Natural rubber is an elastomer formed from the monomer 2-methyl-1,3-butadiene (isoprene).

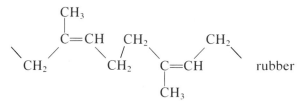

Synthetic rubbers, such as Neoprene and styrene-butadiene rubber (SBR), are made from monomers that are related to isoprene.

Table 18-10 (pages 612–613) summarizes the organic compounds that have been described so far in this chapter.

Part B   Biological chemistry

18.10 CARBOHYDRATES

Carbohydrates are organic molecules that contain only carbon, hydrogen, and oxygen atoms. They are related to either aldehydes or ketones, and contain hydroxyl groups. Thus, they can be defined as polyhydroxyl aldehydes or polyhydroxyl ketones, or their derivatives.

Carbohydrates are classified according to their size. The smallest carbohydrates are called **monosaccharides**. Larger carbohydrates consist of repeating monosaccharide units. For example, a **disaccharide** consists of two monosaccharides joined together. **Polysaccharides**, the largest carbohydrates, are polymers consisting of chains of monosaccharide or disaccharide units.

Glucose, $C_6H_{12}O_6$, is the most important monosaccharide. It is the major carbohydrate present in our bloodstreams. Referred to as blood sugar, its normal concentration is 70 to 90 mg glucose per 100 mL blood. It serves as the basic fuel for most body cells. Glucose contains six carbon atoms.

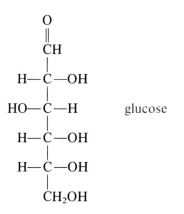

glucose

FIGURE 18-15

Glucose. Three different ways of representing a glucose molecule are shown.

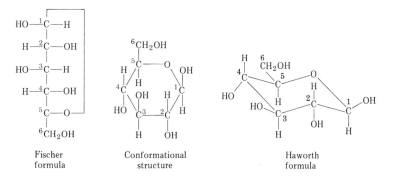

Fischer formula   Conformational structure   Haworth formula

Notice that the first carbon atom of glucose is part of an aldehyde functional group. The remaining carbons have hydroxyl groups attached. Although this straight-chain formula is easy to draw, most glucose molecules in solution actually are cyclic. They exist as six-membered rings. These closed-chain forms of glucose are represented in different ways in Figure 18-15.

Table sugar is a disaccharide called sucrose. Sucrose consists of glucose linked to another monosaccharide, fructose.

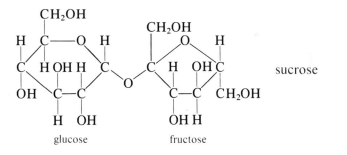

sucrose

glucose   fructose

On the average, each person in the United States eats about 100 pounds of sucrose per year.

Starch is the major carbohydrate in the human diet. It is a mixture of two polysaccharides: amylose, a straight-chain polymer of glucose, and amylopectin, a branched polymer of glucose. When starch is digested in the body, the bonds between the monomer units in these two polymers are broken, producing free glucose. When excess glucose is present in the body, part of it is stored as glycogen, a glucose polymer that is similar in structure to amylopectin.

Cellulose is another straight-chain polymer of glucose. But the monomers in cellulose are bonded in a slightly different manner than they are in amylose, making it impossible for our digestive enzymes to break them apart. Therefore we cannot digest this polymer—it is the roughage or "fiber" in our diet. Cellulose is the most abundant organic substance found in nature. For example, cotton consists almost entirely of cellulose, and wood is 40 to 50% cellulose.

Lipids are organic molecules that dissolve in certain organic solvents such as ether, chloroform, and carbon tetrachloride. Lipids are found in both plants and animals. Many lipids are special types of esters that are formed from an alcohol and **fatty acids**. A fatty acid consists of a straight chain of carbon atoms with a carboxylic acid group at one end of the molecule. The two most common *saturated* fatty acids are palmitic acid ($C_{15}H_{31}\overset{\displaystyle O}{\overset{\|}{C}}$—OH), which contains 16 carbon atoms, and stearic acid ($C_{17}H_{35}\overset{\displaystyle O}{\overset{\|}{C}}$—OH), with 18 carbon atoms.

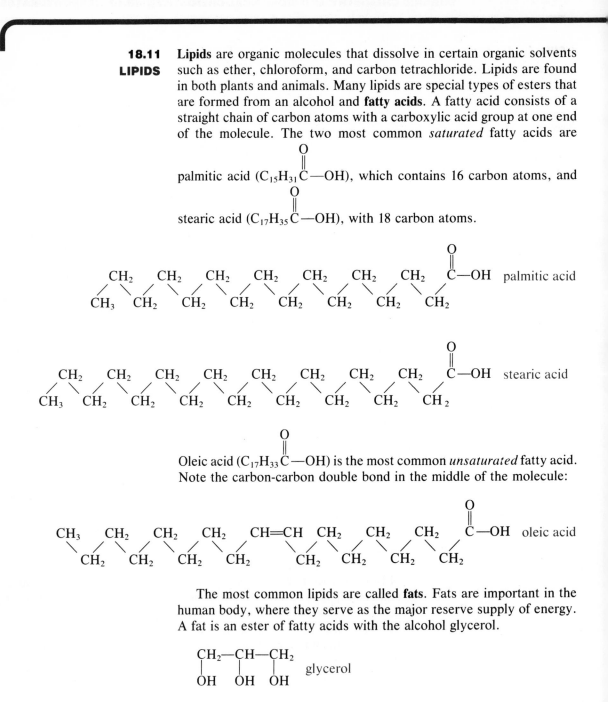

Oleic acid ($C_{17}H_{33}\overset{\displaystyle O}{\overset{\|}{C}}$—OH) is the most common *unsaturated* fatty acid. Note the carbon-carbon double bond in the middle of the molecule:

The most common lipids are called **fats**. Fats are important in the human body, where they serve as the major reserve supply of energy. A fat is an ester of fatty acids with the alcohol glycerol.

$$CH_2\text{—}CH\text{—}CH_2$$
$$\ \ |\qquad |\qquad\ |$$
$$OH\quad OH\quad OH$$ glycerol

For example, if glycerol reacts with three molecules of the fatty acid oleic acid, an ester group, $-\overset{\displaystyle O}{\overset{\|}{C}}-O-$, forms at each of the three hydroxyl groups.

FIGURE 18-16

General structure of a triacylglycerol (triglyceride). The molecule is a triester formed from glycerol and three fatty acids.

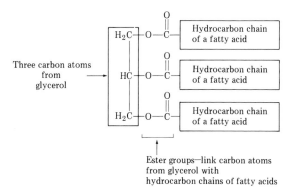

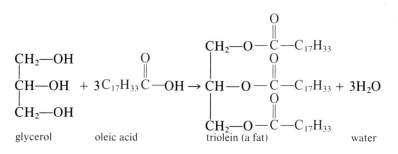

The resulting compound is a fat or **triacylglycerol** (triglyceride). This particular fat is called triolein because it was formed from three molecules of oleic acid.

The general structure of a fat is given in Figure 18-16. A fat can be made from any three fatty acids. In other words, the three fatty acid groups need not all be the same, as they were in triolein. In addition, most naturally occurring fats are actually mixtures of several different triacylglycerols.

Both animals and plants contain fats. Animal fats are generally solid at room temperature and are relatively saturated. For example, butter or milk fat from cows, lard from hogs, and tallow from beef or sheep are solids at room temperature. Vegetable fats, called **oils**, such as corn oil and olive oil, are generally liquid at room temperature and are more unsaturated than animal fats. Vegetable fats are sometimes called polyunsaturated fats because the fatty acid portions of the triacylglycerol molecules contain several double bonds.

Phospholipids, or phosphoglycerides, differ from fats in that one of the hydroxyl groups of glycerol forms an ester with phosphoric acid (H_3PO_4) instead of with a fatty acid. A nitrogen-containing alcohol (such as choline) forms another ester at the free end of the phosphate group. For example, the structural formula of the phospholipid phosphatidylcholine (or lecithin) is the following:

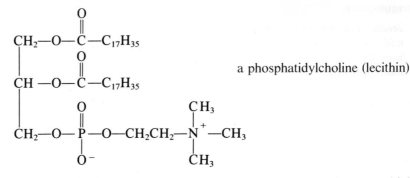

a phosphatidylcholine (lecithin)

Phospholipids are the major component of body cell membranes, which separate cells from their surroundings.

Steroids are another type of lipid. Unlike fats and phospholipids, steroids do not consist of esters. Instead, these molecules are based on a framework of four hydrocarbon rings. Cholesterol is a well-known example of a steroid.

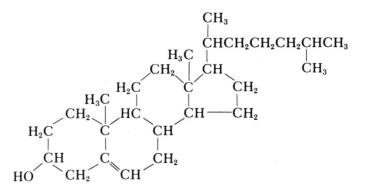

cholesterol (a steroid)

One type of arteriosclerosis or "hardening of the arteries" is associated with deposits of cholesterol inside the walls of the arteries. Other steroids, synthesized from cholesterol, are found throughout the body and are very active biologically. For example, many hormones, which regulate chemical processes in the body, are steroids.

18.12
PROTEINS
AND ENZYMES

Proteins are complex polymers of amino acids. As shown in Figure 18-17, an amino acid contains an amino group and a carboxylic acid group, both bonded to a central carbon atom called the alpha carbon. A side chain consisting of one or more atoms is bonded to the alpha carbon. All of the twenty major amino acids have these structural

FIGURE 18-17

General structure of an amino acid. The side chain differs from one amino acid to another.

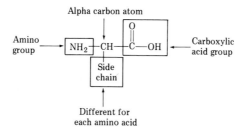

characteristics in common. They differ only in the nature of the side chain, the atom or group of atoms attached to the alpha carbon. These side chains include alkyl and aromatic groups, carboxylic acids, amines, amides, alcohols, and nitrogen heterocycles. For example, the following are the structures of six different amino acids (the side chains are shown in color):

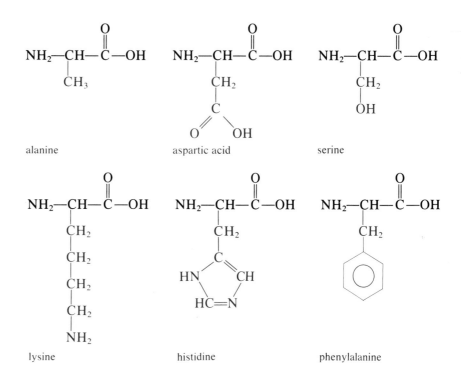

In a protein, amino acids are joined as a result of the reactions between the amino group of one amino acid and the carboxylic acid group of another. The resulting chain of carbon and nitrogen atoms forms the backbone of the protein molecule. These linkages, shown in

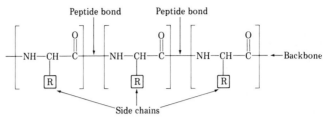

FIGURE 18-18

General structure of a polypeptide. In a polypeptide, amino acids are joined by peptide bonds.

Figure 18-18, are called *peptide bonds*. An amino acid polymer is therefore called a **polypeptide**. Proteins are high molecular weight naturally occurring polypeptides.

The properties of a protein are determined by the nature and order of the side chains that are attached to the backbone of the polypeptide. Therefore, its properties depend on the *sequence* of its amino acids. Since there are 20 different amino acids, and proteins contain hundreds of amino acids linked together, the number of different proteins that are possible is huge. The *primary structure* of a protein is the order of amino acids in the polypeptide chain. Sometimes the replacement of a single amino acid by another in a chain of several hundred amino acids can completely change the properties of the protein. Such a substitution sometimes occurs in hemoglobin, the oxygen-carrying protein of red blood cells, and results in the disease sickle cell anemia.

Do-it-yourself 18

Meringue cookies

If we modify the environment of a protein molecule by changing the temperature, raising or lowering the pH, or adding certain chemical compounds, the protein may lose its normal three-dimensional structure. This process is called protein denaturation. When the protein loses its regular shape, it also loses its characteristic properties.

Egg white contains a protein called albumin. When we boil an egg, the albumin denatures. The molecules clump together, forming a solid mass of egg white. When we beat raw egg white, we create a surface film of denatured albumin that makes the egg white stiff. This second process is the basis for preparing meringue.

To do: Preheat an oven to 300°F. Sift one cup of sugar. Add $\frac{1}{8}$ teaspoon of salt to three egg whites and beat them until they are stiff. If you use a thin wire whisk, beat 300 strokes in 2 minutes. If you use an electric mixer, beat at medium speed for $\frac{1}{2}$ minute, and then at high speed for another $\frac{1}{2}$ minute.

Slowly add the sugar while beating constantly. Add one teaspoon of vanilla and $1\frac{1}{4}$ cup of shredded coconut.

Drop one teaspoon of the mixture at a time onto a cookie sheet that has been greased and floured. Bake for 30 minutes.

Denatured albumin is a major ingredient of these tasty meringue cookies!

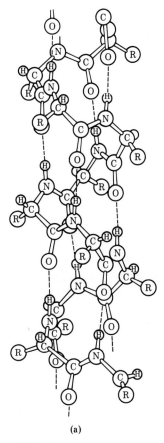

(a)

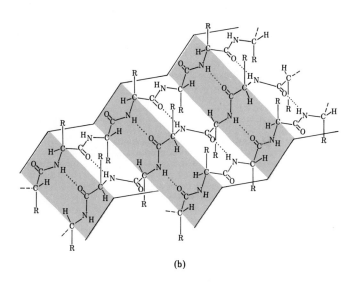

(b)

FIGURE 18-19

Secondary structure in proteins. In both the alpha helix (a) and the pleated sheet (b) arrangements, amino acids are linked by hydrogen bonds.

(This genetic disease is discussed in Section 18.14.)

The *secondary structure* of a protein describes the regular shapes taken by portions of the polypeptide backbone. They can be either helical (like the threads of a screw) or sheet-like, as shown in Figure 18-19. These shapes result from hydrogen bonding (described in Section 12.2) between amino acids. Hydrogen bonds form between the partially positive H atoms that are bonded to nitrogen atoms in the chain and the partially negative oxygen atoms of carbonyl groups in the chain.

The *tertiary structure* of a protein is the overall three-dimensional shape, called the *conformation* of the entire protein. It results from various types of interactions between the amino acids. Most proteins have complex tertiary structures, which give them the special properties needed for their biological roles in a living organism.

Proteins serve many different functions. Structural proteins, like collagen, are fibers that make up most of our connective tissue, cartilage, and bone. Collagen is the most abundant human protein. (It provides the structural framework for the mineral component of bone, hydroxyapatite, which has the formula $Ca_5(PO_4)_3OH$.) Contractile proteins such as actin and myosin are present in muscle, and give it the

FIGURE 18-20

Space-filling model of lysozyme. The arrow indicates the active site in this enzyme molecule. (Photo courtesy of Dr. J.A. Rupley.)

ability to stretch or contract. Transport proteins, like hemoglobin, carry small molecules through the bloodstream. Storage proteins serve as reservoirs for essential chemical substances. For example, iron is stored by the protein ferritin. Protective proteins, such as the antibody globulin, inactivate foreign proteins (called antigens) that invade the body. Some hormones also consist of protein molecules.

Enzymes are proteins that serve as biological catalysts. They increase the rates of chemical reactions in the body. A reactant, which we call the substrate, binds to the enzyme molecule in a special region, called the active site. The active site allows only certain substrates to bind, and therefore makes the enzyme very specific. For example, Figure 18-20 illustrates the active site of the enzyme lysozyme (which breaks the cell walls of bacteria). Once the substrate is attached to the active site of an enzyme, the functional groups of the amino acids in the active site either weaken one or more bonds of the substrate, or participate in a reaction by transferring an electron or proton.

As described in Section 16.4, an enzyme provides an alternate mechanism for the conversion of substrate to product. This alternate mechanism has a lower activation energy than the mechanism for the uncatalyzed reaction. After the reaction takes place, the products are released from the active site. The enzyme is then free to bind another substrate molecule and begin the process again. By repeating this process very rapidly, enzymes catalyze the reactions of millions of substrate molecules per minute.

**18.13
METABOLISM**

In a real sense, we are what we eat. The atoms that make up the molecules of our bodies come from food: carbohydrates, lipids, proteins, and other nutrients. In one year, each of us ingests about 1400 pounds of food. Part of this food is converted into new molecules of our bodies, and part serves as the fuel for our bodies' chemical reactions. Thus food is the source of both matter and energy for the human body. The series of complex chemical steps by which enzymes catalyze all the chemical processes that take place in the body is called **metabolism**.

Metabolism consists of two parts. In **catabolism**, the food molecules that we eat are broken down into smaller molecules, releasing energy. Carbohydrates and proteins produce about 4 kcal/g, and lipids yield about 9 kcal/g. The breakdown of food molecules begins with digestion. During digestion, carbohydrates are broken down from polysaccharides to monosaccharides, lipids (mainly triacylglycerols) to fatty acids and glycerol, and proteins to amino acids. Then each of these types of basic molecules—monosaccharides, fatty acids, and amino acids—can either be reused as they are or broken down further.

Energy is released by the oxidation of these key molecules. For example, the complete oxidation of glucose, which occurs in a complex series of reactions catalyzed by enzymes, can be summarized by the following equation.

$$C_6H_{12}O_6(g) + 6O_2(g) \rightarrow 6CO_2(g) + 6H_2O(l) + \text{energy}$$

glucose   oxygen   carbon   water
                     dioxide

To be useful, the released energy must be stored in a form that can be used later as it is needed. Figure 18-21 shows the energy storage molecule, adenosine triphosphate, or ATP. It consists of a monosaccharide, called ribose, to which a nitrogen heterocycle, called adenine, and a triphosphate group are attached. Energy is stored when ATP is

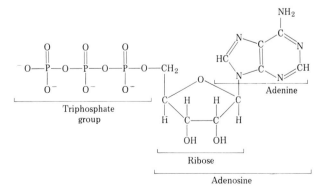

FIGURE 18-21

*Adenosine triphosphate (ATP).
The combination of adenine
and ribose is called adenosine.*

FIGURE 18-22

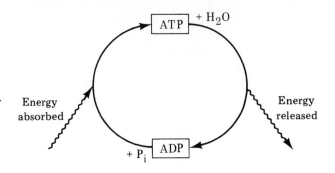

ATP as an energy storage molecule. Energy is stored when adenosine diphosphate (ADP) is converted to adenosine triphosphate (ATP). This energy is released when ATP is converted back to ADP. P_i stands for inorganic phosphate, PO_4^{3-}.

formed from adenosine diphosphate, ADP, and inorganic phosphate, PO_4^{3-}.

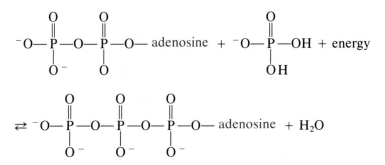

When energy is needed, ATP is broken down by the reverse reaction. The energy stored in the bonds of ATP is released when the terminal phosphate group is removed to form ADP again. Figure 18-22 summarizes this process.

The second part of metabolism, **anabolism**, is the synthesis of the large molecules needed by body cells from the small molecules provided by catabolism. This synthesis requires energy. The ATP formed in catabolism drives the energy-requiring reactions of anabolism. During anabolism, excess glucose is stored as the polysaccharide, glycogen. New triacylglycerols are made, and these are stored as fat, the body's major energy reserve. New proteins, as well as other nitrogen-containing molecules, are synthesized from amino acids. In this way, the atoms from the food molecules eventually reappear as parts of the molecules in our bodies.

18.14
NUCLEIC ACIDS
AND HEREDITY

Nucleic acids are a very important type of biological molecule. They provide the instructions that determine our biological characteristics. The nucleic acid called DNA, deoxyribonucleic acid, contains this information. DNA is present in the central portion, called the nucleus, of every cell in our bodies. It instructs each cell to produce certain proteins. DNA is a tremendously long polymer, much larger than any

FIGURE 18-23

Part of a polynucleotide. A phosphate ester joins carbon 3' of one nucleotide with carbon 5' of the next nucleotide in the chain.

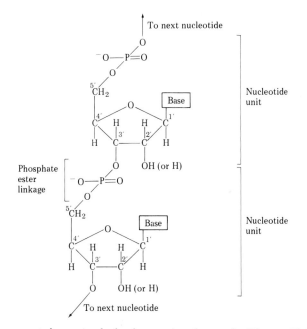

protein or carbohydrate. As shown in Figure 18-23, DNA is a polynucleotide, a polymer of nucleotide units. Each nucleotide consists of a monosaccharide (deoxyribose), a phosphate group, and a "base" that is a nitrogen heterocycle.

Figure 18-24 shows the four different bases found in DNA and the special way that they interact. In a process called *base pairing*, adenine

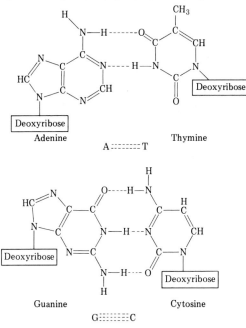

FIGURE 18-24

Base pairing. Hydrogen bonds can form as shown between adenine (A) and thymine (T), and between guanine (G) and cytosine (C).

625

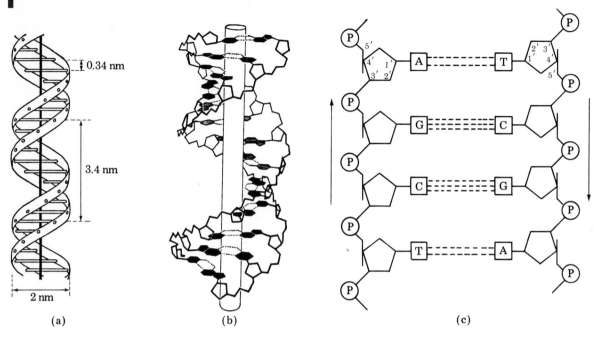

FIGURE 18-25

Deoxyribonucleic acid, DNA. In (a), the phosphate-sugar chains are shown as ribbons and the base pairs as horizontal rods. In (b), the dotted lines show the hydrogen bonds between base pairs. In (c), a portion of the molecule is unfolded (P stands for phosphate group). (Reprinted with permission from J.N. Davidson, The Biochemistry of Nucleic Acids, 7th ed. (London: Chapman and Hall, 1972.)

(A) forms hydrogen bonds to thymine (T), and guanine (G) forms hydrogen bonds to cytosine (C). Because of these specific interactions, adenine is said to be complementary to thymine, and guanine is said to be complementary to cytosine. Adenine and thymine, and guanine and cytosine, are called complementary base pairs.

The DNA molecule contains two parallel polynucleotide chains. These two strands form a spiral arrangement called a double helix, shown in Figure 18-25(a,b). Because of base pairing, the two strands within a DNA molecule are complementary to each other, as shown in Figure 18-25(c).

The order or sequence in which the four nucleotides appear in a DNA molecule is a code that tells the cell how to make its proteins. In a process called *transcription*, this sequence is first copied onto another type of nucleic acid called messenger RNA (ribonucleic acid). RNA molecules are single stranded. They contain the slightly different monosaccharide ribose, instead of deoxyribose, and the related base uracil, instead of thymine. The messenger RNA, formed by base pairing, is a copy of one strand of the DNA molecule. It carries the information obtained from DNA in the nucleus, to the rest of the cell, called the cytoplasm.

Here, in a process called *translation*, the messenger RNA molecules serve as guides for making the body's proteins. Each three nucleotide bases on the messenger RNA form a code word, which signals the

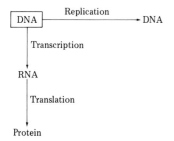

FIGURE 18-26

The central role of DNA. The information contained in DNA can be copied through replication, or used to direct protein synthesis through transcription and translation.

formation of one of the twenty amino acids, or instructs the cell to start or stop protein synthesis. For example, the base sequence guanine-guanine-adenine is the code for the amino acid glycine. Thus *the order of the bases in the messenger RNA, copied from the DNA, makes up a code, which we call the genetic code.* Proteins are put together (synthesized) according to this code. The sequence of amino acids in a protein is thus determined by the base sequence of part of a DNA molecule. Because the proteins of our bodies determine our characteristics, DNA molecules carry the information that controls who we are. The precise sequence of base pairs in the DNA molecules differs from one person to the next. As a result, each person is unique.

All the cells in your body contain the same DNA molecules. Every time new cells form, the DNA molecules must be copied. The process by which a DNA molecule makes a copy of itself is called *replication*. During replication, the two parallel strands of the DNA molecule "unzip." A new strand forms by base pairing along each of the old strands. In this way, two complete molecules form, each identical with the original one. Figure 18-26 compares replication, which produces another DNA molecule, with the processes of transcription and translation, which produce protein molecules.

Our heredity is based on the replication of DNA from our parents. During the process of sexual reproduction, a sperm cell that contains DNA from the father joins an egg cell that contains DNA from the mother. Both the sperm cell and the egg cell contain only half the normal amount of DNA. When they combine to form a fertilized egg, this cell now has the normal amount of DNA. The fertilized egg is the beginning of a new separate organism. Its DNA replicates each time that a cell divides and the organism grows. Thus every cell contains copies of DNA that originally came from both parents.

A *mutation* is a change in the sequence of bases in a DNA molecule. Because the base sequence is changed, the information contained in the DNA molecule changes. Mutations may result from exposure to certain chemicals or to radiation. Once this molecular accident has taken place, the altered DNA is copied through replication. If the organism containing the mutation reproduces, it may pass the mutation on to its offspring. Certain mutations can be beneficial and may result in proteins that give an organism a better chance to survive and reproduce. Evolution is probably the result of beneficial mutations that are passed along in this way.

Mutations can also be harmful. They may result in less effective or missing proteins. Genetic diseases such as sickle cell anemia, result from such mutations. In this disease, a portion of the DNA that controls hemoglobin synthesis contains a single base substitution (adenine for thymine) in one code word. This defect causes a change in one amino acid at a particular point in the protein chain. The modified hemoglobin causes red blood cells to become sickle shaped at low oxygen pressure,

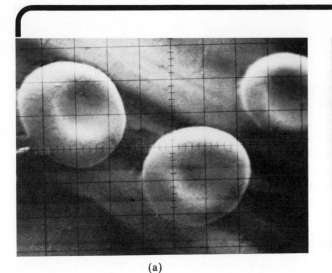

(a)

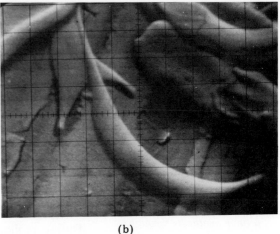

(b)

FIGURE 18-27

Red blood cells (at a magnification of 5000×). The cells in (a) are normal, but those in (b) are sickle cells. (Photos courtesy of Philips Electronic Instruments, Inc.)

as shown in Figure 18-27. These cells get trapped in small blood vessels, causing damage to certain organs. The cells also break easily, resulting in anemia. Our increasing knowledge of nucleic acids and the chemistry of living systems should help us to treat, and perhaps even to cure such diseases. The new field of genetic engineering, which involves techniques for modifying DNA molecules, offers the first real hope for conquering genetic diseases.

Summary

18.1 Carbon forms compounds in which it has four covalent bonds. It can form several million different compounds, because the carbon atom bonds strongly to other carbon atoms as well as to atoms of other nonmetals. Organic chemistry is the study of carbon compounds.

18.2 Hydrocarbons consist only of carbon and hydrogen atoms, and are the simplest organic compounds. Methane, CH_4, is the smallest hydrocarbon.

18.3 The alkanes are a family of saturated hydrocarbons. Each carbon in an alkane is bonded to four other atoms by single bonds. Alkanes have the general formula C_nH_{2n+2}.

18.4 An alkyl group is an alkane that is missing one hydrogen atom. It is a fragment of a molecule and can exist only when attached to another atom.

18.5 Isomers are molecules that have the same molecular formula but different structures. The straight-chain alkanes (with three or more carbon atoms) have branched-chain isomers.

18.6 Alkenes are hydrocarbons that contain a double bond between two carbon atoms. The general formula of an alkene is C_nH_{2n}. Alkynes contain a triple bond between two carbon atoms. The general formula of an alkyne is C_nH_{2n-2}. Alkenes and alkynes are unsaturated, enabling them to undergo addition reactions.

18.7 Aromatic hydrocarbons are cyclic molecules with alternating single and double bonds. The electrons in an aromatic molecule are actually spread out (delocalized) over all the atoms in the ring. Polynuclear (polycyclic) aromatic compounds contain several rings "fused" together.

18.8 A functional group is an atom or group of atoms in a molecule that gives the molecule characteristic properties. Important oxygen-containing functional groups are alcohol ($-OH$),

$$\text{ether } (-O-),\text{ aldehyde } (-\overset{\overset{\displaystyle O}{\|}}{C}H),\text{ ketone } (-\overset{\overset{\displaystyle O}{\|}}{C}-),\text{ carboxylic acid}$$

$$(-\overset{\overset{\displaystyle O}{\|}}{C}-OH),\text{ and ester } (-\overset{\overset{\displaystyle O}{\|}}{C}-O-).$$

18.9 Alkyl halides are hydrocarbons in which halogen atoms have replaced hydrogen atoms. Nitrogen-containing compounds include the amines, amides, and nitrogen heterocycles. Polymers are very large molecules made by chemically joining many small molecules called monomers.

18.10 Carbohydrates are organic molecules that can be defined as polyhydroxyl aldehydes or polyhydroxyl ketones, or their derivatives. The most important carbohydrates are the monosaccharide glucose, the disaccharide sucrose, and the polysaccharides starch, glycogen, and cellulose.

18.11 Lipids are a mixed group of organic compounds that dissolve in certain organic solvents. The major lipids are triacylglycerols (fats or triglycerides), phospholipids, and steroids.

18.12 Proteins are complex polymers of amino acids. The structure and function of a protein molecule depends on the sequence of its amino acids.

18.13 The word metabolism refers to all the enzyme-catalyzed chemical processes that exchange matter and energy between our bodies and the environment. In catabolism, food molecules are broken down, releasing energy. In anabolism, large molcules are synthesized from the smaller molecules formed and the energy released during catabolism.

18.14 Nucleic acids are polynucleotide molecules that contain genetic information in the sequence of their bases. Protein molecules are synthesized by transcription and translation using this information. Heredity is based on the process of DNA replication.

Exercises

KEY-WORD MATCHING EXERCISE

For each term on the left, choose a phrase from the right-hand column that most closely matches its meaning.

1 organic chemistry		**a**	carbon-carbon double bond
2 hydrocarbon		**b**	benzene
3 alkane		**c**	contains —O—
4 alkyl group		**d**	contains $-\overset{\overset{\displaystyle O}{\|}}{C}-OH$
5 isomer			
6 alkene		**e**	carbon-carbon triple bond
7 alkyne		**f**	contains $-\overset{\overset{\displaystyle O}{\|}}{\underset{\|}{C}}-\overset{\|}{\underset{\|}{C}}-\overset{\|}{\underset{\|}{C}}-$
8 aromatic hydrocarbon			
9 alcohol		**g**	different atoms in ring
10 ether		**h**	C_nH_{2n+2}
11 aldehyde		**i**	glucose
12 ketone		**j**	contains —OH
13 carboxylic acid		**k**	chemistry of carbon
14 ester		**l**	includes fats, steroids
15 amine		**m**	triacylglycerol
16 amide			

17 heterocycle

18 polymer

19 carbohydrate

20 lipid

21 fat

22 protein

23 metabolism

24 nucleic acid

n polymer of amino acids

o C and H only

p molecule of heredity

q reactions in body

r same formula, different structure

s contains $-\overset{\overset{\displaystyle O}{\|}}{C}-NH_2$

t many monomers

u contains $-\overset{\overset{\displaystyle O}{\|}}{C}H$

v substituted NH_3

w from alcohol + acid

x missing an H atom

EXERCISES BY TOPIC

Carbon atom (18.1)

25 Why do carbon atoms form four bonds?

26 Why do so many different organic compounds exist?

27 How do organic compounds differ in general from inorganic compounds?

Methane (18.2)

28 What are hydrocarbons?

29 What is the shape of methane?

30 Describe the formation of a **a** sp^3 hybrid orbital **b** sigma bond

Alkanes (18.3)

31 What is an alkane?

32 Why are alkanes said to be saturated?

33 Write the molecular formula of an alkane having **a** 20 carbon atoms **b** 42 carbon atoms

34 Draw the complete structural (straight-chain) formula of **a** butane **b** hexane

Alkyl groups (18.4)

35 How does an alkyl group differ from an alkane?

36 Why can't an alkyl group exist alone?

37 Name the alkyl group based on octane. Draw a possible structure.

Isomers (18.5)

38 What are structural isomers? How are they alike and how do they differ?

39 Draw as many different isomers as you can that have the formula C_5H_{12}.

40 Name the isomers in Exercise 39.

Alkenes and alkynes (18.6)

41 How does each of the following differ from an alkane?
a alkene   **b** alkyne

42 Describe the bonding in propene.

43 Write the molecular formula of   **a** 25-carbon alkene   **b** 18-carbon alkyne

44 Describe the bonding in propyne.

45 Why are alkenes and alkynes considered unsaturated?

Aromatic hydrocarbons (18.7)

46 What are cyclic molecules?

47 Describe the bonding in benzene. Why is it unusual?

48 How does an aromatic hydrocarbon differ from an aliphatic hydrocarbon?

49 What is a polynuclear aromatic hydrocarbon?

Oxygen-containing organic compounds (18.8)

50 What is a functional group?

51 Identify each of the following molecules as an alcohol, ether, aldehyde, ketone, carboxylic acid, or ester:

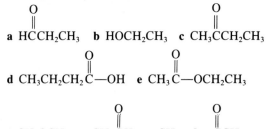

52 Name the compounds in Exercise 51.

53 What are the differences between a carbonyl, a carboxyl, and a carboxylate group?

54 Name the ester formed from acetic acid and ethyl alcohol.

55 Write an equation for the esterification reaction of Exercise 54.

Other organic derivatives and polymers (18.9)

56 Classify the following compounds as an akyl halide, amine, amide, or heterocycle: **a** $CH_3\overset{\displaystyle O}{\overset{\|}{C}}-NH_2$ **b** CCl_2F_2 **c**

d $(CH_3CH_2)_2NH$ **e** $CH_3CH_2NH_2$

57 Why are amines basic?

58 What are polymers?

59 Describe three different polymers and their uses.

60 What is an alkaloid?

Carbohydrates (18.10)

61 What are carbohydrates?

62 Give an example of each: **a** disaccharide   **b** monosaccharide   **c** polysaccharide

63 Name the carbohydrate that is   **a** impossible to digest   **b** known as table sugar   **c** blood sugar   **d** the major food in your diet   **e** stored in your body

64 Describe the composition of starch.

Lipids (18.11)

65 What are lipids?

66 Draw a triacylglycerol containing stearic acid (tristearin).

67 How does a fat differ from an oil?

68 How does a phospholipid differ from a triacylglycerol?

Proteins and enzymes (18.12)

69 How are the 20 major amino acids alike? How do they differ?

70 How are amino acids joined in a protein?

71 Describe the   **a** primary   **b** secondary   **c** tertiary structure of a protein.

72 What functions do proteins serve?

73 How does an enzyme work?

Metabolism (18.13)

74 Why can it be said that "you are what you eat"?

75 What is metabolism? anabolism? catabolism?

76 How does food provide energy?

77 What is the role of ATP and how does it work?

Nucleic acids and heredity (18.14)

78 How do nucleic acids determine heredity?

79 What is base-pairing? Why is it important?

80 Briefly describe how proteins are made.

81 What is the purpose of replication?

82 What are mutations? Why are they inherited?

ADDITIONAL EXERCISES **83** Identify the following compounds by stating their class (such as alkane or alcohol):

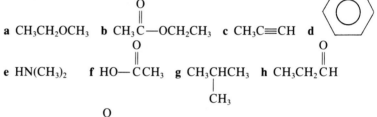

a $CH_3CH_2OCH_3$ **b** $CH_3\overset{\overset{\displaystyle O}{\|}}{C}-OCH_2CH_3$ **c** $CH_3C\equiv CH$ **d**

e $HN(CH_3)_2$ **f** $HO-\overset{\overset{\displaystyle O}{\|}}{C}CH_3$ **g** $CH_3\underset{\underset{\displaystyle CH_3}{|}}{C}HCH_3$ **h** $CH_3CH_2\overset{\overset{\displaystyle O}{\|}}{C}H$

i CH_3Cl **j** $H_2N-\overset{\overset{\displaystyle O}{\|}}{C}CH_2CH_3$

84 Name the compounds in Exercise 83.

85 Draw as many different isomers as you can that have the formula C_8H_{18}.

86 Name the isomers in Exercise 85.

87 Name the following compounds:

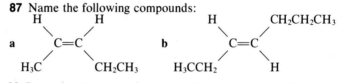

88 Draw the structure of **a** *trans*-2-heptane **b** *cis*-3-hexane

89 Name the following dichlorobenzenes:

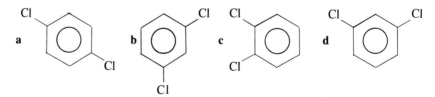

90 Draw the structure of: **a** 2-butanol   **b** dimethyl ether   **c** ethanal **d** trimethylamine   **e** butanone **f** carbon tetrachloride   **g** propanoic acid **h** methyl formate   **i** methanamide **j** 2-hexene

91 What happens when you digest starch?

92 What is a steroid? Give an example.

93 Is a molecular (or genetic) disease contagious? Explain.

94 Since an alcohol contains the —OH group, why isn't it basic?

95 To what does the term "polyunsaturated" refer?

96 What is the relationship between catabolism and anabolism?

97 In Benedict's test for sugars, the following reaction occurs:

$$\text{aldehyde} + Cu^{2+}(aq) \rightarrow \text{carboxylic acid} + Cu_2O(s)$$

Identify the oxidizing agent and reducing agent in the reaction.

98 Draw all the different isomers that could result from the substitution of one Cl atom for one H atom in each of the following compounds:

a $CH_3CH_2CH_2CH_2CH_3$ b
$$CH_3-\overset{\overset{\displaystyle CH_3}{|}}{\underset{\underset{\displaystyle CH_3}{|}}{C}}-CH_3$$

99 Part of one DNA strand has the base sequence T-T-A-G-A-G-T. What is the base sequence on the complementary strand?

100 Identify the functional groups in each of the six amino acids shown in Section 18.12.

101 Draw three isomers that have the formula $C_2H_2Cl_2$.

102 What ester is formed from the reaction of methyl alcohol and propanoic acid? Write its name and structure.

103 Draw the structures of four alcohols that have the formula $C_4H_{10}O$.

104 Identify the class of each of the following compounds:

a CH_3OCH_3 b CCl_4 c $H\overset{\overset{\displaystyle O}{\|}}{C}CH_3$ d $CH_3CH_2CH_2CH_2\overset{\overset{\displaystyle O}{\|}}{C}-OH$

e $CH_3O-\overset{\overset{\displaystyle O}{\|}}{C}CH_2CH_3$ f $(CH_3CH_2)_3N$ g $CH_3CH{=}CHCH_2CH_3$

h $CH_3-\overset{\overset{\displaystyle O}{\|}}{C}-NH_2$ i $CH_3CH_2\overset{\overset{\displaystyle O}{\|}}{C}CH_2CH_3$ j CH_3CH_3

105 Name the compounds in Exercise 104.

106 People with diabetes may need to treat themselves with daily injections of the protein insulin. Why cannot insulin be taken orally?

PROGRESS CHART   Check off each item when completed.

———————   Previewed chapter

Studied each section:

———————   *Part A   Organic chemistry*
———————   **18.1** Carbon atom
———————   **18.2** Methane
———————   **18.3** Alkanes
———————   **18.4** Alkyl groups
———————   **18.5** Isomers
———————   **18.6** Alkenes and alkynes
———————   **18.7** Cyclic and aromatic hydrocarbons
———————   **18.8** Oxygen-containing organic compounds
———————   **18.9** Other organic derivatives and polymers

———————   *Part B   Biological chemistry*
———————   **18.10** Carbohydrates
———————   **18.11** Lipids
———————   **18.12** Proteins and enzymes
———————   **18.13** Metabolism
———————   **18.14** Nucleic acids and heredity

———————   Reviewed chapter

———————   Answered assigned exercises:

#'s ————————————————————————

Overcoming fear of science and math

FIGURE A-1

(Peanuts by Charles M. Schultz. Copyright © 1979 by United Feature Syndicate, Inc.)

Some students are afraid to study chemistry because of bad experiences with previous courses in math or science. (See Figure A-1.) They believe mistakenly that since they did not do well in science and math earlier, they cannot learn these subjects now. You may feel this way yourself. Perhaps you even feel guilty, ashamed, or inadequate because of past difficulties.

The first step in overcoming your fears is to recognize that they exist. Then you can start to deal with them. You may wish to see your college counselor to discuss methods of relaxation. The counselor may suggest that you attend some college workshops designed to help students deal with their fears.

The next step is to begin learning the science and math skills in this book. Sections 1.3 and 1.4 on studying, and Section 3.1 on problem solving should be especially helpful. In many ways, learning a new subject is like learning to engage in a sport or play a musical instrument. You will be tense at first; you will make mistakes, and probably will be embarrassed by them. But by constant practice, you will develop new skills and self-confidence.

While you are learning, do not be afraid to ask questions that may seem elementary, or to say that you don't understand something. These are your rights as a student. Get help from your instructor, from fellow students, and from friends. You may find it helpful to work in a small group with others who have similar problems. Share your frustrations and support each other. You might even keep a diary or an audiotape of your thoughts as you work on assignments. Doing so can help you to identify bad study habits and other stumbling blocks to success.

Keep in mind that present knowledge in science and math took hundreds of years to develop. Don't expect instant understanding. However, learning these subjects does not require a special talent either. In order to succeed in your study of chemistry, you need only the desire to learn and the willingness to work hard.

Arithmetic review

**B.1
BASIC
OPERATIONS
WITH SIGNED
NUMBERS**

Addition is the process of joining two or more numbers. It can be represented as a move to the right on a number line, shown below. For example, 2 + 3 (two plus three, or the sum of two and three) can be shown in the following way:

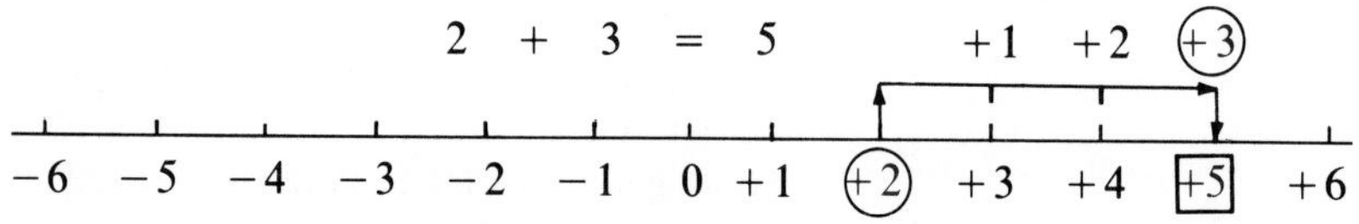

$$2 \ + \ 3 \ = \ 5 \qquad +1 \quad +2 \quad \boxed{+3}$$

$$\overline{-6 \quad -5 \quad -4 \quad -3 \quad -2 \quad -1 \quad 0 \quad +1 \quad (+2) \quad +3 \quad +4 \quad \boxed{+5} \quad +6}$$

Subtraction is the process of finding the difference between two numbers, and corresponds to moving towards the left on the number line. For example, three minus two can be shown as:

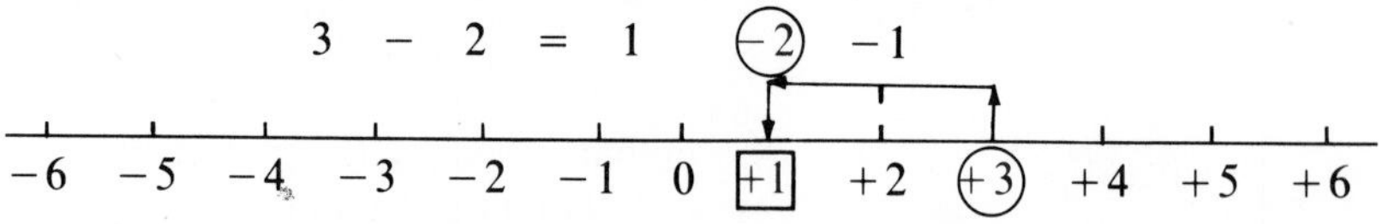

$$3 \ - \ 2 \ = \ 1 \qquad (-2) \quad -1$$

$$\overline{-6 \quad -5 \quad -4 \quad -3 \quad -2 \quad -1 \quad 0 \quad \boxed{+1} \quad +2 \quad (+3) \quad +4 \quad +5 \quad +6}$$

Numbers greater than zero are called **positive numbers**, and may appear with a plus (+) sign. If no sign is given, you can assume that the number is positive. **Negative numbers** (numbers less than zero) always appear with a minus (−) sign.

To add two numbers having the *same* sign, either both positive or both negative, we just add the numbers and keep the sign. For example, the sum of −2 and −3 is found as follows:

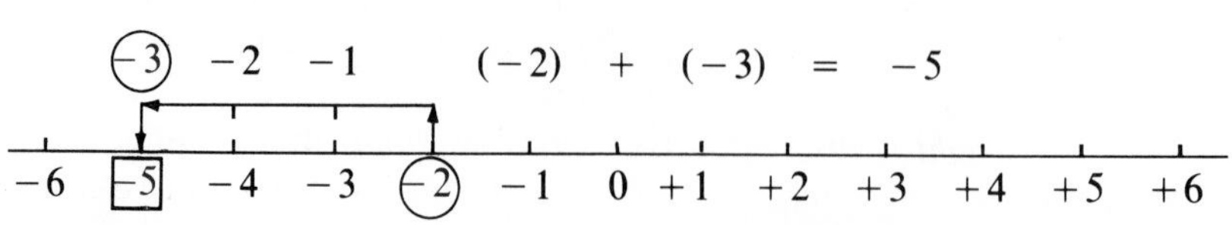

$$(-3) \quad -2 \quad -1 \qquad (-2) \ + \ (-3) \ = \ -5$$

$$\overline{-6 \quad \boxed{-5} \quad -4 \quad -3 \quad (-2) \quad -1 \quad 0 \quad +1 \quad +2 \quad +3 \quad +4 \quad +5 \quad +6}$$

To add two numbers with *different* signs, we first drop the signs, and subtract the smaller number from the larger number. Then we give the answer the sign of the larger number. For example, to find the sum of -3 and $+2$ we drop the signs and then subtract the smaller number 2 from the larger number 3.

$$3 - 2 = 1$$

Then we give the sign of the larger number to the answer. The larger number 3 had a minus $(-)$ sign originally, so our answer is -1.

To subtract numbers, we change the sign of the number being subtracted, and then add the numbers according to the rules for addition. Consider the example $(+2) - (-3)$. We change the sign of the number being subtracted (-3 becomes $+3$) and then add the two numbers: $(+2) + (+3) = +5$. Subtraction, then, is a form of addition in which a sign is reversed. Other examples of subtraction are:

$$(-5) - (+2) = (-5) + (-2) = -7$$
$$(-3) - (-4) = (-3) + (+4) = +1$$
$$(+1) - (+3) = (+1) + (-3) = -2$$

Multiplication is also a form of addition. It is the addition of the same number several times. For example, three times five is the same as $5 + 5 + 5 = 15$. In multiplication, one number (5) is the number being added, and the other (3) represents the number of times it is added.

Division is the repeated subtraction of one number from another. If we subtract one number from another until we get zero, the number of times that we performed the subtraction is the answer. For example, fifteen divided by five means:

$$15 - 5 = 10 \qquad 10 - 5 = 5 \qquad 5 - 5 = 0$$

These subtractions show that fifteen "contains" three fives, so fifteen divided by five is three. When dividing numbers, the number that is divided is called the **dividend**. The number that is doing the dividing is called the **divisor**:

$$\underset{\text{dividend}}{6} \div \underset{\text{divisor}}{3} = 2$$

When multiplying or dividing numbers that have the *same* sign, we carry out the operation and give the answer a *plus* sign. On the other hand, when multiplying or dividing numbers that have *different* signs, we carry out the operation and give the answer a *minus* sign. The following examples illustrate these rules:

$$(-3) \times (-2) = +6 \quad \text{same sign, answer positive}$$
$$(+2) \times (-4) = -8 \quad \text{different signs, answer negative}$$
$$(+8) \div (-4) = -2 \quad \text{different signs, answer negative}$$
$$(-6) \div (-3) = +2 \quad \text{same sign, answer positive}$$

If a calculation involves several different operations, the operations must be carried out in a certain definite order. Consider this example:

$$(6 + 9) \div 5 + 2(-10 + 4)$$

The order is as follows: First work out the expressions within parentheses.

$$15 \div 5 + 2(-6)$$

Then perform the indicated multiplications and divisions, moving from left to right:

$$3 + (-12)$$

Finally, do the additions or subtractions, also from left to right.

$$-9$$

B.2 FRACTIONS

Fractions are used to represent *part* of a whole number. Common fractions are written as one number above another number, with the two numbers separated by a bar. The number on the top is called the **numerator** and the number on the bottom is the **denominator**. The fraction $\frac{1}{2}$, one-half, means one part out of a total of two equal parts.

$\dfrac{1}{2}$ ← numerater (dividend)
← denominator (divisor)

Fractions represent the ratio, or relationship, between two numbers. We can also think of a fraction as a division. For example $\frac{1}{2}$ means one divided by two.

In order to add or subtract fractions, they must have the same denominator, called a *common* denominator. If the denominator of one fraction is a multiple of the denominator of the other, as in

$$\frac{1}{2} + \frac{1}{4}$$

we use the larger number (4) as the common denominator. We convert the other fraction ($\frac{1}{2}$) to one having the common denominator by first dividing its original denominator (2) into the new denominator (4): $4 \div 2 = 2$. Then we multiply this result by the original numerator: $2 \times 1 = 2$. Thus $\frac{1}{2} = \frac{2}{4}$. We can then perform addition or subtraction using the fraction with the new common denominator:

$$\frac{2}{4} + \frac{1}{4} = \frac{3}{4}$$

The numerator of the answer is the sum of the numerators of the fractions being added. The denominator stays the same. If we had been *subtracting* fractions, then the numerator of the answer would be the difference between the numerators being subtracted.

If the original denominators are *not* simple multiples, as in

$$\frac{1}{2} - \frac{1}{3}$$

then we use the product of the denominators ($2 \times 3 = 6$) as the common denominator. We convert each fraction to one that has the common denominator:

$$\frac{1}{2} = \frac{3}{6} \qquad \frac{1}{3} = \frac{2}{6}$$

We can then perform the addition or subtraction.

$$\frac{3}{6} - \frac{2}{6} = \frac{1}{6}$$

Fractions can be multiplied and divided whether or not they have a common denominator. To multiply fractions, we simply multiply the numerators and multiply the denominators:

$$\frac{2}{3} \times \frac{1}{4} = \frac{2 \times 1}{3 \times 4} = \frac{2}{12}$$

Then we simplify the result, if possible. For example, in the fraction, $\frac{2}{12}$, both the numerator and the denominator can be divided by 2:

$$\frac{2}{12} = \frac{2/2}{12/2} = \frac{1}{6}$$

This division simplifies the fraction. Note that dividing (or multiplying) both the numerator and denominator of a fraction by the *same number* does not change the value of the fraction. When a fraction cannot be simplified further, it is said to be in *lowest terms*. For example, the fraction $\frac{1}{6}$ is in lowest terms.

To divide fractions, we invert the second fraction (write it "upside down"), and then multiply the first fraction by the inverted one. For example:

$$\frac{3}{8} \div \frac{1}{4} = \frac{3}{8} \times \frac{4}{1} = \frac{12}{8}$$

The result, $\frac{12}{8}$, can be simplified by dividing the numerator and denominator by the common factor 4. The resulting fraction, $\frac{3}{2}$, is in lowest terms.

A fraction that has a multiple of ten as its denominator is called a decimal fraction or decimal number. As shown in Figure B-1, decimal fractions can be represented by numbers written to the right of a decimal point. The position of a number with respect to the decimal point determines its value. For example, the number 0.5 means five times one-tenth (5 × 0.1, or 5 × 1/10). Its meaning is exactly the same as one-half, since the ratio 1/2 is identical to the ratio 5/10.

Any fraction can be converted to a decimal number, simply by dividing the numerator by the denominator. Table B-1 lists the decimal values of the most common fractions. Note that a zero is usually placed in front of a decimal fraction to emphasize the position of the decimal point.

When *adding* or *subtracting* decimal numbers, we always line up the numbers in columns, placing the decimal points directly under each other:

3	1	8	9	2	1	3
↑	↑	↑	↑	↑	↑	↑
×1000	×100	×10	×1	×0.1	×0.01	×0.001

3000.	thousands
100.	hundreds
80.	tens
9.	units
0.2	tenths
0.01	hundredths
0.003	thousandths
3189.213	

FIGURE B-1

A decimal number. The value of each digit depends on its position with respect to the decimal point.

TABLE B-1 *Decimal values of common fractions*

Fraction	Decimal
$\frac{1}{8}$	0.125
$\frac{1}{6}$	0.167
$\frac{1}{5}$	0.200
$\frac{1}{4}$	0.250
$\frac{1}{3}$	0.333
$\frac{3}{8}$	0.375
$\frac{2}{5}$	0.400
$\frac{1}{2}$	0.500
$\frac{3}{5}$	0.600
$\frac{5}{8}$	0.625
$\frac{2}{3}$	0.667
$\frac{3}{4}$	0.750
$\frac{4}{5}$	0.800
$\frac{5}{6}$	0.833
$\frac{7}{8}$	0.875

$$\begin{array}{r} 56.73 \\ +\,230.15 \\ \hline 286.88 \end{array} \qquad \begin{array}{r} 1.789 \\ -\,0.151 \\ \hline 1.638 \end{array}$$

addition   subtraction

In this way, we add or subtract digits that have the *same place value.*

When *multiplying* decimal numbers, we need not line up the decimal points. The number of decimal places (numbers after the decimal point) in the answer equals the *sum* of the decimal places of the numbers being multiplied. For example:

$$\begin{array}{r} 3.21 \\ \times\,0.8 \\ \hline 2.568 \end{array}$$

Since 3.21 has two decimal places and 0.8 has one, the product must have three (1 + 2 = 3) places.

When *dividing* decimal numbers, we move the decimal point of the divisor (the number doing the dividing) all the way to the right, making it a whole number. Then we shift the decimal point in the dividend (the number being divided) exactly the same number of places, as shown in this example:

$$\begin{array}{r} 2.5 \\ 1.5\,)\overline{3.75} \\ \underline{30} \\ 75 \\ \underline{75} \end{array}$$

The decimal point in the answer lies directly above this new position.

Multiplying or dividing by the number 10 is particularly easy with decimal numbers. Multiplying a decimal number by 10 involves shifting the decimal point one position to the *right* for each multiple of 10. Study these examples:

$$3.145 \times 10 \ = \ 31.45$$

$$3.145 \times 100 \ = \ 314.5$$

$$3.145 \times 1000 = 3145$$

Dividing a decimal number by 10 involves moving the decimal point one place to the *left* for each multiple of 10.

$$18.5 \div 10 \ = 1.85$$

$$18.5 \div 100 \ = 0.185$$

$$18.5 \div 1000 = 0.0185$$

A common fraction, decimal fraction, and percent. Each represents part of a whole.

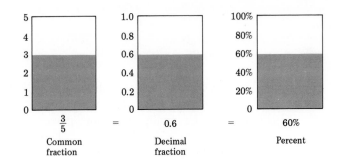

Note that multiplying by 10 makes the number larger and dividing by 10 makes the number smaller. This fact should help you to remember which way to move the decimal point.

The term *percent* stands for a special type of fraction in which the denominator is 100. Fifty percent, written as 50%, means 50 parts out of 100 equal parts, or 50/100. To convert a decimal to percent, we move the decimal point two places to the right. Performing this shift is the same as multiplying by 100. For example:

decimal	percent
0.75	75%
0.062	6.2%
2.51	251%

To convert a percent to a decimal, we move the decimal point two places to the left (divide by 100). Figure B-2 summarizes the relationship between common fractions, decimal fractions, and percent.

Exercises

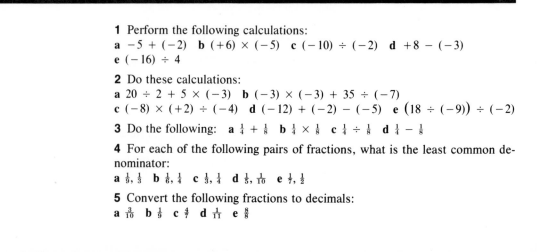

1 Perform the following calculations:
a $-5 + (-2)$ **b** $(+6) \times (-5)$ **c** $(-10) \div (-2)$ **d** $+8 - (-3)$
e $(-16) \div 4$

2 Do these calculations:
a $20 \div 2 + 5 \times (-3)$ **b** $(-3) \times (-3) + 35 \div (-7)$
c $(-8) \times (+2) \div (-4)$ **d** $(-12) + (-2) - (-5)$ **e** $\left(18 \div (-9)\right) \div (-2)$

3 Do the following: **a** $\frac{1}{4} + \frac{1}{8}$ **b** $\frac{1}{4} \times \frac{1}{8}$ **c** $\frac{1}{4} \div \frac{1}{8}$ **d** $\frac{1}{4} - \frac{1}{8}$

4 For each of the following pairs of fractions, what is the least common denominator:
a $\frac{1}{9}, \frac{1}{3}$ **b** $\frac{1}{6}, \frac{1}{4}$ **c** $\frac{1}{3}, \frac{1}{4}$ **d** $\frac{1}{5}, \frac{1}{10}$ **e** $\frac{1}{7}, \frac{1}{2}$

5 Convert the following fractions to decimals:
a $\frac{3}{10}$ **b** $\frac{1}{9}$ **c** $\frac{4}{7}$ **d** $\frac{1}{11}$ **e** $\frac{8}{8}$

6 Perform the following calculations:

a $\begin{array}{r} 89.71 \\ + \ 16.53 \end{array}$ **b** $7.2\overline{)1.95}$ **c** $\begin{array}{r} 1.376 \\ - \ 0.988 \end{array}$ **d** $\begin{array}{r} 1.561 \\ \times \ 3.2 \end{array}$

7 Convert the fractions in Exercise 5 to percent.

8 Carry out these calculations by moving the decimal point:
a 18.95×100 **b** $7.63 \div 1000$ **c** 0.981×10 **d** $253.1 \div 100$
e $0.035 \div 10$

9 Which is larger: $\frac{14}{19}$ or 75%?

10 Do the following calculation: $\dfrac{-(18.1) \times (9.31 - 5.18)}{(-0.731) \times (1.69)}$

Algebra review

**C.1
SOLVING
EQUATIONS**

A **mathematical equation** is a statement that two numerical quantities are equal.

$$2 + 2 = 4$$
$$\underset{\text{of equation}}{\text{left side}} = \underset{\text{of equation}}{\text{right side}}$$

An **algebraic equation** is a special type of mathematical equation in which symbols are used to represent one or more of the numbers. For example

$$x + 5 = 8$$

is an algebraic equation in which the symbol x represents some unknown number. Solving an algebraic equation involves rearranging the terms of the equation so that the unknown is on one side. All the other terms, usually numbers, should appear on the other side of the equation after we have finished rearranging it.

When rearranging the terms of an algebraic equation, *we must perform exactly the same operations on both sides of the equation.* In other words, any addition, subtraction, multiplication, or division carried out on one side of the equation must also be carried out on the other side. For example, in the equation $x + 5 = 8$, the unknown can be found by subtracting 5 from both sides.

$x + 5 = 8$	
$x + 5 - 5 = 8 - 5$	5 subtracted from both sides
$x + 0 = 3$	subtraction performed
$x = 3$	addition performed

The equation $2x = 60$ is solved by dividing both sides of the equation by 2, since $2x$ means two times x.

$$2x = 60$$

$$\frac{2x}{2} = \frac{60}{2} \qquad \text{both sides divided by 2}$$

$$x = 30 \qquad \text{division performed}$$

Note in the final step that the symbol x by itself is understood to mean $1x$.

The equation $4x - 6 = 10$ requires both an addition and a division in order to solve for x.

$$4x - 6 = 10$$

$$4x - 6 + 6 = 10 + 6 \qquad \text{6 added to both sides}$$

$$4x + 0 = 16 \qquad \text{addition performed}$$

$$4x = 16 \qquad \text{addition performed}$$

$$\frac{4x}{4} = \frac{16}{4} \qquad \text{both sides divided by 4}$$

$$x = 4 \qquad \text{division performed}$$

To check the solution to an algebraic equation, we substitute the value found for the unknown back into the original equation. If both sides of the equation are equal, the solution is correct. For example, the previous equation is checked as follows:

$$4x - 6 = 10$$

$$4(4) - 6 \overset{?}{=} 10 \qquad \text{substituted 4 for } x$$

$$16 - 6 \overset{?}{=} 10 \qquad \text{multiplication performed}$$

$$10 = 10 \qquad \text{subtraction performed, answer checks}$$

Additional techniques are sometimes required to solve algebraic equations. For example, if the unknown appears in more than one term, these terms must be combined, as we now show:

$$5x = 3x + 6$$

$$5x - 3x = 3x + 6 - 3x \qquad \text{3x subtracted from both sides}$$

$$2x = 6 \qquad \text{subtraction performed}$$

$$\frac{2x}{2} = \frac{6}{2} \qquad \text{each side divided by 2}$$

$$x = 3 \qquad \text{division performed}$$

If parentheses surround the unknown, they must be removed before the equation can be solved. Study the following example:

$$4(x + 2) = 12$$

$$(4)(x) + (4)(2) = 12 \qquad \text{each term in parentheses multiplied by 4}$$

$$4x + 8 = 12 \qquad \text{multiplications performed}$$

$$4x + 8 - 8 = 12 - 8 \qquad \text{8 subtracted from both sides}$$

$$4x = 4 \qquad \text{subtraction performed}$$

$$\frac{4x}{4} = \frac{4}{4} \qquad \text{each side divided by 4}$$

$$x = 1 \qquad \text{division performed}$$

C.2 PROPORTIONS

A **proportion** is an equation that relates two ratios. For example, the equation

$$\frac{5}{25} = \frac{1}{5}$$

is a proportion. The *products of the diagonal terms in a proportion are always equal:* $5 \times 5 = 25 \times 1$. Therefore, if only three of the terms of a proportion are known, the fourth term can be found by using this relationship.

Consider the following proportion:

$$\frac{1}{2} = \frac{x}{100}$$

To find the value of x, we cross-multiply and set the products equal to each other:

$$1 \times 100 = 2 \times x$$

Then we solve this equation for x using the methods described in the previous section.

$$100 = 2x$$

$$\frac{100}{2} = \frac{2x}{2}$$

$$50 = x$$

The answer can be checked by substituting 50 for x in the original proportion and cross-multiplying.

$$\frac{1}{2} \overset{?}{=} \frac{50}{100}$$

$$1 \times 100 \overset{?}{=} 2 \times 50$$

$$100 = 100$$

The use of proportions to solve problems is often similar to the use of the conversion-factor method described in Section 3.2. For example, in the following equations, if a is the unknown, and b is given, then the ratio c/d corresponds to the conversion factor:

$$\frac{a}{b} = \frac{c}{d}$$

$$ad = bc$$

$$a = \frac{bc}{d} = b \times \frac{c}{d}$$

C.3
FORMULAS
A **formula** is an equation that states the relationship between two or more quantities. A simple example of a formula is:

$$\text{dollars} = \frac{\text{cents}}{100}$$

This formula can be used to convert a given number of cents into dollars. For example, 200 cents could be converted to 2 dollars by using this formula. Terms in a formula that can have different values, such as "dollars" and "cents," are called **variables**. Terms that stay the same, like 100, are called **constants**. Variables, as well as constants, can be represented by symbols. For example, the formula for the area of a circle is:

$$A = \pi r^2$$

The symbol A represents a variable, area. The symbol r represents another variable, the radius. The Greek letter π (pi) stands for a constant, 3.14.

Formulas, like algebraic equations, can be rearranged so that the unknown variable is alone on one side of the equation. For example, if the variable cents is unknown, the formula dollars = cents/100 can be rewritten by multiplying both sides by 100:

$$\text{dollars} = \frac{\text{cents}}{100}$$

$$\text{dollars} \times 100 = \frac{\text{cents}}{\cancel{100}} \times \cancel{100}$$

$$\text{dollars} \times 100 = \text{cents}$$

Problems are often solved by substituting numerical values into a formula. However, it is easy to make a mistake if you "plug" numbers into a formula without thinking. Always be sure that you understand what the formula means and why you are using it. Sometimes it is helpful to know how the formula was derived. For example, in temperature conversion formulas, you may confuse degrees Fahrenheit and Celsius. To avoid this confusion, recall the relative sizes of the degree on each scale, and the different starting points.

C.4
GRAPHS

A **graph** is a drawing that shows how two variables are related to each other. It shows how one variable changes when the other is changed. The quantity being changed is called the *independent* variable; it can have any reasonable value. The other quantity, the *dependent* variable, depends on the value of the independent variable. For example, changing the pressure on a gas will cause the volume of the gas to change. In this example pressure is called the independent variable, and volume is called the dependent variable.

Figure C-1 is a graph of these two quantities. The independent variable (*P*) is marked on the horizontal line or axis, and the dependent variable (*V*) appears on the vertical axis. The table shows the actual values of pressure and volume that were measured. These pairs of values are plotted on the graph, that is, they are placed at positions given by their values. A graph is like a map; each point has a certain

FIGURE C-1

Pressure-volume curve. This graph shows how the pressure of a gas varies when its volume is changed (at constant temperature). Each point on the curve corresponds to a particular pressure and volume.

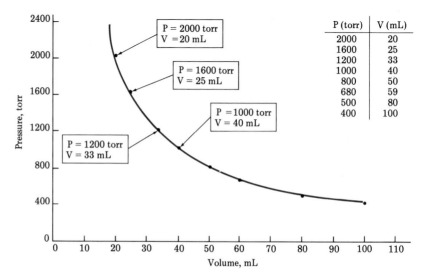

P (torr)	V (mL)
2000	20
1600	25
1200	33
1000	40
800	50
680	59
500	80
400	100

FIGURE C-2

Solubility curves. This graph shows how the solubility of different compounds varies with temperature.

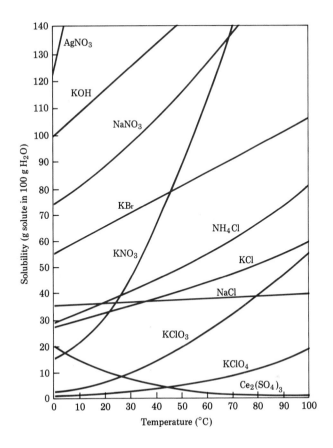

"address" given by the values on the horizontal and vertical axes. Several points on the graph are already labelled. You should be able to label the other points yourself.

When the dependent variable and independent variable change in the same way (both increase or both decrease), they are said to be *directly proportional*. For example, the volume of a gas is directly proportional to its temperature (in kelvins) because as the temperature increases, so does the volume. On the other hand, pressure and volume are *inversely proportional*: as one variable increases, the other decreases.

Graphs show a great deal of information in a compact form. Figure C-2 shows the solubilities of selected compounds in water. Using this graph, we can determine the solubility of any one of these compounds at a particular temperature. For example, the solubility of KCl is about 50 g (in 100 g of H_2O) at 80°C. This graph also indicates *trends* in the way that temperature affects solubility for each compound. The solubility of $KClO_3$, for example, increases dramatically with temperature. This information is much clearer when presented in a graph than when listed in a table.

1 Solve for x: **a** $2 + 3x = 11$ **b** $5x - 8 = 17$ **c** $2x = 18 - x$

2 Find the value of x: **a** $5(x + 1) = 30$ **b** $(3 - x)2 = 4$
c $6(x + 1) = 3(4 + x)$

3 Solve for x: **a** $\dfrac{3}{15} = \dfrac{x}{10}$ **b** $\dfrac{2}{x} = \dfrac{5}{30}$ **c** $\dfrac{9}{24} = \dfrac{27}{x}$

4 Consider the following formula: $\text{zerds} = \dfrac{\text{nudniks}}{\text{plotkins}}$. Rewrite the formula so that you could find plotkins if you are given nudniks and zerds.

5 In the following formula, find the value of P if $Q = 3$, $R = 2$, and $S = 5$.

$$P = \frac{2.5\ QR}{S}$$

6 In the formula in Exercise 5, find R if $P = 10$, $S = 2$, and $Q = 4$.

7 In Figure C-2, which substance is more soluble at 80°C, Na_2SO_4 or KCl?

8 What is the solubility of   **a** $KClO_3$ at 70°C?   **b** $NaCl$ at 10°C?
c $Pb(NO_3)_2$ at 55°C?

9 Solve for x: $2x - 9 = \dfrac{3x + 9}{4}$

10 Solve for x: $\dfrac{4x - a}{b} = \dfrac{cd}{e} + 1$

Using a calculator

Electronic calculators allow us to make arithmetical computations very quickly. Unless we use calculators thoughtfully, however, we may arrive at incorrect answers. It is easy to push the buttons without thinking about what we are doing. The technique of estimating, described in Section 2.11, is essential when using a calculator. By estimating answers first, we can tell immediately if our calculated answer makes sense.

In most calculators we enter the operations in the order in which they are written. Consider the following computation:

$$\boxed{3} \ \boxed{\times} \ \boxed{4} \ \boxed{=} \ 12$$

First we press the "3" key, then the "$\times$" key for multiplication, then the "4" key, followed by the "$=$" (equals) key. The answer 12 appears in the display. Listed below are the results of each of these actions:

Action	Result
enter $\boxed{3}$	3 appears in display
press $\boxed{\times}$	3 stored, multiplication indicated
enter $\boxed{4}$	4 appears in display
press $\boxed{=}$	answer 12 appears in display and is stored

Since the $\boxed{=}$ key stores the result, further calculations can be carried out if necessary by pressing another operation key (such as "divide"). The calculator automatically "clears" to zero, however, if we enter another number.

Most calculators have a "floating" decimal point. Pressing the $\boxed{\cdot}$ (decimal point) key at the proper place while entering the number automatically positions the decimal point correctly in the answer. Note that zeros after the last number to the right of the decimal point are omitted in the display. Thus, 18.100 appears as 18.1.

Although you may never have to do so, it is possible to enter an eight-digit number or a seven-digit decimal number into a calculator. Answers may also be displayed with that many digits. However not all of the numbers shown in the display are necessarily significant. Always follow the rules for making calculations with significant figures. These rules are given in Section 2.8. When performing calculations involving very large or very small numbers, use exponential notation, which is described in Section 2.10.

SI units

Listed below are the official base units of the International System

Quantity	Name	Symbol	Definition or standard
length	meter	m	1,650,763.73 wavelengths in vacuum of the orange-red line of the spectrum of krypton-86
mass	kilogram	kg	cylinder of platinum-iridium alloy kept at the International Bureau of Weights and Measures in Paris, France
time	second	s	duration of 9,192,631,770 cycles of the radiation associated with a specified transition of the cesium-133 atom
electric current	ampere	A	current maintained in two long parallel wires separated by one meter that would produce a certain force (2×10^{-7} newton per meter) due to their magnetic fields
temperature	kelvin	K	1/273.16 of the thermodynamic temperature of the triple point of water (the temperature at which ice, liquid water, and water vapor all co-exist)
amount of substance	mole	mol	amount of substance that contains as many elementary entities as there are atoms in 0.012 kilogram of carbon-12
luminous intensity	candela	cd	luminous intensity of 1/600,000 of a square meter of a black body at the temperature of freezing platinum (2045 K)

Other units can be derived from these base units. For example:

Quantity	Name	Symbol
area	square meter	m^2
volume	cubic meter	m^3
speed	meters per second	m/s
density	kilograms per cubic meter	kg/m^3
concentration	moles per cubic meter	mol/m^3

Some derived units have special names, such as the following:

Quantity	Name	Symbol	Meaning
frequency	hertz	Hz	1/s
force	newton	N	(kg)(m)/s^2
pressure	pascal	Pa	N/m^2
energy	joule	J	(N)(m)

Solubility table

	Al^{3+}	NH_4^+	Ba^{2+}	Ca^{2+}	Cu^{2+}	H^+	Fe^{2+}	Fe^{3+}	Pb^{2+}	Mg^{2+}	Mn^{2+}	Hg_2^{2+}	Hg^{2+}	Ni^{2+}	K^+	Ag^+	Na^+	Zn^{2+}
acetates ($C_2H_3O_2^-$)	W	W	W	W	W	W	W	W	W	W	W	w	W	W	W	W	w	W
bromides (Br^-)	W	W	W	W	W	W	W	W	W	W	W	A	W	W	W	a	W	W
carbonates (CO_3^{2-})	-	W	w	w	-	-	w	-	A	w	w	A	-	w	W	A	W	w
chlorides (Cl^-)	W	W	W	W	W	W	W	W	W	W	W	a	W	W	W	a	W	W
cyanides (CN^-)	-	W	W	W	A	W	a	-	w	W	-	A	W	a	W	a	W	A
fluorides (F^-)	W	W	w	w	w	W	w	w	w	w	A	d	d	w	W	W	W	w
hydroxides (OH^-)	A	W	W	W	A	-	A	A	w	A	A	-	A	w	W	-	W	A
iodides (I^-)	W	W	W	W	a	W	W	W	w	W	W	A	w	W	W	I	W	W
nitrates (NO_3^-)	W	W	W	W	W	W	W	W	W	W	W	W	W	W	W	W	W	W
oxides (O^{2-})	a	-	W	w	A	W	A	A	w	A	A	A	w	A	W	w	d	w
phosphates (PO_4^{3-})	A	W	A	w	A	W	A	w	A	w	w	A	A	A	W	A	W	A
sulfates (SO_4^{2-})	W	W	a	w	W	W	W	W	w	W	W	w	d	W	W	w	W	W
sulfides (S^{2-})	d	W	d	w	A	W	A	d	A	d	A	I	I	A	W	A	W	A

Key: W   soluble in water
    w   sparingly soluble in water but soluble in acids
    A   insoluble in water, but soluble in acids
    a   insoluble in water, and only sparingly soluble in acids
    I   insoluble in both water and acids
    d   decomposes in water

Table of logarithms

	.00	.01	.02	.03	.04	.05	.06	.07	.08	.09
1.0	.000	.004	.009	.013	.017	.021	.025	.029	.033	.037
1.1	.041	.045	.049	.053	.057	.061	.064	.068	.072	.076
1.2	.079	.083	.086	.090	.093	.097	.100	.104	.107	.111
1.3	.114	.117	.121	.124	.127	.130	.134	.137	.140	.143
1.4	.146	.149	.152	.155	.158	.161	.164	.167	.170	.173
1.5	.176	.179	.182	.185	.188	.190	.193	.196	.199	.201
1.6	.204	.207	.210	.212	.215	.217	.220	.223	.225	.228
1.7	.230	.233	.236	.238	.241	.243	.246	.248	.250	.253
1.8	.255	.258	.260	.262	.265	.267	.270	.272	.274	.276
1.9	.279	.281	.283	.286	.288	.290	.292	.294	.297	.299
2.0	.301	.303	.305	.307	.310	.312	.314	.316	.318	.320
2.1	.322	.324	.326	.328	.330	.332	.334	.336	.338	.340
2.2	.342	.344	.346	.348	.350	.352	.354	.356	.358	.360
2.3	.362	.364	.365	.367	.369	.371	.373	.375	.377	.378
2.4	.380	.382	.384	.386	.387	.389	.391	.393	.394	.396
2.5	.398	.400	.401	.403	.405	.407	.408	.410	.412	.413
2.6	.415	.417	.418	.420	.422	.423	.425	.427	.428	.430
2.7	.431	.433	.435	.436	.438	.439	.441	.442	.444	.446
2.8	.447	.449	.450	.452	.453	.455	.456	.458	.459	.461
2.9	.462	.464	.465	.467	.468	.470	.471	.473	.474	.476
3.0	.477	.479	.480	.481	.483	.484	.486	.487	.489	.490
3.1	.491	.493	.494	.496	.497	.498	.500	.501	.502	.504
3.2	.505	.507	.508	.509	.511	.512	.513	.515	.516	.517
3.3	.519	.520	.521	.522	.524	.525	.526	.528	.529	.530
3.4	.531	.533	.534	.535	.537	.538	.539	.540	.542	.543
3.5	.544	.545	.547	.548	.549	.550	.551	.553	.554	.555
3.6	.556	.558	.559	.560	.561	.562	.563	.565	.566	.567
3.7	.568	.569	.571	.572	.573	.574	.575	.576	.577	.579
3.8	.580	.581	.582	.583	.584	.585	.587	.588	.589	.590
3.9	.591	.592	.593	.594	.595	.597	.598	.599	.600	.601
4.0	.602	.603	.604	.605	.606	.607	.609	.610	.611	.612
4.1	.613	.614	.615	.616	.617	.618	.619	.620	.621	.622
4.2	.623	.624	.625	.626	.627	.628	.629	.630	.631	.632
4.3	.633	.634	.635	.636	.637	.638	.639	.640	.641	.642
4.4	.643	.644	.645	.646	.647	.648	.649	.650	.651	.652
4.5	.653	.654	.655	.656	.657	.658	.659	.660	.661	.662
4.6	.663	.664	.665	.666	.667	.667	.668	.669	.670	.671

	.00	.01	.02	.03	.04	.05	.06	.07	.08	.09
4.7	.672	.673	.674	.675	.676	.677	.678	.679	.679	.680
4.8	.681	.682	.683	.684	.685	.686	.687	.688	.688	.689
4.9	.690	.691	.692	.693	.694	.695	.695	.696	.697	.698
5.0	.699	.700	.701	.702	.702	.703	.704	.705	.706	.707
5.1	.708	.708	.709	.710	.711	.712	.713	.713	.714	.715
5.2	.716	.717	.718	.719	.719	.720	.721	.722	.723	.723
5.3	.724	.725	.726	.727	.728	.728	.729	.730	.731	.732
5.4	.732	.733	.734	.735	.736	.736	.737	.738	.739	.740
5.5	.740	.741	.742	.743	.744	.744	.745	.746	.747	.747
5.6	.748	.749	.750	.751	.751	.752	.753	.754	.754	.755
5.7	.756	.757	.757	.758	.759	.760	.760	.761	.762	.763
5.8	.763	.764	.765	.766	.766	.767	.768	.769	.769	.770
5.9	.771	.772	.772	.773	.774	.775	.775	.776	.777	.777
6.0	.778	.779	.780	.780	.781	.782	.782	.783	.784	.785
6.1	.785	.786	.787	.787	.788	.789	.790	.790	.791	.792
6.2	.792	.793	.794	.794	.795	.796	.797	.797	.798	.799
6.3	.799	.800	.801	.801	.802	.803	.803	.804	.805	.806
6.4	.806	.807	.808	.808	.809	.810	.810	.811	.812	.812
6.5	.813	.814	.814	.815	.816	.816	.817	.818	.818	.819
6.6	.820	.820	.821	.822	.822	.823	.823	.824	.825	.825
6.7	.826	.827	.827	.828	.829	.829	.830	.831	.831	.832
6.8	.833	.833	.834	.834	.835	.836	.836	.837	.838	.838
6.9	.839	.839	.840	.841	.841	.842	.843	.843	.844	.844
7.0	.845	.846	.846	.847	.848	.848	.849	.849	.850	.851
7.1	.851	.852	.852	.853	.854	.854	.855	.856	.856	.857
7.2	.857	.858	.859	.859	.860	.860	.861	.862	.862	.863
7.3	.863	.864	.865	.865	.866	.866	.867	.867	.868	.869
7.4	.869	.870	.870	.871	.872	.872	.873	.873	.874	.874
7.5	.875	.876	.876	.877	.877	.878	.879	.879	.880	.880
7.6	.881	.881	.882	.883	.883	.884	.884	.885	.885	.886
7.7	.886	.887	.888	.888	.889	.889	.890	.890	.891	.892
7.8	.892	.893	.893	.894	.894	.895	.895	.896	.897	.897
7.9	.898	.898	.899	.899	.900	.900	.901	.901	.902	.903
8.0	.903	.904	.904	.905	.905	.906	.906	.907	.907	.908
8.1	.908	.909	.910	.910	.911	.911	.912	.912	.913	.913
8.2	.914	.914	.915	.915	.916	.916	.917	.918	.918	.919
8.3	.919	.920	.920	.921	.921	.922	.922	.923	.923	.924
8.4	.924	.925	.925	.926	.926	.927	.927	.928	.928	.929
8.5	.929	.930	.930	.931	.931	.932	.932	.933	.933	.934
8.6	.934	.935	.936	.936	.937	.937	.938	.938	.939	.939
8.7	.940	.940	.941	.941	.942	.942	.943	.943	.943	.944
8.8	.944	.945	.945	.946	.946	.947	.947	.948	.948	.949
8.9	.949	.950	.950	.951	.951	.952	.952	.953	.953	.954
9.0	.954	.955	.955	.956	.956	.957	.957	.958	.958	.959
9.1	.959	.960	.960	.960	.961	.961	.962	.962	.963	.963
9.2	.964	.964	.965	.965	.966	.966	.967	.967	.968	.968
9.3	.968	.969	.969	.970	.970	.971	.971	.972	.972	.973
9.4	.973	.974	.974	.975	.975	.975	.976	.976	.977	.977
9.5	.978	.978	.979	.979	.980	.980	.980	.981	.981	.982
9.6	.982	.983	.983	.984	.984	.985	.985	.985	.986	.986
9.7	.987	.987	.988	.988	.989	.989	.989	.990	.990	.991
9.8	.991	.992	.992	.993	.993	.993	.994	.994	.995	.995
9.9	.996	.996	.997	.997	.997	.998	.998	.999	.999	.999

Using a log table for pH conversions

pH is equal to the negative logarithm of the hydronium ion (H_3O^+) concentration. Therefore, converting hydronium ion concentration to pH involves finding the logarithm of the concentration. This process is most easily carried out by expressing the concentration in exponential notation, and then separately finding the log of the coefficient and the log of the power of 10. As shown in Section 14.5, the log of a power of 10 is simply the exponent. To find the log of the coefficient, however, we must use a log table, such as the one in Appendix G.

Suppose that we want to find the log of the number 0.000500. First we express this number in exponential notation: 5.00×10^{-4}. Now we must find the log of the coefficient 5.00. To do this, we look down the first column of the log table until we reach 5.0, the first two digits of the coefficient. Then, keeping a finger on this number, we look at the headings at the top of the table. They give the possible values of the third digit of the coefficient. Because the coefficient is 5.00, the log lies in the column headed by .00. We look down this column until we come to the number opposite the finger. This number is .699, and the logarithm of 5.00 is therefore .699 or 0.699. Had the coefficient been 5.04 instead, the logarithm would be 0.702 (in the column headed by .04). Note that all logarithms in the table are decimal numbers ranging from 0.000 to 0.999.

Example Find the logarithms of the following numbers using a log table.

number	log
9.85	0.993
1.09	0.037
6.67	0.824
3.16	0.500
7.00	0.845

Now to find the logarithm of the number 5.00×10^{-4}, we *add* the log of the coefficient to the log of the power of ten. The reason for this rule is that the logarithm of a product, $\log (A \times B)$, equals the *sum* of the logarithms of each factor, $\log A + \log B$. We already found that the log of the coefficient is 0.699. The log of the power of ten is simply the exponent -4. Therefore, the logarithm of 5.00×10^{-4} equals

$$\log (5.00) + \log (10^{-4})$$
$$= 0.699 + (-4)$$
$$= -3.301$$

The only additional step involved in finding the pH is to multiply by -1, since $pH = -\log [H_3O^+]$. If, for example, 5.00×10^{-4} represents the hydronium ion concentration of a solution, then the pH is $-\log [H_3O^+] = -\log (5.0 \times 10^{-4}) = -(-3.301) = 3.301$. Because pH is almost never meaningful beyond the hundredths place, the pH of this acid solution should be given as 3.30.

Example The waste water leaving an industrial plant has a hydronium ion concentration of 2.18×10^{-5} moles/liter. Find its pH.

UNKNOWN	GIVEN	CONNECTION
pH	2.18×10^{-5} moles/L $= [H_3O^+]$	$pH = -\log [H_3O^+]$

$$pH = -\log [H_3O^+] = -\log (2.18 \times 10^{-5})$$
$$pH = -(\log (2.18) + \log (10^{-5}))$$
$$pH = -(0.338 + (-5))$$
$$pH = -(0.338 - 5) = -(-4.662)$$
$$pH = 4.66$$

Example A sample of the secretion from a patient's pancreas was found to have a hydronium ion concentration of 1.39×10^{-8}. What is the pH of this solution?

UNKNOWN	GIVEN	CONNECTION
pH	1.39×10^{-8} moles/L $= [H_3O^+]$	$pH = -\log [H_3O^+]$

$$pH = -\log [H_3O^+] = -\log (1.39 \times 10^{-8})$$
$$pH = -(\log (1.39) + \log (10^{-8}))$$
$$pH = -(0.143 + (-8)) = -(0.143 - 8)$$
$$pH = -(-7.857)$$
$$pH = 7.86$$

The reverse calculation, finding the hydronium ion concentration from a given pH, involves what is called "taking the antilogarithm." This phrase simply means determining what number corresponds to the given logarithm. Since $pH = -\log [H_3O^+]$, then $\log [H_3O^+] = -pH$. The hydronium ion concentration is the number whose logarithm has the value $-pH$. For example, if the pH is 9, then $\log [H_3O^+] = -9$. The number whose logarithm is -9 is 10^{-9}, and therefore the H_3O^+ ion concentration is 10^{-9}.

If the pH is not an exact integer, the calculation is slightly more complex. The pH is first separated into two parts, a whole number and a decimal number. For example, if the pH is 9.5, the initial step is $\log [H_3O^+] = -9.5 = -9 - 0.5$. However, the decimal number must be positive so that the log table can be used. We must change the value of the whole number -9 to -10 so that the sum of the whole number and a decimal number is still equal to -9.5. The decimal number part must be $+0.5$. We now have $-9.5 = -10 + 0.5$.

Now we find the antilog of each part, that is, the number whose logarithm has that value. The antilog of -10 is simply 10^{-10}. (To check, note that the log of 10^{-10} is -10.) To find the antilog of 0.5, we look within the table until we find 0.500, or the value closest to it. We then look across to the lefthand column to find the first two digits (3.1) of the antilog, and we look at the top of the column containing 0.500 to find the last digit (.06). Therefore the antilog of 0.500 is 3.16. Combining the two antilogs gives the concentration 3.16×10^{-10} moles/L. We can check the answer by finding its negative logarithm, making sure that it equals the given value 9.5.

Example   When a soap is dissolved in water, the pH becomes 11.1. What is the hydronium ion concentration?

UNKNOWN	GIVEN	CONNECTION
$[H_3O^+]$	pH 11.1	$pH = -\log [H_3O^+]$

$$\log [H_3O^+] = -pH = -11.1$$
$$\log [H_3O^+] = 0.9 - 12$$
$$[H_3O^+] = \text{antilog } (0.9) \times \text{antilog } (-12)$$
$$[H_3O^+] = 7.95 \times 10^{-12} \text{ mole/L}$$

Example   A glass of tomato juice has a pH of 4.2. Find $[H_3O^+]$.

UNKNOWN	GIVEN	CONNECTION
$[H_3O^+]$	pH 4.2	$pH = -\log [H_3O^+]$

$$\log [H_3O^+] = -pH = -4.2$$
$$\log [H_3O^+] = 0.8 - 5$$
$$[H_3O^+] = \text{antilog } (0.8) \times \text{antilog } (-5)$$
$$[H_3O^+] = 6.3 \times 10^{-5} \text{ mole/L}$$

Exercises

1 Find the pH of a solution whose hydronium ion concentration is:
a 3.5×10^{-8} M   **b** 7.1×10^{-3} M   **c** 5.0×10^{-5} M

2 Find the hydronium ion concentration of a solution whose pH is: **a** 2.3
b 13.5   **c** 6.7

3 A 0.1 M solution of sodium carbonate has a pH of 11.5. Find the hydronium ion concentration of this solution.

4 If pOH is defined as $-\log [OH^-]$, find the pOH of a solution whose hydroxide ion concentration is 8.3×10^{-4} M.

5 A solution contains 10.0 g of KOH in 500. mL. Find its pH.

Glossary

acid an electrolyte that furnishes H$^+$ ions (Arrhenius definition); a proton donor (Brönsted definition)

activated complex an unstable, short-lived molecule that is intermediate between the reactants and products

activation energy the amount of energy needed in order for colliding molecules to react

adhesion the attraction of molecules of one substance for other substances

alcohol a compound containing a hydroxyl group (—OH) covalently bonded to a carbon atom

aldehyde an organic compound containing a carbonyl group bonded on one side to a hydrogen atom $\left(\begin{array}{c} O \\ \parallel \\ -CH \end{array}\right)$

alkane a hydrocarbon with only single bonds between carbon atoms

alkene a hydrocarbon containing a double bond between carbon atoms

alkyl group a group of atoms formed from an alkane by the removal of a hydrogen atom

alkyne a hydrocarbon containing a triple bond between carbon atoms

amide a nitrogen-containing organic compound formed by the reaction of an amine with a carboxylic acid, and containing the group

$$\begin{array}{c} O \\ \parallel \\ -C-N\diagup \\ \diagdown \end{array}$$

amine a nitrogen-containing organic compound formed by replacing one or more hydrogens from ammonia with hydrocarbon groups

amphoteric able to act either as an acid or as a base

anion a negatively charged ion

anode the electrode at which oxidation takes place

aromatic hydrocarbon a molecule containing carbon and hydrogen atoms that has electrons evenly distributed in a ring system like benzene

atom the basic building block of matter; consists of an inner region called the nucleus surrounded by particles called electrons

atomic mass unit (amu) a unit of mass equal to 1/12th the mass of an atom of carbon-12

atomic number the number of protons in the nucleus of an atom

atomic weight an average of the atomic masses of the isotopes of an element as found in nature

Avogadro's number 6.02×10^{23}, the number of units (atoms, ions, formula units, or molecules) in one mole of a substance

aufbau process the procedure by which electron configurations of atoms are determined; electrons are added to orbitals having lowest energy, according to certain rules

base an electrolyte that furnishes OH$^-$ ions (Arrhenius definition); a proton acceptor (Brönsted definition)

boiling point the temperature at which a liquid is converted to a gas, because its vapor pressure equals the atmospheric pressure

buffer solution a solution whose pH changes only very slightly upon addition of an acid or base

carbohydrate an organic molecule containing carbon, hydrogen, and oxygen atoms that is a polyhydroxyl aldehyde or polyhydroxyl ketone

carbonyl group an oxygen atom double bonded

$$\text{to carbon,} \quad -\overset{\overset{\displaystyle O}{\|}}{C}-$$

catalyst a substance that causes a chemical reaction to take place more rapidly and can be recovered unchanged at the end of a reaction

cathode the electrode at which reduction takes place

cation a positively charged ion

chain reaction a process in which the products of a reaction cause further reactions to take place

charge a fundamental property of matter that exists in two forms, positive and negative, depending on whether there is a shortage or excess of electrons

chemical bond the force that holds atoms together in a chemical compound, resulting from the attraction of the nucleus of one atom for the electrons of another

chemical equation a shorthand representation of a chemical reaction, summarizing the chemical change that takes place

chemical equilibrium a condition in which the forward and reverse processes of a reversible reaction take place at the same rate so that the concentrations of reactants and products do not change

chemical kinetics the study of the relationship between reaction rates and reaction mechanisms

chemical reaction the process by which one or more substances are converted to different substances by the making and breaking of chemical bonds

chemistry the study of the composition, structure, and properties of substances, and the changes that they undergo

cohesion the attraction of molecules for each other

colligative property a property of a solution that depends only on the number of solute particles present

colloid a mixture containing relatively large particles (1 to 1000 nm in diameter) dispersed in another substance

combining capacity the number of single bonds that an atom can form; also called valence

compound a substance formed by the chemical combination of two or more different elements

concentration the amount of solute present in a given quantity of solvent or solution

conjugate acid-base pair an acid and a base that differ only by the presence or absence of a proton

constant a mathematical quantity that has a fixed value in a formula or equation

conversion factor method a method for performing calculations that is based on multiplying by ratios containing related quantities in the numerator and denominator

covalent bond the attraction of two nonmetal atoms to each other through the sharing of outer electrons

deliquescence the property of dissolving by absorbing moisture from the air

density the ratio of an object's mass to its volume; usually expressed in g/cm^3

diffusion the movement of molecules away from a region of high concentration to a region of lower concentration

dissociation the process by which an ionic compound separates into ions

efflorescence the loss of water of hydration upon exposure to air

electrochemical cell a set-up in which a flow of electrons can be produced by or result in a redox reaction

electrochemistry the study of oxidation reduction reactions that produce or use electrical energy

electrolyte a substance that produces ions in solution and whose solution therefore conducts electricity

electrolytic cell a set-up in which electrical energy from an outside source causes a redox reaction to take place

electron a negatively charged subatomic particle found in regions surrounding the nucleus of an atom

electron affinity a measure of an atom's tendency to attract electrons; the amount of energy released upon addition of an electron to an atom of an element (in its gaseous state)

electron configuration the arrangement of electrons in the orbitals of an atom

electronegativity the attraction of an atom for the electrons in its chemical bonds

element a collection of atoms with the same atomic number

empirical formula a representation of the simplest whole number ratios of the atoms in a compound

energy the ability or capacity to do work

energy level an allowed value of energy for electrons in a certain region of an atom

equilibrium a condition in which opposing processes balance so that no observable change takes place

equilibrium constant (K) a number that indicates the extent of a reversible reaction based on the ratio of concentrations of products and reactants at equilibrium (at a given temperature)

ester an organic compound containing a

$$-\overset{\overset{\textstyle O}{\|}}{C}-O-$$ group formed from the reaction of a carboxylic acid and an alcohol

ether an organic compound containing an oxygen atom covalently bonded to two carbon atoms

evaporation the escape of molecules from the surface of a liquid in an open container

exponential notation a shorthand way of writing numbers as powers of 10

fat see triacylglycerol

fission the splitting of a heavy atomic nucleus into two or more pieces

formula (chemical) a shorthand way to express the numbers and kinds of atoms in a molecule

formula (mathematical) an equation that states the relationship between two or more quantities

formula unit the smallest number of ions that represent the composition of an ionic compound

formula weight the sum of the atomic weights of all the atoms (or ions) present in a formula unit of a compound

fusion the combination of small nuclei at high temperature to form a larger, more stable nucleus

galvanic cell a set-up in which a redox reaction results in a flow of electricity; also called a voltaic cell

gas matter in the state in which it has neither definite shape nor definite volume

gram (g) the basic unit for expressing measurements of mass in the metric system

graph a drawing that shows the relationship between two variables

group a vertical column of the periodic table

half-cell an electrode and its accompanying solution containing the oxidized and reduced forms of a substance

half-life (t$_{1/2}$) the time required for one-half of a given number of nuclei to decay

half-reaction either the reduction or the oxidation step in a redox reaction

heat the transfer of energy that takes place between a hotter substance and a cooler one

heat capacity the ratio of the amount of heat absorbed by a substance to the resulting change in temperature, expressed in cal (or J) per mole per °C

heat of reaction the amount of energy transferred as heat during a reaction; also known as enthalpy change

heterocycle an organic ring compound that contains atoms of two or more different elements in the ring

hydration a process in which water molecules react as single units, remaining as H_2O

hydrocarbon a compound containing only carbon and hydrogen atoms

hydrogen bond the weak attraction of a partially positive hydrogen atom by a partially negative atom such as oxygen or nitrogen

hydrolysis a reaction with water in which water molecules break up into hydrogen ions and hydroxide ions

hydronium ion a hydrated proton, H_3O^+

hygroscopic capable of absorbing water vapor from the air

hypothesis a tentative explanation or model

ideal gas law the equation that relates the pressure, volume, temperature, and amount of an ideal gas: $PV = nRT$; R is the ideal gas constant

immiscible a term used to describe liquids that do not dissolve in each other

indicator a dye whose color changes over a specific pH range

ion an electrically charged atom or group of atoms

ionic bond the attraction of oppositely charged ions

ionization the process by which a molecule breaks apart into ions

ionization energy the amount of energy needed to remove an electron from an atom of an element in its gaseous state

isomer a molecule with the same molecular formula as another but with a different structure

isotope an atom related to another atom by having the same atomic number but different mass number

ketone an organic compound containing a carbonyl group $\left(\begin{array}{c} O \\ \| \\ -C- \end{array} \right)$ covalently bonded to two carbon atoms

kinetic molecular theory the model used to describe the motion of molecules (or atoms) within matter

law of conservation of energy a scientific law which states that energy is neither created nor destroyed in chemical changes

law of conservation of mass a scientific law which states that the total mass of the reactants equals the total mass of the products in a chemical reaction

law of definite proportions a law of chemistry, stating that a compound always contains its component elements in fixed proportions by weight

length the distance between two points

Lewis symbol a way to represent atoms using dots or crosses for outermost s and p electrons; Lewis symbols are combined in Lewis structures to represent the bonding in covalent molecules

limiting reagent the reactant that is used up first, and thus limits the amount of product that can form in a chemical reaction

lipid an organic compound, found in plants and animals, that dissolves in certain organic solvents like ether, chloroform, and carbon tetrachloride

liquid matter in the state in which it has constant volume and takes the shape of its container, filling it up from the bottom

liter (L) the basic unit for expressing measurements of volume in the metric system, equivalent to the cubic decimeter (dm^3)

main group element an element in the main part of the periodic table where s and p orbitals are being filled; also called a representative element

mass the amount of matter an object contains

mass number the sum of the numbers of protons and neutrons in the nucleus of an atom

matter anything that has mass and occupies space; the stuff from which everything is made

measurement   a number (and unit) that indicates the amount of something

mechanism   the series of steps by which a reaction is thought to take place

meniscus   the curved upper surface of a liquid in a column

metabolism   the complex chemical processes that take place in the body in order to exchange matter and energy with the body's surroundings

metal   an element whose atoms tend to lose their outer electrons; metals are generally shiny, and conduct heat and electricity well

meter (m)   the basic unit for expressing measurements of length in the metric (SI) system

metric system   a system of measurement based on the number 10 (see SI)

miscible   the ability of liquids to mix to form a solution in any proportion

mixture   a combination of two or more substances whose atoms have not been joined by chemical bonding; the composition of a mixture is variable

molality (*m*)   the number of moles of solute per kilogram of solvent in a solution

molarity (M)   the number of moles of solute per liter of solution

molar mass   the mass, in grams per mole, of a substance

mole   a unit used to express the amount of a substance; it contains 6.02×10^{23} atoms, molecules, or formula units, and has a mass in grams equal to the atomic, formula, or molecular weight

molecular weight   the sum of the atomic weights of all the atoms present in a molecule

molecule   the unit formed by the covalent bonding of two or more atoms

mole ratio   a conversion factor that relates the number of moles of two substances involved in a chemical reaction

net ionic equation   a simplified chemical equation that shows only those ions, atoms, or molecules that take part in a chemical reaction

neutralization   the reaction of an acid with a base

neutron   a neutral subatomic particle present in the nuclei of atoms

nonmetal   an element whose atoms tend to gain electrons to complete their outermost level; nonmetals are generally dull in appearance, and conduct heat and electricity poorly

normality (N)   the number of equivalents of solute per liter of solution

nucleic acid   a long polymer of nucleotides that provides information for the synthesis of proteins and the transmission of heredity from one generation to the next

nucleus   central portion of an atom, containing protons and neutrons

octet rule   the statement that atoms usually have a filled outer level of eight electrons (s^2p^6) after forming chemical bonds with other atoms

orbital   a specific region around the nucleus of an atom in which the probability of finding an electron is high

organic chemistry   the study of carbon compounds

osmosis   the diffusion of water molecules through a semi-permeable membrane toward the more concentrated solution

oxidation   the loss of electrons by the atoms of a substance

oxidation number   a number assigned to an atom according to certain rules based on the distribution of electrons

oxidation-reduction reaction   see redox reaction

oxidizing agent   a substance that accepts electrons, becoming reduced in the process

partial pressure   the pressure exerted by a particular gas in a mixture of gases

parts per million (ppm)   parts of solute in 10^6 parts of solution

percent (%)   a fraction whose denominator is 100; parts per hundred

percentage composition   the percentage by weight of each element in a compound

percent yield the ratio of the actual yield (grams of product formed) to the theoretical yield (the number of grams predicted by stoichiometry)

period a horizontal row of the periodic table

periodic law the observation that the properties of elements repeat in a regular way when the elements are arranged by increasing atomic number

periodic table a systematic arrangement of the elements that reveals trends in their properties

pH a measure of the acidity or basicity of a solution; the negative logarithm of the hydronium ion concentration

polar having an uneven distribution of electrical charge

polyatomic ion a group of atoms that has an electrical charge

polymer a very large molecule formed by chemically joining many small molecules called monomers

pressure force applied to a certain area; the units may be torr (mm Hg), atm, pascals, or psi

product a substance produced by a chemical reaction

proportion a mathematical equation that relates two ratios

protein a naturally occurring polymer of amino acids

proton a positively charged subatomic particle present in the nuclei of atoms

quantum mechanics a branch of science based on the theory that some quantities such as energies of electrons in atoms can have only certain allowed values

radiation a means by which energy is emitted or absorbed

radioactivity the process of nuclear disintegration or decay in which nuclei break down, releasing radiation

radioisotope a radioactive isotope of an element

rate constant (k) the proportionality factor in a rate equation that relates the concentrations to the measured rate of a reaction

rate-determining step the slowest step in a reaction that takes place in a series of steps

rate equation (law) an experimentally determined expression that relates the reaction rate to the concentrations of reactants

reactant a starting substance in a chemical reaction

reaction rate the change in concentration of a reactant or product with time

reduction the gain of electrons by the atoms of a substance

reduction potential (E) a measure of the tendency of the oxidized form of a redox couple to gain electrons and become the reduced form of the substance; expressed in volts

redox couple the oxidized and reduced forms of a substance that differ by one or more electrons

redox reaction a chemical change involving the processes of oxidation and reduction

reducing agent a substance that donates electrons, becoming oxidized in the process

reversible reaction a reaction that occurs in both the forward direction, to form products, and in the reverse direction, to form reactants

salt an ionic compound formed during neutralization; contains the cation from the base and the anion from the acid

saturated solution a solution in which the rate of solute dissolving equals the rate of solute returning from solution

saturated hydrocarbon an organic compound to which no more hydrogen atoms can be added

science a systematic and logical organization of facts that describe nature

scientific law a statement about the way that nature behaves

scientific method the process of creating and testing hypotheses

scientific notation see exponential notation

scientific theory a hypothesis that has been tested and confirmed many times under a wide variety of conditions

shell a region of space around an atomic nucleus, which can hold a certain number of electrons; a set of orbitals at a certain average distance from the nucleus (see energy level)

SI abbreviation for International System of Units, the official version of the modern metric system

significant figures the number of meaningful digits in a measured value

solid matter in the state that has definite shape and volume

solubility the mass of solute that can be dissolved in a certain amount of solvent at a given temperature; usually expressed as g solute per 100 mL (or 100 g) solvent

solute the substance that is being dissolved or the component that is present in smaller amount, in the formation of a solution

solution a homogeneous mixture of two or more substances

solvent the substance doing the dissolving, or present as the major component, in the formation of a solution

specific gravity the ratio of the density of a substance to the density of water

specific heat the amount of heat needed to raise the temperature of one gram of a substance by one degree Celsius; given in cal (or J) per g per °C

spectrum the specific pattern of frequences of radiation that are absorbed or emitted by a substance

stoichiometry the methods used to calculate relationships between the quantities of elements or compounds involved in a chemical reaction

STP standard temperature and pressure; 273 K (or 0°C) and 1 atm (or 760 torr or 101.3 kPa)

supersaturated containing more solute than necessary to reach solubility equilibrium at a particular temperature

surface tension the inward attraction experienced by the molecules at the surface of a liquid

suspension a mixture consisting of relatively large particles temporarily distributed in another substance

temperature describes how hot or cold a substance is; a measure of the average kinetic energy of the molecules of a substance

titration a procedure, based on a neutralization reaction, used to determine the unknown concentration of an acid or base

triacylglycerol a triester of glycerol and three fatty acids; also called fat or triglyceride

unsaturated solution a solution that contains less solute than is indicated by the solubility of that solute at the temperature of the solution

unsaturated hydrocarbon an organic compound to which additional hydrogen atoms can be added

valence the combining capacity of an atom; the number of single bonds that an atom can form

valence electrons electrons in the outermost shell of an atom

vapor the gaseous form of a substance that normally exists as a liquid or solid

vaporization the escape of molecules from the surface of a liquid

variable a mathematical quantity that can have different values in a formula or equation

voltaic cell see galvanic cell

volume the amount of space an object occupies

volume-volume percent (v/v%) the volume of a liquid solute in milliliters present in 100 mL of solution

weight percent the mass of solute in grams that is present in 100 g of solution

weight-volume percent (w/v%) the mass of solute in grams that is present in 100 mL of solution

Numerical
answers
to exercises
by topic

Chapter Two

23 $1000

25 10^6

28 9.3×10^6 miles

29 6.72×10^2

30 9.843×10^{-7} inch

31 185,000,000 bottles

32 0.000004 inch

33 a 9.22×10^1 **b** 5.55×10^{-3} **c** 3.894×10^5 **d** 1.5×10^1 **e** 2.5×10^{-2}

34 a 2800 **b** 0.0152 **c** 80,000 **d** 760,000,000,000 **e** 0.000031

38 a 10,000 g **b** 0.010 g **c** 1 μg

40 180 m = 196.9 yds

41 10^{-2} m

42 10^6

43 10 pm, 1 Å, 1000 μm, 100 mm, 100 m, 10 km

45 1/1000

47 10^3

55 5

56 a 4 **b** 3 **c** 1 **d** 3 **e** 2

57 a 1.50×10^3 **b** 1.5×10^3

58 a 2 **b** 1 **c** 2 **d** 2 **e** 4

59 a 1.1 **b** 3.6×10^2 **c** 6.7×10^{-2} **d** 0.048 **e** 59

60 a 4.59 **b** 0.00836 **c** 34.5 **d** 0.0665 **e** 1.26×10^3

61 1675.8 lb

62 15 g

63 a 19.0 **b** 2.3

64 a 0.2 **b** 2 **c** 6.0 **d** 3.0×10^2

65 a 4.11 **b** 8.1×10^2

66 3.6×10^4 m^2

67 a 2×10^2 **b** 6×10^{11} **c** 8×10^{-8} **d** 4×10^{-1}

68 a 6×10^{-1} **b** 2×10^1 **c** 3×10^1 **d** 4×10^{-6} **e** 2×10^5

69 a $4. \times 10^3$ **b** 9×10^4 **c** 4.3×10^{-2} **d** 8.1×10^{-1}

70 a 5.6×10^7 **b** 3.5×10^{-1} **c** -1.5×10^6 **d** 7.6×10^8

71 a 1.0×10^{-2} **b** 6.6×10^{-8}

72 4×10^3 ft

73 a 9×10^2 **b** 8×10^{-2} **c** 3×10^1 **d** 1 **e** 4×10^5

74 a 4×10^3 **b** 3×10^4 **c** 1×10^3 **d** 1×10^2

75 a 6 **b** 2 **c** 1×10^{-1}

76 a 1×10^{-2} **b** 1×10^{-7}

Chapter Three

9 a 0.25 dollar **b** 120 months

11 5.5 acres

12 0.20 dollar

13 1.9×10^2 ziggles

14 0.20 dollar

15 3.0×10^3 mg

16 a 4×10^{-2} m **b** 43 mg **c** 1.55×10^4 mm **d** 0.328 L **e** 9.8×10^2 g

17 1.5×10^{13} cm

18 1.63×10^5 g

19 1.0×10^4 m

20 0.025 L

21 2.134×10^6 mg

22 207 mL

23 6.804×10^7 mg

24 0.025 mL

25 33 mm

26 a 1.52×10^{-5} kg **b** 3.51×10^{-6} cm **c** 6.5×10^4 μL **d** 7.5×10^7 μg

27 5×10^{-5} cm

28 2.234×10^{-2} km

29 0.305 kg

30 485.3 kg

31 3.6×10^2 mL

32 2 oz

33 a 7.3 kg **b** 5×10^3 mL **c** 0.43 m **d** 5.29 oz **e** 0.633 qt

35 109 yd

36 160. kg

37 3.3×10^2 mm

38 291 K, 64°F

39 16°C

40 38.6°C

41 61°F

42 −273°C, −459°F

43 a −9.4°C, 264 K **b** 84°F, 302 K **c** 126°F, 52°C **d** −34°C, 239 K **e** 14°F, 263 K

44 −23°C

46 4.0×10^2 kJ

47 1638 kcal/hr

48 669 kJ

49 79.8 kcal

50 965 g

51 500. cm³

52 7.9 g/cm³

53 379 cm³

54 1.03 g/cm³

55 1.2 g

Chapter Four

33 15, 32

34 19

35 a 9, 10, 9 **b** 16, 34, 16 **c** 25, 27, 25 **d** 35, 73, 35 **e** 182, 90, 90 **f** 33, 33, 35 **g** 40, 41, 40 **h** 55, 27, 27

45 6.941 amu

Chapter Eight

9 a 186.2 **b** 54.9 **c** 102.9 **d** 200.6 **e** 27.0

10 354.6 amu

11 334.2 amu

12 a 294.2 amu **b** 265.8 amu **c** 59.1 amu **d** 183.5 amu

13 a 115.7 amu **b** 342.0 amu **c** 46.1 amu **d** 164.1 amu **e** 84.0 amu

17 a 197.0 g/mole **b** 64.1 g/mole **c** 164.1 g/mole

18 180.1 g/mole

19 a 164.0 g Na_3PO_4 **b** 79.1 g C_5H_5N **c** 205.7 g $Sr(C_2H_3O_2)_2$

20 0.810 g Ni

21 a 1.82×10^3 g $Pb(NO_3)_2$ **b** 32.1 g KH_2PO_4 **c** 181 g $NaClO_4$

22 53.5 g C_3H_8O

23 a 1.58×10^3 g $KMnO_4$ **b** 71.1 g $KMnO_4$ **c** 0.245 g $KMnO_4$

24 a 1.41×10^3 g $C_8H_{16}O_2$ **b** 1.47 g $HgCl_2$ **c** 24.2 g $Ca(C_2H_3O_2)_2$

25 1.18×10^3 g $C_3H_5N_3O_9$

26 a 0.935 mole NH_4Cl **b** 0.0935 mole NH_4Cl **c** 5.50 moles NH_4Cl
d 1.87×10^{-4} mole NH_4Cl

27 0.0642 mole $C_{10}H_{14}N_2$

28 0.474 mole $Na_2S_2O_3$

29 a 0.343 mole $Mg(OH)_2$ **b** 0.0487 mole C_6H_6 **c** 1.20 moles $Al_2(SO_4)_3$

30 a 0.173 mole H_3PO_4 **b** 3.81×10^{-3} mole $KAl(SO_4)_2$
c 2.28×10^{-3} mole CH_4O

31 a 0.0532 mole I_2 **b** 0.0102 mole H_3PO_4

32 1.81×10^{24} atoms O

33 1.93×10^{24} formula units K_3PO_4

34 1.48×10^{24} Li^+ ions

35 9.42×10^{23} formula units Fe_2O_3

36 6.02×10^{22} molecules $C_6H_{12}O_6$

37 1.23×10^3 g U

39 79.2% C, 9.82% H, 11.1% O

40 63.3% Hg, 7.58% C, 20.2% S, 8.84% N

41 84.8% C, 11.2% H, 4.04% O

42 73.9% C, 8.14% H, 2.75% Mg, 6.27% N, 8.96% O

45 C_5H_{12}

46 $NaIO_3$

47 $C_{20}H_{12}O_5$

48 $C_{12}H_8N_2$

49 $K_2S_2O_8$

Chapter Ten

9 a 0.459 mole Mg **b** 6.30 moles N

10 a 2.63 moles H_2S **b** 1.55 moles Al_2S_3

11 a 0.0549 mole HBr   **b** 0.2745 mole H_3PO_3

12 a 63.0 g C   **b** 2.71 moles HCl

13 a 0.612 mole KNO_3 **b** 413 g $KClO_3$

14 a 5.39 moles H_2O **b** 804 g H_3PO_4

15 a 165 kg O_2 **b** 2.92×10^4 g S

16 a 209 g KI   **b** 0.122 g I_2

17 a 143 g $SnCl_2$ **b** 3.35×10^3 g $SnCl_4$

18 a 64.4 g SO_2 **b** 17.0 kg Pb

19 a 15.8 g H_3PO_4 **b** 1.17 g Na_3PO_4

20 a 3.61×10^{23} formula units $Mg(OH)_2$ **b** 3.06×10^3 g H_2O

21 a 1.56 g NO_2 **b** 4.04×10^{22} atoms S

22 a 7.45×10^{24} molecules H_2O **b** 2.27×10^{24} atoms Mn

24 a $FeCl_2$ **b** 160. g $FeCl_3$

25 a 160. g C   **b** 5.38 kg P_4

26 a 6.03 g HNO_3 **b** 0.358 g NO_2

28 29.5%

29 a 52.8%   **b** 95.7%

Chapter Eleven

14 a 1.22×10^4 kPa   **b** 9.12×10^4 torr   **c** 1.76×10^3 psi

15 a 0.987 atm   **b** 100. kPa

16 a 0.953 atm   **b** 96.6 kPa

17 7.44 L

18 0.587 atm

19 5.23 L

20 0.34 atm

23 927 torr

24 $-47°C$

25 109 kPa

28 2.12 L

29 529 mL

30 $-161°C$

31 27.8 L

32 895 mL

33 2.46 L

34 40.7 kPa

35 349°C

38 19.6 L

39 0.817 mole NH_3

40 2.08 L

41 **a** 37.4 L SO_2 **b** 1.45×10^3 L O_2

42 **a** 692 g $CaCO_3$ **b** 483 L CO_2

43 **a** 376 L CO_2 **b** 4.14 g NH_3

46 0.587 mole BF_3

47 124 L

48 6.1×10^2 atm

49 $-219°C$

52 890 torr

53 723 torr

54 20.0 L

55 173 torr

Chapter Twelve

25 0.791

32 0.088 cal/(g)(°C)

33 1.00×10^5 cal

34 114 kJ

39 **a** 540. kcal **b** 80. kcal

40 218 kcal

Chapter Thirteen

34 18 g/L

35 2.30×10^3 g $NaNO_3$

49 **a** 1.25 g KCl **b** 8.00 g KCl

50 290. g NH_3

51 67.0 g HCl

52 15.8 mL

53 **a** 1.4 ppm **b** 1.4×10^3 ppb **c** 1.4×10^{-4}% (w/w)

54 0.025 g Hg

55 **a** 6.0 g $KMnO_4$ **b** 0.38 g $KMnO_4$ **c** 1.3×10^{-3} g $KMnO_4$

56 0.753 M KI

57 0.085 M Na_2SO_4

58 **a** 2.41 g $LiNO_3$ **b** 85.3 g $LiNO_3$

59 **a** 13.9 g CsCl **b** 1.24 g CsCl

60 320. mL

62 25 mL

63 0.96 M

64 63 mL

65 22 mL

66 dilute 0.13 L of 3.0 M $AgNO_3$

74 $-0.37°C$

75 $-0.61°C$

76 100.17°C

Chapter Fourteen

33 1×10^{-10} M

35 **a** 9 **b** 3 **c** 7 **d** 1 **e** 5

37 **a** 1×10^{-4} M H_3O^+, 1×10^{-10} M OH^- **b** 1×10^{-8} M H_3O^+, 1×10^{-6} M OH^- **c** 1×10^{-13} M H_3O^+, 1×10^{-1} M OH^-

44 0.0227 M H_3BO_3

45 0.173 M $Ca(OH)_2$

46 0.467 M H_3PO_4

55 **a** 63.0 g **b** 85.7 g **c** 32.7 g

56 2.78 g $Ca(OH)_2$

57 **a** 0.20 N H_2SO_4 **b** 18 N H_3PO_4 **c** 2.1 N KOH

58 0.224 N

59 0.181 N NaOH

60 **a** 0.0050 M H_2CO_3 **b** 1.0 M H_3BO_3 **c** 2.5×10^{-3} M $Mg(OH)_2$

Chapter Sixteen

40 2.0

41 1×10^{-53}

43 **a** 1×10^{-11} M OH^- **b** 1.8×10^{-7} M OH^- **c** 1.1×10^{-13} M OH^-

45 2×10^{-11}

50 1.6×10^{-3} mole/liter

51 58 moles/dm³

52 2.2

53 11

54 1.65×10^{-4} mole/L [9.62×10^{-4} g $Mg(OH)_2$ per 100 g H_2O]

55 1.1×10^{-7} g FeS

Chapter Seventeen

20 **a** 2.88×10^{17} J **b** 1.0×10^{-7} g

28 5.0 μg

29 5.0 μg

Numerical answers to review exercises

Review Exercises for Chapters Two–Five

4 a 2.91×10^1 **b** 7.3×10^{-2} **c** 1.0541×10^4

7 $1\ cm^3$, $10\ mL$, $0.1\ dm^3$, $1\ L$

8 0.011

9 3

10 2.9

11 a $1.5 \times 10^8\ km$ **b** $4.9 \times 10^{11}\ ft$

12 a $3.49\ kg$ **b** $3.49 \times 10^3\ g$

13 a 2.0×10^4 **b** 1.9×10^4

14 $114\ g$

15 $500\ cm^3$

17 $310\ K$

18 $6.28 \times 10^3\ kJ$

19 $69\ cm^3$

20 $3.95 \times 10^{-22}\ g$

21 24 protons, 29 neutrons, 24 electrons

Review Exercises for Chapters Six–Eight

13 a $96.1\ amu$ **b** $294.2\ amu$ **c** $136.1\ amu$

14 a $70.2\ g\ (NH_4)_2CO_3$ **b** $215\ g\ K_2Cr_2O_7$ **c** $99.5\ g\ CaHPO_4$

15 a $0.605\ mole\ (NH_4)_2CO_3$ **b** $0.197\ mole\ K_2Cr_2O_7$ **c** $0.427\ mole\ CaHPO_4$

16 a 6.26×10^{24} formula units $(NH_4)_2CO_3$ **b** 2.05×10^{24} formula units $K_2Cr_2O_7$ **c** 4.42×10^{24} formula units $CaHPO_4$

17 69.6% C, 8.4% H, 22.1% O

18 $C_9H_{11}NO_2$

19 $C_{10}H_{14}N_2$

Review Exercises for Chapters Nine–Eleven

8 2.11 kg HCN

9 6.21×10^{24} molecules O_2

10 1.62 kg O_2

11 a 30.2 g HCN   **b** 2.5 g NH_3

12 33.6%

14 a 1.01 atm   **b** 102 kPa

15 1.01 g C_2H_2

16 715 torr

17 0.959 kPa

18 830 L

19 5.36 L

20 5.3 L

Review Exercises for Chapters Twelve–Fourteen

4 922 kcal, 3.86×10^3 kJ

10 a 3.8 g $NH_4C_2H_3O_2$ **b** 87 g $NH_4C_2H_3O_2$

11 a 0.0289 M $Hg(NO_3)_2$ **b** 0.935% (w/v)   **c** 9.35×10^3 ppm

12 0.101 M $KMnO_4$

16 pH 1

17 pH 8

19 0.0765 M $HClO_4$

Index

A
B
C
D
E
F
G
H
I
J

Common metric units in chemistry

	Unit		Meaning
Mass	kilogram	kg	1000 g
	gram	g	454 g weighs one pound
	milligram	mg	0.001 g
Length	meter	m	equal to 39.37 inches
	centimeter	cm	0.01 m
	millimeter	mm	0.001 m
Volume	cubic decimeter	dm^3	$0.001\ m^3$
	cubic centimeter	cm^3	$0.001\ dm^3$
	liter	L	equal to 1.037 quarts
	milliliter	mL	0.001 L, same as cm^3
Temperature	kelvin	K	$°C + 273$
	degree Celsius	°C	$(°F - 32)/1.8$
Energy	joule	J	equal to 0.239 calorie
Pressure	pascal	Pa	equal to 0.00750 torr (mm Hg)
Amount	mole	mol	contains 6.02×10^{23} particles
Concentration	molarity	M	moles solute/liter of solution
	molality	m	moles solute/kilogram of solvent

Common metric prefixes in chemistry

Prefix	Symbol	Meaning
kilo-	k	10^3
centi-	c	10^{-2}
milli-	m	10^{-3}
micro-	μ	10^{-6}

Common constants in chemistry

Avogadro's number $\qquad 6.02 \times 10^{23}$ particles/mole

Gas constant, R $\qquad 0.0821\ \dfrac{(atm)\ (L)}{(mole)\ (K)}$

$\qquad\qquad\qquad 8.31\ \dfrac{(kPa)\ (dm^3)}{(mole)\ (K)}$

Molar gas volume at STP $\qquad$ 22.4 liters/mole

Planck's constant, h $\qquad 6.63 \times 10^{-34}$ (joule) (sec)

Speed of light, c $\qquad 3.00 \times 10^{-8}$ meters/sec